D0933730

APPLIED
LONGITUDINAL
DATA ANALYSIS

Applied Longitudinal Data Analysis

Modeling Change and Event Occurrence

Judith D. Singer
John B. Willett

OXFORD
UNIVERSITY PRESS

2003

OXFORD
UNIVERSITY PRESS

Oxford   New York
Auckland   Bangkok   Buenos Aires   Cape Town   Chennai
Dar es Salaam   Delhi   Hong Kong   Istanbul   Karachi   Kolkata
Kuala Lumpur   Madrid   Melbourne   Mexico City   Mumbai   Nairobi
São Paulo   Shanghai   Taipei   Tokyo   Toronto

Published by Oxford University Press, Inc.
198 Madison Avenue, New York, New York 10016

www.oup.com

Oxford is a registered trademark of Oxford University Press

Library of Congress Cataloging-in-Publication Data
Singer, Judith D.
   Applied longitudinal data analysis : modeling change and event
occurrence/by Judith D. Singer and John B. Willett.
      p.   cm.
Includes bibliographical references and index.
   ISBN 978-0-19-515296-8
   1. Longitudinal methods.   2. Social sciences—Research.
I. Willett, John B.   II. Title.
   H62 .S47755 2002
   001.4′2—dc21   2002007055

20   19   18   17

Printed in the United States of America
on acid-free paper

Preamble

Time, occasion, chance and change.
To these all things are subject.
—Percy Bysshe Shelley

Questions about change and event occurrence lie at the heart of much empirical research. In some studies, we ask how people mature and develop; in others, we ask whether and when events occur. In their two-week study of the effects of cocaine exposure on neurodevelopment, Espy, Francis, and Riese (2000) gathered daily data from 40 premature infants: 20 had been exposed to cocaine, 20 had not. Not only did the cocaine-exposed infants have slower rates of growth, but the effect of exposure was greater the *later* the infant was delivered. In his 23-year study of the effects of wives' employment on marital dissolution, South (2001) tracked 3523 couples to examine whether and, if so, when they divorced. Not only did the effect of wives' employment become larger over time (the risk differential was greater in the 1990s than in the 1970s), it increased the longer a couple stayed married.

In this book, we use concrete examples and careful explanation to demonstrate how research questions about change and event occurrence can be addressed with longitudinal data. In doing so, we reveal research opportunities unavailable in the world of cross-sectional data. In fact, the work of Espy and colleagues was prompted, at least in part, by the desire to improve upon an earlier cross-sectional study. Brown, Bakeman, Coles, Sexson, and Demi (1998) found that gestational age moderated the effects of cocaine exposure. But with only one wave of data, they could do little more than establish that babies born later had poorer functioning. They could not describe infants' *rates* of development, nor establish whether change trajectories were linear or nonlinear, nor determine whether gestational age affected infants' functioning at birth. With 14 waves of data, on the other hand, Espy and colleagues could do this and more. Even though their study was brief—covering just the two weeks immediately after birth—they found that growth trajectories were non-linear and that the trajectories of later-born babies began lower, had shallower slopes, and had lower rates of acceleration.

South (2001), too, laments that many researchers fail to capitalize on the richness of longitudinal data. Even among those who do track individuals over time, "relatively few . . . have attempted to ascertain whether the critical socioeconomic and demographic determinants of divorce and separation vary across the marital life course" (p. 230). Researchers are too quick to assume that the effects of predictors like wives' employment remain constant over time. Yet as South points out, why should they? The predictors of divorce among newlyweds likely differ from those among couples who have been married for years. And concerning secular trends, South offers two cogent, but conflicting, arguments about how the effects of wives' employment might change over time. First, he argues that the effects might diminish, as more women enter the labor force and working becomes normative. Next, he argues that the effects might increase, as changing mores weaken the link between marriage and parenthood. With rich longitudinal data on thousands of couples in different generations who married in different years, South carefully evaluates the evidence for, and against, these competing theories in ways that cross-sectional data do not allow.

Not all longitudinal studies will use the same statistical methods—the method must be matched to the question. Because these two studies pose different types of research questions, they demand different analytic approaches. The first focuses on a continuous outcome—neurological functioning—and asks how this attribute changes over time. The second focuses on a specific event—divorce—and asks about its occurrence and timing. Conceptually, we say that in the first study, *time is a predictor* and our analyses assess how a continuous outcome varies as a function of time and other predictors. In the second study, time is an object of study in its own right and we want to know whether, and when, events occur and how their occurrence varies as a function of predictors. Conceptually, then, *time is an outcome.*

Answering each type of research question requires a different statistical approach. We address questions about change using methods known variously as *individual growth modeling* (Rogosa, Brandt, & Zimowski, 1982; Willett, 1988), *multilevel modeling* (Goldstein, 1995), *hierarchical linear modeling* (Raudenbush & Bryk, 2002), *random coefficient regression* (Hedeker, Gibbons, & Flay, 1994), and *mixed modeling* (Pinheiro & Bates, 2000). We address questions about event occurrence using methods known variously as *survival analysis* (Cox & Oakes, 1984), *event history analysis* (Allison, 1984; Tuma & Hannan, 1984), *failure time analysis* (Kalbfleish & Prentice, 1980), and *hazard modeling* (Yamaguchi, 1991). Recent years have witnessed major advances in both types of methods. Descriptions of these advances appear throughout the technical literature and their strengths are well documented. Statistical software is abundant, in the form of dedicated packages and preprogrammed routines in the large multipurpose statistical packages.

But despite these advances, application lags behind. Inspection of substantive papers across many disciplines, from psychology and education to criminology and public health, suggests that—with exceptions, of course—these methods have yet to be widely and wisely used. In a review of over 50 longitudinal studies published in *American Psychological Association* journals in 1999, for example, we found that only four used individual growth modeling (even though many wanted to study change in a continuous outcome) and only one used survival analysis (even though many were interested in event occurrence; Singer & Willett, 2001). Certainly, one cause for this situation is that many popular applied statistics books fail to describe these methods, creating the misimpression that familiar techniques, such as regression analysis, will suffice in these longitudinal applications.

Failure to use new methods is one problem; failure to use them *well* is another. Without naming names, we find that even when individual growth modeling and survival analysis are used in appropriate contexts, they are too often implemented by rote. These methods are complex, their statistical models sophisticated, their assumptions subtle. The default options in most computer packages do *not* automatically generate the statistical models you need. Thoughtful data analysis requires diligence. But make no mistake; hard work has a payoff. If you learn how to analyze longitudinal data well, your approach to empirical research will be altered fundamentally. Not only will you frame your research questions differently but you will also change the kinds of effects that you can detect.

We are not the first to write on these topics. For each method we describe, there are many excellent volumes well worth reading and we urge you to consult these resources. Current books on growth modeling tend to be somewhat technical, assuming advanced knowledge of mathematical statistics (a topic that itself depends on probability theory, calculus, and linear algebra). That said, Raudenbush and Bryk (2002) and Diggle, Liang, and Zeger (1994) are two classics we are proud to recommend. Goldstein (1995) and Longford (1993) are somewhat more technical but also extremely useful. Perhaps because of its longer history, there are several accessible books on survival analysis. Two that we especially recommend are Hosmer and Lemeshow (1999) and Collett (1994). For more technically oriented readers, the classic Kalbfleisch and Prentice (1980) and the newer Therneau and Grambsch (2000) extend the basic methods in important ways.

Our book is different from other books in several ways. To our knowledge, no other book at this level presents growth modeling and survival analysis within a single, coherent framework. More often, growth modeling is treated as a special case of multilevel modeling (which it is), with repeated measurements "grouped" within the individual. Our book stresses the primacy of the sequential nature of the empirical growth record, the repeated observations on an individual over time. As we will show, this structure has far-reaching ramifications for statistical models and their assumptions. Time is not just "another" predictor; it has unique properties that are key to our work. Many books on survival analysis, in contrast, treat the method itself as an object of study in its own right. Yet isolating one approach from all others conceals important similarities among popular methods for the analysis of longitudinal data, in everything from the use of a *person-period data set* to ways of interpreting the effects of *time-varying predictors*. If you understand both growth modeling and survival analysis, and their complementarities, you will be able to apply both methods synergistically to different research questions in the *same* study.

Our targeted readers are our professional colleagues (and their students) who are comfortable with traditional statistical methods but who have yet to fully exploit these longitudinal approaches. We have written this book as a tutorial—a structured conversation among colleagues. In its pages, we address the questions that our colleagues and students ask us when they come for data analytic advice. Because we have to start somewhere, we assume that you are comfortable with linear and logistic regression analysis, as well as with the basic ideas of decent data analysis. We expect that you know how to specify and compare statistical models, test hypotheses, distinguish between main effects and interactions, comprehend the notions of linear and nonlinear relationships, and can use residuals and other diagnostics to examine your assumptions. Many of you may also be comfortable with multilevel modeling or structural equation modeling, although we assume no familiarity with either. And although our methodological colleagues are not our prime audience, we hope they, too, will find much of interest.

Our orientation is data analytic, not theoretical. We explain how to use growth modeling and survival analysis via careful step-by-step analysis of real data. For each method, we emphasize five linked phases: identifying research questions, postulating an appropriate model and understanding its assumptions, choosing a sound method of estimation, interpreting results, and presenting your findings. We devote considerable space—over 150 tables and figures—to illustrating how to present your work not just in words but also in displays. But ours is not a cookbook filled with checklists and flowcharts. The craft of good data analysis cannot be prepackaged into a rote sequence of steps. It involves more than using statistical computer software to generate reams of output. Thoughtful analysis can be difficult and messy, raising delicate problems of model specification and parameter interpretation. We confront these thorny issues directly, offering concrete advice for sound decision making. Our goal is to provide the short-term guidance you need to quickly start using the methods in your own work, as well as sufficient long-term advice to support your work once begun.

Many of the topics we discuss are rooted in complex statistical arguments. When possible, we do not delve into technical details. But if we believe that understanding these details will improve the quality of your work, we offer straightforward conceptual explanations that do not sacrifice intellectual rigor. For example, we devote considerable space to issues of estimation because we believe that you should not fit a statistical model and interpret its results without understanding intuitively what the model stipulates about the underlying population and how sample data are used to estimate parameters. But instead of showing you how to maximize a likelihood function, we discuss heuristically what maximum likelihood methods of estimation are, why they make sense, and how the computer applies them. Similarly, we devote considerable attention to explicating the assumptions of our statistical models so that you can understand their foundations and limitations. When deciding whether to include (or exclude) a particular topic, we asked ourselves: Is this something that empirical researchers need to know to be able to conduct their analyses wisely? This led us to drop some topics that are discussed routinely in other books (for example, we do not spend time discussing what *not* to do with longitudinal data) while we spend considerable time discussing some topics that other books downplay (such as how to include and interpret the effects of time-varying predictors in your analyses).

All the data sets analyzed in this book—and there are many—are *real* data from *real* studies. To provide you with a library of resources that you might emulate, we also refer to many other published papers. Dozens of researchers have been extraordinarily generous with their time, providing us with data sets in psychology, education, sociology, political science, criminology, medicine, and public health. Our years of teaching convince us that it is easier to master technical material when it is embedded in real-world applications. But we hasten to add that the methods are unaware of the substance involved. Even if your discipline is not repre-sented in the examples in these pages, we hope you will still find much of analytic value. For this reason, we have tried to choose examples that require little disciplinary knowledge so that readers from other fields can appreciate the subtlety of the substantive arguments involved.

Like all methodologists writing in the computer age, we faced a dilemma: how to balance the competing needs of illustrating the use of statistical software with the inevitability that specific advice about any particular computer package would soon be out of date. A related concern that we shared was a sense that the ability to program a statisti-cal package does not substitute for understanding what a statistical model is, how it represents relationships among variables, how its parameters are estimated, and how to interpret its results. Because we have no vested interest in any particular statistical package, we decided to use a variety of them throughout the book. But instead of presenting unadulterated computer output for your perusal, we have reformatted the results obtained from each program to provide templates you can use when reporting findings. Recognizing that empirical researchers must be able to use software effectively, however, we have provided an associated website that lists the data sets used in the book, as well as a library of com-puter programs for analyzing them, and selected additional materials of interest to the data analyst.

The book is divided into two major parts: individual growth modeling in the first half, survival analysis in the second. Throughout each half, we stress the important connections between the methods. Each half has its own introduction that: (1) discusses when the method might be used; (2) distinguishes among the different types of research questions in that domain; and (3) identifies the major statistical features of empirical studies that lend themselves to the specified analyses. Both types of analy-ses require a sensible metric for clocking time, but in growth modeling, you need multiple waves of data and an outcome that changes systemati-cally, whereas in survival analysis, you must clearly identify the beginning of time and the criteria used to assess event occurrence. Subsequent chap-ters in each half of the book walk you through the details of analysis. Each begins with a chapter on data description and exploratory analysis, fol-lowed by a detailed discussion of model specification, model fitting, and parameter interpretation. Having introduced a basic model, we then con-sider extensions. Because it is easier to understand the path that winds through the book only after important issues relevant for each half have been introduced, we defer discussion of each half's outline to its associated introductory chapter.

Acknowledgments

We have spent the last eighteen years working closely together in the most productive, mutually supportive, and personally enjoyable collaboration of our professional lives. We offer this book as testament to that collaboration.

We first met in January 1985. The previous academic year, we had each applied for a single position as an Assistant Professor of Quantitative Methods at the Harvard Graduate School of Education (HGSE). When the chair of the search committee announced that he was leaving Harvard for the University of Chicago, the School discovered it had two vacancies to fill and decided to hire us both. We had never met, and everyone told us they expected us to compete. Instead, we began meeting regularly for lunch—first for mutual support, then to coordinate courses, and ultimately to link our scholarship. Despite the popular image of the competitive lone scholar, we've found that by working together, we're more imaginative, productive, and effective than either of us is working apart. And perhaps more importantly, we have more fun.

As junior academics, we had to weather the usual storms of promotion and review. For this, we owe our sincere thanks to colleagues at Harvard and elsewhere who encouraged us to pursue our own interests and scholarship above all else. Initially, we were set on our path by our doctoral advisors: Fred Mosteller and Dick Light at Harvard (for Judy), David Rogosa and Ingram Olkin at Stanford (for John). Tony Bryk, the chair of the Harvard search committee that hired us, inadvertently laid the foundation for our collaboration by bringing us together and then leaving us alone. Over our years at Harvard, we benefited greatly from the active help and gentle advice of colleagues. Dick Light and Dick

Murnane nurtured and guided us by unselfish example and personal friendship. Catherine Snow and Susan Johnson led the way by exploring the promotional pathway at HGSE, just ahead of us. Two far-sighted HGSE Deans, Pat Graham and Jerry Murphy, found ways to help an institution steeped in tradition entertain the unusual—a pair of quantitative methodologists working together.

We trace our planful collaboration to a conversation one warm spring afternoon in April 1987, on a bench along the Mississippi River in New Orleans. With youthful hubris, we hatched the first of several "five-year" plans: together we would become the "great communicators of statistical methods," bringing powerful new quantitative techniques to empirical researchers throughout education and the social sciences. A former B-movie actor had carried that banner into the Oval Office, so why couldn't a nice Jewish girl from Brooklyn and an expatriate Yorkshire lad do the academic equivalent? We decided right there to give it a shot.

Part of our strategy was to make our collaboration seamless. We would never divulge who wrote what; if one of us was invited to give a talk or contribute a paper, s/he would insist that the other participate as well; we would never compete with each other for any opportunity; and all our papers would include the disclaimer: "The order of the authors has been determined by randomization."

The majority of our joint scholarly activity has focused on the analytic issues and problems that arise when modeling change and event occurrence. Like any intellectual endeavor, our understanding of the field has grown more nuanced over time, largely as a consequence of interactions not only with one and other but with others as well. This book draws together and organizes our own thoughts in light of the many understandings we have derived from the pioneering work of others. Too numerous to count, the list includes: Paul Allison, Mark Appelbaum, Carl Bereiter, Tony Bryk, Harris Cooper, Dennis Cox, Lee Cronbach, Art Dempster, Brad Efron, Jan de Leeuw, Harvey Goldstein, Larry Hedges, Dave Hoaglin, Fred Lord, Jack Kalbfleisch, Nan Laird, Bob Linn, Jack McArdle, Bill Meredith, Rupert Miller, Fred Mosteller, Bengt Muthen, John Nesselrode, Ross Prentice, Steve Raudenbush, Dave Rindskopf, David Rogosa, John Tisak, John Tukey, Nancy Tuma, Jim Ware, Russ Wolfinger, and Marvin Zelen. To all of these, and to the many others not listed here, we offer our sincere thanks.

We would also like to thank the many people who contributed directly to the genesis, production, and completion of the book. Our first thanks go to the Spencer Foundation, which under then-President Pat Graham, provided the major grant that permitted us to buy back time from our teaching schedules to begin assembling this manuscript. Anonymous reviewers and board members at the Spencer Foundation provided early feedback on our original proposal and helped refine our notions of the book's content, audience, and organization. Other friends, particularly Steve Raudenbush and Dave Rindskopf, read early drafts of the book and gave us detailed comments. Our colleague Suzanne Graham tested out earlier versions of the book in her class on longitudinal data analysis at HGSE. Suzanne, and the cohorts of students who took the class, provided helpful feedback on everything from typos to conceptual errors to writing style.

We could not have written a book so reflective of our pedagogic philosophy without access to many real longitudinal data sets. To provide the data for this book, we surveyed the research literature across a wide array of substantive domains and contacted the authors of papers that caught our collective eye. In this search, we were very ably assisted by our colleague, Librarian John Collins, and his team at HGSE's Monroe C. Gutman Library.

The empirical researchers that we contacted—often out of the blue— were unfailingly generous and helpful with our requests to use their data. Many of these scholars are themselves pioneers in applying innovative analytic methods. We are grateful for their time, their data, and their willingness to allow us to capitalize on their work in this book. Specifically, we would like to thank the following colleagues (in alphabetical order), who made a direct contribution of data to our work: Niall Bolger; Peg Burchinal; Russell Burton; Deborah Capaldi and Lynn Crosby; Ned Cooney; Patrick Curran and Laurie Chassin; Andreas Diekmann; Al Farrell; Michael Foster; Beth Gamse; Elizabeth Ginexi; Suzanne Graham; James Ha; Sharon Hall; Kris Henning; Margaret Keiley; Dick Murnane and Kathy Boudett; Steve Raudenbush; Susan Sorenson; Terry Tivnan; Andy Tomarken; Blair Wheaton; Christopher Zorn. In the text and bibliography, we provide citations to exemplary papers by these authors in which the data were originally reported. These citations list both the scholars who were responsible for providing us with the data and also the names of their collaborating colleagues, many of whom were also important in granting permission to use the data. And, while we cannot list everyone here in the brief space allowed for our acknowledgments, we recognize them all explicitly in the text and bibliography in our citation of their scholarship, and we thank them enormously for their support.

Of course, the data will always remain the intellectual property of the original authors, but any mistakes in the analyses reported here are ours alone. We must emphasize that we used these data examples strictly for the illustration of statistical methods. In many of our examples, we modified the original data to suit our pedagogic purposes. We may have selected specific variables from the original dataset for re-analysis, perhaps combining several into a single composite. We transformed variables as we saw fit. We selected subgroups of individuals, or particular cohorts, from the original sample for re-analysis. We also eliminated specific waves of data and individual cases from our analyses, as necessary. Consequently, any substantive results that we present may not necessarily match those of the original published studies. The original researchers retain the rights to the substantive findings of the studies from which our data-examples were drawn and their results naturally take precedence over ours. For this reason, if you are interested in those findings explicitly, you must consult the original empirical papers.

We decided early on that this book would describe the ideas behind analyses, not the programming of statistical software. Computer software for analyzing longitudinal data is now ubiquitous. The major statistical packages include routines for modeling change and event occurrence, and there are dedicated software packages available as well. Software packages differ not so much in their core purpose as in their implementation; they generally fit the same statistical models but offer different user interfaces, methods of estimation, ancillary statistics, graphics and diagnostics. We therefore decided not to feature any particular piece of software but to employ a sampling of what was readily available at the time. We thank the SAS Institute, Scientific Software International, SPSS, and the STATA Corporation for their support, and we appreciate the willingness of the authors and publishers of the HLM, MLwiN, and LISREL software for providing us with up-to-the minute versions.

Needless to say, software continues to change rapidly. Since we began this book, all the packages we initially used have been improved and revamped, and new software has been written. This process of steady improvement is a great benefit to empirical researchers and we fully expect it to continue unabated. We suggest that researchers use whatever software is most convenient at any given moment rather than committing permanently to any single piece of software. While analytic processes may differ with different software, findings will probably not.

We would like to comment specifically on the help, feedback and support that we have received from the Statistical Training and Consulting Division (STCD) of the Academic Technology Services at UCLA, under the directorship of Michael Mitchell. The STCD has graciously written computer programs to execute all the analyses featured in this book, using several major statistical packages (including HLM, MLwiN, SAS, SPSS, SPLUS, and STATA), and they have posted these programs along with selected output to a dedicated website (http://www.ats.ucla.edu/stat/examples/alda/). This website is a terrific practical companion to our book and we recommend it: access is free and open to all. We would like to thank Michael and his dedicated team of professionals for the foresight and productivity they have displayed in making this service available to us and to the rest of the scholarly community.

It goes without saying that we owe an immense debt to all members of the production team at Oxford University Press. We are particularly grateful to: Joan Bossert, Vice President and Acquiring Editor; Lisa Stallings, Managing Editor; Kim Robinson and Maura Roessner, Assistant Editors. There are also many others who touched the book during its long journey and we thank them as well for all the energy, care, and enthusiasm they devoted to this effort.

Finally, we want to recognize our love for those who gave us life and who provide us with a reason to live—our parents, our families, and our partners.

P. S.: The order of the authors *was* determined by randomization.

Contents

PART I

1

A Framework for Investigating Change over Time

Change is inevitable. Change is constant.
—Benjamin Disraeli

Change is pervasive in everyday life. Infants crawl and walk, children learn to read and write, the elderly become frail and forgetful. Beyond these natural changes, targeted interventions can also cause change: cholesterol levels may decline with new medication; test scores might rise after coaching. By measuring and charting changes like these—both naturalistic and experimentally induced—we uncover the temporal nature of development.

The investigation of change has fascinated empirical researchers for generations. Yet it is only since the 1980s, when methodologists developed a class of appropriate statistical models—known variously as *individual growth models*, *random coefficient models*, *multilevel models*, *mixed models*, and *hierarchical linear models*—that researchers have been able to study change well. Until then, the technical literature on the measurement of change was awash with broken promises, erroneous half-truths, and name-calling. The 1960s and 1970s were especially rancorous, with most methodologists offering little hope, insisting that researchers should not even attempt to measure change because it could not be done well (Bereiter, 1963; Linn & Slinde, 1977). For instance, in their paper, "How should we measure change? Or should we?," Cronbach and Furby (1970) tried to end the debate forever, advising researchers interested in the study of change to "frame their questions in other ways."

Today we know that it is possible to measure change, and to do it well, *if you have longitudinal data* (Rogosa, Brandt, & Zimowski, 1982; Willett, 1989). Cross-sectional data—so easy to collect and so widely available—will not suffice. In this chapter, we describe why longitudinal data are necessary for studying change. We begin, in section 1.1, by introducing three longitudinal studies of change. In section 1.2, we distinguish between the two types of question these examples address, questions about: (1) *within-individual change*—How does *each person* change over time?—and (2) *interindividual differences in change*—What predicts differences among people in their changes? This distinction provides an appealing heuristic for framing research questions and underpins the statistical models we ultimately present. We conclude, in section 1.3, by identifying three requisite *methodological* features of any study of change: the availability of (1) multiple waves of data; (2) a substantively meaningful metric for time; and (3) an outcome that changes systematically.

1.1 When Might You Study Change over Time?

Many studies lend themselves to the measurement of change. The research design can be experimental or observational. Data can be collected prospectively or retrospectively. Time can be measured in a variety of units—months, years, semesters, sessions, and so on. The data collection schedule can be fixed (everyone has the same periodicity) or flexible (each person has a unique schedule). Because the phrases "growth models" and "growth curve analysis" have become synonymous with the measurement of change, many people assume that outcomes must "grow" or *increase* over time. Yet the statistical models that we will specify care little about the direction (or even the functional form) of change. They lend themselves equally well to outcomes that *decrease* over time (e.g., weight loss among dieters) or exhibit complex trajectories (including plateaus and reversals), as we illustrate in the following three examples.

1.1.1 Changes in Antisocial Behavior during Adolescence

Adolescence is a period of great experimentation when youngsters try out new identities and explore new behaviors. Although most teenagers remain psychologically healthy, some experience difficulty and manifest antisocial behaviors, including aggressive *externalizing behaviors* and depressive *internalizing behaviors*. For decades, psychologists have postulated a variety of theories about why some adolescents develop problems and others do not, but lacking appropriate statistical methods, these suppositions went untested. Recent advances in statistical methods have allowed empirical exploration of developmental trajectories and assessment of their predictability based upon early childhood signs and symptoms.

Coie, Terry, Lenox, Lochman, and Hyman (1995) designed an ingenious study to investigate longitudinal patterns by capitalizing on data gathered routinely by the Durham, North Carolina, public schools. As part of a systemwide screening program, every third grader completes a battery of sociometric instruments designed to identify classmates who are overly aggressive (who start fights, hit children, or say mean things) or extremely rejected (who are liked by few peers and disliked by many). To investigate the link between these early assessments and later antisocial behavioral trajectories, the researchers tracked a random sample of 407 children, stratified by their third-grade peer ratings. When they were in sixth, eighth, and tenth grade, these children completed a battery of instruments, including the Child Assessment Schedule (CAS), a semistructured interview that assesses levels of antisocial behavior. Combining data sets allowed the researchers to examine these children's patterns of change between sixth and tenth grade and the predictability of these patterns on the basis of the earlier peer ratings.

Because of well-known gender differences in antisocial behavior, the researchers conducted separate but parallel analyses by gender. For simplicity here, we focus on boys. Nonaggressive boys—regardless of their peer rejection ratings—consistently displayed few antisocial behaviors between sixth and tenth grades. For them, the researchers were unable to reject the null hypothesis of no systematic change over time. Aggressive nonrejected boys were indistinguishable from this group with respect to patterns of externalizing behavior, but their sixth-grade levels of internalizing behavior were temporarily elevated (declining linearly to the nonaggressive boys' level by tenth grade). Boys who were both aggressive *and* rejected in third grade followed a very different trajectory. Although they were indistinguishable from the nonaggressive boys in their sixth-grade levels of either outcome, over time they experienced significant linear increases in both. The researchers concluded that adolescent boys who will ultimately manifest increasing levels of antisocial behavior can be identified as early as third grade on the basis of peer aggression and rejection ratings.

1.1.2 Individual Differences in Reading Trajectories

Some children learn to read more rapidly than others. Yet despite decades of research, specialists still do not fully understand why. Educators and pediatricians offer two major competing theories for these interindividual differences: (1) the *lag* hypothesis, which assumes that every child can become a proficient reader—children differ only in the *rate* at which they acquire skills; and (2) the *deficit* hypothesis, which assumes that some children will never read well because they lack a crucial skill. If the lag hypothesis were true, all children would eventually become proficient; we need only follow them for sufficient time to see their mastery. If the deficit hypothesis were true, some children would never become proficient no matter how long they were followed—they simply lack the skills to do so.

Francis, Shaywitz, Stuebing, Shaywitz, and Fletcher (1996) evaluated the evidence for and against these competing hypotheses by following 363 six-year-olds until age 16. Each year, children completed the Woodcock-Johnson Psycho-educational Test Battery, a well-established measure of reading ability; every other year, they also completed the Wechsler Intelligence Scale for Children (WISC). By comparing third-grade reading scores to expectations based upon concomitant WISC scores, the researchers identified three distinct groups of children: 301 "normal readers"; 28 "discrepant readers," whose reading scores were much different than their WISC scores would suggest; and 34 "low achievers," whose reading scores, while not discrepant from their WISC scores, were far below normal.

Drawing from a rich theoretical tradition that anticipates complex trajectories of development, the researchers examined the tenability of several alternative nonlinear growth models. Based upon a combination of graphical exploration and statistical testing, they selected a model in which reading ability increases nonlinearly over time, eventually reaching an asymptote—the maximum reading level the child could be expected to attain (if testing continued indefinitely). Examining the fitted trajectories, the researchers found that the two groups of disabled readers were indistinguishable statistically, but that both differed significantly from the normal readers in their eventual plateau. They estimated that the average child in the normal group would attain a reading level 30 points higher than that of the average child in either the discrepant or low-achieving group (a large difference given the standard deviation of 12). The researchers concluded that their data were more consistent with the deficit hypothesis—that some children will *never* attain mastery—than with the lag hypothesis.

1.1.3 Efficacy of Short-Term Anxiety-Provoking Psychotherapy

Many psychiatrists find that short-term anxiety-provoking psychotherapy (STAPP) can ameliorate psychological distress. A methodological strength of the associated literature is its consistent use of a well-developed instrument: the Symptom Check List (SCL-90), developed by

Derogatis (1994). A methodological weakness is its reliance on two-wave designs: one wave of data pretreatment and a second wave posttreatment. Researchers conclude that the treatment is effective when the decrease in SCL-90 scores among STAPP patients is lower than the decrease among individuals in a comparison group.

Svartberg, Seltzer, Stiles, and Khoo (1995) adopted a different approach to studying STAPP's efficacy. Instead of collecting just two waves of data, the researchers examined "the course, rate and correlates of symptom improvement as measured with the SCL-90 during and after STAPP" (p. 242). A sample of 15 patients received approximately 20 weekly STAPP sessions. During the study, each patient completed the SCL-90 up to seven times: once or twice at referral (before therapy began), once at mid-therapy, once at termination, and three times after therapy ended (after 6, 12, and 24 months). Suspecting that STAPP's effectiveness would vary with the patients' abilities to control their emotional and motivational impulses (known as *ego rigidity*), two independent psychiatrists reviewed the patients' intake files and assigned ego rigidity ratings.

Plotting each patient's SCL-90 data over time, the researchers identified two distinct temporal patterns, one during treatment and another after treatment. Between intake and treatment termination (an average of 8.5 months later), most patients experienced relatively steep linear declines in SCL-90 scores—an average decrease of 0.060 symptoms per month (from an initial mean of 0.93). During the two years after treatment, the rate of linear decline in symptoms was far lower—only 0.005 per month—although still distinguishable from 0. In addition to significant differences among individuals in their rates of decline before and after treatment termination, ego rigidity was associated with rates of symptom decline during therapy (but not after). The researchers concluded that: (1) STAPP can decrease symptoms of distress *during* therapy; (2) gains achieved during STAPP therapy *can* be maintained; but (3) major gains *after* STAPP therapy ends are rare.

1.2 Distinguishing Between Two Types of Questions about Change

From a substantive point of view, each of these studies poses a unique set of research questions about its own specific outcomes (antisocial behavior, reading levels, and SCL-90 scores) and its own specific predictors (peer ratings, disability group, and ego rigidity ratings). From a statistical point of view, however, each poses an identical pair of questions: (1)

How does the outcome change over time? and (2) Can we predict differences in these changes? From this perspective, Coie and colleagues (1995) are asking: (1) How does each adolescent's level of antisocial behavior change from sixth through tenth grade?; and (2) Can we predict differences in these changes according to third grade peer ratings? Similarly, Francis and colleagues (1996) are asking: (1) How does reading ability change between ages 6 and 16?; and (2) Can we predict differences in these changes according to the presence or absence of a reading disability?

These two kinds of question form the core of every study about change. The first question is descriptive and asks us to characterize each person's pattern of change over time. Is individual change linear? Nonlinear? Is it consistent over time or does it fluctuate? The second question is relational and asks us to examine the association between predictors and the patterns of change. Do different types of people experience different patterns of change? Which predictors are associated with which patterns? In subsequent chapters, we use these two questions to provide the conceptual foundation for our analysis of change, leading naturally to the specification of a pair of statistical models—one per question. To develop your intuition about the questions and how they map onto subsequent studies of change, here we simply emphasize their sequential and hierarchical nature.

In the first stage of an analysis of change, known as *level-1*, we ask about *within-individual change* over time. Here, we characterize the individual pattern of change so that we can describe each person's *individual growth trajectory*—the way his or her outcome values rise and fall over time. Does this child's reading skill grow rapidly, so that she begins to understand complex text by fourth or fifth grade? Does another child's reading skill start out lower and grow more slowly? The goal of a level-1 analysis is to describe the *shape* of each person's individual growth trajectory.

In the second stage of an analysis of change, known as *level-2*, we ask about *interindividual differences in change*. Here, we assess whether different people manifest different patterns of within-individual change and ask what predicts these differences. We ask whether it is possible to predict, on the basis of third-grade peer ratings, which boys will remain psychologically healthy during adolescence and which will become increasingly antisocial? Can ego rigidity ratings predict which patients will respond most rapidly to psychotherapy? The goal of a level-2 analysis is to detect heterogeneity in change across individuals and to determine the *relationship* between predictors and the *shape* of each person's individual growth trajectory.

In subsequent chapters, we map these two research questions onto a pair of statistical models: (1) a level-1 model, describing within-individual change over time; and (2) a level-2 model, relating predictors to any interindividual differences in change. Ultimately, we consider these two models to be a "linked pair" and refer to them jointly as the *multilevel model for change.* But for now, we ask only that you learn to distinguish the two types of questions. Doing so helps clarify why research studies of change must possess certain methodological features, a topic to which we now turn.

1.3 Three Important Features of a Study of Change

Not every longitudinal study is amenable to the analysis of change. The studies introduced in section 1.1 share three methodological features that make them particularly well suited to this task. They each have:

- Three or more waves of data
- An outcome whose values change systematically over time
- A sensible metric for clocking time

We comment on each of these features of research design below.

1.3.1 Multiple Waves of Data

To model change, you need longitudinal data that describe how each person in the sample changes over time. We begin with this apparent tautology because too many empirical researchers seem willing to leap from cross-sectional data that describe differences among individuals of different ages to making generalizations about change over time. Many developmental psychologists, for example, analyze cross-sectional data sets composed of children of differing ages, concluding that outcome differences between age groups—in measures such as antisocial behavior—reflect real change over time. Although change is a compelling explanation of this situation—it might even be the *true* explanation—cross-sectional data can never confirm this possibility because equally valid competing explanations abound. Even in a sample drawn from a single school, a random sample of older children may differ from a random sample of younger children in important ways: the groups began school in different years, they experienced different curricula and life events, and if data collection continues for a sufficient period of time, the older sample omits age-mates who dropped out of school. Any observed differences in outcomes between grade-separated cohorts may be due to these explanations and not to systematic individual change. In statistical terms, cross-sectional studies confound age and cohort effects (and age and history effects) and are prone to selection bias.

Studies that collect two waves of data are only marginally better. For decades, researchers erroneously believed that two-wave studies were sufficient for studying change because they narrowly conceptualized change as an *increment*: the simple difference between scores assessed on two measurement occasions (see Willett, 1989). This limited perspective views change as the acquisition (or loss) of the focal increment: a "chunk" of achievement, attitude, symptoms, skill, or whatever. But there are two reasons an increment's size cannot describe the *process of change*. First, it cannot tell us about the *shape* of each person's individual growth trajectory, the focus of our level-1 question. Did all the change occur immediately after the first assessment? Was progress steady or delayed? Second, it cannot distinguish true change from measurement error. If measurement error renders pretest scores too low and posttest scores too high, you might conclude erroneously that scores increase over time when a longer temporal view would suggest the opposite. In statistical terms, two-waves studies cannot describe individual trajectories of change and they confound true change with measurement error (see Rogosa, Brandt, & Zimowski, 1982).

Once you recognize the need for multiple waves of data, the obvious question is, How many waves are enough? Are three sufficient? Four? Should you gather more? Notice that Coie's study of antisocial behavior included just three waves, while Svartberg's STAPP study included at least six and Francis's reading study included up to ten. In general, more waves are always better, within cost and logistical constraints. Detailed discussion of this design issue requires clear understanding of the statistical models presented in this book. So for now, we simply note that more waves allow you to posit more elaborate statistical models. If your data set has only three waves, you must fit simpler models with stricter assumptions— usually assuming that individual growth is *linear* over time (as Coie and colleagues did in their study of antisocial behavior). Additional waves allow you to posit more flexible models with less restrictive assumptions; you can assume that individual growth is nonlinear (as in the reading study) or linear in chunks (as in the STAPP study). In chapters 2–5, we assume that individual growth is linear over time. In chapter 6, we extend these basic ideas to situations in which level-1 growth is discontinuous or nonlinear.

1.3.2 A Sensible Metric for Time

Time is the fundamental predictor in every study of change; it must be measured reliably and validly in a sensible metric. In our examples, reading scores are associated with particular *ages*, antisocial behavior is associated with particular *grades*, and SCL-90 scores are associated with particular *months since intake*. Choice of a time metric affects several interrelated decisions about the number and spacing of data collection waves. Each of these, in turn, involves consideration of costs, substantive needs, and statistical benefits. Once again, because discussion of these issues requires the statistical models that we have yet to develop, we do not delve into specifics here. Instead we discuss general principles.

Our overarching point is that there is no single answer to the seemingly simple question about the most sensible metric for time. You should adopt whatever scale makes most sense for your outcomes and your research question. Coie and colleagues used *grade* because they expected antisocial behavior to depend more on this "social" measure of time than on chronological age. In contrast, Francis and colleagues used age because each reading score was based on the child's age at testing. Of course, these researchers also had the option of analyzing their data using grade as the time metric; indeed, they present tables in this metric. Yet when it came to data analysis, they used the child's age at testing so as to increase the precision with which they measured each child's growth trajectory.

Many studies possess several plausible metrics for time. Suppose, for example, your interest focuses on the longevity of automobiles. Most of us would initially assess time using the vehicle's *age*—the number of weeks (or months) since purchase (or manufacture). And for many automotive outcomes—particularly those that assess appearance qualities like rust and seat wear—this choice seems appropriate. But for other outcomes, other metrics may be better. When modeling the depth of tire treads, you might measure time in *miles*, reasoning that tire wear depends more on actual use, not years on the road. The tires of a one-year-old car that has been driven 50,000 miles will likely be more worn than those of a two-year-old car that has been driven only 20,000 miles. Similarly, when modeling the health of the starter/igniter, you might measure time in *trips*, reasoning that the starter is used only once each drive. The condition of the starters in two cars of identical age and mileage may differ if one car is driven infrequently for long distances and the other is driven several times daily for short hops. So, too, when modeling the life of the engine, you might measure time in *oil changes*, reasoning that lubrication is most important in determining engine wear.

Our point is simple: choose a metric for time that reflects the cadence you expect to be most useful for your outcome. Psychotherapy studies can clock time in *weeks* or *number of sessions*. Classroom studies can clock time in *grade* or *age*. Studies of parenting behavior can clock time using *parental age* or *child age*. The only constraint is that, like time itself, the temporal variable can change only monotonically—in other words, it cannot reverse direction. This means, for example, that when studying child outcomes, you could use height, but not weight, as a gauge of time.

Having chosen a metric for time, you have great flexibility concerning the *spacing* of the waves of data collection. The goal is to collect sufficient data to provide a reasonable view of each individual's growth trajectory. *Equally spaced waves* have a certain appeal, in that they offer balance and symmetry. But there is nothing sacrosanct about equal spacing. If you expect rapid nonlinear change during some time periods, you should collect more data at those times. If you expect little change during other periods, space those measurements further apart. So in their STAPP study, Svartberg and colleagues (1995) spaced their early waves more closely together—at approximately 0, 4, 8, and 12 months—because they expected greater change during therapy. Their later waves were further apart—at 18 and 30 months—because they expected fewer changes.

A related issue is whether everyone should share the same data collection schedule—in other words, whether everyone needs an identical distribution of waves. If everyone is assessed on an identical schedule—whether the waves are equally or unequally spaced—we say that the data set is *time-structured*. If data collection schedules vary across individuals, we say the data set is *time-unstructured*. Individual growth modeling is flexible enough to handle both possibilities. For simplicity, we begin with time-structured data sets (in chapters 2, 3, and 4). In chapter 5, we show how the same multilevel model for change can be used to analyze time-unstructured data sets.

Finally, the resultant data set need not be *balanced*; in other words, each person need not have the same number of waves. Most longitudinal studies experience some attrition. In Coie and colleagues' (1995) study of antisocial behavior, 219 children had three waves, 118 had two, and 70 had one. In Francis and colleagues' (1996) reading study, the total number of assessments per child varied between six and nine. While non-random attrition can be problematic for drawing inferences, individual growth modeling does not require balanced data. Each individual's empirical growth record can contain a unique number of waves collected at unique occasions of measurement—indeed, as we will see in chapter 5, some individuals can even contribute fewer than three waves!

1.3.3 A Continuous Outcome That Changes
Systematically Over Time

Statistical models care little about the substantive meaning of the individual outcomes. The same models can chart changes in standardized test scores, self-assessments, physiological measurements, or observer ratings. This flexibility allows individual growth models to be used across diverse disciplines, from the social and behavioral sciences to the physical and natural sciences. The *content* of measurement is a substantive, not statistical, decision.

How to measure a given construct, however, *is* a statistical decision, and not all variables are equally suitable. Individual growth models are designed for continuous outcomes whose values change systematically over time.[1] This focus allows us to represent individual growth trajectories using meaningful parametric forms (an idea we introduce in chapter 2). Of course, it must make conceptual and theoretical sense for the outcome to follow such a trajectory. Francis and colleagues (1996) invoke developmental theory to argue that reading ability will follow a logistic trajectory as more complex skills are layered upon basic building blocks and children head toward an upper asymptote. Svartberg and colleagues (1995) invoke psychiatric theory to argue that patients' trajectories of symptomatology will differ when they are in therapy and after therapy ends.

Continuous outcomes support all the usual manipulations of arithmetic: addition, subtraction, multiplication, and division. Differences between pairs of scores, equidistantly spaced along the scale, have identical meanings. Scores derived from standardized instruments developed by testing companies—including the Woodcock Johnson Psychoeducational Test Battery—usually display these properties. So, too, do arithmetic scores derived from most public-domain instruments, like Hodges's Child Assessment Schedule and Derogatis's SCL-90. Even home-grown instruments can produce scores with the requisite measurement properties as long as they include a large enough number of items, each scored using a large enough number of response categories.

Of course, your outcomes must also possess decent psychometric properties. Using well-known or carefully piloted instruments can ensure acceptable standards of validity and precision. But longitudinal research imposes three additional requirements because the metric, validity, and precision of the outcome must also be preserved across time.

When we say that the metric in which the outcome is measured must be preserved across time, we mean that the outcome scores must be equatable over time—a given value of the outcome on any occasion must represent the same "amount" of the outcome on every occasion. Outcome equatability is easiest to ensure when you use the identical instrument for measurement repeatedly over time, as did Coie and colleagues (1995) in their study of antisocial behavior and Svartverg and colleagues (1995) in their study of STAPP. Establishing outcome equatability when the measures differ over time—like the Woodcock Johnson test battery used by Francis and colleagues (1996)—requires more effort. If the instrument has been developed by a testing organization, you can usually find support for equatability over time in the testing manuals. Francis and colleagues (1996) note that:

> The Rasch-scaled score reported for the reading-cluster score is a transformation of the number correct for each subtest that yields a score with interval scale properties and a constant metric. The transformation is such that a score of 500 corresponds to the average performance level of fifth graders. Its interval scale and constant metric properties make the Rasch-scaled score ideal for longitudinal studies of individual growth. (p. 6)

If outcome measures are not equatable over time, the longitudinal equivalence of the score meanings cannot be assumed, rendering the scores useless for measuring change.

Note that measures cannot be made equatable simply by standardizing their scores on each occasion to a common standard deviation. Although occasion-by-occasion standardization appears persuasive—it seems to let you talk about children who are "1 (standard deviation) unit" above the mean at age 10 and "1.2 units" above the mean at age 11, say—the "units" from which these scores are derived (i.e., the underlying age-specific standard deviations used in the standardization process) are themselves unlikely to have had either the same size or the same meaning.

Second, your outcomes must be equally valid across all measurement occasions. If you suspect that cross-wave validity might be compromised, you should replace the measure *before* data collection begins. Sometimes, as in the psychotherapy study, it is easy to argue that validity is maintained over time because the respondents have good reason to answer honestly on successive occasions. But in other studies, such as Coie and colleagues' (1996) antisocial behavior study, instrument validity over time may be more difficult to assert because young children may not understand all the questions about antisocial behavior included in the measure and older children may be less likely to answer honestly. Take the time to be cautious even when using instruments that appear valid on the surface. In his landmark paper on dilemmas in the measurement of change, Lord (1963) argued that, just because a measurement was valid on one occasion, it would not necessarily remain so on all subsequent occasions even when administered to the same individuals under the same conditions. He argued that a multiplication test may be a valid measure of mathematical skill among young children, but becomes a measure of memory among teenagers.

Third, you should try to preserve your outcome's precision over time, although precision need not be identical on every occasion. Within the logistical constraints imposed by data collection, the goal is to minimize errors introduced by instrument administration. An instrument that is "reliable enough" in a cross-sectional study—perhaps with a reliability of .8 or .9—will no doubt be sufficient for a study of change. So, too, the measurement error variance can vary across occasions because the methods we introduce can easily accommodate heteroscedastic error variation. Although the reliability of change measurement depends directly on outcome reliability, the precision with which you estimate individual change depends more on the number and spacing of the waves of data collection. In fact, by carefully choosing and placing the occasions of measurement, you can usually offset the deleterious effects of measurement error in the outcome.

2

Exploring Longitudinal Data
on Change

Change is the nursery of music, joy, life, and Eternity.
—John Donne

Wise researchers conduct descriptive exploratory analyses of their data before fitting statistical models. As when working with cross-sectional data, exploratory analyses of longitudinal data can reveal general patterns, provide insight into functional form, and identify individuals whose data do not conform to the general pattern. The exploratory analyses presented in this chapter are based on numerical and graphical strategies already familiar from cross-sectional work. Owing to the nature of longitudinal data, however, they are inevitably more complex in this new setting. For example, before you conduct even a single analysis of longitudinal data, you must confront a seemingly innocuous decision that has serious ramifications: how to store your longitudinal data efficiently. In section 2.1, we introduce two different data organizations for longitudinal data—the "person-level" format and the "person-period" format—and argue in favor of the latter.

We devote the rest of this chapter to describing exploratory analyses that can help you learn how different individuals in your sample change over time. These analyses serve two purposes: to identify important features of your data and to prepare you for subsequent model-based analyses. In section 2.2, we address the *within-person* question—How does each person change over time?—by exploring and summarizing *empirical growth records*, which list each individual's outcome values over time. In section 2.3, we address the *between-person* question—How does individual change differ across people?—by exploring whether different people change in similar or different ways. In section 2.4, we show how to ascertain descriptively whether observed differences in change across people (*interindividual differences in change*) are associated with individual characteristics. These between-person explorations can help identify variables that may ultimately prove to be important predictors of change. We conclude, in section 2.5, by examining the reliability and precision of exploratory estimates of change and commenting on their implications for the design of longitudinal studies.

2.1 Creating a Longitudinal Data Set

Your first step is to organize your longitudinal data in a format suitable for analysis. In cross-sectional work, data-set organization is so straightforward as to not warrant explicit attention—all you need is a "standard" data set in which each individual has his or her own record. In longitudinal work, data-set organization is less straightforward because you can use two very different arrangements:

- A *person-level data set*, in which each person has one record and multiple variables contain the data from each measurement occasion
- A *person-period data set*, in which each person has multiple records—one for each measurement occasion

A person-level data set has as many records as there are people in the sample. As you collect additional waves, the file gains new variables, not new cases. A person-period data set has many more records—one for each person-period combination. As you collect additional waves of data, the file gains new records, but no new variables.

All statistical software packages can easily convert a longitudinal data set from one format to the other. The website associated with our book presents illustrative code for implementing the conversion in a variety of statistical packages. If you are using SAS, for example, Singer (1998, 2001) provides simple code for the conversion. In STATA, the "reshape" command can be used. The ability to move from one format to the other means that you can enter, and clean, your data using whichever format is most convenient. But as we show below, when it comes to data analysis—either exploratory or inferential—you need to have your data in a person-period format because this most naturally supports meaningful analyses of change over time.

We illustrate the difference between the two formats in figure 2.1, which presents five waves of data from the *National Youth Survey* (NYS; Raudenbush & Chan, 1992). Each year, when participants were ages 11, 12, 13, 14, and 15, they filled out a nine-item instrument designed to assess their tolerance of deviant behavior. Using a four-point scale

"Person-Level" data set

ID	TOL11	TOL12	TOL13	TOL14	TOL15	MALE	EXPOSURE
9	2.23	1.79	1.9	2.12	2.66	0	1.54
45	1.12	1.45	1.45	1.45	1.99	1	1.16
268	1.45	1.34	1.99	1.79	1.34	1	0.9
314	1.22	1.22	1.55	1.12	1.12	0	0.81
442	1.45	1.99	1.45	1.67	1.9	0	1.13
514	1.34	1.67	2.23	2.12	2.44	1	0.9
569	1.79	1.9	1.9	1.99	1.99	0	1.99
624	1.12	1.12	1.22	1.12	1.22	1	0.98
723	1.22	1.34	1.12	1	1.12	0	0.81
918	1	1	1.22	1.99	1.22	0	1.21
949	1.99	1.55	1.12	1.45	1.55	1	0.93
978	1.22	1.34	2.12	3.46	3.32	1	1.59
1105	1.34	1.9	1.99	1.9	2.12	1	1.38
1542	1.22	1.22	1.99	1.79	2.12	0	1.44
1552	1	1.12	2.23	1.55	1.55	0	1.04
1653	1.11	1.11	1.34	1.55	2.12	0	1.25

"Person-Period" data set

ID	AGE	TOL	MALE	EXPOSURE
9	11	2.23	0	1.54
9	12	1.79	0	1.54
9	13	1.9	0	1.54
9	14	2.12	0	1.54
9	15	2.66	0	1.54
45	11	1.12	1	1.16
45	12	1.45	1	1.16
45	13	1.45	1	1.16
45	14	1.45	1	1.16
45	15	1.99	1	1.16
.	.	.	.	.
.	.	.	.	.
1653	11	1.11	0	1.25
1653	12	1.11	0	1.25
1653	13	1.34	0	1.25
1653	14	1.55	0	1.25
1653	15	2.12	0	1.25

Figure 2.1. Conversion of a person-level data set into a person-period data set for selected participants in the tolerance study.

(1 = very wrong, 2 = wrong, 3 = a little bit wrong, 4 = not wrong at all), they indicated whether it was wrong for someone their age to: (a) cheat on tests, (b) purposely destroy property of others, (c) use marijuana, (d) steal something worth less than five dollars, (e) hit or threaten someone without reason, (f) use alcohol, (g) break into a building or vehicle to steal, (h) sell hard drugs, or (i) steal something worth more than fifty dollars. At each occasion, the outcome, *TOL*, is computed as the respondent's average across the nine responses. Figure 2.1 also includes two potential predictors of change in tolerance: *MALE*, representing respondent gender, and *EXPOSURE*, assessing the respondent's self-reported exposure to deviant behavior at age 11. To obtain values of this latter predictor, participants estimated the proportion of their close friends who were involved in each of the same nine activities on a five-point scale (ranging from 0 = none, to 4 = all). Like *TOL*, each respondent's value of *EXPOSURE* is the average of his or her nine responses. Figure 2.1 presents data for a random sample of 16 participants from the larger NYS data set. Although the exploratory methods of this chapter apply in data sets of all sizes, we have kept this example purposefully small to enhance manageability and clarity. In later chapters, we apply the same methods to larger data sets.

2.1.1 The Person-Level Data Set

Many people initially store longitudinal data as a *person-level* data set (also known as the *multivariate format*), probably because it most resembles the familiar cross-sectional data-set format. The top panel of figure 2.1 displays the NYS data using this arrangement. The hallmark feature of a person-level data set is that each person has only one row (or "record") of data, regardless of the number of waves of data collection. A 16-person data set has 16 records; a 20,000-person data set has 20,000. Repeated measurements of each outcome appear as additional variables (hence the alternate "multivariate" label for the format). In the person-level data set of figure 2.1, the five values of tolerance appear in columns 2 through 6 (*TOL11, TOL12, . . . TOL15*). Suffixes attached to column headings identify the measurement occasion (here, respondent's age) and additional variables—here, *MALE* and *EXPOSURE*—appear in additional columns.

The primary advantage of a person-level data set is the ease with which you can examine visually each person's *empirical growth record*, his or her temporally sequenced outcome values. Each person's empirical growth record appears compactly in a single row making it is easy to assess quickly the way he or she is changing over time. In examining the top panel of figure 2.1, for example, notice that change differs considerably across

Table 2.1: Estimated bivariate correlations among tolerance scores assessed on five measurement occasions ($n = 16$)

	TOL11	TOL12	TOL13	TOL14	TOL15
TOL11	1.00				
TOL12	0.66	1.00			
TOL13	0.06	0.25	1.00		
TOL14	0.14	0.21	0.59	1.00	
TOL15	0.26	0.39	0.57	0.83	1.00

adolescents. Although most become more tolerant of deviant behavior over time (e.g., subjects 514 and 1653), many remain relatively stable (e.g., subjects 569 and 624), none of the 16 becomes much less tolerant (although subject 949 declines for a while before increasing).

Despite the ease with which you can examine each person's empirical growth record visually, the person-level data set has four disadvantages that render it a poor choice for most longitudinal analyses: (1) it leads naturally to noninformative summaries; (2) it omits an explicit "time" variable; (3) it is inefficient, or useless, when the number and spacing of waves varies across individuals; and (4) it cannot easily handle the presence of time-varying predictors. Below, we explain these difficulties; in section 2.1.2, we demonstrate how each is addressed by a conversion to a person-period data set.

First, let us begin by examining the five separate tolerance variables in the person-level data set of figure 2.1 and asking how you might analyze these longitudinal data. For most researchers, the instinctive response is to examine wave-to-wave relationships among *TOL11* through *TOL15* using bivariate correlation analyses (as shown in table 2.1) or companion bivariate plots. Unfortunately, summarizing the bivariate relationships between waves tells us little about change over time, for either individuals or groups. What, for example, does the weak but generally positive correlation between successive assessments of *TOLERANCE* tell us? For any pair of measures, say *TOL11* and *TOL12*, we know that adolescents who were more tolerant of deviant behavior at one wave tend to be more tolerant at the next. This indicates that the *rank order* of adolescents remains relatively stable across occasions. But it does not tell us *how* each person changes over time; it does not even tell us about the *direction* of change. If everyone's score declined by one point between age 11 and age 12, but the rank ordering was preserved, the correlation between waves would be positive (at +1)! Tempting though it is to infer a direct link between the wave-to-wave correlations and change, it is a futile exercise. Even with a small data set—here just five waves of data for 16 people—wave-to-wave correlations and plots tell us nothing about change over time.

Second, the person-level data set has no explicit numeric variable identifying the occasions of measurement. Information about "time" appears in the variable names, not in the data, and is therefore unavailable for statistical analysis. Within the actual person-level data set of figure 2.1, for example, information on *when* these *TOLERANCE* measures were assessed—the numeric values 11, 12, 13, 14, and 15—appears nowhere. Without including these values in the dataset, we cannot address within-person questions about the relationship between the outcome and "time."

Third, the person-level format is inefficient if either the number, or spacing, of waves varies across individuals. The person-level format is best suited to research designs with *fixed* occasions of measurement— each person has the same number of waves collected on the same exact schedule. The person-level data set of figure 2.1 is compact because the NYS used such a design—each adolescent was assessed on the same five annual measurement occasions (at ages 11, 12, 13, 14, and 15). Many longitudinal data sets do not share this structure. For example, if we reconceptualized "time" as the adolescent's *specific* age (say, in months) at each measurement occasion, we would need to expand the person-level data set in some way. We would need either five additional columns to record the respondent's precise age on each measurement occasion (e.g., variables with names like *AGE11, AGE12, AGE13, AGE14*, and *AGE15*) or even more additional columns to record the respondent's tolerance of deviant behavior on each of the many *unique* measurement occasions (e.g., variables with names like *TOL11.1, TOL11.2, . . . TOL15.11*). This latter approach is particularly impractical. Not only would we add 55 variables to the data set, we would have missing values in the cells corresponding to each month not used by a particular individual. In the extreme, if each person in the data set has his or her own unique data collection schedule—as would be the case were *AGE* recorded in days— the person-level format becomes completely unworkable. Hundreds of columns would be needed and most of the data entries would be missing!

Finally, person-level data sets become unwieldy when the values of *predictors* can vary over time. The two predictors in this data set are *time-invariant*—the values of *MALE* and *EXPOSURE* remain the same on every occasion. This allows us to use a single variable to record the values of each. If the data set contained *time-varying predictors*—predictors whose values vary over time—we would need an additional *set* of columns for each—one per measurement occasion. If, for example, exposure to deviant behavior were measured each year, we would need four additional columns. While the data could certainly be recorded in this way, this leads to the same disadvantages for time-varying predictors as we have just described for time-varying outcomes.

Taken together, these disadvantages render the person-level format, so familiar in cross-sectional research, ill suited to longitudinal work. Although we will return to the multivariate format in chapter 8, when we introduce a covariance structure analysis approach to modeling change (known as *latent growth modeling*), for now we suggest that longitudinal data analysis is facilitated—and made more meaningful—if you use the "person-period" format for your data.

2.1.2 The Person-Period Data Set

In a person-period data set, also known as *univariate format*, each individual has multiple records, one for each period in which he or she was observed. The bottom panel of figure 2.1 presents illustrative entries for the NYS data. Both panels present identical information; they differ only in *structure*. The person-period data set arrays each person's empirical growth record vertically, not horizontally. Person-period data sets therefore have fewer columns than person-level data sets (here, five instead of eight), but many more rows (here, 80 instead of 16). Even for this small example, the person-period data set has so many rows that figure 2.1 displays only a small subset.

All person-period data sets contain four types of variables: (1) a subject identifier; (2) a time indicator; (3) outcome variable(s); and (4) predictor variable(s). The *ID* number, which identifies the participant that each record describes, typically appears in the first column. Time-invariant by definition, *IDs* are identical across each person's multiple records. Including an *ID* number is more than good record keeping; it is an integral part of the analysis. Without an *ID*, you cannot sort the data set into person-specific subsets (a first step in examining individual change trajectories in section 2.2).

The second column in the person-period data set typically displays a *time indicator*—usually labeled *AGE, WAVE,* or *TIME*—which identifies the specific occasion of measurement that the record describes. For the NYS data, the second column of the person-period data set identifies the respondent's *AGE* (in years) on each measurement occasion. A dedicated time variable is a fundamental feature of every person-period data set; it is what renders the format amenable to recording longitudinal data from a wide range of research designs. You can easily construct a person-period data set even if each participant has a unique data collection schedule (as would be the case if we clocked time using each adolescent's precise age on the date of interview). The new *AGE* variable would simply record each adolescent's age on that particular date (e.g., 11.24, 12.32, 13.73, 14.11, 15.40 for one case; 11.10, 12.32, 13.59, 14.21, 15.69 for the next, etc.). A dedicated *TIME* variable also allows person-period data sets to accommodate research designs in which the number of measurement occasions differs across people. Each person simply has as many records as he or she has waves of data in the design. Someone with three waves will have three records; someone with 20 will have 20.

Each outcome in a person-period data set—here, just *TOL*—is represented by a single variable (hence the alternate "univariate" label for the format) whose values represent that person's score on each occasion. In figure 2.1, every adolescent has five records, one per occasion, each containing his or her tolerance of deviant behavior at the age indicated.

Every predictor—whether time-varying or time-invariant—is also represented by a single variable. A person-period data set can include as many predictors of either type as you would like. The person-period data set in figure 2.1 includes two time-invariant predictors, *MALE* and *EXPOSURE*. The former is time-invariant; the latter is time-invariant only because of the way it was constructed (as exposure to deviant behavior at one point in time, age 11). Time-invariant predictors have identical values across each person's multiple records; time-varying predictors have potentially differing values. We defer discussion of time-varying predictors to section 5.3. For now, we simply note how easy it is to include them in a person-period data set.

We hope that this discussion convinces you of the utility of storing longitudinal data in a person-period format. Although person-period data sets are typically longer than their person-level cousins, the ease with which they can accommodate any data collection schedule, any number of outcomes, and any combination of time-invariant and time-varying predictors outweigh the cost of increased size.

2.2 Descriptive Analysis of Individual Change over Time

Having created a person-period data set, you are now poised to conduct exploratory analyses that describe how individuals in the data set change over time. Descriptive analyses can reveal the nature and idiosyncrasies of each person's temporal pattern of growth, addressing the question: How does each person change over time? In section 2.2.1, we present a simple graphical strategy; in section 2.2.2, we summarize the observed trends by superimposing rudimentary fitted trajectories.

2.2.1 Empirical Growth Plots

The simplest way of visualizing how a person changes over time is to examine an *empirical growth plot*, a temporally sequenced graph of his or her empirical growth record. You can easily obtain empirical growth plots from any major statistical package: sort the person-period data set by subject identifier (*ID*) and separately plot each person's outcome vs. time (e.g., *TOL* vs. *AGE*). Because it is difficult to discern similarities and differences among individuals if each page contains only a single plot, we recommend that you cluster sets of plots in smaller numbers of panels.

Figure 2.2 presents empirical growth plots for the 16 adolescents in the NYS study. To facilitate comparison and interpretation, we use identical axes across panels. We emphasize this seemingly minor point because many statistical packages have the annoying habit of automatically expanding (or contracting) scales to fill out a page or plot area. When this happens, individuals who change only modestly acquire seemingly steep trajectories because the vertical axis expands to cover their limited outcome range; individuals who change dramatically acquire seemingly shallow trajectories because the vertical axis shrinks to accommodate their wide outcome range. If your axes vary inadvertently, you may draw erroneous conclusions about any similarities and differences in individual change.

Empirical growth plots can reveal a great deal about how each person changes over time. You can evaluate change in both absolute terms (against the outcome's overall scale) and in relative terms (in comparison to other sample members). Who is increasing? Who is decreasing? Who is increasing the most? The least? Does anyone increase and then decrease (or vice versa)? Inspection of figure 2.2 suggests that tolerance of deviant behavior generally increases with age (only subjects 314, 624, 723, and 949 do not fit this trend). But we also see that most adolescents remain in the lower portion of the outcome scale—here shown in its full extension from 1 to 4—suggesting that tolerance for deviant behavior never reaches alarming proportions (except, perhaps, for subject 978).

Should you examine every possible empirical growth plot if your data set is large, including perhaps thousands of cases? We do not suggest that you sacrifice a ream of paper in the name of data analysis. Instead, you can randomly select a subsample of individuals (perhaps stratified into groups defined by the values of important predictors) to conduct these exploratory analyses. All statistical packages can generate the random numbers necessary for such subsample selection; in fact, this is how we selected these 16 individuals from the NYS sample.

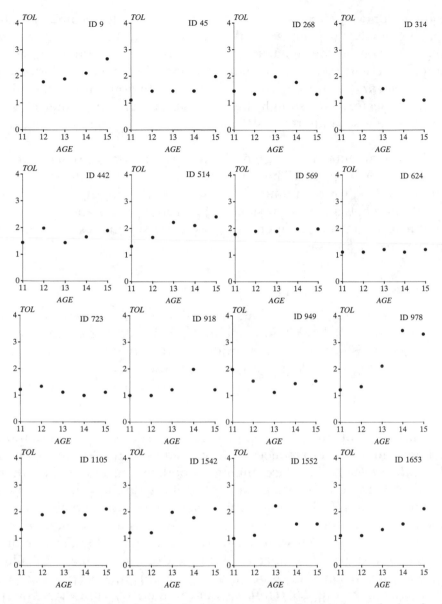

Figure 2.2. Exploring how individuals change over time. Empirical growth plots for 16 participants in the tolerance study.

2.2.2 Using a Trajectory to Summarize Each Person's Empirical Growth Record

It is easy to imagine summarizing the plot of each person's empirical growth record using some type of smooth trajectory. Although we often begin by drawing freehand trajectories, we strongly recommend that you also apply two standardized approaches. With the *nonparametric* approach, you let the "data speak for themselves" by smoothing across temporal idiosyncrasies without imposing a specific functional form. With the *parametric* approach, you select a common functional form for the trajectories—a straight line, a quadratic or some other curve—and then fit a separate regression model to each person's data, yielding a fitted trajectory.

The fundamental advantage of the nonparametric approach is that it requires no assumptions. The parametric approach requires assumptions but, in return, provides numeric summaries of the trajectories (e.g., estimated intercepts and slopes) suitable for further exploration. We find it helpful to begin nonparametrically—as these summaries often inform the parametric analysis.

Smoothing the Empirical Growth Trajectory Nonparametrically

Nonparametric trajectories summarize each person's pattern of change over time graphically without committing to a specific functional form. All major statistical packages provide several options for assumption-free smoothing, including the use of splines, loess smoothers, kernel smoothers, and moving averages. Choice of a particular smoothing algorithm is primarily a matter of convenience; all are adequate for the exploratory purposes we intend here.

Figure 2.3 plots the NYS empirical growth records and superimposes a smooth nonparametric trajectory (obtained using the "curve" option in *Harvard Graphics*). When examining smoothed trajectories like these, focus on their elevation, shape, and tilt. Where do the scores hover—at the low, medium, or high end of the scale? Does everyone change over time or do some people remain the same? What is the overall pattern of change? Is it linear or curvilinear; smooth or steplike? Do the trajectories have an inflection point or plateau? Is the rate of change steep or shallow? Is this rate of change similar or different across people? The trajectories in figure 2.3 reinforce our preliminary conclusions about the nature of individual change in the tolerance of deviant behavior. Most adolescents experience a gentle increase between ages 11 and 15, except for subject 978, who registers a dramatic leap after age 13.

After examining the nonparametric trajectories individually, stare at the entire set together as a group. Group-level analysis can help inform decisions that you will soon need to make about a functional form for the trajectory. In our example, several adolescents appear to have linear trajectories (subjects 514, 569, 624, and 723) while others have

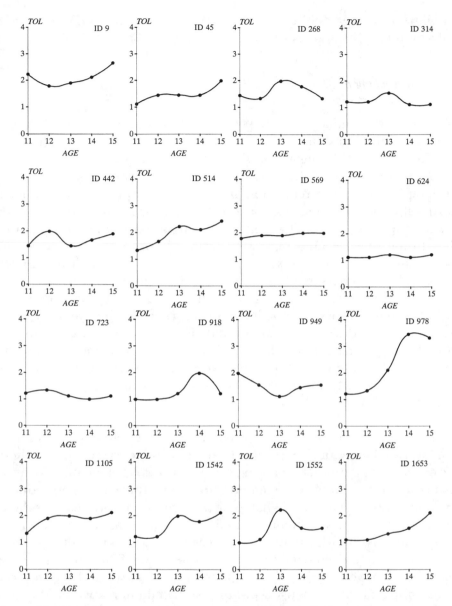

Figure 2.3. Smooth nonparametric summaries of how individuals change over time. Smooth nonparametric trajectories superimposed on empirical growth plots for participants in the tolerance study.

curvilinear ones that either accelerate (9, 45, 978, and 1653) or rise and fall around a central peak or trough (268, 314, 918, 949, 1552).

Smoothing the Empirical Growth Trajectory Using OLS Regression

We can also summarize each person's growth trajectory by fitting a separate parametric model to each person's data. Although many methods of model fitting are possible, we find that ordinary least squares (OLS) regression is usually adequate for exploratory purposes. Of course, fitting person-specific regression models, one individual at a time, is hardly the most efficient use of longitudinal data; that's why we need the multilevel model for change that we will soon introduce. But because the "fitting of little OLS regression models" approach is intuitive and easy to implement in a person-period data set, we find that it connects empirical researchers with their data in a direct and intimate way.

To fit an exploratory OLS regression model to each person's data, you must first select a specific functional form for that model. Not only is this decision crucial during exploratory analysis, it becomes even more important during formal model fitting. Ideally, substantive theory and past research will guide your choice. But when you observe only a restricted portion of the life span—as we do here—or when you have only three or four waves of data, model selection can be difficult.

Two factors further complicate the choice of a functional form. First, exploratory analyses often suggest that different people require different functions—change might appear linear for some, curvilinear for others. We observe this pattern, to some extent, in figure 2.3. Yet the simplification that comes from adopting a common functional form across everyone in the data set is so compelling that its advantages totally outweigh its disadvantages. Adopting a common functional form across everyone in the sample allows you to distinguish people easily using the same set of numerical summaries derived from their fitted trajectories. This process is especially simple if you adopt a linear change model, as we do here; you can then compare individuals using just the estimated intercepts and slopes of their fitted trajectories. Second, measurement error makes it difficult to discern whether compelling patterns in the empirical growth record really reflect true change or are simply due to random fluctuation. Remember, each *observed score* is just a fallible operationalization of an underlying *true score*—depending upon the sign of the error, the observed score can be inappropriately high or low. The empirical growth records do not present a person's true pattern of change over time; they present the fallible observed reflection of that change. Some of what we see in the empirical growth records and plots is nothing more than measurement error.

These complications argue for parsimony when selecting a functional form for exploratory analysis, driving you to adopt the simplest trajectory that can do the job. Often the best choice is simply a straight line. In this example, we adopted a linear individual change trend because it provides a decent description of the trajectories for these 16 adolescents. In making this decision, of course, we assume implicitly that any deviations from linearity in figure 2.3 result from either the presence of outliers or measurement error. Use of an individual linear change model simplifies our discussion enormously and has pedagogic advantages as well. We devote chapter 6 to a discussion of models for discontinuous and nonlinear change.

Having selected an appropriate parametric form for summarizing the empirical growth records, you obtain fitted trajectories using a three-step process:

1. Estimate a within-person regression model for each person in the data set. With a linear change model, simply regress the outcome (here *TOL*) on some representation of time (here, *AGE*) in the person-period data set. Be sure to conduct a separate analysis for each person (i.e., conduct the regression analyses "by *ID*").
2. Collect summary statistics from all the within-person regression models into a separate data set. For a linear-change model, each person's estimated intercept and slope summarize their growth trajectory; the R^2 and residual variance statistics summarize their goodness of fit.
3. Superimpose each person's fitted regression line on a plot of his or her empirical growth record. For each person, plot selected predicted values and join them together smoothly.

We now apply this three-step process to the NYS data.

We begin by fitting a separate linear change model to each person's empirical growth record. Although we can regress *TOL* on *AGE* directly, we instead regress *TOL* on (*AGE* − *11*) years, providing a *centered* version of *AGE*. Centering the temporal predictor is optional, but doing so improves the interpretability of the intercept. Had we not centered *AGE*, the fitted intercept would estimate the adolescent's tolerance of deviant behavior at age 0—an age beyond the range of these data and hardly one at which a child can report an attitude. Subtracting 11 years from each value of *AGE* moves the origin of the plot so that each intercept now estimates the adolescent's tolerance of deviant behavior at the more reasonable age of 11 years.

Centering *AGE* has no effect on the interpretation of each person's slope: it still estimates his or her annual rate of change. Adolescents with positive slopes grow more tolerant of deviant behavior as they age; those with the largest slopes become more tolerant the most rapidly. Adoles-

Table 2.2: Results of fitting separate within-person exploratory OLS regression models for *TOLERANCE* as a function of linear time

| | Initial status | | Rate of change | | Residual | | | |
ID	Estimate	se	Estimate	se	variance	R²	*MALE*	*EXPOSURE*
0009	1.90	0.25	0.12	0.10	0.11	0.31	0	1.54
0045	1.14	0.13	0.17	0.05	0.03	0.77	1	1.16
0268	1.54	0.26	0.02	0.11	0.11	0.02	1	0.90
0314	1.31	0.15	−0.03	0.06	0.04	0.07	0	0.81
0442	1.58	0.21	0.06	0.09	0.07	0.14	0	1.13
0514	1.43	0.14	0.27	0.06	0.03	0.88	1	0.90
0569	1.82	0.03	0.05	0.01	0.00	0.88	0	1.99
0624	1.12	0.04	0.02	0.02	0.00	0.33	1	0.98
0723	1.27	0.08	−0.05	0.04	0.01	0.45	0	0.81
0918	1.00	0.30	0.14	0.13	0.15	0.31	0	1.21
0949	1.73	0.24	−0.10	0.10	0.10	0.25	1	0.93
0978	1.03	0.32	0.63	0.13	0.17	0.89	1	1.59
1105	1.54	0.15	0.16	0.06	0.04	0.68	1	1.38
1542	1.19	0.18	0.24	0.07	0.05	0.78	0	1.44
1552	1.18	0.37	0.15	0.15	0.23	0.25	0	1.04
1653	0.95	0.14	0.25	0.06	0.03	0.86	0	1.25

cents with negative slopes grow less tolerant of deviant behavior over time; those with the most negative slopes become less tolerant the most rapidly. Because the fitted slopes estimate the annual rate of change in the outcome, they are the parameter of central interest in an exploratory analysis of change.

Table 2.2 presents the results of fitting 16 linear-change OLS regression models to the NYS data. The table displays OLS-estimated intercepts and slopes for each person along with associated standard errors, residual variance, and R^2 statistics. Figure 2.4 presents a stem-and-leaf display of each summary statistic. Notice that both the fitted intercepts and slopes vary considerably, reflecting the heterogeneity in trajectories observed in figure 2.3. Although most adolescents have little tolerance for deviant behavior at age 11, some—like subjects 9 and 569—are more tolerant. Notice, too, that many adolescents register little change over time. Comparing the estimated slopes to their associated standard errors, we find that the slopes for nine people (subjects 9, 268, 314, 442, 624, 723, 918, 949, and 1552) are indistinguishable from 0. Three have moderate increases (514, 1542, and 1653) and one extreme case (978) increases three times faster than his closest peer.

Figure 2.5 superimposes each adolescent's fitted OLS trajectory on his or her empirical growth plot. All major statistical packages can generate

Fitted initial status	
1.9	0
1.8	2
1.7	3
1.6	
1.5	4 4 8
1.4	3
1.3	1
1.2	7
1.1	2 4 8 9
1	0 3
0.9	5

Fitted rate of change	
0.6	3
0.5	
0.4	
0.3	
0.2	4 5 7
0.1	2 4 5 6 7
0	2 2 5 6
⁻0	3 5
-0.1	0

Residual variance	
.2 lo	3
.1 hi	5 7
.1 lo	0 1 1
.0 hi	5 7
.0 lo	0 0 1 3 3 3 4 4

R^2 statistic	
0.8	6 8 8 9
0.7	7 8
0.6	8
0.5	
0.4	5
0.3	1 1 3
0.2	5 5
0.1	4
0	2 7

Figure 2.4. Observed variation in fitted OLS trajectories. Stem and leaf displays for fitted initial status, fitted rate of change, residual variance, and R^2 statistic resulting from fitting separate OLS regression models to the tolerance data.

such plots. For example, because the estimated intercept and slope for subject 514 are 1.43 and 0.27, the fitted values at ages 11 and 15 are: 1.43 (computed as $1.43 + 0.27(11 - 11)$) and 2.51 (computed as $1.43 + 0.27(15 - 11)$). To prevent extrapolation beyond the temporal limits of the data, we plot this trajectory only between ages 11 and 15.

Comparing the exploratory OLS-fitted trajectories with the observed data points allows us to evaluate how well the chosen linear change model fits each person's growth record. For some adolescents (such as 569 and 624), the linear change model fits well—their observed and fitted values nearly coincide. A linear change trajectory may also be reasonable for many other sample members (including subjects 45, 314, 442, 514, 723, 949, 1105, and 1542) if we are correct in regarding the observed deviations from the fitted trajectory as random error. For five adolescents (subjects 9, 268, 918, 978, and 1552), observed and fitted values are more disparate. Inspection of their empirical growth records suggests that their change may warrant a curvilinear model.

Table 2.2 presents two simple ways of quantifying the quality of fit for each person: an individual R^2 statistic and an individual estimated residual variance. Even in this small sample, notice the striking variability in

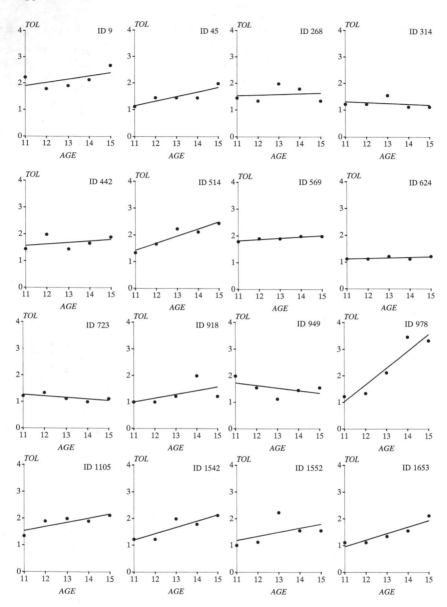

Figure 2.5. OLS summaries of how individuals change over time. Fitted OLS trajectories superimposed on empirical growth plots for participants in the tolerance study.

the individual R^2 statistics. They range from a low of 2% for subject 268 (whose trajectory is essentially flat and whose data are widely scattered) to highs of 88% for subjects 514 and 569 (whose empirical growth records show remarkable linearity in change) and 89% for subject 978 (who has the most rapid rate of growth). The individual estimated residual variances mirror this variability (as you might expect, given that they are an element in the computation of the R^2 statistic). Skewed by definition (as apparent in figure 2.4), they range from a low near 0 for subjects 569 and 624 (whose data are predicted nearly perfectly) to highs of 0.17 and 0.23 for subjects 978 and 1552 (who each have an extreme observation). We conclude that the quality of exploratory model fit varies substantially from person to person; the linear change trajectory works well for some sample members and poorly for others.

By now you may be questioning the wisdom of using OLS regression methods to conduct even exploratory analyses of these data. OLS regression methods assume independence and homoscedasticity of residuals. Yet these assumptions are unlikely to hold in longitudinal data where residuals tend to be autocorrelated and heteroscedastic over time within person. Despite this concern, OLS estimates can be very useful for exploratory purposes. Although they are less efficient when the assumption of residual independence is violated (i.e., their sampling variance is too high), they still provide unbiased estimates of the intercept and slope of the individual change (Willett, 1989). In other words, these exploratory estimates of the key features of the individual change trajectory—each person's intercept and slope—will be on target, if a little noisy.

2.3 Exploring Differences in Change across People

Having summarized how each individual changes over time, we now examine similarities and differences in these changes across people. Does everyone change in the same way? Or do the trajectories of change differ substantially across people? Questions like these focus on the assessment of *interindividual differences* in change.

2.3.1 Examining the Entire Set of Smooth Trajectories

The simplest way of exploring interindividual differences in change is to plot, on a single graph, the entire set of smoothed individual trajectories. The left panel of figure 2.6 presents such a display for the NYS data using the nonparametric smoother; the right panel presents a similar display using OLS regression methods. In both, we omit the observed data to decrease clutter.

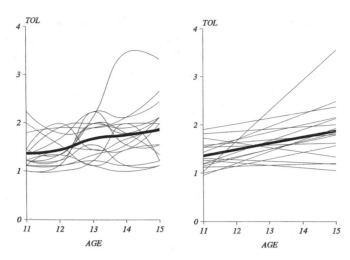

Figure 2.6. Examining the collection of smooth nonparametric and OLS trajectories across participants in the tolerance study. Panel A presents the collection of smooth nonparametric trajectories; Panel B presents the collection of fitted OLS trajectories. Both panels also present an *average change trajectory* for the entire group.

Each panel in figure 2.6 also includes a new summary: an *average change trajectory* for the entire group. Depicted in bold, this summary helps us compare individual change with group change. Computing an average change trajectory is a simple two-step process. First, sort the person-period data set by time (here, *AGE*), and separately estimate the mean outcome (here, *TOLERANCE*) for each occasion of measurement. Second, plot these time-specific means and apply the same smoothing algorithm, nonparametric or parametric, used to obtain the individual trajectories.

Both panels in figure 2.6 suggest that, on average, the change in tolerance of deviant behavior between ages 11 and 15 is positive but modest, rising by one to two-tenths of a point per year (on this 1 to 4 scale). This suggests that as adolescents mature, they gradually tolerate more deviant behavior. Note that even the nonparametrically smoothed average trajectory seems approximately linear. (The slight curvature or discontinuity between ages 12 and 13 disappears if we set aside the extreme case, subject 978.) Both panels also suggest substantial interindividual heterogeneity in change. For some adolescents, tolerance increases moderately with age; for others, it remains stable; for some, it declines. This heterogeneity creates a "fanning out" of trajectories as increasing age engenders greater diversity in tolerance. Notice that the OLS regression panel is somewhat easier to interpret because of its greater structure.

Although the average change trajectory is a valuable summary, we inject a note of caution: the shape of the average change trajectory may not mimic the shape of the individual trajectories from which it derives. We see this disconcerting behavior in figure 2.6, where the nonparametrically smoothed trajectories manifest various curvilinear shapes but the average trajectory is nearly linear. This means that you should never infer the shape of the individual change trajectories from the shape of their average. As we explain in section 6.4, the only kind of trajectory for which the "average of the curves" is identical to the "curve of the averages" is one whose mathematical representation is *linear in the parameters* (Keats, 1983). All polynomials—including linear, quadratic, and cubic trajectories —are linear in the parameters; their average trajectory will always be a polynomial of the same order as the individual trajectories. The average of a set of straight lines will be a straight line; the average of a set of quadratics will be a quadratic. But many other common curves do not share this property. The average of a set of logistic curves, for example, is usually a smoothed-out step function. This means that you must exercise extreme caution when examining an average growth trajectory. We display the average simply for comparison, not to learn anything about underlying shapes of the individual trajectories.

2.3.2 Using the Results of Model Fitting to Frame Questions about Change

Adopting a parametric model for individual change allows us to re-express *generic* questions about interindividual differences in "change" as *specific* questions about the behavior of parameters in the individual models. If we have selected our parametric model wisely, little information is lost and great simplification is achieved. If you adopt a linear individual change model, for instance, you are implicitly agreeing to summarize each person's growth using just two parameter estimates: (1) the fitted intercept; and (2) the fitted slope. For the NYS data, variation in fitted intercepts across adolescents summarizes observed interindividual differences in tolerance at age 11. If these intercepts describe fitted values at the first wave of data collection, as they do here, we say that they estimate someone's "initial status." Variation in the fitted slopes describes observed interindividual differences in the rates at which tolerance for deviant behavior changes over time.

Greater specificity and simplification accrues if we reframe general questions about interindividual heterogeneity in change in terms of key parameters of the individual change trajectory. Rather than asking "Do individuals differ in their changes, and if so, how?" we can now ask "Do individuals differ in their intercepts? In their slopes?" To learn about the observed *average* pattern of change, we examine the sample averages of the fitted intercepts and slopes; these tell us about the average initial status and the average annual rate of change in the sample as a whole. To learn about the observed *individual differences* in change, we examine the sample *variances* and *standard deviations* of the intercepts and slopes; these tell us about the observed variability in initial status and rates of change in the sample. And to learn about the observed relationship between initial status and the rate of change, we can examine the sample *covariance* or *correlation* between intercepts and slopes.

Formal answers to these questions require the multilevel model for change of chapter 3. But we can presage this work by conducting simple descriptive analyses of the estimated intercepts and slopes. In addition to plotting their distribution (as in figure 2.4), we can examine standard descriptive statistics (means and standard deviations) and bivariate summaries (correlation coefficients) obtained using the data set that describes the separate fitted regression results in table 2.2.

We find it helpful to examine three specific quantities, the:

- *Sample means of the estimated intercepts and slopes.* The level-1 OLS-estimated intercepts and slopes are unbiased estimates of initial status and rate of change for each person. Their sample means are therefore unbiased estimates of the key features of the average observed change trajectory.
- *Sample variances (or standard deviations) of the estimated intercepts and slopes.* These measures quantify the amount of observed interindividual heterogeneity in change.
- *Sample correlation between the estimated intercepts and slopes.* This correlation summarizes the association between fitted initial status and fitted rate of change and answers the question: Are observed initial status and rate of change related?

Results of these analyses for the NYS data appear in table 2.3.

Across this sample, we find an average estimated intercept of 1.36 and an average estimated slope of 0.13. We therefore conclude that the average adolescent in this sample has an observed tolerance level of 1.36 at age 11 and that this increases by an estimated 0.13 points per year. The magnitude of the sample standard deviations (in comparison to their means) suggests that adolescents are scattered widely around both these averages. This tells us that the adolescents differ considerably in their fitted initial status and fitted rates of change. Finally, the correlation coefficient of −0.45 indicates a negative relationship between fitted initial status and fitted rate of change, suggesting that adolescents with greater

Table 2.3: Descriptive statistics for the individual growth parameters obtained by fitting separate within-person OLS regression models for *TOLERANCE* as a function of linear time ($n = 16$)

	Initial status (intercept)	Rate of change (slope)
Mean	1.36	0.13
Standard deviation	0.30	0.17
Bivariate correlation		−0.45

initial tolerance tend to become more tolerant less rapidly over time (although we must be cautious in our interpretation because of negative bias introduced by the presence of measurement error).

2.3.3 Exploring the Relationship between Change and Time-Invariant Predictors

Evaluating the impact of predictors helps you uncover systematic patterns in the individual change trajectories corresponding to interindividual variation in personal characteristics. For the NYS data, we consider two time-invariant predictors: *MALE* and *EXPOSURE*. Asking whether the observed tolerance trajectories differ by gender allows us to explore whether boys (or girls) are initially more tolerant of deviant behavior and whether they tend to have different annual rates of change. Asking whether the observed tolerance trajectories differ by early exposure to deviant behavior (at age 11) allows us to explore whether a child's fitted initial level of tolerance is associated with early exposure and whether the fitted rate of change in tolerance is related as well. All of these questions focus on *systematic interindividual differences in change*.

Graphically Examining Groups of Smoothed Individual Growth Trajectories

Plots of smoothed individual growth trajectories, displayed separately for groups distinguished by important predictor values, are valuable exploratory tools. If a predictor is categorical, display construction is straightforward. If a predictor is continuous, you can temporarily categorize its values. For example, we split *EXPOSURE* at its median (1.145) for the purposes of display. For numeric analysis, of course, we continue to use its continuous representation.

Figure 2.7 presents smoothed OLS individual growth trajectories separately by gender (upper pair of panels) and exposure (lower pair of

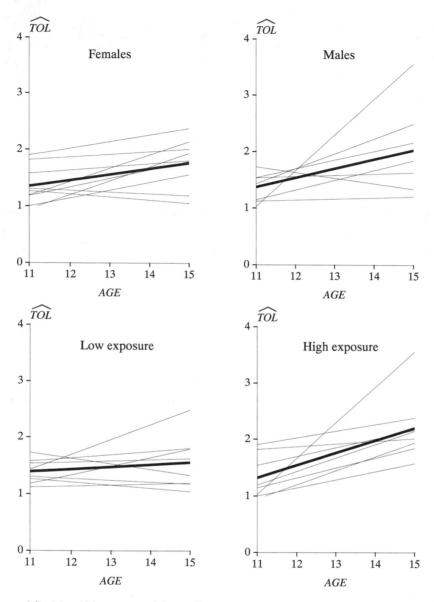

Figure 2.7. Identifying potential predictors of change by examining OLS fitted trajectories separately by levels of selected predictors. Fitted OLS trajectories for the tolerance data displayed separately by gender (upper panel) and exposure (lower panel).

panels). The bold trajectory in each panel depicts the average trajectory for the subgroup. When you examine plots like these, look for systematic patterns: Do the observed trajectories differ across groups? Do observed differences appear more in the intercepts or in the slopes? Are some groups' observed trajectories more heterogeneous than others'? Setting aside subject 978, who had extremely rapid growth, we find little difference in the distribution of fitted trajectories by gender. Each group's average observed trajectory is similar in intercept, slope, and scatter. We also find little difference in fitted initial status by exposure, but we do discern a difference in the fitted rate of change. Even discounting subject 978, those with greater initial exposure to deviant behavior seem to become tolerant more rapidly as they age.

The Relationship between OLS-Estimated Trajectories and Substantive Predictors

Just as we described the distribution of fitted intercepts and slopes in section 2.3, we can also use them as objects of further exploratory analysis. To investigate whether fitted trajectories vary systematically with predictors, we can treat the estimated intercepts and slopes as outcomes and explore the relationship between them and predictors. For the NYS data, these analyses explore whether the initial tolerance of deviant behavior or the annual rate of change in tolerance is observed to differ by: (1) gender or (2) early exposure to deviant behavior.

Because these analyses are exploratory—soon to be replaced in chapter 3 by the fitting of a multilevel model for change—we restrict ourselves to the simplest of approaches: the use of bivariate plots and sample correlations. Figure 2.8 plots the fitted intercepts and slopes versus the two predictors: *MALE* and *EXPOSURE*. Accompanying each plot is a sample correlation coefficient. All signs point to little or no gender differential in either fitted initial status or rate of change. But with respect to *EXPOSURE*, it does appear that adolescents with greater early exposure to deviant behavior become more tolerant at a faster rate than peers who were less exposed.

Despite their utility for descriptive and exploratory analyses, OLS estimated intercepts and slopes are hardly the final word in the analysis of change. Estimates are not true values—they are imperfect measures of each person's true initial status and true rate of change. They have biases that operate in known directions; for example, their sample variances are inflated by the presence of measurement error in the outcome. This means that the variance in the true rate of change will necessarily be smaller than the variance of the fitted slope because part of the latter's

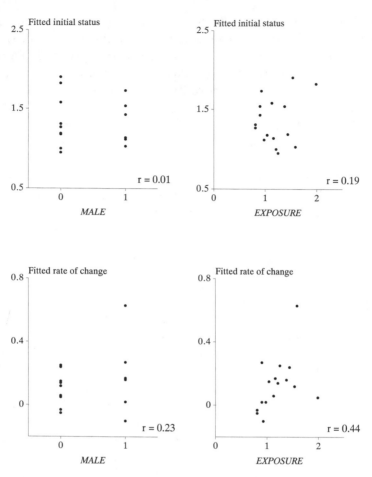

Figure 2.8. Examining the relationship between OLS parameter estimates (for initial status and rates of change) and potential predictors. Fitted OLS intercepts and slopes for the tolerance data plotted vs. two predictors: *MALE* and *EXPOSURE*.

variability is error variation. So, too, the sample correlation between the fitted intercept and slope is negatively biased (it underestimates the population correlation) because the measurement error in fitted initial status is embedded, with opposite sign, in the fitted rate of change.

These biases suggest that you should use the descriptive analyses of this chapter for exploratory purposes only. They can help you get your feet wet and in touch with your data. Although it is technically possible to improve these estimates—for example, we can deflate the sample variances of OLS estimates and we can correct the correlation coefficient for measurement error (Willett, 1989)—we do not recommend expending this extra effort. The need for ad hoc corrections has been effectively replaced by the widespread availability of computer software for fitting the multilevel model for change directly.

2.4 Improving the Precision and Reliability of OLS-Estimated Rates of Change: Lessons for Research Design

Before introducing the multilevel model for change, let us examine another feature of the within-person exploratory OLS trajectories introduced in this chapter: the precision and reliability of the estimated rates of change. We do so not because we will be using these estimates for further analysis, but because it allows us to comment on—in a particularly simple arena—some fundamental principles of longitudinal design. As you would hope, these same basic principles also apply directly to the more complex models we will soon introduce.

Statisticians assess the precision of a parameter estimate in terms of its *sampling variation*, a measure of the variability that would be found across infinite resamplings from the same population. The most common measure of sampling variability is an estimate's *standard error*, the square root of its estimated sampling variance. Precision and standard error have an inverse relationship; the smaller the standard error, the more precise the estimate. Table 2.2 reveals great variability in the standard errors of the individual slope estimates for the NYS data. For some, the estimated rate of change is very precise (e.g., subjects 569 and 624); for others, it is not (e.g., subject 1552).

Understanding why the individual slope estimates vary in precision provides important insights into how you can improve longitudinal studies of change. Standard results from mathematical statistics tell us that the precision of an OLS-estimated rate of change depends upon an individual's: (1) residual variance, the vertical deviations of observed values around the fitted line; and (2) number and spacing of the waves of longitudinal data. If individual i has T waves of data, gathered at times $t_{i1}, t_{i2}, \ldots, t_{iT}$, the sampling variance of the OLS-estimated rate of change is[1]:

$$\begin{pmatrix} \text{Sampling variance} \\ \text{of the OLS rate of change} \\ \text{for individual } i \end{pmatrix} = \frac{\sigma_{\varepsilon_i}^2}{\sum\limits_{j=1}^{T}(t_{ij}-\bar{t}_i)^2} = \frac{\sigma_{\varepsilon_i}^2}{CSST_i}, \tag{2.1}$$

where $\sigma_{\varepsilon_i}^2$ represents the residual variance for the *ith* individual and $CSST_i$ represents his or her corrected sum of squares for *TIME*, the sum of squared deviations of the time values around the average time, $\bar{t}_i$.

Equation 2.1 suggests two ways of increasing the precision of OLS estimated rates of change: (1) decrease the residual variance (because it appears in the numerator); or (2) increase variability in measurement times (because the corrected sums of squares for time appears in the denominator). Of course, the magnitude of the residual variance is largely outside your control; strictly speaking, you cannot directly modify its value. But because at least some of the residual variance is nothing more than measurement error, you can improve precision by using outcome measures with better psychometric properties.

Greater improvements in precision accrue if you work to increase the corrected sum of squares for time by modifying your research design. Inspection of equation 2.1 indicates that the greater the variability in the timing of measurement, the more precise the assessment of change. There are two simple ways of achieving increased variability in the timing of measurement: (1) redistribute the timing of the planned measurement occasions to be further away from their average; and (2) increase the number of waves. Both strategies yield substantial payoffs because it is the *squared* deviations of the measurement times about their average in the denominator of equation 2.1. A change as simple as adding another wave of data to your research design, far afield from the central set of observations, can reap dramatic improvements in the precision with which change can be measured.

We can reach similar conclusions by examining the reliability of the OLS estimated rates of change. Even though we believe that precision is a better criterion for judging measurement quality, we have three reasons for also examining reliability. First, the issue of reliability so dominates the literature on the measurement of change that it may be unwise to avoid all discussion. Second, it is useful to define reliability explicitly so as to distinguish it mathematically from precision. Third, even though reliability and precision are different criteria for evaluating measurement quality, they do, in this case, lead to similar recommendations about research design.

Unlike precision, which describes how well an individual slope estimate measures that person's true rate of change, reliability describes how much the rate of change varies across people. Precision has meaning for the individual; reliability has meaning for the group. Reliability is defined in terms of interindividual variation: it is the proportion of a measure's observed variance that is true variance. When test developers claim that a test has a reliability of .90 in a population, they mean that 90% of the person-to-person variation in observed scores across the population is variability in true scores.

Reliability of change is defined similarly. The population reliability of the OLS slope is the proportion of population variance in observed rate of change that is variance in true rate of change (see Rogosa et al., 1982; Willett, 1988, 1989). If reliability is high, a large portion of the interindividual differences in observed rate of change will be differences in true rate of change. Were we to rank everyone in the population on their observed changes, we would then be pretty confident that the rankings reflect the rank order of the true changes. If reliability is low, the rankings on observed change might not reflect the true underlying rankings at all.

Improvements in precision generally lead to improvements in reliability —when you measure individual change more accurately, you can better distinguish individuals on the basis of these changes. But as a group-level parameter, reliability's magnitude is also affected by the amount of variability in true change in the population. If everyone has an identical value of true rate of change, you will be unable to effectively distinguish among people even if their observed rates of change are precise, so reliability will be zero. This means that you can simultaneously enjoy excellent individual precision for the rate of change and poor reliability for detecting interindividual differences in change; you can measure everyone's change well, but be unable to distinguish people because everyone's changes are identical. For a constant level of measurement precision, as population heterogeneity in true change increases, so does reliability.

The disadvantage of reliability as a gauge of measurement quality is that it confounds the effect of within-person precision with the effect of between-person heterogeneity in true change. When individual precision is poor or when interindividual heterogeneity in true change is small, reliability tends to 0. When precision is high or when heterogeneity in true change is large, reliability tends to 1. This means that reliability does not tell you uniquely about either precision or heterogeneity in true change; instead, it tells you about both simultaneously, impairing its value as an indicator of measurement quality.

We can confirm these inadequacies algebraically, albeit under a pair of limiting assumptions: (1) that the longitudinal data are fully balanced—everyone in the population is observed on the same set of occasions, t_1, t_2, ..., t_T; and (2) that each person's residuals are drawn identically and independently from a common distribution with variance σ_ε^2. The population reliability of the OLS estimate of individual rate of change is then:

$$\text{Reliability of the OLS rate of change} = \frac{\sigma_{True\,Slope}^2}{\sigma_{True\,Slope}^2 + \dfrac{\sigma_\varepsilon^2}{CSST}}, \qquad (2.2)$$

where $\sigma^2_{True\,Slope}$ is the population variance of the true rate of change and *CSST* is the corrected sum-of-squares-time, now common across individuals (Willett, 1988). Because $\sigma^2_{True\,Slope}$ appears in both the numerator and denominator, it plays a central role in determining reliability. If everyone is growing at the same true rate, all true growth trajectories will be parallel and there will be no variability in the true rate of change across people. When this happens, both $\sigma^2_{True\,Slope}$ and the reliability of change will be 0, no matter how precisely the individual change is measured. Ironically, this means that the OLS slope can be a very precise yet completely unreliable measure of change. If there are large differences in the true rate of change across people, the true growth trajectories will crisscross considerably. When this happens, $\sigma^2_{True\,Slope}$ will be large, dominating both numerator and denominator, and the reliability of the OLS slope will tend to 1, regardless of its precision. This means that the OLS slope can be an imprecise yet reliable measure of change. The conclusion: you can be fooled about the quality of your change measurement if you use reliability as your sole criterion.

We can also use equation 2.2 to reinforce our earlier conclusions about longitudinal research design. First, for a given level of interindividual difference in true change in the population, the reliability of the OLS slope depends solely on the residual variance. Once again, the better the quality of your outcome measurement, the better the reliability with which change can be measured because at least part of the residual variance is simply measurement error. Second, reliability can be improved through design, by manipulating the number and spacing of the measurement occasions. Anything that you can do to increase corrected sum-of-squares time, *CSST*, will help. As you add waves of data or move the existing waves further away from the center of the data collection period, the reliability with which change can be measured will improve.

3

Introducing the Multilevel Model for Change

When you're finished changing, you're finished
—Benjamin Franklin

In this chapter, we introduce the multilevel model for change, demonstrating how it allows us to address within-person and between-person questions about change simultaneously. Although there are several ways of writing the statistical model, here we adopt a simple and common approach that has much substantive appeal. We specify the multilevel model for change by simultaneously postulating a pair of subsidiary models—a level-1 submodel that describes how each person changes over time, and a level-2 model that describes how these changes differ across people (Bryk & Raudenbush, 1987; Rogosa & Willett, 1985).

We begin, in section 3.1, by briefly reviewing the rationale and purpose of statistical models in general and the multilevel model for change in particular. We then introduce the level-1 model for individual change (section 3.2) and the level-2 model for interindividual heterogeneity in change (section 3.3). In section 3.4, we provide an initial foray into the world of estimation, introducing the method of maximum likelihood. (We discuss other methods of estimation in subsequent chapters.) We close, in sections 3.5 and 3.6, by illustrating how the resultant parameter estimates can be interpreted and how key hypotheses can be tested.

We do not intend this chapter to present a complete and general account of the multilevel model for change. Our goal is to provide a single "worked" example—from beginning to end—that illustrates all the steps you must go through when specifying the model, fitting it to data, and interpreting its results. We proceed in this way because we believe it is easier to learn about the model by first walking through a simple, but complete, analysis in a constrained, yet realistic, context. This minimizes notational and analytic complexity and lets us focus on interpretation and understanding. As a result, this chapter is limited to: (1) a linear change model for individual growth; (2) a time-structured data set in which everyone shares an identical data collection schedule; (3) an evaluation of the impact of a single dichotomous time-invariant predictor; and (4) the use of one piece of dedicated statistical software, HLM. In subsequent chapters, we extend this basic model in many ways, generalizing it to situations in which growth is curvilinear or discontinuous; the timing, spacing, and number of waves of data differ across individuals; interest centers on the effects of many predictors, both discrete and continuous, time-invariant and time-varying; distributional assumptions differ; and other methods of estimation and statistical software are used.

3.1 What Is the Purpose of the Multilevel Model for Change?

Even though you have surely fit many types of statistical models in your data analytic career, experience tells us that when researchers get caught up in a novel and complex analysis, they often need to be reminded just what a statistical model is and what it is not. So before presenting the multilevel model for change itself, we briefly review the purpose of statistical models.

Statistical models are mathematical representations of population behavior; they describe salient features of the hypothesized process of interest among individuals in the target population. When you use a particular statistical model to analyze a particular set of data, you implicitly declare that *this* population model gave rise to *these* sample data. Statistical models are not statements about sample behavior; they are statements about the *population process* that generated the data.

To provide explicit statements about population processes, statistical models are expressed using parameters—intercepts, slopes, variances, and so on—that represent specific population quantities of interest. Were you to use the following simple linear regression model to represent the relationship between infant birth weight (in pounds) and neurological functioning on a single occasion in a cross-sectional data set (with the usual notation) $NEURO_i = \beta_0 + \beta_1 (BWGT_i - 3) + \varepsilon_i$, you would be declaring implicitly that, in the population from which your sample was drawn: (1) β_0 is an unknown intercept parameter that represents the expected level of neurological functioning for a three-pound newborn; and (2) β_1 is an unknown slope parameter that represents the expected difference in functioning between newborns whose birth weights differ by one pound. Even an analysis as simple as a one-sample *t*-test invokes a statistical model expressed in terms of an unknown population parameter: the population mean, μ. In conducting this test, you use sample data to evaluate the evidence concerning μ's value: Is μ equal to zero (or some other prespecified value)? Analyses may differ in form and function, but a statistical model underpins every inference.

In whatever context, having postulated a statistical model, you then fit the model to sample data and estimate the population parameters' unknown values. Most methods of estimation provide a measure of "goodness-of-fit"—such as an R^2 statistic or a residual variance—that quantifies the correspondence between the fitted model and sample data. If the model fits well, you can use the estimated parameter values to draw conclusions about the direction and magnitude of hypothesized effects in the population. Were you to fit the simple linear regression model just specified above, and find that $\widehat{NEURO}_i = 80 + 5(BWGT_i - 3)$, you would be able to predict that an average three-pound newborn has a functional level of 80 and that functional levels are five points higher for each extra pound at birth. Hypothesis tests and confidence intervals could then be used to make inferences from the sample back to the population.

The simple regression model above is designed for cross-sectional data. What kind of statistical model is needed to represent change processes in longitudinal data? Clearly, we seek a model that embodies two types of research questions: level-1 questions about *within-person change* and level-2 questions about *between-person differences in change*. If the hypothetical study of neurological functioning just described were longitudinal, we might ask: (1) How does each child's neurological functioning change over time? and (2) Do children's trajectories of change vary by birth weight? The distinction between the within-person and the between-person questions is more than cosmetic—it provides the core rationale for specifying a statistical model for change. It suggests that a model for change must include components at two levels: (1) a level-1 submodel that describes how individuals change over time; and (2) a level-2 sub-model that describes how these changes vary across individuals. Taken together, these two components form what is known as a multilevel statistical model (Bryk & Raudenbush, 1987; Rogosa & Willett, 1985).

In this chapter, we develop and explain the multilevel model for change using an example of three waves of data collected by Burchinal and colleagues (1997). As part of a larger study of the effects of early intervention on child development, these researchers tracked the cognitive performance of 103 African-American infants born into low-income families. When the children were 6 months old, approximately half ($n = 58$) were randomly assigned to participate in an intensive early intervention program designed to enhance their cognitive functioning; the other

Table 3.1: Excerpts from the person-period data set for the early intervention study

ID	AGE	COG	PROGRAM
68	1.0	103	1
68	1.5	119	1
68	2.0	96	1
70	1.0	106	1
70	1.5	107	1
70	2.0	96	1
71	1.0	112	1
71	1.5	86	1
71	2.0	73	1
72	1.0	100	1
72	1.5	93	1
72	2.0	87	1
...	...	...	...
902	1.0	119	0
902	1.5	93	0
902	2.0	99	0
904	1.0	112	0
904	1.5	98	0
904	2.0	79	0
906	1.0	89	0
906	1.5	66	0
906	2.0	81	0
908	1.0	117	0
908	1.5	90	0
908	2.0	76	0
...	...	...	...

half ($n = 45$) received no intervention and constituted a control group. Each child was assessed 12 times between ages 6 and 96 months. Here, we examine the effects of program participation on changes in cognitive performance as measured by a nationally normed test administered three times, at ages 12, 18, and 24 months.

Table 3.1 presents illustrative entries from the person-period data set for this example. Each child has three records, one per wave of data collection. Each record contains four variables: (1) *ID*; (2) *AGE*, the child's age (in years) at each assessment (1.0, 1.5, or 2.0); (3) *COG*, the child's cognitive performance score at that age; and (4) *PROGRAM*, a dichotomy that describes whether the child participated in the early intervention program. Because children remained in their group for the duration of data collection, this predictor is time-invariant. Notice that all eight empirical growth records in table 3.1 suggest a decline in cognitive performance over time. As a result, although we might wish that we would be determining whether program participants experience a faster rate of *growth*, it appears that we will actually be determining whether they experience a slower rate of *decline*.

3.2 The Level-1 Submodel for Individual Change

The *level-1* component of the multilevel model, also known as the *individual growth model*, represents the change we expect each member of the population to experience during the time period under study. In the current example, the level-1 submodel represents the individual change in cognitive performance that we hypothesize will occur during each child's second year of life.

Whatever level-1 submodel we specify, we must believe that the observed data could reasonably have come from a population in which the model is functioning. To align expectations with reality, we usually precede level-1 submodel specification with visual inspection of the empirical growth plots (although purists might question the wisdom of "peeking"). Figure 3.1 presents empirical growth plots of *COG vs AGE* for the 8 children whose data appear in table 3.1. We also examined plots for the 95 other children in the sample but we do not present them here, to conserve space. The plots reinforce our perception of declining cognitive performance over time. For some, the decline appears smooth and systematic (subjects 71, 72, 904, 908); for others, it appears scattered and irregular (subjects 68, 70, 902, 906).

When examining empirical growth plots like these, with an eye toward ultimate model specification, we ask global questions such as: What type of population individual growth model might have generated these sample data? Should it be linear or curvilinear with age? Smooth or jagged? Continuous or disjoint? As discussed in chapter 2, try and look beyond inevitable sample zigs and zags because plots of observed data confound information on true change with the effects of random error. In these plots, for example, the slight nonlinearity with age for subjects 68, 70, 902, 906, and 908 might be due to the imprecision of the cognitive assessment. Often, and especially when you have few waves of data, it is difficult to argue for anything except a linear-change individual-growth model. So when we determine which trajectory to select for modeling change, we often err on the side of parsimony and postulate a simple linear model.[1]

Adopting an individual growth model in which change is a linear function of *AGE*, we write the level-1 submodel as:

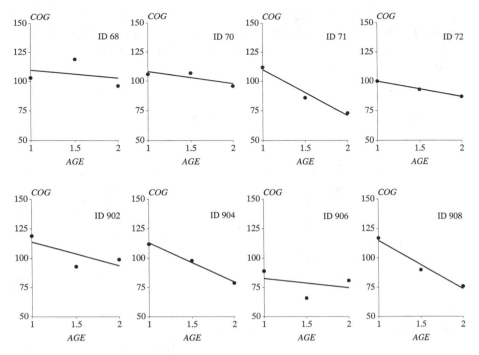

Figure 3.1. Identifying a suitable functional form for the level-1 submodel. Empirical growth plots with superimposed OLS trajectories for 8 participants in the early intervention study.

$$Y_{ij} = [\pi_{0i} + \pi_{1i}(AGE_{ij} - 1)] + [\varepsilon_{ij}].\qquad(3.1)$$

In postulating this submodel, we assert that, in the population from which this sample was drawn, Y_{ij}, the value of *COG* for child i at time j, is a linear function of his or her age on that occasion (AGE_{ij}). This model assumes that a straight line adequately represents each person's true change over time and that any deviations from linearity observed in sample data result from random measurement error (ε_{ij}).

Equation 3.1 uses two subscripts, i and j, to identify individuals and occasions, respectively. For these data, i runs from 1 through 103 (for the 103 children) and j runs from 1 through 3 (for the three waves of data). Although everyone in this data set was assessed on the same three occasions (ages 1.0, 1.5, and 2.0), the level-1 submodel in equation 3.1 is not limited in application to *time-structured* designs. The identical submodel could be used for data sets in which the timing and spacing of waves differs across people.[2] For now, we work with this time-structured

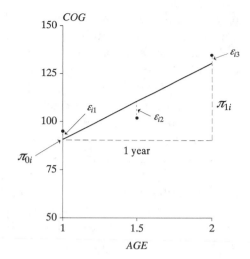

Figure 3.2. Understanding the structural and stochastic features of the level-1 individual growth model. Mapping the model in equation 3.1 onto imaginary data for child i, an arbitrarily selected member of the population.

example; in chapter 5, we extend our presentation to data sets in which data collection schedules vary across people.

In writing equation 3.1, we use brackets to distinguish two parts of the submodel: the *structural* part (in the first set of brackets) and the *stochastic* part (in the second). This distinction parallels the classical psychometric distinction between "true scores" and "measurement error," but as we discuss below, its implications are much broader.

3.2.1 The Structural Part of the Level-1 Submodel

The structural part of the level-1 submodel embodies our hypotheses about the shape of each person's *true trajectory of change* over time. Equation 3.1 stipulates that this trajectory is linear with age and has *individual growth parameters* π_{0i} and π_{1i} that characterize its shape for the ith child in the population. Harkening back to section 2.2.2, these individual growth parameters are the population parameters that lie beneath the individual intercepts and slopes obtained when we fit OLS-estimated individual change trajectories in our exploratory analyses.

To clarify what the individual growth model says about the population, examine figure 3.2, which maps the model onto imaginary data for an arbitrarily selected member of the population, child i. First notice the intercept. Because we specify the level-1 submodel using the predictor $(AGE-1)$, the intercept, π_{0i}, represents child i's true cognitive performance at age 1. We concretize this interpretation in figure 3.2 by showing that the child's hypothesized trajectory intersects the Y axis at π_{0i}. Because we hypothesize that each child in the population has his or her own intercept, this growth parameter includes the subscript i. Child 1's intercept is π_{01}, child 2's intercept is π_{02}, and so on.

Notice that equation 3.1 uses a special representation for the predictor, *AGE*. We used a similar approach in chapter 2, when we subtracted 11 from each adolescent's age before fitting exploratory OLS change trajectories to the tolerance data. This practice, known as *centering*, facilitates parameter interpretation. By using (*AGE-1*) as a level-1 predictor, instead of *AGE*, the intercept in equation 3.1 represents child i's true value of *Y* at age 1. Had we simply used *AGE* as a level-1 predictor, with no centering, π_{0i} would represent child i's true value of *Y* at age 0, an age that precedes the onset of data collection. This representation is less attractive because: (1) we would be predicting beyond the data's temporal limits; and (2) we don't know whether the trajectory extends back to birth linearly with age.

As you become adept at positing level-1 submodels, you will find that it is wise to consider empirical and interpretive issues like these when chosing the scale of your temporal predictor. In section 5.4, we explore other temporal representations, including those in which we center time on its *middle* and *final* values. The approach we adopt here—centering time on the first wave of data collection—is usually a good way to start. Aligning π_{0i} with the first wave of data collection allows us to interpret its value using simple nomenclature: it is child i's true *initial status*. If π_{0i} is large, child i has a high true initial status; if π_{0i} is small, child i has low true initial status. We summarize this interpretation in the first row of the top panel of table 3.2, which defines all parameters in equation 3.1.

The second parameter in equation 3.1, π_{1i}, represents the *slope* of the postulated individual change trajectory. The slope is the most important parameter in a level-1 linear change submodel because it represents the rate at which individual i changes over time. Because *AGE* is clocked in years, π_{1i} represents child i's true annual rate of change. We represent this parameter in figure 3.2 using the right triangle whose hypotenuse is the child's hypothesized trajectory. During the single year under study in our example—as child i goes from age 1 to 2—the trajectory rises by π_{1i}. Because we hypothesize that each individual in the population has his (or her) own rate of change, this growth parameter is subscripted by i. Child 1's rate of change is π_{11}, child 2's rate of change is π_{12}, and so on. If π_{1i} is positive, child i's true outcome increases over time; if π_{1i} is negative, child i's true outcome decreases over time (this latter case prevails in our example).

In specifying a level-1 submodel that attempts to describe everyone (all the i's) in the population, we implicitly assume that all the true individual change trajectories have a common algebraic form. But we do not assume that everyone has the same exact trajectory. Because each person

Table 3.2: Definition and interpretation of parameters in the multilevel model for change

	Symbol	Definition	Illustrative interpretation
Level-1 Model (See Equation 3.1)			
Individual growth parameters	π_{0i}	*Intercept* of the true change trajectory for individual i in the population.	Individual i's true value of *COG* at age 1 (i.e., his *true initial status*).
	π_{1i}	*Slope* of the true change trajectory for individual i in the population.	Individual i's yearly rate of change in true *COG* (i.e., his *true annual rate of change*).
Variance component	σ_ε^2	*Level-1 residual variance* across all occasions of measurement, for individual i in the population.	Summarizes the net (vertical) scatter of the observed data around individual i's hypothesized change trajectory.
Level-2 Model (See Equation 3.3)			
Fixed effects	γ_{00}	Population average of the level-1 intercepts, π_{0i}, for individuals with a level-2 predictor value of 0.	Population average true initial status for nonparticipants.
	γ_{01}	Population average difference in level-1 intercept, π_{0i}, for a 1-unit difference in the level-2 predictor.	Difference in population average true initial status between participants and nonparticipants.
	γ_{10}	Population average of the level-1 slopes, π_{1i}, for individuals with a level-2 predictor value of 0.	Population average annual rate of true change for nonparticipants.
	γ_{11}	Population average difference in level-1 slope, π_{1i}, for a 1-unit difference in the level-2 predictor.	Difference in population average annual rate of true change between participants and non-participants.
Variance components	σ_0^2	Level-2 residual variance in true intercept, π_{0i}, across all individuals in the population.	Population residual variance of true initial status, controlling for program participation.
	σ_1^2	*Level-2 residual variance in true slope*, π_{1i}, across all individuals in the population.	Population residual variance of true rate of change, controlling for program participation.
	σ_{01}	Level-2 residual covariance between true intercept, π_{0i}, and true slope, π_{1i}, across all individuals in the population.	Population residual covariance between true initial status and true annual rate of change, controlling for program participation.

has his or her own individual growth parameters (intercepts and slopes), different people can have their own distinct change trajectories.

Positing a level-1 submodel allows us to distinguish the trajectories of different people using just their individual growth parameters. This leap is the cornerstone of individual growth modeling because it means that we can study interindividual differences in change by studying interindividual variation in the growth parameters. Imagine a population in which each member dips into a well of possible individual growth parameter values and selects a pair—a personal intercept and a slope. These values then determine his or her true change trajectory. Statistically, we say that each person has drawn his or her individual growth parameter values from an underlying bivariate distribution of intercepts and slopes. Because each individual draws his or her coefficients from an unknown *random* distribution of parameters, statisticians often call the multilevel model for change a *random coefficients model*.

3.2.2 The Stochastic Part of the Level-1 Submodel

The *stochastic* part of the level-1 submodel appears in the second set of brackets on the right-hand side of equation 3.1. Composed of just one term, the stochastic part represents the effect of random error, ε_{ij}, associated with the measurement of individual i on occasion j. The level-1 errors appear in figure 3.2 as ε_{i1}, ε_{i2} and ε_{i3}. Each person's *true* change trajectory is determined by the structural component of the submodel. But each person's *observed* change trajectory also reflects the measurement errors. Our level-1 submodel accounts for these perturbations—the differences between the true and observed trajectories—by including random errors: ε_{i1} for individual i's first measurement occasion, ε_{i2} for individual i's second measurement occasion, and so on.

Psychometricians consider random errors a natural consequence of measurement fallibility and the vicissitudes of data collection. We think it wise to be less specific, labeling the ε_{ij} as *level-1 residuals*. For these data, each residual represents that part of child i's value of *COG* at time j not predicted by his or her age. We adopt this vaguer interpretation because we know that we can reduce the magnitude of the level-1 residuals by introducing selected time-varying predictors other than *AGE* into the level-1 submodel (as we show in section 5.3). This suggests that the stochastic part of the level-1 submodel is not just measurement error.

Regardless of how you conceptualize the level-1 errors, one thing is incontrovertible: they are *unobserved*. In ultimately fitting the level-1 submodel to data, we must invoke assumptions about the distribution of the level-1 residuals, from occasion to occasion and from person to person.

Traditional OLS regression invokes "classical" assumptions: that residuals are independently and identically distributed, with homoscedastic variance across occasions and individuals. This implies that, regardless of individual and occasion, each error is drawn independently from an underlying distribution with zero mean and an unknown residual variance. Often, we also stipulate the form of the underlying distribution, usually claiming normality. When we do, we can embody our assumptions about the level-1 residuals, ε_{ij}, by writing:

$$\varepsilon_{ij} \sim N(0, \sigma_\varepsilon^2), \qquad (3.2)$$

where the symbol ~ means "is distributed as," N stands for a normal distribution, and the first element in parentheses identifies the distribution's mean (here, 0) and the second element identifies its variance (here, σ_ε^2). As documented in table 3.2, the residual variance parameter σ_ε^2 captures the scatter of the level-1 residuals around each person's true change trajectory.

Of course, classical assumptions like these may be less credible in longitudinal data. When individuals change, their level-1 error structure may be more complex. Each person's level-1 residuals may be autocorrelated and heteroscedastic over time, not independent as equation 3.2 stipulates. Because the same person is measured on several occasions, any unexplained person-specific time-invariant effect in the residuals will create a correlation across occasions. So, too, the outcome may have a different precision (and reliability) for individuals at different times, perhaps being more suitable at some occasions than at others. When this happens, the error variance may differ over time and the level-1 residuals will be heteroscedastic over occasions within person. How does the multilevel model for change account for these possibilities? Although this is an important question, we cannot address it fully without further technical work. We therefore delay addressing the issues of residual autocorrelation and heteroscedasticity until chapter 4, where we show, in section 4.2, how the full multilevel model for change accommodates automatically for certain kinds of complex error structure. Later, in chapter 8, we go further and demonstrate how using covariance structure analysis to conduct analyses of change lets you hypothesize, implement, and evaluate other alternative error structures.

3.2.3 Relating the Level-1 Submodel to the OLS Exploratory Methods of Chapter 2

The exploratory OLS-fitted trajectories of section 2.2.2 may now make more sense. Although they are not fully efficient because they do not properly exploit all the information present in longitudinal data, they do provide invaluable insights into the functioning of the hypothesized individual growth model. The top panel of figure 3.3 presents the results of using OLS methods to fit the level-1 submodel in equation 3.1 to the data for all 103 children (regressing *COG* on (*AGE-1*), separately by *ID*). The bottom panel presents stem and leaf displays for three summary statistics from these models: the fitted intercepts, the fitted slopes, and the estimated residual variances.

For most children, cognitive performance declines over time. For some, the decline is rapid; for others, less so. Few children show any improvement. Each fitted intercept estimates that child's true initial status; each fitted slope estimates that child's true annual rate of change during the second year of life. The fitted intercepts are centered near 110; the fitted slopes are centered near −10. This suggests that at age 1, the average child has a true cognitive level slightly above the national norm (of 100 for this test). Over time, however, most children decline (we estimate that only 7 improve).

The stem-and-leaf displays in the bottom left panel of figure 3.3 reveal great heterogeneity in fitted intercept and slope across children in the sample and suggest that not all children have identical trajectories of change. Of course, you must be cautious when interpreting the interindividual heterogeneity in change trajectories evident in figure 3.3. The between-person variation in the estimated change trajectories that you observe is necessarily inflated over the underlying interindividual variability in the unknown true change trajectories because the fitted trajectories, having been estimated from observed data, are *fallible* representations of true change. The actual variability in underlying true change will always be somewhat less than what you observe in exploratory analysis, with the magnitude of the difference depending on the quality of your outcome measurement and the efficacy of your hypothesized individual growth model.

The skewed distribution of residual variances in the bottom right panel of figure 3.3 suggests great variation in the *quality* of the OLS summaries across children (we expect the distribution of these statistics to be skewed, as they are "squared" quantities and are therefore bounded by zero below). When the residual variance is near 0, as it is for many children, the fitted trajectories are reasonable summaries of the observed data for those children. When the residual variance is larger, as it often is here, the fitted trajectories are poorer summaries: the observed values of *COG* are further away from the fitted lines, making the magnitude of the estimated level-1 residuals, and therefore the residual variance, large.

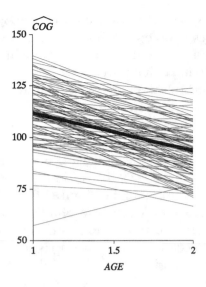

Figure 3.3. Observed variation in fitted OLS trajectories. Fitted OLS trajectories for participants in the early intervention study as well as stem and leaf displays for fitted initial status, fitted rate of change, residual variance.

Fitted initial status		Fitted rate of change		Residual variance	
14	0	2.	0	46	8
13*	5568	1*		44	
13.	00134	1.	0	42	
12*	5556778999	0*	79	40	00
12.	02233344	0.	134	38	
11*	55667777888889	-0*	4444332	36	8
11.	000111112222233334444	-0.	99998888777765	34	
10*	55666688999	-1*	4333322211000	32	3
10.	0012222244	-1.	99888877666655	30	
9*	6666677799	-2*	44322211110000	28	4
9.	344	-2.	9999877776655	26	7
8*	89	-3*	443322100000	24	1444
8.	34	-3.	987	22	8
7*	7	-4*	443111	20	
7.				18	3
6*				16	00011
6.				14	
5*	7			12	21
				10	44433
				8	1118886666
				6	77744
				4	333844
				2	04444888833338888888
				0	0000111122233334444444466668111114447

3.3 The Level-2 Submodel for Systematic Interindividual Differences in Change

The level-2 submodel codifies the relationship between interindividual differences in the change trajectories and time-invariant characteristics of the individual. The ability to formulate this relationship using a level-2 submodel stems from the realization that adoption of a common level-1 submodel forces people to differ only in the values of their individual growth parameters. When we use a level-1 linear change model, people can differ only in their intercepts and slopes. This allows us to recast vague questions about the relationship between "change" and predictors as specific questions about the relationship between the individual growth parameters and predictors.

Like all statistical models, the level-2 submodel describes hypothesized population processes, not sample behavior. But insights gleaned from sample data can often provide valuable insight into model formulation. In this spirit, examine the top panel of figure 3.4, which separately plots fitted OLS trajectories according to the child's program participation (program participants in the right panel, nonparticipants in the left). The average change trajectory for each group is shown in bold. Program participants tend to have higher scores at age 1 and decline less precipitously over time. This suggests that their intercepts are higher but their slopes are shallower. Also note the substantial interindividual heterogeneity *within* groups. Not all participants have higher intercepts than nonparticipants; not all nonparticipants have steeper slopes. Our level-2 model must simultaneously account for both the general patterns (here, the between-group differences in intercepts and slopes) *and* interindividual heterogeneity in patterns within groups.

What kind population model might have given rise to these patterns? The preceding discussion suggests four specific features for the level-2 submodel. First, its outcomes must be the individual growth parameters (here, π_{0i} and π_{1i} from equation 3.1). As in regular regression, where we model the population distribution of a random variable by making it an outcome, here, where we model the population distribution of the individual growth parameters, they, too, must be the outcomes. Second, the level-2 submodel must be written in separate parts, one for each level-1 growth parameter. When we use a linear change individual growth model at level-1 (as in equation 3.1), we need two level-2 submodels: one for the intercept, π_{0i}, another for the slope, π_{1i}. Third, each part must specify a relationship between an individual growth parameter and the predictor (here, *PROGRAM*). As you move across the panels in the top of figure 3.4, the value of the predictor, *PROGRAM*, shifts from 0 to 1. This suggests that each level-2 model should ascribe differences in either π_{0i} or π_{1i} to *PROGRAM* just as in a regular regression model. Fourth, each model must allow individuals who share common predictor values to vary in their individual change trajectories. This means that each level-2 submodel must allow for stochastic variation in the individual growth parameters.

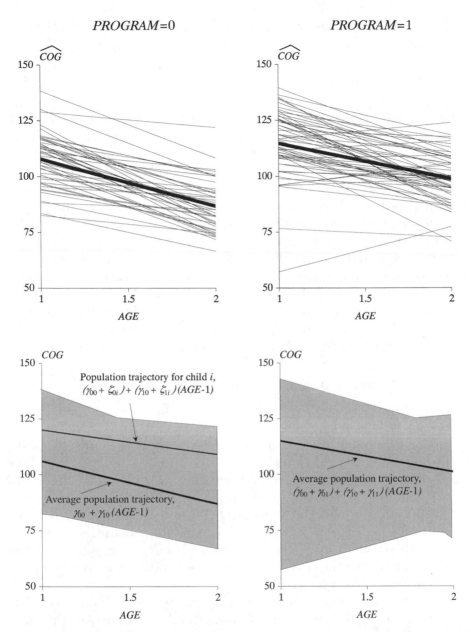

Figure 3.4. Understanding the structural and stochastic features of the level-2 submodel for inter-individual differences in change. Top panel presents fitted OLS trajectories separately by levels of the predictor *PROGRAM*. Bottom panel maps the model in equation 3.3 onto imaginary data for an arbitrary child *i* and the average population trajectory. The shaded portion in each of the lower panels is designed to suggest the existence of many distinct population trajectories for different children.

These considerations lead us to postulate the following level-2 sub-model for these data:

$$\pi_{0i} = \gamma_{00} + \gamma_{01}PROGRAM_i + \zeta_{0i}$$
$$\pi_{1i} = \gamma_{10} + \gamma_{11}PROGRAM_i + \zeta_{1i}.$$

(3.3)

Like all level-2 submodels, equation 3.3 has more than one component, each resembling a regular regression model. Taken together, the two components treat the intercept (π_{0i}) and the slope (π_{1i}) of an individual's growth trajectory as level-2 outcomes that may be associated with the predictor, *PROGRAM*. Each component also has its own residual—here, ζ_{0i} and ζ_{1i}—that permits the level-1 parameters (the π's) of one person to differ stochastically from those of others.

Although not yet apparent, the two components of this level-2 sub-model have *seven* population parameters: the four regression parameters (the γ's) shown in equation 3.3 and three residual variance/covariance parameters we will soon define. All are estimated when we fit the multi-level model for change to data. We list, label, and define these parameters in the second section of table 3.2 and illustrate their action in the bottom panel of figure 3.4. We discuss their interpretation below.

3.3.1 Structural Components of the Level-2 Submodel

The structural parts of the level-2 submodel contain four level-2 parameters—γ_{00}, γ_{01}, γ_{10}, and γ_{11}—known collectively as the *fixed effects*. The fixed effects capture systematic interindividual differences in change trajectory according to values of the level-2 predictor(s). In equation 3.3, two of the fixed effects, γ_{00} and γ_{10}, are level-2 intercepts; two, γ_{01} and γ_{11}, are level-2 slopes. As in regular regression, the slopes are of greater interest because they represent the effect of predictors (here, the effect of *PROGRAM*) on the individual growth parameters. You can interpret the level-2 parameters much as you do regular regression coefficients, except that you must remember that they describe variation in "outcomes" that are themselves level-1 individual growth parameters.

The easiest way to unravel the meaning of the level-2 fixed effects is to identify a *prototypical individual* distinguished by particular predictor values, substitute those values into the level-2 submodel, and examine the consequences. To derive the postulated level-2 submodel for a prototypical nonparticipant, for example, we set *PROGRAM* to 0 in both parts of equation 3.3 to find: when $PROGRAM = 0$, $\pi_{0i} = \gamma_{00} + \zeta_{0i}$ and $\pi_{1i} = \gamma_{10} + \zeta_{1i}$. This model hypothesizes that, in the population of nonparticipants, the values of initial status and annual rate of change, π_{0i} and π_{1i}, are centered around the level-2 parameters γ_{00}, and γ_{10}. γ_{00} represents the average true initial status (cognitive score at age 1); γ_{10} represents the average true annual rate of change. By fitting the multilevel model for change to data and estimating these parameters, we address the question: What is the average true trajectory of change in the population for children who did not receive the early intervention program? The lower left panel of figure 3.5 depicts this average population trajectory. Its intercept is γ_{00}; its slope is γ_{10}.

We repeat this process for program participants by setting *PROGRAM* to 1: in this case, $\pi_{0i} = (\gamma_{00} + \gamma_{01}) + \zeta_{0i}$ and $\pi_{1i} = (\gamma_{10} + \gamma_{11}) + \zeta_{1i}$. In the population of program participants, the values of initial status and annual rate of change, π_{0i} and π_{1i}, are centered around $(\gamma_{00} + \gamma_{01})$ and $(\gamma_{10} + \gamma_{11})$. Comparing these centers to those for nonparticipants illustrates that the level-2 parameters γ_{01} and γ_{11} capture the effects of *PROGRAM*. γ_{01} represents the hypothesized difference in average true initial status between groups; γ_{11} represents the hypothesized difference in average true annual rate of change. This allows us to think of the level-2 slopes, γ_{01} and γ_{11}, as "shifts" associated with program participation. The lower right panel of figure 3.4 depicts these shifts. If γ_{01} and γ_{11} are non-zero, the average population trajectories in the two groups differ; if they are both 0, they do not. These two level-2 slope parameters therefore address the question: What is the difference in the average trajectory of true change associated with program participation?

3.3.2 Stochastic Components of the Level-2 Submodel

Each part of the level-2 submodel contains a residual that allows the value of each person's growth parameters to be scattered around the relevant population averages. These residuals, ζ_{0i} and ζ_{1i} in equation 3.3, represent those portions of the level-2 outcomes—the individual growth parameters—that remain "unexplained" by the level-2 predictor(s). As is true for most residuals, we are interested less in their specific values than in their population variances and covariance, which we label σ_0^2, σ_1^2, and σ_{01}. You should know that labeling conventions for these population variances vary considerably across authors and statistical packages. For example, Raudenbush and Bryk (2002) label them τ_{00}, τ_{11}, and τ_{01}, while Goldstein (1995) labels them σ_{u0}^2, σ_{u1}^2, and σ_{u01}.

If child i is a member of the population of nonparticipants, *PROGRAM* takes on the value 0 and the level-2 residuals in equation 3.3 represent deviations between his or her true initial status and annual rate of change from the population average intercept and slope for nonparticipants (γ_{00} and γ_{10}). We display a trajectory for this prototypical child in the lower left panel of figure 3.4. The trajectory begins at a true initial status of $(\gamma_{00} + \zeta_{0i})$ and has a (declining) true annual rate of change of $(\gamma_{10} + \zeta_{1i})$.

Trajectories for other children can be constructed similarly by combining parameters γ_{00} and γ_{10} with other child-specific residuals. The shaded area in this panel is designed to suggest the existence of many different true trajectories, one for each nonparticipant in the population (if they could be fully enumerated). Similarly, if child i is a member of the population of participants, *PROGRAM* takes on the value 1 and the level-2 residuals in equation 3.3 represent deviations between his true initial status and annual rate of change and the population average intercept and slope for participants $(\gamma_{00} + \gamma_{01})$ and $(\gamma_{10} + \gamma_{11})$. To illustrate the heterogeneity in change for this group, the lower right panel of figure 3.4 also includes a shaded area.

Because the level-2 residuals represent deviations between the individual growth parameters and their respective population averages, their variances, σ_0^2 and σ_1^2, summarize the population variation in true individual intercept and slope around these averages. Because they describe those portions of the intercepts and slopes *left over* after accounting for the effect(s) of the model's predictor(s), they are actually *conditional* residual variances. Conditional on the presence of the model's predictors, σ_0^2 represents the population residual variance in true initial status and σ_1^2 represents the population residual variance in true annual rate of change. These variance parameters allow us to address the question: How much heterogeneity in true change remains after accounting for the effects of program participation?

When we posit a level-2 submodel, we also allow for a possible association between individual initial status and individual rates of change. Children who begin at a higher level may have higher (or lower) rates of change. To account for this possibility, we permit the level-2 residuals to be correlated. Since ζ_{0i} and ζ_{1i} represent the deviations of the individual growth parameters from their population averages, their population covariance summarizes the association between true individual intercepts and slopes. Again because of their conditional nature, the population covariance of the level-2 residuals, σ_{01}, summarizes the magnitude and direction of the association between true initial status and true annual rate of change, controlling for program participation. This parameter allows us to address the question: Controlling for program participation, are true initial status and true rate of change related?

To fit the multilevel model for change to data, we must make some assumptions about the level-2 residuals (just as we did for the level-1 residuals in equation 3.2). But because we have two level-2 residuals, we describe their underlying behavior using a *bivariate distribution*. The standard assumption is that the two level-2 residuals, ζ_{0i} and ζ_{1i}, are bivariate normal with mean 0, unknown variances, σ_0^2, and σ_1^2, and unknown covariance, σ_{01}. We can express these assumptions compactly using matrix notation by writing:

$$\begin{bmatrix} \zeta_{0i} \\ \zeta_{1i} \end{bmatrix} \sim N\left(\begin{bmatrix} 0 \\ 0 \end{bmatrix}, \begin{bmatrix} \sigma_0^2 & \sigma_{01} \\ \sigma_{10} & \sigma_1^2 \end{bmatrix} \right). \tag{3.4}$$

Matrix notation greatly simplifies the way in which we codify the model's assumptions. In broad outline, we interpret equation 3.4 in the same way we interpret the assumptions about the level-1 residuals in equation 3.2. The first matrix on the right of the equals sign in parentheses specifies the bivariate distribution's mean vector; here, we assume it to be 0 for each residual (as usual). The second matrix specifies the bivariate distribution's variance-covariance matrix, also known as the *level-2 error covariance matrix* because it captures the covariation among the level-2 residuals (or errors). Two variances, σ_0^2 and σ_1^2, appear along the diagonal, the covariance, σ_{01}, appears on the off-diagonal. Because the covariance between ζ_{0i} and ζ_{1i} is the same as the covariance between ζ_{1i} and ζ_{0i}, the off-diagonal elements are identical—that is, $\sigma_{01} = \sigma_{10}$. The complete set of residual variances and covariances—both the level-2 error variance-covariance matrix and the level-1 residual variance, σ_ε^2—is known collectively as the model's *variance components*.

3.4 Fitting the Multilevel Model for Change to Data

Until the widespread availability of software for fitting multilevel models, researchers used ad hoc strategies like those presented in chapter 2 to analyze longitudinal data: they fitted individual growth trajectories in separate within-person OLS-regression analyses and then they regressed the individual growth parameter estimates obtained on selected level-2 predictors (Willett, 1989). But as previously discussed, this approach has at least two flaws: (1) it ignores information about the individual growth parameter estimates' precision, even though we know that it varies (as seen in the varying residual variances in the bottom panel of figure 3.3); and (2) it replaces *true* individual growth parameters—the real outcomes in a level-2 submodel—with their fallible estimates. The level-2 submodels do not describe the relationship between the parameter *estimates* and predictors, but between the parameters' *true values* and predictors.

Beginning in the 1980s, several teams of statisticians began developing specialized software for fitting the multilevel model for change to data. By the early 1990s, four major packages were widely used: HLM (Bryk, Raudenbush, & Congdon, 1988), MLn (Rasbash & Woodhouse, 1995), GENMOD (Mason, Anderson, & Hayat, 1988), and VARCL (Longford, 1993).

Although the latter two are no longer supported, HLM (Raudenbush, Bryk, Cheong, & Congdon, 2001, available from http://www.ssicentral.com) and MLwiN (Goldstein, 1998, available from http://multilevel.ioe.ac.uk) continue to be modified, expanded, and upgraded regularly to handle an increasing variety of multilevel models. Several multipurpose software packages have also added multilevel routines, including SAS PROC MIXED and PROC NLMIXED (SAS Institute, 2001, http://www.sas.com), the STATA "xt" routines, such as xtreg (Stata, 2001, http://www.stata.com), and SPLUS' NLME library (Pinheiro & Bates, 2001, available from http://cm.bell-labs.com/cm/ms/departments/sia/project/nlme/). So, too, teams of statisticians continue to develop new specialty programs including BUGS (Gilks, Richardson, & Spiegelhalter, 1996, available from http://www.mrcbsu.cam.ac.uk/bugs) and MIXREG (Hedeker & Gibbons, 1996; available from http://www.uic.edu/~hedeker).

As this list suggests, you have a wide and growing array of model fitting options in the investigation of change. We ourselves have no vested interest in any particular software program and do not promote any one above the others. All have their strengths, and we use many of them in our research and in this book. At their core, each program does the same job: it fits the multilevel model for change to data and provides parameter estimates, measures of precision, diagnostics, and so on. There is also some evidence that all the different packages produce the same, or similar, answers to a given problem (Kreft & de Leeuw, 1990). So, in one sense, it does not matter which program you choose. But the packages do differ in many important ways including the "look and feel" of their interfaces, their ways of entering and preprocessing data, their model specification process, their estimation methods, their strategies for hypothesis testing, and the provision of diagnostics. These differences may lead you to decide that one piece of software is especially convenient for your work.

For now, we focus on one particular method of estimation—*maximum likelihood*—as implemented in one program, HLM (Raudenbush, Bryk, Cheong, & Congdon, 2001). In subsequent chapters, we describe other methods of estimation and we apply other statistical software, allowing us to provide advice and compare the competing approaches and packages.

3.4.1 The Advantages of Maximum Likelihood Estimation

The method of maximum likelihood (ML) is currently the most popular approach to statistical estimation. Its popularity results, in part, from its excellent performance in large random samples from well-defined target populations. As sample size increases, ML estimates have three desirable properties: (1) they are *asymptotically unbiased (consistent)*—they converge on the unknown true values of population parameters; (2) they are *asymptotically normally distributed*—their sampling distributions are approximately normal with known variance; and (3) they are *asymptotically efficient*—their standard errors are smaller than those derived by other methods. Another advantage is that any function of ML estimates is also an ML estimate. This means that predicted growth trajectories (constructed from ML estimates of initial status and rates of change) are ML estimates of the true trajectories. All else being equal, statisticians prefer estimates that are consistent and efficient, that make use of well-established normal theory, and that can generate decent estimates of more complex quantities. Hence the appeal of ML methods.

Notice that the attractive properties of ML estimates are *asymptotic*. This means that in practice—in any actual analysis of a real sample—the properties hold only *approximately*. In large samples, they are likely to hold; in small samples, they may not.[3] To enjoy these advantages, you need a relatively large sample, and the question, how large is large, has no simple answer. Although 10 is certainly small and 100,000 is certainly large, no one can say definitively how large is large enough. In cross-sectional work, Long (1997), for example, recommends a minimum of 100 individuals and he labels sample sizes of 500 "adequate." For a general multilevel model, Snijders and Bosker (1999) consider samples of 30 or more large. Although "rules of thumb" like these provide broad guidelines, we tend to distrust them. The answer to the question "How large?" differs by context, by the particularities of different types of ML estimation, by features of the data, and by the requirements of the tests conducted. Instead we simply offer practical advice: if you use ML methods in "small" samples, treat *p*-values and confidence intervals circumspectly.

Derivation of computational formulas for ML estimation is beyond our scope or intent here. Below, we offer a heuristic explanation of what happens when you use ML methods to fit a multilevel model for change. Our goal is to lay the conceptual foundation for future chapters by explaining why ML estimates make sense and why they have such useful properties. Readers interested in mathematical details should consult Raudenbush and Bryk (2002), Goldstein (1995), or Longford (1993).

3.4.2 Using Maximum Likelihood Methods to Fit a Multilevel Model

Conceptually, maximum likelihood estimates are those guesses for the values of the unknown population parameters that maximize the probability of observing a particular sample of data. In the early intervention study, they are those estimates of the fixed effects and variance components that make it most likely we would have observed the specific patterns of change found for these 103 children.

To derive an ML estimate for a population parameter, a statistician must first construct a *likelihood function*—an expression that describes the probability of observing the sample data as a function of the model's unknown parameters. Then, he, she, or more accurately, a computer, numerically examines the relative performance of potentially competing estimates until those that maximize the likelihood function are found. The likelihood function for the early intervention data is a function of the probability that we would observe the particular temporal pattern of *COG* values found in the person-period data set. We seek estimates of the fixed effects and variance components whose values maximize the probability of observing this specific pattern.

All likelihood functions are expressed as the product of probabilities (or probability densities). For cross-sectional data, each sample member usually contributes just one term, related to the probability that *that* person has his or her observed data. But because longitudinal data consist of several observations, one per measurement occasion, each person contributes several terms to the likelihood function, which contains as many terms as there are records in the person-period data set.

The particular term that each person contributes on each occasion depends on the specification and assumptions of the hypothesized model. The multilevel model contains structural parts (as shown in, for example, in equations 3.1 and 3.3) and stochastic parts (whose behavior is described in equations 3.2 and 3.4). The structural portion describes the true outcome value for person i on occasion j for his or her particular predictor values. It depends on the unknown values of the fixed effects. The stochastic portion—the level-1 and level-2 residuals—introduce an element of randomness into the proceedings, scattering the observations for person i on occasion j from the structurally specified value.

To derive a maximum likelihood estimate, we must also make assumptions about the *distribution* of the residuals. We have already stated assumptions in equation 3.2 for the level-1 residual, ε_{ij}, and in equation 3.4 for the two-level-2 residuals, ζ_{0i} and ζ_{1i}. Each is assumed to be normally distributed with mean 0; ε_{ij} has unknown variance, σ_ε^2; ζ_{0i} and ζ_{1i} have unknown variances, σ_0^2 and σ_1^2, and covariance, σ_{01}. We also assume that the level-2 residuals are independent of the level-1 residual and that all residuals are independent of the model's predictors.

Given a model and its underlying assumptions, a statistician can write a mathematical expression for the distribution, or *probability density*, of the outcome. This expression has a mean determined by the model's structural parts and a variance determined by its stochastic parts. As a probability density function, it also describes the likelihood that a person with particular values of the predictors—only *PROGRAM* in equation 3.3—could have particular outcome values using a set of unknown fixed effects and variance components whose values we would like to estimate. That is, it also contains the actual data values observed for that person on that occasion.

It is a short step from here to the full sample likelihood, which we reach by exploiting the well-known multiplicative property of independent probabilities. If you toss one coin, there is a probability of .5 that it will turn up heads. If you independently toss two coins, the probability that each will turn up heads is still .5. But taken together, the probability that you will obtain two heads is only .25 (.5 × .5). If you independently toss three coins, the probability of three heads declines to 0.125 (.5 × .5 × .5). Statisticians use this principle to create a full sample likelihood from the separate person-period likelihoods just developed. First they write down the value of the probability density of the outcome for each person in the data set on every occasion, thereby describing the likelihood that he or she obtained his or her particular value of the outcome on that occasion. Then they multiply these terms together, yielding an expression for the likelihood of simultaneously observing *all* the data in the person-period data set. Because each person-period likelihood is a function of the data and the unknown parameters, so is their product the full sample likelihood.

To find ML estimates of the unknown population parameters, we identify those values of the unknown parameters that maximize this product of probabilities. Conceptually, imagine a computer trying out billions of alternative estimates, multiplying them together as specified in the sample likelihood function to yield a numeric value for the likelihood, and comparing those numeric values across all of the billions of tries until those estimates that yield the maximum value of the likelihood function are found. These would be the maximum likelihood estimates for this particular problem.

Of course, an enormous numerical search like this is daunting, even with fast computers. Calculus can facilitate the search, but it cannot eliminate the difficulty of working with the products of probability densities that make up the sample likelihood function. To facilitate the search, statisticians use a simple strategy: instead of finding those values of the unknown parameters that maximize the likelihood function, they find those that maximize its logarithm. Working with this new function, known as the *log-likelihood function*, sacrifices nothing because the values that maximize it also maximize the raw likelihood function. The transformation to logarithms simplifies the intensive numerical calculations involved because (1) the logarithm of a product is a *sum* of the separate logarithms, and (2) the logarithm of a term raised to a power is the power multiplied by the logarithm of the term. And so, since the sample likelihood contains both multiplicative and exponentiated terms, the logarithmic transformation moves the numerical maximization into a more tractable sphere, computationally speaking.

Although simpler than maximizing the likelihood function itself, maximizing the log-likelihood function also involves iteration. All software programs that provide ML estimates for the multilevel model for change use an iterative procedure. To begin, the program generates reasonable "starting" values for all model parameters, usually by applying something like the OLS methods we just rejected in chapter 2! In successive iterations, the program gradually refines these estimates as it searches for the log-likelihood function's maximum. When this search converges—and the difference between successive estimates is trivially small—the resultant estimates are output. If the algorithm does not converge (and this happens more often than you might like), you must repeat the search allowing more iterations or you must improve your model specification. (We discuss these issues in section 5.2.2.)

Once the ML estimates are found, it is relatively easy for a computer to estimate their associated sampling variation in the form of *asymptotic standard errors* (*ase*). We use the adjective "asymptotic" because, as noted earlier, ML standard errors are accurate only in large samples. Like any standard error, the *ase* measures the precision with which an estimate has been obtained—the smaller the *ase*, the more precise the estimate.

We now use maximum likelihood methods to fit the multilevel model in equations 3.1 and 3.3 to the early intervention data. Table 3.3 presents results obtained using the HLM software.[4] We first discuss the estimated fixed effects in the first four rows; in section 3.6, we discuss the estimated variance components shown in the next four rows.

3.5   Examining Estimated Fixed Effects

Empirical researchers usually conduct hypothesis tests before scrutinizing parameter estimates to determine whether an estimate warrants inspection. If an estimate is consistent with a null hypothesis of no population effect, it is unwise to interpret its direction or magnitude.

Table 3.3: Results of fitting a multilevel model for change to the early intervention data ($n = 103$)

		Parameter	Estimate	*ase*	z
Fixed Effects					
Initial status, π_{0i}	Intercept	γ_{00}	107.84***	2.04	52.97
	PROGRAM	γ_{01}	6.85*	2.71	2.53
Rate of change, π_{1i}	Intercept	γ_{10}	−21.13***	1.89	−11.18
	PROGRAM	γ_{11}	5.27*	2.52	2.09
Variance Components					
Level 1:	Within-person, ε_{ij}	σ_ε^2	74.24***	10.34	7.17
Level 2:	In initial status, ζ_{0i}	σ_0^2	124.64***	27.38	4.55
	In rate of change, ζ_{1i}	σ_1^2	12.29	30.50	0.40
	Covariance between ζ_{0i} and ζ_{1i}	σ_{01}	−36.41	22.74	−1.60

$\sim p < .10$; $* p < .05$; $** p < .01$; $*** p < .001$.
This model predicts cognitive functioning between ages 1 and 2 years as a function of (*AGE*-1) (at level-1) and *PROGRAM* (at level-2).

Note: Full ML, HLM.

Although we agree that it is wise to test hypotheses before interpreting parameters, here we reverse this sequence for pedagogic reasons, discussing interpretation in section 3.5.1 and testing in section 3.5.2. Experience convinces us that when learning a new statistical method, it is easier to understand what you are doing if you interpret parameters first and conduct tests second. This sequence emphasizes conceptual understanding over up-or-down decisions about "statistical significance" and ensures that you understand the hypotheses you test.

3.5.1 Interpreting Estimated Fixed Effects

The fixed effects parameters of the level-2 submodel—the γ's of equation 3.3—quantify the effects of predictors on the individual change trajectories. In our example, they quantify the relationship between the individual growth parameters and program participation. We interpret these estimates much as we do any regression coefficient, with one key difference: the level-2 "outcomes" that these fixed effects describe are the level-1 individual growth parameters themselves.

Until you are comfortable directly interpreting the output from software programs, we strongly recommend that you take the time to actually write down the structural portion of the fitted model before attempting to interpret the fixed effects. Although some software programs facilitate the linkage between model and estimates through structured displays (e.g., MlwiN), others (e.g., SAS PROC MIXED) use somewhat esoteric conventions for labeling output. Substituting estimates $\hat{\gamma}$ in table 3.3 into the level-2 submodel in equation 3.3, we have:

$$\hat{\pi}_{0i} = 107.84 + 6.85 PROGRAM_i$$
$$\hat{\pi}_{1i} = -21.13 + 5.27 PROGRAM_i \tag{3.5}$$

The first part of the fitted submodel describes the effects of *PROGRAM* on initial status; the second part describes its effects on the annual rates of change.

Begin with the first part of the fitted submodel, for initial status. In the population from which this sample was drawn, we estimate the true initial status (*COG* at age 1) for the average nonparticipant to be 107.84; for the average participant, we estimate that it is 6.85 points higher (114.69). The means of both groups are higher than national norms (100 for this test). The age 1 performance of participants is 6.85 points higher than that of nonparticipants. Before concluding that this differential in initial status casts doubt on the randomization mechanism, remember that the intervention started *before* the first wave of data collection, when the children were already 6 months old. This modest seven-point elevation in initial status may reflect early treatment gains attained between ages 6 months and 1 year.

Next, examine the second part of the fitted submodel, for the annual rate of change. In the population from which this sample was drawn, we estimate the true annual rate of change for the average nonparticipant to be −21.13; for the average participant, we estimate it to be 5.27 points higher (−15.86). The average nonparticipant dropped over 20 points during the second year of life; the average participant dropped over 15. The cognitive functioning of both groups of children declines over time. As we suspected when we initially examined these data, the intervention slows the rate of decline.

Another way of interpreting fixed effects is to plot fitted trajectories for prototypical individuals. Even in a simple analysis like this, which involves just one dichotomous predictor, we find it invaluable to inspect prototypical trajectories visually. For this particular multilevel model, only two prototypes are possible: a program participant (*PROGRAM* = 1) and a nonparticipant (*PROGRAM* = 0). Substituting these values into equation 3.5 yields the estimated initial status and annual growth rates for each:

When *PROGRAM* = 0:
$$\hat{\pi}_{0i} = 107.84 + 6.85(0) = 107.84$$
$$\hat{\pi}_{1i} = -21.13 + 5.27(0) = -21.13.$$

When *PROGRAM* = 1:
$$\hat{\pi}_{0i} = 107.84 + 6.85(1) = 114.69$$
$$\hat{\pi}_{1i} = -21.13 + 5.27(1) = -15.86.$$

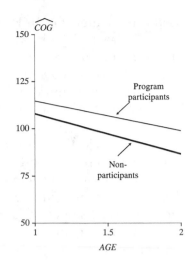

Figure 3.5. Displaying the results of a fitted multi-level model for change. Prototypical trajectories for an average program participant and nonparticipant in the early intervention data.

We use these estimates to plot the fitted individual change trajectories in figure 3.5. These plots reinforce the numeric conclusions just articulated. In comparison to nonparticipants, the average participant has a higher score at age 1 and a slower annual rate of decline.

3.5.2 Single Parameter Tests for the Fixed Effects

As in regular regression, you can conduct a hypothesis test on each fixed effect (each γ) using a single parameter test. Although you can equate the parameter value to any pre-specified value in your hypothesis test, most commonly you examine the null hypothesis that, controlling for all other predictors in the model, the population value of the parameter is 0, $H_0: \gamma = 0$, against the two-sided alternative that it is not, $H_1: \gamma \neq 0$. When you use ML methods, this test's properties are known only asymptotically (for exceptions, see note 3). You test this hypothesis for each fixed effect by computing the familiar z-statistic:

$$z = \frac{\hat{\gamma}}{ase(\hat{\gamma})}. \tag{3.7}$$

Most multilevel modeling programs provide z-statistics; if not, you can easily compute them by hand. However, care is needed because there is much looseness and inconsistency in output labels; terms like z-statistic, z-ratio, quasi-t-statistic, t-statistic, and t-ratio, which are not the same, are used interchangeably. In HLM, the package we used here, this statistic is labeled a "t-ratio." Most programs also output either an associated p-value or confidence interval to facilitate testing.[5]

Table 3.3 presents z-statistics (column 6) and approximate p-values (as superscripts in column 4) for testing hypotheses about the fixed effects. We reject all four null hypotheses, suggesting that each parameter plays a role in the story of the program's effect on children's cognitive development. In rejecting (at the .001 level) the null hypotheses for the two level-2 intercepts, γ_{00} and γ_{10}, we conclude that the average nonparticipant had a non-zero cognitive score at age 1 (hardly surprising!) which declined over time. In rejecting (at the .05 level) the null hypotheses for the two level-2 slopes, γ_{01} and γ_{11}, we conclude that differences between program participants and nonparticipants—in both initial status and annual rates of change—are statistically significant.

3.6 Examining Estimated Variance Components

Estimated variance and covariance components are trickier to interpret as their numeric values have little absolute meaning and there are no graphic aids to fall back on. Interpretation for a single fitted model is especially difficult as you lack benchmarks for evaluating the components' magnitudes. This increases the utility of hypothesis testing, for at least the tests provide some benchmark (against the null value of 0) for comparison.

3.6.1 Interpreting Estimated Variance Components

Variance components assess the amount of outcome variability left—at either level-1 or level-2—after fitting the multilevel model. The level-1 residual variance, σ_ε^2, summarizes the population variability in an average person's outcome values around his or her own true change trajectory. Its estimate for these data is 74.24, a number that is difficult to evaluate in absolute terms. In chapter 4, we provide strategies making relative comparisons to residual variances in other models.

The level-2 variance components summarize the between-person variability in change trajectories that remains after controlling for predictors (here, *PROGRAM*). Using the matrix notation of equation 3.4, we write:

$$\begin{bmatrix} 124.64 & -36.41 \\ -36.41 & 12.29 \end{bmatrix}.$$

Because hypothesis tests, discussed below, reveal that only one of these elements, σ_0^2, is significantly different from 0, it is the only parameter we discuss here. But because we have no point of comparison, it is difficult to say whether its value, 124.64, is small or large. All we can say is that it quantifies the amount of residual variation in true initial status remaining after we control for program participation.

3.6.2 Single Parameter Tests for the Variance Components

Tests for variance components evaluate whether there is any remaining *residual* outcome variation that could potentially be explained by other predictors. The level of the particular variance component—either level-1 or level-2—dictates the type of predictor that might be added. In general, all the tests are similar in that they assess the evidence concerning the null hypothesis that the parameter's population value is 0, $H_0 : \sigma^2 = 0$, against the alternative that it is not, $H_1 : \sigma^2 \neq 0$.

There are two very different methods for conducting these hypothesis tests. In this chapter, we offer the simpler approach—the *single parameter test*. Some programs provide this test as a z-statistic—the ratio of the estimated variance component to its asymptotic standard error. Others offer the identical test by squaring the z-statistic and labeling it a χ^2 statistic on one degree of freedom. The appeal of a single parameter hypothesis test is simple. Even if you fit just one statistical model, as we have here, you can garner some insight into the variance components' relative values—at least in comparison to 0.

Unfortunately, statisticians disagree as to the nature, form, and effectiveness of these tests. Miller (1986), Raudenbush and Bryk (2002), and others have long questioned their utility because of their sensitivity to departures from normality. Longford (1999) describes their sensitivity to sample size and imbalance (unequal numbers of observations per person) and argues that they are so misleading that they should be abandoned completely. Because they can be useful for quick, albeit imprecise, assessment, we suggest you examine them only with extreme caution. In section 4.6, we present a superior method for testing hypotheses about variance components, an approach whose use we normally recommend.

Table 3.3 presents single-parameter hypothesis tests for the model's four variance/covariance components. The first three test the null hypothesis that the population variance of the level-1 residuals, σ_ε^2, is 0, that the population variance of the level-2 residuals for initial status, σ_0^2, is 0 and that the population variance of the level-2 residuals for the annual rate of change, σ_1^2, is 0. The last tests whether the covariance between the level-2 residuals for initial status and annual rates of change, σ_{01}, is 0, indicating whether true initial status and true annual rate of change are correlated, after participation in the intervention program is accounted for.

For these data, we reject only two of these null hypotheses (each at the .001 level). The test for the level-1 residual, on σ_ε^2, suggests the existence of additional outcome variation at level-1, which may be predictable. To explain some of this remaining within-person variation, we might add suitable time-varying predictors such as the number of books in the child's home or the amount of parent-child interaction to the level-1 submodel.

The test for the level-2 residual for initial status, on σ_0^2, suggests the existence of additional variation in true initial status, π_{0i}, after accounting for the effects of program participation. This again suggests the need for additional predictors, but because this is a level-2 variance component (describing residual variation in true initial status), we would consider adding both time-invariant *and* time-varying predictors to the multilevel model.

We cannot reject the null hypotheses for the two remaining variance components. Failure to reject the null hypothesis for σ_1^2 indicates that *PROGRAM* explains all the potentially predictable variation between children in their true annual rates of change. Failure to reject the null hypothesis for σ_{01} indicates that the intercepts and slopes of the individual true change trajectories are uncorrelated—that there is no association between true initial status and true annual rates of change (once the effects of *PROGRAM* are removed). As we discuss in subsequent chapters, the results of these two tests might lead us to drop the second level-2 residual, ζ_{1i}, from our model, for neither its variance nor covariance with ζ_{0i}, is significantly different from 0.

4

Doing Data Analysis with the Multilevel Model for Change

We are restless because of incessant change, but we would be frightened if change were stopped.

—Lyman Bryson

In chapter 3, we used a pair of linked statistical models to establish the multilevel model for change. Within this representation, a level-1 submodel describes how each person changes over time and a level-2 submodel relates interindividual differences in change to predictors. To introduce these ideas in a simple context, we focused on just one method of estimation (maximum likelihood), one predictor (a dichotomy), and a single multilevel model for change.

We now delve deeper into the specification, estimation, and interpretation of the multilevel model for change. Following introduction of a new data set (section 4.1), we present a *composite* formulation of the model that combines the level-1 and level-2 submodels together into a single equation (section 4.2). The new composite model leads naturally to consideration of alternative methods of estimation (section 4.3). Not only do we describe two new methods—*generalized least squares* (GLS) and *iterative generalized least squares* (IGLS)—within each, we distinguish further between two types of approaches, the *full* and the *restricted.*

The remainder of the chapter focuses on real-world issues of data analysis. Our goal is to help you learn how to articulate and implement a coherent approach to model fitting. In section 4.4, we present two "standard" multilevel models for change that you should always fit initially in any analysis—the *unconditional means* model and the *unconditional growth* model—and we discuss how they provide invaluable baselines for subsequent comparison. In section 4.5, we discuss strategies for adding time-invariant predictors to the multilevel model for change. We then discuss methods for testing complex hypotheses (sections 4.6 and 4.7) and examining model assumptions and residuals (section 4.8). We conclude, in section 4.9, by recovering "model-based" estimates of the individual growth trajectories that improve upon the exploratory person-by-person OLS estimates introduced in chapter 3. To highlight concepts and strategies rather than technical details, we continue to limit our presentation in several ways, by using: (1) a linear individual growth model; (2) a time-structured data set in which everyone shares the same data collection schedule; and (3) a single piece of statistical software (MLwiN).

4.1 Example: Changes in Adolescent Alcohol Use

As part of a larger study of substance abuse, Curran, Stice, and Chassin (1997) collected three waves of longitudinal data on 82 adolescents. Each year, beginning at age 14, the teenagers completed a four-item instrument assessing their alcohol consumption during the previous year. Using an 8-point scale (ranging from 0 = "not at all" to 7 = "every day"), adolescents described the frequency with which they (1) drank beer or wine, (2) drank hard liquor, (3) had five or more drinks in a row, and (4) got drunk. The data set also includes two potential predictors of alcohol use: *COA*, a dichotomy indicating whether the adolescent is a child of an alcoholic parent; and *PEER*, a measure of alcohol use among the adolescent's peers. This latter predictor was based on information gathered during the initial wave of data collection. Participants used a 6-point scale (ranging from 0 = "none" to 5 = "all") to estimate the proportion of their friends who drank alcohol occasionally (one item) or regularly (a second item).

In this chapter, we explore whether individual trajectories of alcohol use during adolescence differ according to the history of parental alcoholism and early peer alcohol use. Before proceeding, we note that the values of the outcome we analyze, *ALCUSE*, and of the continuous predictor, *PEER*, are both generated by computing the *square root* of the sum of participants' responses across each variable's constituent items. Transformation of the outcome allows us to assume linearity with *AGE* at level-1; transformation of the predictor allows us to assume linearity with *PEER* at level-2. Otherwise, we would need to posit nonlinear models at both levels in order to avoid violating the necessary linearity assumptions. If you find these transformations unsettling, remember that each item's original scale was arbitrary, at best. As in regular regression, analysis is often clearer if you fit a linear model to transformed variables instead of a nonlinear model to raw variables. We discuss this issue further when we introduce strategies for evaluating the tenability of the multilevel model's assumptions in section 4.8, and we explicitly introduce models that relax the linearity assumption in chapter 6.

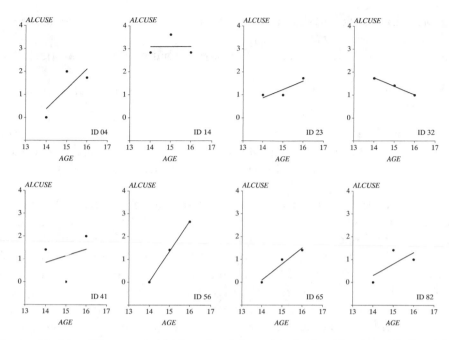

Figure 4.1. Identifying a suitable functional form for the level-1 submodel. Empirical growth plots with superimposed OLS trajectories for 8 participants in the alcohol use study.

To inform model specification, figure 4.1 presents empirical change plots with superimposed OLS-estimated linear trajectories for 8 adolescents randomly selected from the larger sample. For them, and for most of the other 74 not shown, the relationship between (the now-transformed) *ALCUSE* and *AGE* appears linear between ages 14 and 16. This suggests that we can posit a level-1 individual growth model that is linear with adolescent age $Y_{ij} = \pi_{0i} + \pi_{1i}(AGE_{ij} - 14) + \varepsilon_{ij}$, where Y_{ij} is adolescent *i*'s value of *ALCUSE* on occasion *j* and AGE_{ij} is his or her age (in years) at that time. We have centered *AGE* on 14 years (the age at the first wave of data collection) to facilitate interpretation of the intercept.

As you become comfortable with model specification, you may find it easier to write the level-1 submodel using a generic variable $TIME_{ij}$ instead of a specific temporal predictor like $(AGE_{ij} - 14)$:

$$Y_{ij} = \pi_{0i} + \pi_{1i}TIME_{ij} + \varepsilon_{ij}. \tag{4.1}$$

This representation is general enough to apply to all longitudinal data sets, regardless of outcome or time scale. Its parameters have the usual interpretations. In the population from which this sample was drawn:

- π_{0i} represents individual i's true initial status, the value of the outcome when $TIME_{ij} = 0$.
- π_{1i} represents individual i's true rate of change during the period under study.
- ε_{ij} represents that portion of individual i's outcome that is unpredicted on occasion j.

We also continue to assume that the ε_{ij} are independently drawn from a normal distribution with mean 0 and variance σ_ε^2. They are also uncorrelated with the level-1 predictor, $TIME$, and are homoscedastic across occasions.

To inform specification of the level-2 submodel, figure 4.2 presents exploratory OLS-fitted linear change trajectories for a random sample of 32 of the adolescents. To construct this display, we twice divided this subsample into two groups: once by COA (top panel) and again by $PEER$ (bottom panel). Because $PEER$ is continuous, the bottom panel represents a split at the sample mean. Thicker lines represent coincident trajectories—the thicker the line, the more trajectories. Although each plot suggests considerable interindividual heterogeneity in change, some patterns emerge. In the top panel, ignoring a few extreme trajectories, children of alcoholic parents have generally higher intercepts (but no steeper slopes). In the bottom panel, adolescents whose young friends drink more appear to drink more themselves at age 14 (that is, they tend to have higher intercepts), but their alcohol use appears to increase at a slower rate (they tend to have shallower slopes). This suggests that both COA and $PEER$ are viable predictors of change, each deserving further consideration.

We now posit a level-2 submodel for interindividual differences in change. For simplicity, we focus only on COA, representing its hypothesized effect using the two parts of the level-2 submodel, one for true initial status (π_{0i}) and a second for true rate of change (π_{1i}):

$$\begin{aligned} \pi_{0i} &= \gamma_{00} + \gamma_{01}COA_i + \zeta_{0i} \\ \pi_{1i} &= \gamma_{10} + \gamma_{11}COA_i + \zeta_{1i}. \end{aligned} \qquad (4.2)$$

In the level-2 submodel:

- γ_{00} and γ_{10}, the level-2 intercepts, represent the population average initial status and rate of change, respectively, for the child of a non-alcoholic ($COA = 0$). If both parameters are 0, the average child whose parents are non-alcoholic uses no alcohol at age 14 and does not change his or her alcohol consumption between ages 14 and 16.

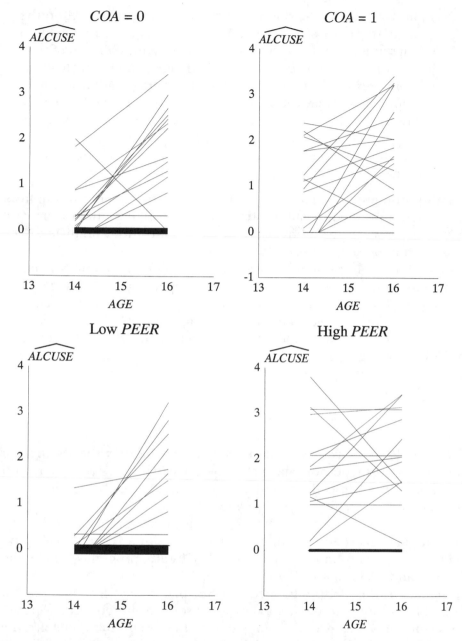

Figure 4.2. Identifying potential predictors of change by examining OLS fitted trajectories separately by levels of selected predictors. Fitted OLS trajectories for the alcohol use
data displayed separately by *COA* status (upper panel) and *PEER* alcohol use (lower
panel).

- γ_{01} and γ_{11}, the level-2 slopes, represent the effect of *COA* on the change trajectories, providing increments (or decrements) to initial status and rates of change, respectively, for children of alcoholics. If both parameters are 0, the average child of an alcoholic initially uses no more alcohol than the average child of a non-alcoholic and the rates of change in alcohol use do not differ as well.
- ζ_{0i} and ζ_{1i}, the level-2 residuals, represent those portions of initial status or rate of change that are unexplained at level-2. They represent deviations of the individual change trajectories around their respective group average trends.

We also continue to assume that ζ_{0i} and ζ_{1i} are independently drawn from a bivariate normal distribution with mean 0, variances σ_0^2 and σ_1^2, and covariance σ_{01}. They are also uncorrelated with the level-2 predictor, *COA*, and are homoscedastic over all values of *COA*.

As in regular regression analysis, we can modify the level-2 submodel to include other predictors—for example, replacing *COA* with *PEER* or adding *PEER* to the current model. We illustrate these modifications in section 4.5. For now, we continue with a single level-2 predictor so that we can introduce a new idea: the creation of the *composite* multilevel model for change.

4.2 The Composite Specification of the Multilevel Model for Change

The level-1/level-2 representation above is not the only specification of the multilevel model for change. A more parsimonious representation arises if you collapse the level-1 and level-2 submodels together algebraically into a single *composite* model. The composite representation, while identical to the level-1/level-2 specification mathematically, provides an alternative way of codifying hypotheses and is the specification required by many multilevel statistical software programs (including MLwiN and SAS PROC MIXED).

To derive the composite specification, first notice that any pair of linked level-1 and level-2 submodels share some common terms. Specifically, the individual growth parameters of the level-1 submodel are the outcomes of the level-2 submodel. We can therefore collapse the submodels together by substituting for π_{0i} and π_{1i} from the level-2 submodel (in equation 4.2, say) into the level-1 submodel (equation 4.1), as follows:

$$Y_{ij} = \pi_{0i} + \pi_{1i}TIME_{ij} + \varepsilon_{ij}$$
$$= (\gamma_{00} + \gamma_{01}COA_i + \zeta_{0i}) + (\gamma_{10} + \gamma_{11}COA_i + \zeta_{1i})TIME_{ij} + \varepsilon_{ij}.$$

The first parenthesis contains the level-2 specification for the level-1 intercept, π_{0i}; the second parenthesis contains the level-2 specification for the level-1 slope, π_{1i}. Multiplying out and rearranging terms then yields the *composite multilevel model for change*:

$$Y_{ij} = [\gamma_{00} + \gamma_{10}TIME_{ij} + \gamma_{01}COA_i + \gamma_{11}(COA_i \times TIME_{ij})]$$
$$+ [\zeta_{0i} + \zeta_{1i}TIME_{ij} + \varepsilon_{ij}], \tag{4.3}$$

where we once again use brackets to distinguish the model's structural and stochastic components.

Even though the composite specification in equation 4.3 appears more complex than the level-1/level-2 specification, the two forms are logically and mathematically equivalent. Each posits an identical set of links between an outcome (Y_{ij}) and predictors (here, *TIME* and *COA*). The specifications differ only in how they organize the hypothesized relationships, each providing valuable insight into what the multilevel model represents. The advantage of the level-1/level-2 specification is that it reflects our conceptual framework directly: we focus first on individual change and next on interindividual differences in change. It also provides an intuitive basis for interpretation because it directly identifies which parameters describe interindividual differences in initial status (γ_{00} and γ_{01}) and which describe interindividual differences in change (γ_{10} and γ_{11}). The advantage of the composite specification is that it clarifies which statistical model is actually being fit to data when the computer begins to iterate.

In introducing the composite model, we do not argue that its representation is uniformly superior to the level-1/level-2 specification. In the remainder of this book, we use both representations, adopting whichever best suits our purposes at any given time. Sometimes we invoke the substantively appealing level-1/level-2 specification; other times we invoke the algebraically parsimonious composite specification. Because both are useful, we recommend that you take the time to become equally facile with each. To aid in this process, below, we now delve into the structural and stochastic components of the composite model itself.

4.2.1 The Structural Component of the Composite Model

The structural portion of the composite multilevel model for change, in the first set of brackets in equation 4.3, may appear unusual, at least at first. Comfortingly, it contains all the original predictors—here, *COA* and *TIME*—as well as the now familiar fixed effects, γ_{00}, γ_{01}, γ_{10}, and γ_{11}. In chapter 3, we demonstrated that the γ's describe the average change trajectories for individuals distinguished by their level-2 predictor values: γ_{00} and γ_{10} are the intercept and slope of the average trajectory for the children of parents who are not alcoholic; $(\gamma_{00} + \gamma_{01})$ and $(\gamma_{10} + \gamma_{11})$ are the intercept and slope of the average trajectory for the children of alcoholics.

The γ's retain these interpretations in the composite model. To demonstrate this equivalence, let us substitute different values of COA into the model's structural portion and recover the population average change trajectories. As COA has only two values, 0 and 1, recovery is easy. For the children of non-alcoholic parents, we substitute 0 into equation 4.3 to find:

$$\begin{pmatrix} \text{Population average} \\ \text{trajectory for the children} \\ \text{of non-alchoholic parents} \end{pmatrix} = \gamma_{00} + \gamma_{10}TIME_{ij} + \gamma_{01}0 + \gamma_{11}(0 \times TIME_{ij})$$

$$= \gamma_{00} + \gamma_{10}TIME_{ij},$$

$$(4.4a)$$

a trajectory with intercept γ_{00} and slope γ_{10}, as indicated in the previous paragraph. For the children of alcoholic parents, we substitute in 1 to find:

$$\begin{pmatrix} \text{Population average} \\ \text{trajectory for the children} \\ \text{of alchoholic parents} \end{pmatrix} = \gamma_{00} + \gamma_{10}TIME_{ij} + \gamma_{01}1 + \gamma_{11}(1 \times TIME_{ij})$$

$$= (\gamma_{00} + \gamma_{01}) + (\gamma_{10} + \gamma_{11})TIME_{ij},$$

$$(4.4b)$$

a trajectory with intercept $(\gamma_{00} + \gamma_{01})$ and slope $(\gamma_{10} + \gamma_{11})$ also as just described.

Although their interpretation is identical, the γ's in the composite model describe patterns of change in a different way. Rather than postulating first how $ALCUSE$ is related to $TIME$ and the individual growth parameters, and second how the individual growth parameters are related to COA, the composite specification in equation 4.3 postulates that $ALCUSE$ depends *simultaneously* on: (1) the level-1 predictor, $TIME$; (2) the level-2 predictor, COA; and (3) the *cross-level* interaction, COA by $TIME$. From this perspective, the composite model's structural portion strongly resembles a regular regression model with predictors, $TIME$ and COA, appearing as main effects (associated with γ_{10} and γ_{01}, respectively) and in a *cross-level* interaction (associated with γ_{11}).

How did this cross-level interaction arise, when the level-1/level-2 specification appears to have no similar term? Its appearance arises from the "multiplying out" procedure used to generate the composite model. When we substitute the level-2 submodel for π_{1i} into its appropriate position in the level-1 submodel, the parameter γ_{11}, previously associated only with *COA*, gets multiplied by *TIME*. In the composite model, then, this parameter becomes associated with the interaction term, *COA* by *TIME*. This association makes sense if you consider the following logic. When γ_{11} is non-zero in the level-1/level-2 specification, the *slopes* of the change trajectories differ according to values of *COA*. Stated another way, the effect of *TIME* (whose effect is represented by the slopes of the change trajectories) differs by levels of *COA*. When the effects of one predictor (here, *TIME*) differ by the levels of another predictor (here, *COA*), we say that the two predictors *interact*. The cross-level interaction in the composite specification codifies this effect.

4.2.2 The Stochastic Component of the Composite Model

The *random effects* of the composite model appear in the second set of brackets in equation 4.3. Their representation is more mysterious than that of the fixed effects and differs dramatically from the simple error terms in the separate submodels. But as you would expect, ultimately, they have the same meaning under both the level-1/level-2 and composite representations. In addition, their structure in the composite model provides valuable insight into our assumptions about the behavior of residuals over time in longitudinal data.

To understand how to interpret this stochastic portion, recall that in chapter 3, we described how the random effects allow each person's true change trajectory to be scattered around the relevant population average trajectory. For example, given that the population average change trajectory for the children of non-alcoholic parents (in equation 4.4a has intercept γ_{00} and slope γ_{10}, the level-2 residuals, ζ_{0i} and ζ_{1i}, allow individual i's trajectory to differ from this average. The true trajectory for individual i, a specific child of non-alcoholic parents, therefore has intercept $(\gamma_{00} + \zeta_{0i})$ and slope $(\gamma_{10} + \zeta_{1i})$. Once this trajectory has been determined, the level-1 residuals, ε_{ij}, then allow his or her data for occasion j to be scattered randomly about it.

We can see how the composite model represents this conceptualization by deriving the true trajectories for different individuals with specific predictor values. Using equation (4.3), we note that if adolescent i has nonalcoholic parents ($COA = 0$):

$$Y_{ij} = [\gamma_{00} + \gamma_{10}TIME_{ij} + \gamma_{01}0 + \gamma_{11}(0 \times TIME_{ij})] + [\zeta_{0i} + \zeta_{1i}TIME_{ij} + \varepsilon_{ij}]$$
$$= [\gamma_{00} + \gamma_{10}TIME_{ij}] + [\zeta_{0i} + \zeta_{1i}TIME_{ij} + \varepsilon_{ij}]$$
$$= (\gamma_{00} + \zeta_{0i}) + (\gamma_{10} + \zeta_{1i})TIME_{ij} + \varepsilon_{ij},$$

leading to a true trajectory with intercept ($\gamma_{00} + \zeta_{0i}$) and slope ($\gamma_{10} + \zeta_{1i}$) as described above. If adolescent i has an alcoholic parent ($COA = 1$):

$$
\begin{aligned}
Y_{ij} &= [\gamma_{00} + \gamma_{10}TIME_{ij} + \gamma_{01}1 + \gamma_{11}(1 \times TIME_{ij})] + [\zeta_{0i} + \zeta_{1i}TIME_{ij} + \varepsilon_{ij}] \\
&= [(\gamma_{00} + \gamma_{01}) + (\gamma_{10} + \gamma_{11})TIME_{ij}] + [\zeta_{0i} + \zeta_{1i}TIME_{ij} + \varepsilon_{ij}] \\
&= (\gamma_{00} + \gamma_{01} + \zeta_{0i}) + (\gamma_{10} + \gamma_{11} + \zeta_{1i})TIME_{ij} + \varepsilon_{ij},
\end{aligned}
$$

leading to a true trajectory with intercept ($\gamma_{00} + \gamma_{01} + \zeta_{0i}$) and slope ($\gamma_{10} + \gamma_{11} + \zeta_{1i}$).

A distinctive feature of the composite multilevel model is its "composite residual," the three terms in the second set of brackets on the right of equation 4.3 that combine together the level-1 residual and the two level-2 residuals:

$$
\text{Composite residual: } [\zeta_{0i} + \zeta_{1i}TIME_{ij} + \varepsilon_{ij}].
$$

The composite residual is not a simple sum. Instead, the second level-2 residual, ζ_{1i}, is multiplied by the level-1 predictor, *TIME*, before joining its siblings. Despite its unusual construction, the interpretation of the composite residual is straightforward: it describes the difference between the observed and the expected value of Y for individual i on occasion j.

The mathematical form of the composite residual reveals two important properties about the occasion-specific residuals not readily apparent in the level-1/level-2 specification: they can be both *autocorrelated* and *heteroscedastic* within person. As we describe briefly below, and more elaborately explain in chapter 7, these are exactly the kinds of properties that you would expect among residuals for repeated measurements of a changing outcome.

When residuals are heteroscedastic, the unexplained portions of each person's outcome have unequal variances across occasions of measurement. Although heteroscedasticity has many roots, one major cause is the effects of omitted predictors—the consequences of failing to include variables that are, in fact, related to the outcome. Because their effects have nowhere else to go, they bundle together, by default, into the residuals. If their impact differs across occasions, the residual's magnitude may differ as well, creating heteroscedasticity. The composite model allows for heteroscedasticity via the level-2 residual ζ_{1i}. Because ζ_{1i} is multiplied by *TIME* in the composite residual, its magnitude can differ (linearly, at least, in a linear level-1 submodel) across occasions. If there are systematic differences in the *magnitudes* of the composite residuals across occasions, there will be accompanying differences in residual *variance*, hence heteroscedasticity.

When residuals are autocorrelated, the unexplained portions of each person's outcome are correlated with each other across repeated occasions. Once again, omitted predictors, whose effects are bundled into the residuals, are a common cause. Because their effects may be present identically in each residual over time, an individual's residuals may become linked across occasions. The presence of the time-invariant ζ_{0i}'s and ζ_{1i}'s in the composite residual of equation 4.3 allows the residuals to be autocorrelated. Because they have only an "i" subscript (and no "j"), they feature identically in each individual's composite residual on every occasion, creating the potential for autocorrelation across time.

4.3 Methods of Estimation, Revisited

When we discussed estimation in section 3.4, we focused on the method of maximum likelihood (ML). As we suggested then, there are other ways of fitting the multilevel model for change. Below, in section 4.3.1, we describe two other methods that are extensions of the popular OLS estimation method, with which you are already familiar: *generalized least squares (GLS)* estimation and *iterative generalized least squares (IGLS)* estimation. In section 4.3.2, we delve deeper into ML methods themselves and distinguish further between two important types of ML estimation—called *full* and *restricted* maximum-likelihood estimation. Finally, in section 4.3.3, we comment on the various methods and how you might choose among them.

4.3.1 Generalized Least-Squares Estimation

Generalized least-squares (GLS) estimation is an extension of ordinary least-squares estimation that allows you to fit statistical models under more complex assumptions on the residuals. Like OLS, GLS seeks parameter estimates that minimize the sum of squared residuals.[1] But instead of requiring the residuals to be independent and homoscedastic, as OLS does, GLS allows them to be autocorrelated and heteroscedastic, as in the composite multilevel model for change.

To understand how you can use GLS to fit the composite multilevel model for change, first reconsider the inefficient exploratory OLS analyses of chapter 2. In section 2.3, our exploratory analyses actually mirrored our later level-1/level-2 specification of the multilevel model for change. To fit the model, we used OLS methods twice. First, in a set of exploratory level-1 analyses, we divided the person-period data set into person-specific chunks (by *ID*) and fit separate within-person regressions of the outcome on *TIME*. Then, in an exploratory level-2 analysis, we regressed the resultant individual growth parameter estimates on predictors. The existence and form of the composite multilevel model for change suggests that, instead of this piecewise analysis, you could keep the person-period data set intact and regress the outcome (here, *ALCUSE*) on the predictors in the structural portion of the composite model for change (here, *TIME, COA*, and *COA* by *TIME*). This would allow you to estimate the fixed effects of greatest interest (γ_{00}, γ_{10}, γ_{01}, γ_{11}) without dividing the data set into person-specific chunks.

Were you to use OLS to conduct this regression analysis in the full person-period data set, the resultant regression coefficients (estimates of γ_{00}, γ_{10}, γ_{01}, γ_{11}) would indeed be unbiased estimates of the composite model's fixed effects. Unfortunately, their standard errors would not possess the optimal properties needed for testing hypotheses efficiently because the residuals in the stochastic portion of the composite model do not possess the "classical" assumptions of independence and homoscedasticity. In other words, the OLS approach is simply inappropriate in the full person-period data set. To estimate the fixed effects efficiently by fitting the composite model directly in the person-period data set requires the methods of GLS estimation.

This leads to a conundrum. In reality, to estimate the fixed effects in the composite model by a regression analysis in the entire person-period data set, we need GLS methods. But to conduct a GLS analysis, we need to know the shape and contents of the *true error covariance* matrix— specifically we need to know the degree of autocorrelation and heteroscedasticity that actually exists among the residuals in the population so that we can account for this error structure during GLS estimation. We cannot know these population values explicitly, as they are hidden from view; we only possess information on the sample, not the population. Hence the conundrum: to conduct an appropriate analysis of the composite multilevel model for change directly in the person-period data set we need information that we do not, indeed cannot, know.

GLS addresses this conundrum using a two-stage approach. First, fit the composite model by regressing *ALCUSE* on predictors *TIME, COA*, and *COA* by *TIME* in the full person-period data set using OLS methods and *estimate* the error covariance matrix using residuals from the OLS-fitted model. Then, refit the composite model using GLS treating the *estimated* error covariance matrix as though it were the *true* error covariance matrix. In this process, the first stage uses OLS to provide *starting values* (initial estimates) of the fixed effects. These starting values then yield predicted outcome values that allow computation of the residuals for each person on each occasion. The population error covariance matrix is then estimated using these residuals. In the second stage, compute *revised* GLS estimates of the fixed effects and associated standard errors under the assumption that the estimated error covariance matrix from the first stage is a correct representation of the population error covariance matrix of the composite model. All of this, of course, is hidden from view because the computer does it for you.

If GLS estimation with two steps is good, could GLS estimation with many steps be better? This simple question leads to an extension of GLS known as IGLS (*iterative* generalized least squares). Instead of stopping after one round of estimation and refitting, you ask the computer to implement the approach repeatedly, each time using the previous set of estimated fixed effects to re-estimate the error covariance matrix, which then leads to GLS estimates of the fixed effects that are further refined. After each round, you can ask the computer to check whether the current set of estimates is an improvement over the last. If they have not improved (as judged by criteria that you define, or the software package specifies by default), then declare that the process has *converged* and stop, out-putting the estimates, their standard errors, and model goodness-of-fit statistics for your perusal.

As with all iterative procedures, the convergence of IGLS is not guar-anteed. If your data set is small or severely unbalanced, or if your hypoth-esized model is too complex, IGLS may iterate indefinitely. To prevent this, all software packages invoke an upper limit on the number of itera-tions for each analysis (that you can modify, if you wish). If an IGLS analy-sis fails to converge after a pre-specified number of iterations, you can try again, increasing this upper limit. If it still fails to converge, the estimates may be incorrect and should be treated with caution. We illustrate the use of IGLS methods later in this chapter and discuss issues of noncon-vergence in section 5.2.

4.3.2 Full and Restricted Maximum-Likelihood Estimation

Statisticians distinguish between two types of maximum likelihood esti-mation: *full* (FML) and *restricted* (RML). These two variants on a common theme differ in how the likelihood function is formed, which affects parameter estimation and the strategies used to test hypotheses. You must select a particular ML method *before* fitting models. Perhaps more impor-tantly, you should understand which method your software package selects as its default (although this can usually be overridden).

Although we were not specific in chapter 3, the ML method that we described there was FML. The likelihood function described in section 3.4 assesses the joint probability of simultaneously observing all the sample data actually obtained. The sample likelihood, a function of the data and the hypothesized model and its assumptions, contains all the unknown parameters, both the fixed effects (the γ's) and the variance components (σ_ε^2, σ_0^2, σ_1^2, and σ_{01}). Under FML, the computer computes those estimates of these population parameters that jointly maximize this likelihood.

FML estimation is not without problems. Because of the way we construct and maximize the likelihood function, FML estimates of the variance components ($\hat{\sigma}_\varepsilon^2$, $\hat{\sigma}_0^2$, $\hat{\sigma}_1^2$, and $\hat{\sigma}_{01}$) contain FML estimates of the fixed effects (the $\hat{\gamma}$'s). This means that we ignore uncertainty about the fixed effects when estimating the variance components, treating their values as known. By failing to allocate some degrees of freedom to the estimation of fixed effects, FML overstates the degrees of freedom left for estimating variance components and underestimates the variance components themselves, leading to biased estimates when samples are small (they are still asymptotically unbiased).

These concerns led statisticians to develop restricted maximum likelihood (RML; Dempster Laird & Rubin, 1977). Because both FML and RML require intensive numerical iteration when used to fit the multilevel model for change, we cannot illustrate their differences algebraically. But because similar issues arise when these methods are used to fit simpler models, including the linear regression model for cross-sectional data, we can illustrate their differences in this context where closed-form estimates *can* be written down.

We begin by describing what happens when we use FML to fit a linear regression model to cross-sectional data. Imagine using the following simple regression model to predict an outcome, Y, on the basis of p predictors, X_1 through X_p, in a sample of size n, $Y_i = \beta_0 + \beta_1 X_{1i} + \beta_2 X_{2i} + \cdots + \beta_p X_{pi} + \varepsilon_i$, where i indexes individuals and ε_i represents the usual independent, normally distributed residual with zero mean and homoscedastic variance, σ_ε^2. If it were somehow possible to know the *true population values* of the regression parameters, the residual for individual i would be: $\varepsilon_i = Y_i - (\beta_0 + \beta_1 X_{1i} + \beta_2 X_{2i} + \cdots + \beta_p X_{pi})$. The FML estimator of the unknown residual variance σ_ε^2, would then be the sum of squared residuals divided by the sample size, n:

$$\hat{\sigma}_\varepsilon^2 = \frac{\sum_{i=1}^{n} \varepsilon_i^2}{n}. \tag{4.5a}$$

Because we imagine that we *know* the population values of the regression coefficients, we need not estimate them to compute residuals, leaving n degrees of freedom for the residual variance calculation.

In practice, of course, we never know the true population values of the regression parameters; we estimate them using sample data, and so:

$$\hat{\varepsilon}_i = Y_i - \left(\hat{\beta}_0 + \hat{\beta}_1 X_{1i} + \hat{\beta}_2 X_{2i} + \cdots + \hat{\beta}_p X_{pi} \right).$$

Substituting these estimates into equation (4.5a) yields an FML estimate of the residual variance:

$$\hat{\sigma}_\varepsilon^2 = \frac{\sum_{i=1}^{n} \hat{\varepsilon}_i^2}{n}, \qquad (4.5b)$$

because functions of FML estimators, the $\hat{\beta}$'s, are themselves FML estimators.

Notice that the denominator of the FML estimated residual variance in equation 4.5b is the sample size n. Use of this denominator assumes that we still have all the original degrees of freedom in the sample to estimate this parameter. But because we estimated $(p + 1)$ regression parameters to compute the residuals, and did so with uncertainty, we used up $(p + 1)$ degrees of freedom. An *unbiased* estimate of the residual variance decreases the denominator of equation 4.5b to account for this loss:

$$\hat{\sigma}_\varepsilon^2 = \frac{\sum_{i=1}^{n} \hat{\varepsilon}_i^2}{n - (p+1)}. \qquad (4.5c)$$

The distinction between the estimated residual variances in equations 4.5b and 4.5c is exactly the same as that between *full* and *restricted* ML estimation in the multilevel model for change. Like RML, equation 4.5c accounts for the uncertainty associated with estimating the regression parameters (the fixed effects) before estimating the residual variance (the variance components); like FML, equation (4.5b) does not.

How are RML estimates computed? Technical work by Patterson and Thompson (1971) and Harville (1974) provides a conceptually appealing strategy. RML estimates of the variance components are those values that maximize the likelihood of observing the sample *residuals* (not the sample data). Once again, an iterative process is used. First, we estimate the fixed effects, the γ's, using some other method, often OLS or GLS. Next, as in regular regression analysis, we use the $\hat{\gamma}$'s to estimate a residual for each person on each occasion (by subtracting observed and predicted values). Under the usual assumptions about the level-1 and level-2 residuals—independence, homoscedasticity, and normality—we can write down the likelihood of observing this particular collection of "data" (that is, *residuals*), in terms of the residuals and the unknown variance components that govern their distributions. We then take the logarithm of the restricted likelihood and maximize it to yield RML estimates of the variance components, the only unknown parameters remaining (as we have assumed that the fixed effects, the γ's, are known).

For decades, controversy has swirled around the comparative advantages of these two methods. Although Dempster et al. (1977, p. 344) declared RML to be "intuitively more correct," it has not proved to be unilaterally better than FML in practice. In their review of simulation studies that compare these methods for fitting multilevel models, Kreft and deLeeuw (1998) find no clear winner. They suggest that some of the ambiguity stems from the decreased precision that accompanies the decreased small sample bias of RML estimation.

If neither approach is uniformly superior, why belabor this distinction? An important issue is that goodness-of-fit statistics computed using the two methods (introduced in section 4.6) refer to different portions of the model. Under FML, they describe the fit of the entire model; under RML, they describe the fit of only the *stochastic* portion (the random effects). This means that the goodness-of-fit statistics from FML can be used to test hypotheses about any type of parameter, either a fixed effect or a variance component, but those from RML can be used only to test hypotheses about variance components (not the fixed effects). This distinction has profound implications for hypothesis testing as a component of model building and data analysis (as we will soon describe). When we compare models that differ only in their variance components, we can use either method. When we compare models that differ in both fixed effects and variance components, we must use full information methods. To further complicate matters, different software programs use different methods as their default option (although all can use either approach). SAS PROC MIXED, for example, uses RML by default, whereas MLwiN and HLM use FML. This means that when you use a particular statistical computer program, you must be sure to ascertain which method of ML estimation is used by default; if you prefer the alternative method—for reasons of potentially increased precision or the ability to conduct a wider array of hypothesis tests—be sure you are obtaining the desired estimates.

4.3.3 Practical Advice about Estimation

Generalized least squares and maximum likelihood estimation are not identical methods of estimation. They use different procedures to fit the model and they allow us to make different assumptions about the distribution of the random effects. We obtain GLS estimates by *minimizing* a weighted function of the residuals; we obtain ML estimates by *maximizing* a log-likelihood. Only ML estimation requires that the residuals be normally distributed. These differences imply that GLS and ML estimates of the same parameters in the same model using the same data may differ. Although you might find this disturbing, we note that two methods can yield unbiased estimates of the same population parameter but that the estimates themselves can differ. While extensive simulation studies comparing methods are still underway (Draper, 1995; Browne & Draper, 2000), limited data-based comparisons suggest that, in practice, both methods lead to similar conclusions (Kreft, de Leeuw & Kim, 1990).

There is one condition under which the correspondence between GLS and ML methods is well known: if the usual normal distribution assumptions required for ML estimation hold, GLS estimates *are* ML estimates.[2] This equivalence means that, if you are prepared to assume normality for ε and the ζ's, as we did in chapter 3, GLS estimates usually enjoy the same asymptotic unbiasedness, efficiency, and normality that ML estimates do. And since you must invoke normal theory assumptions to conduct hypothesis tests anyway, most data analysts find them compelling and easy to accept. In the remainder of the book, we therefore continue to invoke the standard normal theory assumptions when specifying the multilevel model for change.

GLS and ML are currently the dominant methods of fitting multilevel models to data. They appear in a variety of guises in different packages. Both FML and RML appear in HLM and SAS PROC MIXED. STATA xtreg uses a GLS approach. MLwiN uses IGLS and an extension of it, restricted IGLS (RIGLS), which is the GLS equivalent of RML. And new estimation approaches appear each year. This suggests that whatever we write about a particular method of estimation, or its implementation in a particular package, will soon be out of date. But if your goal is data analysis (not the development of estimation strategies), these modifications of the software are unproblematic. The educated user needs to understand the statistical model, its assumptions, and how it represents reality; the mathematical details of the method of estimation are less crucial. That said, we have three reasons for recommending that you take the time to become comfortable with both ML and GLS methods, at least at the heuristic level presented here. First, you cannot conduct credible analyses nor interpret parameter estimates without at least a conceptual understanding how the model is fit. Second, under the assumptions for which they were designed, these methods have decent statistical properties. Third, most new methods will ultimately descend from, or seek to rectify weaknesses in, these methods. In other words, the ML and GLS methods are here to stay.

4.4 First Steps: Fitting Two Unconditional Multilevel Models for Change

You've articulated your research questions, created a person-period data set, conducted exploratory analyses, chosen an estimation approach, and selected a software package. Although you might be tempted to begin by fitting models that include your substantive predictors, we suggest that you first fit the two simpler models presented in this section: the *unconditional means model* (section 4.4.1) and the *unconditional growth model* (section 4.4.2). These unconditional models partition and quantify the outcome variation in two important ways: first, across people without regard to time (the unconditional means model), and second, across both people *and* time (the unconditional growth model). Their results allow you to establish: (1) whether there is systematic variation in your outcome that is worth exploring; and (2) *where* that variation resides (within or between people). They also provide two valuable baselines against which you can evaluate the success of subsequent model building, as we discuss in section 4.4.3.

4.4.1 The Unconditional Means Model

The *unconditional means model* is the first model you should always fit. Instead of describing *change* in the outcome over time, it simply describes and partitions the outcome *variation*. Its hallmark is the absence of predictors at every level:

$$Y_{ij} = \pi_{0i} + \varepsilon_{ij}$$
$$\pi_{0i} = \gamma_{00} + \zeta_{0i}, \tag{4.6a}$$

where we assume, as usual, that:

$$\varepsilon_{ij} \sim N(0, \sigma_\varepsilon^2) \text{ and } \zeta_{0i} \sim N(0, \sigma_0^2). \tag{4.6b}$$

Notice that because there is only one level-2 residual, ζ_{0i}, we assume *univariate* normality at level-2 (not *bivariate* normality, as we do when we have two level-2 residuals).

The unconditional means model stipulates that, at level-1, the true individual change trajectory for person i is completely flat, sitting at elevation π_{0i}. Because the trajectory lacks a slope parameter associated with a temporal predictor, it cannot tilt. The single part of the level-2 submodel stipulates that while these flat trajectories may differ in elevation, their average elevation, across everyone in the population, is γ_{00}. Any interindividual variation in elevation is not linked to predictors. Even though you hope that this model did *not* give rise to your sample data —for it is not really about *change* at all—we recommend that you always fit it first because it partitions the total *variation* in the outcome meaningfully.

To understand how this variance partition operates, notice that flat individual change trajectories are really just *means*. The true mean of Y for individual i is π_{0i}; the true mean of Y across everyone in the population is γ_{00}. Borrowing terminology from analysis of variance, π_{0i} is the *person-specific mean* and γ_{00} is the *grand mean*. The unconditional means model postulates that the *observed* value of Y for individual i on occasion j is composed of deviations about these means. On occasion j, Y_{ij} deviates from individual i's true mean (π_{0i}) by ε_{ij}. The level-1 residual is thus a "within-person" deviation that assesses the "distance" between Y_{ij} and π_{0i}. Then, for person i, his or her true mean (π_{0i}) deviates from the population average true mean (γ_{00}) by ζ_{0i}. This level-2 residual is thus a "between-person" deviation that assesses the "distance" between π_{0i} and γ_{00}.

The variance components of equation 4.6b summarize the variability in these deviations across everyone in the population: σ_ε^2 is the "within-person" variance, the pooled scatter of each person's data around his or her own mean; σ_0^2 is the "between-person" variance, the pooled scatter of the person-specific means around the grand mean. The primary reason we fit the unconditional means model is to estimate these variance components, which assess the amount of outcome variation that exists at each level. Associated hypothesis tests help determine whether there is sufficient variation at that level to warrant further analysis. If a variance component is zero, there is little point in trying to predict outcome variation *at that level*—there is too little variation to explain. If a variance component is non-zero, then there is some variation at that level that could potentially be explained.

Model A of table 4.1 presents the results of fitting the unconditional means model to the alcohol use data. Its one fixed effect, $\hat{\gamma}_{00}$, estimates the outcome's grand mean across all occasions and individuals. Rejection of its associated null hypothesis ($p < .001$) confirms that the average alcohol consumption of the average adolescent between ages 14 and 16 is non-zero. Squaring 0.922 (which yields 0.85) to obtain its value on the instrument's original scale, we conclude that the average adolescent does drink during these years, but not very much.

Next, examine the random effects, the major purpose for fitting this model. The estimated within-person variance, $\hat{\sigma}_\varepsilon^2$, is 0.562; the estimated

Table 4.1: Results of fitting a taxonomy of multilevel models for change to the alcohol use data ($n = 82$)

	Parameter	Model A	Model B	Model C	Model D	Model E	Model F (CPEER)	Model G (CCOA & CPEER)
Fixed Effects								
Initial status, π_{0i}								
Intercept	γ_{00}	0.922***	0.651***	0.316***	−0.317***	−0.314***	0.394***	0.651***
		(0.096)	(0.105)	(0.131)	(0.148)	(0.146)	(0.104)	(0.080)
COA	γ_{01}			0.743***	0.579***	0.571***	0.571***	0.571***
				(0.195)	(0.162)	(0.146)	(0.146)	(0.146)
PEER	γ_{02}				0.694***	0.695***	0.695***	0.695***
					(0.112)	(0.111)	(0.111)	(0.111)
Rate of change, π_{1i}								
Intercept	γ_{10}		0.271***	0.293***	0.429***	0.425***	0.271***	0.271***
			(0.062)	(0.084)	(0.114)	(0.106)	(0.061)	(0.061)
COA	γ_{11}			−0.049	−0.014			
				(0.125)	(0.125)			
PEER	γ_{12}				−0.150~	−0.151~	−0.151~	−0.151~
					(0.086)	(0.085)	(0.085)	(0.085)

Variance Components

		(1)	(2)	(3)	(4)	(5)	(6)	(7)
Level 1	Within-person σ_ε^2	0.562***	0.337***	0.337***	0.337***	0.337***	0.337***	0.337***
		(0.062)	(0.053)	(0.053)	(0.053)	(0.053)	(0.053)	(0.053)
Level 2	In initial status σ_0^2	0.564***	0.624***	0.488**	0.241**	0.241**	0.241**	0.241**
		(0.119)	(0.148)	(0.128)	(0.093)	(0.093)	(0.093)	(0.093)
	In rate of change σ_1^2		0.151**	0.151*	0.139*	0.139*	0.139*	0.139*
			(0.056)	(0.056)	(0.055)	(0.055)	(0.055)	(0.055)
	Covariance σ_{01}		−0.068	−0.059	−0.006	−0.006	−0.006	−0.006
			(0.070)	(0.066)	(0.055)	(0.055)	(0.055)	(0.055)

Pseudo R^2 Statistics and Goodness-of-fit

	(1)	(2)	(3)	(4)	(5)	(6)	(7)
$R_{y\hat{y}}^2$		.043	.150	.291	.291	.291	.291
R_ε^2		.40	.40	.40	.40	.40	.40
R_0^2			.218	.614	.614	.614	.614
R_1^2			.000	.079	.079	.079	.079
Deviance	670.2	636.6	621.2	588.7	588.7	588.7	588.7
AIC	676.2	648.6	637.2	608.7	606.7	606.7	606.7
BIC	683.4	663.0	656.5	632.8	628.4	628.4	628.4

$\sim p < .10$; $* p < .05$; $** p < .01$; $*** p < .001$

These models predict *ALCUSE* between ages 14 and 16 as a function of *AGE*-14 (at level-1) and various combinations of *COA* and *PEER* (at level-2). Models C, D, and E enter the level-2 predictors in their raw form; Models F and G enter the level-2 predictors in *centered* forms as indicated.

Note: MLwiN, full IGLS.

between-person variance, $\hat{\sigma}_0^2$, is 0.564. Using the single parameter hypothesis tests of section 3.6, we can reject both associated null hypotheses at the .001 level. (Although these tests can mislead—(see section 3.6.2), we use them in table 4.1 because it turns out—for these data, at least—that the conclusions are supported by the superior methods of testing presented in section 4.6.) We conclude that the average adolescent's alcohol consumption varies over time and that adolescents differ from each other in alcohol use. Because each variance component is significantly different from 0, there is hope for linking both within-person and between-person variation in alcohol use to predictors.

The unconditional means model serves another purpose: it allows us to evaluate numerically the relative magnitude of the within-person and between-person variance components. In this data set, they happen to be almost equal. A useful statistic for quantifying their relative magnitude is the *intraclass correlation coefficient*, ρ, which describes the proportion of the total outcome variation that lies "between" people. Because the total variation in Y is just the sum of the within and between-person variance components, the population intraclass correlation coefficient is:

$$\rho = \frac{\sigma_0^2}{\sigma_0^2 + \sigma_\varepsilon^2} \tag{4.7}$$

We can estimate ρ by substituting the two estimated variance components from table 4.1 into equation (4.7). For these data, we find:

$$\hat{\rho} = \frac{0.564}{0.564 + 0.562} = 0.50,$$

indicating that half the total variation in alcohol use is attributable to differences among adolescents.

The intraclass correlation coefficient has another role as well: it summarizes the size of the residual autocorrelation in the composite unconditional means model. To understand how it does this, substitute the level-2 submodel in equation 4.6a into its level-1 submodel to yield the following composite unconditional means model:

$$Y_{ij} = \gamma_{00} + (\zeta_{0i} + \varepsilon_{ij}). \tag{4.8}$$

In this representation, Y_{ij} is composed of one fixed effect, γ_{00}, and one composite residual $(\zeta_{0i} + \varepsilon_{ij})$. Each person has a different composite residual on each occasion of measurement. But notice the difference in the subscripts of the pieces of the composite residual: while the level-1 residual, ε_{ij}, has two subscripts (i and j), the level-2 residual, ζ_{0i}, has only one (i). Each person can have a different ε_{ij} on each occasion, but has only one ζ_{0i} across every occasion. The repeated presence of ζ_{0i} in individual i's composite residual links his or her composite residuals across occasions. The error autocorrelation coefficient quantifies the magnitude of this linkage; in the unconditional means model, the error autocorrelation coefficient *is* the intraclass correlation coefficient. Thus, we estimate that, for each person, the average correlation between any pair of composite residuals—between occasions 1 and 2, or 2 and 3, or 1 and 3—is 0.50. This is quite large, and far from the zero residual autocorrelation that an OLS analysis of these data would require. We discuss the intraclass correlation coefficient further in chapter 7.

4.4.2 The Unconditional Growth Model

The next logical step is the introduction of predictor *TIME* into the level-1 submodel. Based on the exploratory analyses of section 4.1, we posit a linear change trajectory:

$$Y_{ij} = \pi_{0i} + \pi_{1i} TIME_{ij} + \varepsilon_{ij}$$
$$\pi_{0i} = \gamma_{00} + \zeta_{0i} \tag{4.9a}$$
$$\pi_{1i} = \gamma_{10} + \zeta_{1i},$$

where we assume that

$$\varepsilon_{ij} \sim N(0, \sigma_\varepsilon^2) \text{ and } \begin{bmatrix} \zeta_{0i} \\ \zeta_{1i} \end{bmatrix} \sim N\left(\begin{bmatrix} 0 \\ 0 \end{bmatrix}, \begin{bmatrix} \sigma_0^2 & \sigma_{01} \\ \sigma_{10} & \sigma_1^2 \end{bmatrix} \right). \tag{4.9b}$$

Because the only predictor in this model is *TIME*, we call equation 4.9 the *unconditional growth model*.

Begin by comparing the unconditional growth model in equation 4.9a to the unconditional means model in equation 4.6a. We facilitate this comparison in table 4.2, which presents these models as well as several others we will soon fit. Instead of postulating that individual i's observed score on occasion j, Y_{ij}, deviates by ε_{ij} from his or her person-specific mean, it specifies that Y_{ij} deviates by ε_{ij} from his or her *true change trajectory*. In other words, altering the level-1 specification alters what the level-1 residuals represent. In addition, we now have a second part to the level-2 submodel that depicts interindividual variation in the rates of change (π_{1i}). But because the model includes no *substantive* predictors, each part of the level-2 submodel simply stipulates that an individual growth parameter (either π_{0i} or π_{1i}) is the sum of an intercept (either γ_{00} or γ_{10}) and a level-2 residual (ζ_{0i} or ζ_{1i}).

An important consequence of altering the level-1 specification is that the meaning of the variance components changes as well. The level-1

Table 4.2: Taxonomy of multilevel models for change fitted to the alcohol use data

Model	Level-1/level-2 specification		Composite model
	level-1 model	level-2 model	
A	$Y_{ij} = \pi_{0i} + \varepsilon_{ij}$	$\pi_{0i} = \gamma_{00} + \zeta_{0i}$	$Y_{ij} = \gamma_{00} + (\varepsilon_{ij} + \zeta_{0i})$
B	$Y_{ij} = \pi_{0i} + \pi_{1i}TIME_{ij} + \varepsilon_{ij}$	$\pi_{0i} = \gamma_{00} + \zeta_{0i}$ $\pi_{1i} = \gamma_{10} + \zeta_{1i}$	$Y_{ij} = \gamma_{00} + \gamma_{10}TIME_{ij}$ $+ (\varepsilon_{ij} + \zeta_{0i} + \zeta_{1i}TIME_{ij})$
C	$Y_{ij} = \pi_{0i} + \pi_{1i}TIME_{ij} + \varepsilon_{ij}$	$\pi_{0i} = \gamma_{00} + \gamma_{01}COA_i + \zeta_{0i}$ $\pi_{1i} = \gamma_{10} + \gamma_{11}COA_i + \zeta_{1i}$	$Y_{ij} = \gamma_{00} + \gamma_{01}COA_i + \gamma_{10}TIME_{ij} + \gamma_{11}COA_i \times TIME_{ij}$ $+ (\varepsilon_{ij} + \zeta_{0i} + \zeta_{1i}TIME_{ij})$
D	$Y_{ij} = \pi_{0i} + \pi_{1i}TIME_{ij} + \varepsilon_{ij}$	$\pi_{0i} = \gamma_{00} + \gamma_{01}COA_i + \gamma_{02}PEER_i + \zeta_{0i}$ $\pi_{1i} = \gamma_{10} + \gamma_{11}COA_i + \gamma_{12}PEER_i + \zeta_{1i}$	$Y_{ij} = \gamma_{00} + \gamma_{01}COA_i + \gamma_{02}PEER_i + \gamma_{10}TIME_{ij}$ $+ \gamma_{11}COA_i \times TIME_{ij} + \gamma_{12}PEER_i \times TIME_{ij}$ $+ (\varepsilon_{ij} + \zeta_{0i} + \zeta_{1i}TIME_{ij})$
E	$Y_{ij} = \pi_{0i} + \pi_{1i}TIME_{ij} + \varepsilon_{ij}$	$\pi_{0i} = \gamma_{00} + \gamma_{01}COA_i + \gamma_{02}PEER_i + \zeta_{0i}$ $\pi_{1i} = \gamma_{10} + \gamma_{12}PEER_i + \zeta_{1i}$	$Y_{ij} = \gamma_{00} + \gamma_{01}COA_i + \gamma_{02}PEER_i + \gamma_{10}TIME_{ij}$ $+ \gamma_{12}PEER_i \times TIME_{ij}$ $+ (\varepsilon_{ij} + \zeta_{0i} + \zeta_{1i}TIME_{ij})$
F	$Y_{ij} = \pi_{0i} + \pi_{1i}TIME_{ij} + \varepsilon_{ij}$	$\pi_{0i} = \gamma_{00} + \gamma_{01}COA_i + \gamma_{02}CPEER_i + \zeta_{0i}$ $\pi_{1i} = \gamma_{10} + \gamma_{12}CPEER_i + \zeta_{1i}$	$Y_{ij} = \gamma_{00} + \gamma_{01}COA_i + \gamma_{02}CPEER_i + \gamma_{10}TIME_{ij}$ $+ \gamma_{12}CPEER_i \times TIME_{ij}$ $+ (\varepsilon_{ij} + \zeta_{0i} + \zeta_{1i}TIME_{ij})$
G	$Y_{ij} = \pi_{0i} + \pi_{1i}TIME_{ij} + \varepsilon_{ij}$	$\pi_{0i} = \gamma_{00} + \gamma_{01}\left(COA_i - \overline{COA}\right)$ $\quad + \gamma_{02}CPEER_i + \zeta_{0i}$ $\pi_{1i} = \gamma_{10} + \gamma_{12}CPEER_i + \zeta_{1i}$	$Y_{ij} = \gamma_{00} + \gamma_{01}\left(COA_i - \overline{COA}\right) + \gamma_{02}CPEER_i$ $+ \gamma_{10}TIME_{ij} + \gamma_{12}CPEER_i \times TIME_{ij}$ $+ (\varepsilon_{ij} + \zeta_{0i} + \zeta_{1i}TIME_{ij})$

These models predict *ALCUSE* between ages 14 and 16 as a function of *AGE-14* (at level-1) and various combinations of *COA* and *PEER* (at level-2). Models C, D, and E enter the level-2 predictors in their raw form; Models F and G enter the level-2 predictors in *centered* forms as indicated. Results of model fitting appear in Table 4.1.

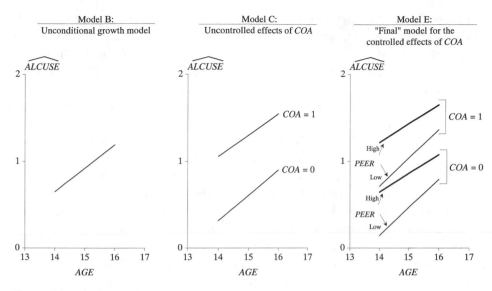

Figure 4.3. Displaying the results of fitted multilevel models for change. Prototypical trajectories from three models presented in table 4.1: Model B: the unconditional growth model, Model C: the uncontrolled effect of *COA*, Model E: the effect of *COA* controlling for *PEER*.

residual variance, σ_ε^2, now summarizes the scatter of each person's data *around his or her own linear change trajectory* (not his or her person-specific mean). The level-2 residual variances, σ_0^2 and σ_1^2, now summarize between-person variability in initial status and rates of change. Estimating these variance components allows us to distinguish level-1 variation from the two different kinds of level-2 variation and to determine whether interindividual differences in change are due to interindividual differences in true initial status or true rate of change.

Model B in table 4.1 presents the results of fitting the unconditional growth model to the alcohol use data. The fixed effects, $\hat{\gamma}_{00}$ and $\hat{\gamma}_{10}$, estimate the starting point and slope of the population average change trajectory. We reject the null hypothesis for each ($p < .001$), estimating that the average true change trajectory for *ALCUSE* has a non-zero intercept of 0.651 and a non-zero slope of +0.271. Because there are no level-2 predictors, it is simple to plot this trajectory, as we do in the left panel of figure 4.3. Although alcohol use for the average adolescent remains low, we estimate that *ALCUSE* rises steadily between ages 14 and 16, from 0.65 to 1.19. We will soon determine whether these trajectories differ systematically by parental alcoholism history or early peer alcohol use.

To assess whether there is hope for future analyses—whether there is statistically significant variation in individual initial status or rate of change that level-2 predictors could explain—examine the variance components. By now, we hope you are beginning to see that variance components are often more interesting than fixed effects. The level-1 residual variance, σ_ε^2, summarizes the average scatter of an individual's observed outcome values around his or her own true change trajectory. If the true change trajectory is linear with age, the unconditional growth model will do a better job of predicting the observed outcome data than the unconditional means model, resulting in smaller level-1 residuals and a smaller level-1 residual variance. Comparing $\hat{\sigma}_\varepsilon^2$ in Model B to that of Model A, we find a decline of .40 (from 0.562 to 0.337). We conclude that 40% of the within-person variation in *ALCUSE* is systematically associated with linear *TIME*. Because we can reject the null hypothesis for this variance component in Model B, we also know that some important within-person variation still remains at level-1 ($p < .001$). This suggests that it might be profitable to introduce substantive predictors into the level-1 submodel. We defer discussion of level-1 substantive predictors until section 5.3 because they must be *time-varying* (not *time-invariant* like the level-2 predictors in this data set).

The level-2 variance components quantify the amount of unpredicted variation in the individual growth parameters. σ_0^2 assesses the unpredicted variability in true initial status (the scatter of the π_{0i} around γ_{00}); σ_1^2 assesses the unpredicted variability in true rates of change (the scatter of the π_{1i} around γ_{10}). Because we reject each associated null hypothesis (at $p < .001$ and $p < .01$, respectively), we conclude that there is non-zero variability in both true initial status and true rate of change. This suggests that it worth trying to use level-2 predictors to explain heterogeneity in each parameter. When we do so, these variance components—0.624 and 0.151—will provide benchmarks for quantifying the predictors' effects. We do not compare these variance components with estimates from the unconditional means model because introduction of *TIME* into the model changes their interpretation.

The population covariance of the level-2 residuals σ_{01}, has an important interpretation in the unconditional growth model. It not only assesses the relationship between the level-2 residuals, it quantifies the population covariance between true initial status and true change. This means that we can assess whether adolescents who drink more at age 14 increase their drinking more (or less) rapidly over time. Interpretation is easier if we re-express the covariance as a correlation coefficient, dividing it by the square root of the product of its associated variance components:

$$\hat{\rho}_{\pi_0\pi_1} = \hat{\rho}_{01} = \frac{\hat{\sigma}_{01}}{\sqrt{\hat{\sigma}_0^2 \hat{\sigma}_1^2}} = \frac{-0.068}{\sqrt{(0.624)(0.151)}} = -0.22.$$

We conclude that the relationship between true rate of change in *ALCUSE* and its level at age 14 is negative and weak and, because we cannot reject its associated null hypothesis, possibly zero.

We can learn more about the residuals in the unconditional growth model by examining the composite specification of the multilevel model:

$$Y_{ij} = \gamma_{00} + \gamma_{10}TIME_{ij} + (\zeta_{0i} + \zeta_{1i}TIME_{ij} + \varepsilon_{ij}). \tag{4.10}$$

Each person has j composite residuals, one per occasion of measurement. The structure of the composite residual, which combines the original level-1 and level-2 residuals (with ζ_{1i} multiplied by *TIME* before being bundled into the sum), provides the anticipated heteroscedasticity and autocorrelation that longitudinal data analysis may demand.

First, we examine the variances of the composite residual. Mathematical results not presented here allow us to write the population variance of the composite residual on the jth occasion of measurement as:

$$\sigma^2_{Residual_j} = \sigma_0^2 + \sigma_1^2 TIME_j^2 + 2\sigma_{01}TIME_j + \sigma_\varepsilon^2. \tag{4.11}$$

Substituting the estimated variance components from Model B in table 4.1 we have:

$$(0.624 + 0.151TIME_j^2 - 0.136TIME_j + 0.337).$$

Substituting values for *TIME* at ages 14 ($TIME_1 = 0$), 15 ($TIME_2 = 1$) and 16 ($TIME_3 = 2$), we find estimated composite residual variances of 0.961, 0.976, and 1.293, respectively. While not outrageously heteroscedastic, especially for ages 14 and 15, this is beyond the bland homoscedasticity we assume of residuals in cross-sectional data.

Further mathematical results not shown here allow us to write the autocorrelation between composite residuals on occasions j and j' as:

$$\rho_{Residual_j Residual_{j'}} = \frac{\sigma_0^2 + \sigma_{01}(TIME_j + TIME_{j'}) + \sigma_1^2 TIME_j TIME_{j'}}{\sqrt{\sigma^2_{Residual_j}\sigma^2_{Residual_{j'}}}}, \tag{4.12}$$

where the residual variances in the denominator are given by equation (4.11). Substituting the estimated variance components and *TIME* into equation 4.12 yields a residual autocorrelation of 0.57 between occasions 1 and 2, 0.64 between occasions 2 and 3, and 0.44 between occasions 1 and 3. We conclude that there is substantial autocorrelation between the residuals across successive measurement occasions. We explore this behavior further in chapter 7.

4.4.3 Quantifying the Proportion of Outcome Variation "Explained"

The two unconditional models assess whether there is potentially predictable outcome variation and, if so, where it resides. For these data, the unconditional means model suggests roughly equal amounts of within-person and between-person variation. The unconditional growth model suggests that some of the within-person variation is attributable to linear *TIME* and that there is between-person variation in both true initial status and true rate of change that level-2 predictors might explain.

In multiple regression analysis, we quantify the proportion of outcome variation that a model's predictors "explain" using an R^2 (or adjusted R^2) statistic. In the multilevel model for change, definition of a similar statistic is trickier because total outcome variation is partitioned into several variance components: here, σ_ε^2, σ_0^2 and σ_1^2. As a result, statisticians have yet to agree on appropriate summaries (Kreft & deLeeuw, 1998; Snidjers & Bosker, 1994). Below, we present several *pseudo-R^2 statistics* that quantify how much outcome variation is "explained" by a multilevel model's predictors. First, we assess the proportion of *total* variation explained using a statistic similar to the traditional R^2 statistic; second, we dissect the level-1 and level-2 outcome variation using statistics similar to traditional *adjusted-R^2* statistics. These pseudo-R^2 statistics can be useful data analytic tools, as long as you construct and interpret them carefully.

An Overall Summary of Total Outcome Variability Explained

In multiple regression, one simple way of computing a summary R^2 statistic is to square the sample correlation between observed and predicted values of the outcome. The same approach can be used in the multilevel model for change. All you need do is: (1) compute a predicted outcome value for each person on each occasion of measurement; and (2) square the sample correlation between observed and predicted values. The resultant pseudo-R^2 statistic assesses the proportion of total outcome variation "explained" by the multilevel model's specific combination of predictors.

The bottom panel of table 4.1 presents this pseudo-R^2 statistic (labeled $R_{y,\hat{y}}^2$) for each model fit. We calculate these statistics by correlating predicted and observed values of *ALCUSE* for each person on each occasion of measurement. For Model B, for example, the predicted values for individual i on occasion j are: $\hat{Y}_{ij} = 0.651 + 0.271\,TIME_{ij}$. As everyone in this data set has the identical set of measurement occasions (0, 1, and 2), Model B yields only three distinct predicted values:

$$\hat{Y}_{i1} = 0.651 + 0.271(0) = 0.651$$
$$\hat{Y}_{i2} = 0.651 + 0.271(1) = 0.922$$
$$\hat{Y}_{i3} = 0.651 + 0.271(2) = 1.193.$$

Across the entire person-period data set, the sample correlation between these predicted values and the observed values is 0.21, which yields a pseudo-R^2 statistic of .043. We conclude that 4.3% of the total variability in *ALCUSE* is associated with linear time. As we add substantive predictors to this model, we examine whether, and by how much, this pseudo-R^2 statistic increases.

Pseudo-R^2 Statistics Computed from the Variance Components

Residual variation—that portion of the outcome variation *unexplained* by a model's predictors—provides another criterion for comparison. When you fit a series of models, you hope that added predictors further explain unexplained outcome variation, causing residual variation to decline. The magnitude of this decline quantifies the improvement in fit. A large decline suggests that the predictors make a big difference; a small, or zero, decline suggests that they do not. To assess these declines on a common scale, we compute the *proportional reduction in residual variance* as we add predictors.

Each unconditional model yields residual variances that serve as yardsticks for comparison. The unconditional means model provides a baseline estimate of σ_ε^2; the unconditional growth model provides baseline estimates of σ_0^2 and σ_1^2. Each leads to its own pseudo-R^2 statistic.

Let us begin by examining the decrease in within-person residual variance (σ_ε^2) between the unconditional means model and unconditional growth model. As shown in table 4.1, our initial level-1 residual variance estimate, 0.562, drops to .337 in the initial model for change. As the fundamental difference between these models is the introduction of *TIME*, this pseudo-R^2 statistic assesses the proportion of within-person variation "explained by time." We compute the statistic as:

$$\text{Pseudo } R_\varepsilon^2 = \frac{\hat{\sigma}_\varepsilon^2(\text{unconditional means model}) - \sigma_\varepsilon^2(\text{unconditional growth model})}{\hat{\sigma}_\varepsilon^2(\text{unconditional means model})}.$$

(4.13)

For the alcohol use data, we have $(.562 - .337)/.562 = 0.400$. We conclude that 40.0% of the within-person variation in *ALCUSE* is explained by linear *TIME*. The only way of reducing this variance component further is to add time-varying predictors to the level-1 submodel. As this data set has no such predictors, $\hat{\sigma}_\varepsilon^2$ remains unchanged in every subsequent model in table 4.1.

We can use a similar approach to compute pseudo-R^2 statistics quantifying the proportional reduction in level-2 residual variance on the addition of one or more level-2 predictors. Each level-2 residual variance component has its own pseudo-R^2 statistic. A level-1 linear change model, with two level-2 variance components, σ_0^2 and σ_1^2, has two pseudo-R^2s. Baseline estimates of these components come from the unconditional growth model. For any subsequent model, we compute a pseudo-R^2 statistic as:

$$\text{Pseudo-}R_\zeta^2 = \frac{\hat{\sigma}_\zeta^2(\text{unconditional growth model}) - \sigma_\zeta^2(\text{subsequent model})}{\hat{\sigma}_\zeta^2(\text{unconditional growth model})}. \quad (4.14)$$

Estimates of these statistics for each of the models in table 4.1 appear in the bottom of the table. We will examine these proportional declines in the next section when we evaluate the results of subsequent model fitting.

Before doing so, however, we close by identifying a potentially serious flaw with the pseudo-R^2 statistics. Unlike traditional R^2 statistics, which will always be positive (or zero), some of these statistics can be *negative*! In ordinary regression, additional predictors generally reduce the residual variance and increase R^2. Even if every added predictor is worthless, the residual variance will not change and R^2 will not change. In the multilevel model for change, additional predictors generally reduce variance components and increase pseudo-R^2 statistics. But because of explicit links among the model's several parts, you can find yourself in extreme situations in which the addition of predictors *increases* the variance components' magnitude. This is most likely to happen when all, or most, of the outcome variation is exclusively either within-individuals or between-individuals. Then, a predictor added at one level reduces the residual variance at that level but potentially *increases* the residual variance(s) at the other level. This yields negative pseudo-R^2 statistics, a disturbing result to say the least. Kreft and de Leeuw (1998, pp. 117–118) and Snijders and Bosker (1999, pp. 99–109) provide mathematical accounts of this phenomenon, explicitly calling for caution when computing and interpreting pseudo-R^2 statistics.

4.5 Practical Data Analytic Strategies for Model Building

A sound statistical model includes all necessary predictors and no unnecessary ones. But how do you separate the wheat from the chaff? We suggest you rely on a combination of substantive theory, research questions, and statistical evidence. *Never* let a computer select predictors mechanically. The computer does not know your research questions nor the literature upon which they rest. It cannot distinguish predictors of direct substantive interest from those whose effects you want to control.

In this section, we describe one data analytic path through the alcohol use data, distilling general principles from this specific case. We begin, in section 4.5.1, by introducing the notion of a *taxonomy* of statistical models, a systematic path for addressing your research questions. In section 4.5.2, we compare fitted models in the taxonomy, interpreting parameter estimates, their associated tests and pseudo-R^2 statistics. In section 4.5.3, we demonstrate how to display analytic results graphically. In section 4.5.4, we discuss alternative strategies for representing the effects of predictors. In the remaining sections of the chapter, we use these basic principles to introduce other important topics related to model building.

4.5.1 A Taxonomy of Statistical Models

A *taxonomy* of statistical models is a systematic sequence of models that, as a set, address your research questions. Each model in the taxonomy extends a prior model in some sensible way; inspection and comparison of its elements tell the story of predictors' individual and joint effects. Most data analysts iterate toward a meaningful path; good analysis does not proceed in a rigidly predetermined order.

We suggest that you base decisions to enter, retain, and remove predictors on a combination of logic, theory, and prior research, supplemented by judicious hypothesis testing and comparison of model fit. At the outset, you might examine the effect of each predictor individually. You might then focus on predictors of primary interest (while including others whose effects you want to control). As in regular regression, you can add predictors singly or in groups and you can address issues of functional form using interactions and transformations. As you develop the taxonomy, you will progress toward a "final model" whose interpretation addresses your research questions. We place quotes around this term to emphasize that we believe no statistical model is *ever* final; it is simply a placeholder until a better model is found.

When analyzing longitudinal data, be sure to capitalize on your intuition and skills cultivated in the cross-sectional world. But longitudinal analyses are more complex because they involve: (1) *multiple level-2 outcomes* (the individual growth parameters), *each* of which can be related to predictors; and (2) *multiple kinds of effects*, both fixed effects and variance components. A level-1 linear change submodel has two level-2 outcomes; a more complex level-1 submodel may have more. The simplest strategy is to initially include each level-2 predictor simultaneously in all level-2 submodels, but as we show below, they need not remain. Each individual growth parameter can have its own predictors, and one goal of model building is to identify which predictors are important for which level-1 parameters. So, too, although each level-2 submodel can contain fixed and random effects, both are not necessarily required. Sometimes a model with fewer random effects will provide a more parsimonious representation and clearer substantive insights.

Before fitting models, take the time to distinguish between: (1) *question* predictors, whose effects are of primary substantive interest; and, (2) *control* predictors, whose effects you would like to remove. Substantive and theoretical concerns usually support the classification. For the alcohol use data, our classifications and analytic path will differ depending on our research questions. If interest centers on parental influences, *COA* is a question predictor and *PEER* a control. We would then evaluate the effect of *COA* on its own and after control for *PEER*. But if interest centers on peer influences, *PEER* is a question predictor and *COA* a control. We would then evaluate the effect of *PEER* on its own and after control for *COA*. Different classification schemes may lead to the same "final model," but they would arrive there via different paths. Sometimes, they lead to different "final models," each designed to answer its own research questions.

In what follows, we assume that research interest centers on the effects of parental alcoholism; *PEER* is a control. This allows us to adopt the analytic path illustrated in tables 4.1 and 4.2. Model C includes *COA* as a predictor of both initial status and change. Model D adds *PEER* to both level-2 models. Model E is a simplification of Model D in which the effect of *COA* on one of the individual growth parameters (the rate of change) is removed. We defer discussion of Models F and G until section 4.5.4.

4.5.2 Interpreting Fitted Models

You need not interpret every model you fit, especially those designed to guide interim decision making. When writing up findings for presentation and publication, we suggest that you identify a manageable subset of models that, taken together, tells a persuasive story parsimoniously. At a minimum, this includes the unconditional means model, the unconditional growth model, and a "final model." You may also want to present intermediate models that either provide important building blocks or tell interesting stories in their own right.

Columns 4–8 of table 4.1 present parameter estimates and associated single parameter hypothesis tests for five models in our taxonomy. (We discuss the last two models in section 4.5.4.) We recommend that you always construct a table like this because it allows you to compare fitted models systematically, describing what happens as you add and remove predictors. Sequential inspection and comparison of estimated fixed effects and variance components and their associated tests allows you to: (1) ascertain whether, and how, the variability in initial status and rate of change is gradually "explained"; and (2) identify which predictors explain what variation. Tests on the fixed effects help identify the predictors to retain; tests on the variance components help assess whether there is additional outcome variation left to predict. Integrating these conclusions helps identify the sources of outcome variation available for prediction and those predictors that are most effective in explaining that variation. As we have discussed Models A and B in section 4.3, we turn now to Model C.

Model C: The Uncontrolled Effects of COA

Model C includes *COA* as a predictor of both initial status and change. Interpretation of its four fixed effects is straightforward: (1) the estimated initial *ALCUSE* for the average child of non-alchoholic parents is 0.316 ($p < .001$); (2) the estimated differential in initial *ALCUSE* between children of alchoholic and non-alchoholic parents is 0.743 ($p < .001$); (3) the estimated rate of change in *ALCUSE* for an average child of non-alchoholic parents is 0.293 ($p < .001$); and (4) the estimated differential in the rate of change in *ALCUSE* between children of alchoholic and non-alcoholic parents is indistinguishable from 0 (-0.049, *ns*). This model provides uncontrolled answers to our research questions, suggesting that while children of alchoholic parents initially drink more than children of non-alchoholic parents, their rate of change in alcohol consumption between ages 14 and 16 does not differ.

Next examine the variance components. The statistically significant within-person variance component ($\hat{\sigma}_\varepsilon^2$) for Model C is identical to that of Model B, reinforcing the need to explore the effects of time-varying predictors (if we had some). Stability like this is expected because we added no additional level-1 predictors (although estimates can vary because of uncertainties arising from iterative estimation). The level-2 variance components, however, do change: $\hat{\sigma}_0^2$ declines by 21.8% from Model B. Because it is still statistically significant, potentially explainable residual variation in initial status remains. While $\hat{\sigma}_1^2$ is unchanged, it, too, is still statistically significant, suggesting the continued presence of potentially explainable residual variation in rates of change. These variance components are now called *partial* or *conditional* variances because they quantify the interindividual differences in change that remain unexplained by the model's predictors. We conclude that we should explore the effects of a level-2 predictor like *PEER* because it might help explain some of the level-2 residual variation.

Failure to find a relationship between *COA* and the rate of change might lead some analysts to immediately remove this term. We resist this temptation because *COA* is our focal question predictor and we want to evaluate the full spectrum of its effects. If subsequent analyses continue to suggest that this term be removed, we can always do so (as we do, in Model E).

Model D: The Controlled Effects of COA

Model D evaluates the effects of *COA* on initial status and rates of change in *ALCUSE*, controlling for the effects of *PEER* on initial status and rate of change. Notice that the level-2 intercepts change substantially from Model C: $\hat{\gamma}_{00}$ reverses sign, from +0.316 to −0.317; $\hat{\gamma}_{10}$ increases by 50%, from 0.293 to 0.429. We expect changes like these when we add level-2 predictors to our model. This is because each level-2 intercept represents the value of the associated individual growth parameter—π_{0i} or π_{1i}—when *all* predictors in each level-2 model are 0. In Model C, which includes only one predictor, *COA*, the intercepts describe initial status and rate of change for children of non-alchoholic parents. In Model D, which includes two predictors, the intercepts describe initial status and rate of change for a subset of children of non-alchoholic parents—those for whom *PEER* also equals 0. Because we can reject the null hypothesis associated with each parameter ($p < .001$), we might conclude that children of non-alchoholic parents whose early peers do not drink have non-zero levels of alcohol consumption themselves. But this conclusion is incorrect because the fitted intercept for initial status (−0.317) is *negative* suggesting that the confidence interval for the parameter does not even reach zero from below! As *ALCUSE* cannot be negative, this interval is implausible. As in regular regression, fitted intercepts may be implausible even when they correspond to observable combinations of predictor values. We discuss strategies for improving the interpretability of the level-2 intercepts in section 4.5.4.

The remaining parameters in Model D have expected interpretations: γ_{01} and γ_{11} describe the differential in *ALCUSE* between children of alchoholic and non-alchoholic parents controlling for the effects of *PEER* and γ_{02} and γ_{12} describe the differential in *ALCUSE* for a one-unit difference in *PEER* controlling for the effect of *COA*. Given our focus on the effects of *COA*, we are more interested in the former effects than the latter. We therefore conclude that, controlling for the effects of *PEER*: (1) the estimated differential in initial *ALCUSE* between children of alchoholic and non-alchoholic parents is 0.579 ($p < .001$); and (2) the estimated differential in the rate of change in *ALCUSE* between children of alchoholic and non-alchoholic parents is indistinguishable from 0 (−0.014, *ns*). This model provides *controlled* answers to our research questions. As before, we conclude that children of alchoholic parents initially drink more than children of non-alchoholic parents but their annual rate of change in consumption between ages 14 and 16 is no different. The magnitude of the early differential in *ALCUSE* is lower after *PEER* is controlled. At least some of the differential initially found between the two groups may be attributable to this predictor.

Next examine the associated variance components. Comparing Model D to the unconditional growth model B, we find that while $\hat{\sigma}_\varepsilon^2$ remains stable (as expected), $\hat{\sigma}_0^2$ and $\hat{\sigma}_1^2$ both decline. Taken together, *PEER* and *COA* explain 61.4% of the variation in initial status and 7.9% of the variation in rates of change. Notice that we *can* compare these random effects across models even though we *cannot* compare their fixed effects ($\hat{\gamma}_{00}$ and $\hat{\gamma}_{10}$). This is because the random effects describe the residual variance of the level-1 growth parameters—π_{0i} or π_{1i}—which retain their meaning across successive models even though the corresponding fixed effects (at level-2) do not.

Rejection of the null hypotheses associated with σ_0^2 and σ_1^2 suggests that there is further unpredicted variation in both initial status and rates of change. If our data set had included other person-level predictors, we would introduce them into the level-2 model to explain this variation. But we have no such predictors. And hypothesis tests for the parameter associated with the effect of *COA* on rate of change (γ_{11}) suggest that it need not be included in Models C or D as a predictor of change. In comparison to all other fixed effects, it is the only one whose null hypothesis cannot be rejected. We conclude that even though *COA* is our focal question predictor, we should remove this term to obtain a more parsimonious model.

Model E: A Tentative "Final Model" for the Controlled Effects of COA

Model E includes *PEER* as a predictor of both initial status and change but *COA* as a predictor of only initial status. For ease of exposition, we tentatively label this our "final model," but we hasten to add that our decision to temporarily stop here is based on many other analyses not shown. In particular, we examined issues of functional form, including nonlinearity and interactions, and found no evidence of either (beyond that which we addressed by transforming the original outcome and predictor). We discuss issues like these in section 4.8 and in subsequent chapters as we extend the multilevel model for change.

By now, you should be able to interpret the fixed effects in Model E directly. Controlling for the effects of *PEER*, the estimated differential in initial *ALCUSE* between children of alchoholic and non-alchoholic parents is 0.571 ($p < .001$) and controlling for the effect of parental alcoholism, for each 1-point difference in *PEER*: the average initial *ALCUSE* is 0.695 higher and the average rate of change in *ALCUSE* is .151 lower. We conclude that children of alchoholic parents drink more alcohol initially than children of non-alchoholic parents but their rate of change in consumption between ages 14 and 16 is no different. We also conclude that *PEER* is positively associated with early consumption but negatively associated with the rate of change in consumption. Fourteen-year-olds whose friends drink more tend to drink more at that age, but they have a slower rate of increase in consumption over time.

Examining the random effects for Model E in comparison to Model D, we find no differences in $\hat{\sigma}_\varepsilon^2$, $\hat{\sigma}_0^2$ or $\hat{\sigma}_1^2$. This confirms that we lose little by eliminating the effect of *COA* on change. As before, rejection of all three associated null hypotheses suggests the presence of unpredicted variation that we might be able to explain with additional predictors. The population covariance of the level-2 residuals, σ_{01}, summarizes the bivariate relationship between initial status and change, controlling for the specified effects of *COA* and *PEER*; in other words, the *partial* covariance between true initial status and change. Its estimate, -0.006, is even smaller than the unconditional estimate of -0.068 in the initial model for change and its associated hypothesis test indicates that it may well be zero in the population. We conclude that, after accounting for the effects of *PEER* and *COA*, initial status and rate of change in alcohol use are unrelated.

4.5.3 Displaying Prototypical Change Trajectories

Numerical summaries are just one way of describing the results of model fitting. For longitudinal analyses, we find that graphs of fitted trajectories for prototypical individuals are more powerful tools for communicating results. These plots are especially helpful when fitted intercepts in level-2 submodels refer to unlikely or implausible combinations of predictors, as they do for Model E (as evidenced by the negative fitted intercept for the initial status model). Some multilevel software packages provide these plots; if not, the calculations are simple and can be executed in any spreadsheet or graphics program, as shown below.

Let us begin with Model C, which includes the effect of *COA* on both initial status and change. From table 4.1, we have the following two level-2 fitted models:

$$\hat{\pi}_{0i} = 0.316 + 0.743COA_i$$
$$\hat{\pi}_{1i} = 0.293 - 0.049COA_i.$$

We can obtain fitted values for each group by substituting 0 and 1 for *COA*:

$$\text{When } COA_i = 0 \quad \begin{cases} \hat{\pi}_{0i} = 0.316 + 0.743(0) = 0.316 \\ \hat{\pi}_{1i} = 0.293 - 0.049(0) = 0.293 \end{cases}$$

$$\text{When } COA_i = 1 \quad \begin{cases} \hat{\pi}_{0i} = 0.316 + 0.743(1) = 1.059 \\ \hat{\pi}_{1i} = 0.293 - 0.049(1) = 0.244. \end{cases}$$

The average child of a non-alchoholic parent has a fitted trajectory with an intercept of 0.316 and a slope of 0.293; the average child of an alchoholic parent has a fitted trajectory with an intercept of 1.059 and a slope of 0.244.

We plot these fitted trajectories in the middle panel of figure 4.3. Notice the dramatic difference in level and trivial (nonsignificant) difference in slope. Unlike the numeric representation of these effects in table 4.1, the graph depicts both how much higher the *ALCUSE* level is at each age among children of alchoholic parents and it emphasizes the similarity in slopes.

We can also obtain fitted trajectories by working directly with the composite specification. From Model C's composite specification $\hat{Y}_{ij} = 0.316 + 0.743COA_i + 0.293TIME_{ij} - 0.049COA_i \times TIME_{ij}$, we obtain the following two trajectories by substituting in the two values of *COA*:

$$\text{When } COA_i = 0 \quad \begin{cases} \hat{Y}_{ij} = 0.316 + 0.743(0) + 0.293TIME_{ij} - 0.049(0)TIME_{ij} \\ \hat{Y}_{ij} = 0.316 + 0.293TIME_{ij} \end{cases}$$

$$\text{When } COA_i = 1 \quad \begin{cases} \hat{Y}_{ij} = 0.316 + 0.743(1) + 0.293TIME_{ij} - 0.049(1)TIME_{ij} \\ \hat{Y}_{ij} = 1.059 + 0.244TIME_{ij}. \end{cases}$$

By working with composite model directly, we obtain fitted trajectories expressed as a function of *TIME*.

It is easy to extend these strategies to models with multiple predictors, some of which may be continuous. Instead of obtaining a fitted function for *each* predictor value, we recommend that you select *prototypical* values of the predictors and derive fitted functions for *combinations* of these predictor values. Although you may be tempted to select many prototypical values for each predictor, we recommend that you limit yourself lest the displays become crowded, precluding the very interpretation they were intended to facilitate.

Prototypical values of predictors can be selected using one (or more) of the following strategies:

- *Choose substantively interesting values.* This strategy is best for categorical predictors or those with intuitively appealing values (such as 8, 12, and 16 for years of education in the United States).
- *Use a range of percentiles.* For continuous predictors without well-known values, consider using a range of percentiles (either the 25th, 50th, and 75th or the 10th, 50th, and 90th).
- *Use the sample mean ± .5 (or 1) standard deviation.* Another strategy useful for continuous predictors without well-known values.
- *Use the sample mean.* If you just want to control for the impact of a predictor rather than displaying its effect, set its value to the sample mean, yielding the "average" fitted trajectory controlling for that predictor.

Exposition is easier if you select whole number values (if the scale permits) or easily communicated fractions (e.g., $\frac{1}{4}$, $\frac{1}{2}$, and $\frac{3}{4}$). When using sample data to obtain prototypical values, be sure to do the calculations on the time-invariant predictors in the original person data set, *not* the person-period data set. If you are interested in every substantive predictor in a model, display fitted trajectories for all combinations of prototypical predictor values. If you want to focus on certain predictors while statistically controlling for others, eliminate clutter by setting the values of these latter variables to their means.

The right panel of figure 4.3 presents fitted trajectories for four prototypical adolescents derived from Model E. To construct this display we needed to select prototypical values for *PEER*. Based on its standard deviation of 0.726, we chose 0.655 and 1.381, values positioned a half a standard deviation from the sample mean (1.018). For ease of exposition, we label these "low" and "high" *PEER*. Using the level-1/level-2 specification, we calculate the fitted values as follows:

PEER	COA	Initial status ($\hat{\pi}_{0i}$)	Rate of change ($\hat{\pi}_{1i}$)
Low	No	$-0.314 + 0.695(0.655) + 0.571(0) = 0.142$	$0.425 - 0.151(0.655) = 0.326$
Low	Yes	$-0.314 + 0.695(0.655) + 0.571(1) = 0.713$	$0.425 - 0.151(0.655) = 0.326$
High	No	$-0.314 + 0.695(1.381) + 0.571(0) = 0.646$	$0.425 - 0.151(1.381) = 0.216$
High	Yes	$-0.314 + 0.695(1.381) + 0.571(1) = 1.217$	$0.425 - 0.151(1.381) = 0.216$

The fitted trajectories of alcohol use differ by both parental history of alcoholism and peer alcohol use. At each level of *PEER*, the trajectory for children of alchoholic parents is consistently above that of children of non-alchoholic parents. But *PEER* also plays a role. Fourteen-year-olds whose friends drink more tend to drink more at that age. Regardless of parental history, the fitted change trajectory for high *PEER* is above that of low *PEER*. But *PEER* has an inverse effect on the *change* in *ALCUSE* over time. The slope of the prototypical change trajectory is about 33% lower when *PEER* is high, regardless of parental history. We note that this negative impact is not sufficient to counteract the positive early effect of *PEER*. Despite the lower rates of change, the change trajectories when *PEER* is high never approach, let alone fall below, that of adolescents whose value of *PEER* is low.

4.5.4 Recentering Predictors to Improve Interpretation

When introducing the level-1 submodel in chapter 2, we discussed the interpretive benefits of recentering the predictor used to represent time. Rather than entering time as a predictor in its raw form, we suggested that you subtract a constant from each observed value, creating variables like *AGE-11* (in chapter 2), *AGE-1* (in chapter 3), and *AGE-14* (here in chapter 4). The primary rationale for temporal recentering is that it simplifies interpretation. If we subtract a constant from the temporal predictor, the intercept in the level-1 submodel, π_{0i}, refers to the true value of *Y* at that particular age—11, 1, or 14. If the constant chosen represents a study's first wave of data collection, we can simplify interpretation even further by referring to π_{0i} as individual *i*'s true "initial status."

We now extend the practice of rescaling to time-invariant predictors like *COA* and *PEER*. To understand why we might want to recenter time-invariant predictors, reconsider Model E in tables 4.1 and 4.2. When it came to the level-2 fitted intercepts, $\hat{\gamma}_{00}$ and $\hat{\gamma}_{10}$, interpretation was difficult because each represents the value of a level-1 individual growth parameter—π_{0i} or π_{1i}—when *all* predictors in the associated level-2 model are 0. If a level-2 model includes many substantive predictors or if zero is not a valid value for one or more of them, interpretation of its fitted intercepts can be difficult. Although you can always construct prototypical change trajectories in addition to direct interpretation of parameters, we often find it easier to recenter the substantive predictors *before* analysis so that direct interpretation of parameters is possible.

The easiest strategy for recentering a time-invariant predictor is to subtract its sample mean from each observed value. When we center a predictor on its sample mean, the level-2 fitted intercepts represent the *average* fitted values of initial status (or rate of change). We can also recenter a time-invariant predictor by subtracting another meaningful value—for example, 12 would be a suitable centering constant for a predictor representing years of education among U.S. residents; 100 may be a suitable centering constant for scores on an IQ test. Recentering works best when the centering constant is substantively meaningful—either because it has intuitive meaning for those familiar with the predictor *or* because it corresponds to the sample mean. Recentering can be equally beneficial for continuous and dichotomous predictors.

Models F and G in tables 4.1 and 4.2 demonstrate what happens when we center the time-invariant predictors *PEER* and *COA* on their sample means. Each of these models is equivalent to Model E, our tentative "final" model, in that all include the effect of *COA* on initial status and the effect of *PEER* on both initial status and rate of change. The difference between models is that before fitting Model F, we centered *PEER* on its sample mean of 1.018 and before fitting Model G, we also centered *COA* on its sample mean of .451. Some software packages (e.g., HLM) allow you to center predictors by toggling a switch on an interactive menu; others (e.g., MLwiN and SAS PROC MIXED) require you to create a new variable using computer code (e.g., by computing *CPEER* = *PEER* − 1.018). Our only word of caution is that you should compute the sample mean in the *person-level* data set. Otherwise, you may end up giving greater weight to individuals who happen to have more waves of data (unless the person-period data set is fully balanced, as it is here).

To evaluate empirically how recentering affects interpretation, compare the last three columns of table 4.1 and notice what remains the same and what changes. The parameter estimates for *COA* and *PEER* remain identical, regardless of recentering. This means that conclusions about the effects of predictors like *PEER* and *COA* are unaffected: $\hat{\gamma}_{01}$ remains at 0.571, $\hat{\gamma}_{02}$ remains at 0.695, and $\hat{\gamma}_{12}$ remains at −0.151 (as do their standard errors). Also notice that each of the variance components remains unchanged. This demonstrates that our conclusions about the variance components for the level-1 and level-2 residuals are also unaffected by recentering level-2 predictors.

What *does* differ across Models E, F and G are the parameter estimates (and standard errors) for the *intercepts* in each level-2 submodel. These estimates change because they represent different parameters:

- If neither *PEER* nor *COA* are centered (Model E), the intercepts represent a child of non-alchoholic parents whose peers at age 14 were totally abstinent (*PEER* = 0 and *COA* = 0).

- If *PEER* is centered and *COA* is not (Model F), the intercepts represent a child of non-alchoholic parents with an *average* value of *PEER* (*PEER* = 1.018 and *COA* = 0).
- If *both PEER* and *COA* are centered (Model G), the intercepts represent an *average* study participant—someone with *average* values of *PEER* and *COA* (*PEER* = 1.018 and *COA* = 0.451).

Of course, this last individual does not really exist because only two values of *COA* are possible: 0 and 1. Conceptually, though, the notion of an *average* study participant has great intuitive appeal.

When we center *PEER* and not *COA* in Model F, the level-2 intercepts describe an "average" child of non-alchoholic parents: $\hat{\gamma}_{00}$ estimates his or her true initial status (0.394, $p < .001$) and $\hat{\gamma}_{10}$ estimates his or her true rate of change (0.271, $p < .001$). Notice that the latter estimate is unchanged from Model B, the unconditional growth model. When we go further and center both *PEER* and *COA* in Model G, each level-2 intercept is numerically identical to the corresponding level-2 intercept in the unconditional growth model (B).[3]

Given that Models E, F, and G are substantively equivalent, which do we prefer? The advantage of Model G, in which both *PEER* and *COA* are centered, is that its level-2 intercepts are comparable to those in the unconditional growth model (B). Because of this comparability, many researchers routinely center *all* time-invariant predictors—even dichotomies—around their grand means so that the parameter estimates that result from the inclusion of additional predictors hardly change. Model E has a different advantage: because each predictor retains its original scale, we need not remember which predictors are centered and which are not. The predictor identified is the predictor included.

But both of these preferences are context free; they do not reflect our specific research questions. When we consider not just algebra but research interests—which here focus on parental alcoholism—we find ourselves preferring Model F. We base this decision on the easy interpretability of parameters for the dichotomous predictor *COA*. Not only is zero a valid value, it is an especially meaningful one (it represents children of non-alchoholic parents). We therefore see little need to center its values to yield consistency in parameter estimates with the unconditional growth model. When it comes to *PEER*, however, we have a different preference. Because it is of less substantive interest—we view it as a control predictor—we see no need *not* to center its values. Our goal is to evaluate the effects of *COA* controlling for *PEER*. By centering *PEER* at its mean, we achieve the goal of statistical control and interpretations of the level-2 intercepts are reasonable and credible. For the remainder of this chapter, we therefore adopt Model F as our "final model." (We continue to use quotes to emphasize that even this model might be set aside in favor of an alternative in subsequent analyses.)

4.6 Comparing Models Using Deviance Statistics

In developing the taxonomy in tables 4.1 and 4.2, we tested hypotheses on fixed effects and variance components using the single parameter approach of chapter 3. This testing facilitated our decision making and helped us determine whether we should render a simpler model more complex (as when moving from Model B to C) or a more complex model simpler (as when moving from Model D to E). As noted in section 3.6, however, statisticians disagree as to the nature, form, and effectiveness of these tests. The disagreement is so strong that some multilevel software packages do not routinely output these tests, especially for variance components. We now introduce an alternative method of inference—based on the *deviance statistic*—which statisticians seem to prefer. The major advantages of this approach are that it: (1) has superior statistical properties; (2) permits composite tests on several parameters simultaneously; and (3) conserves the reservoir of Type I error (the probability of incorrectly rejecting H_0 when it is true).

4.6.1 The Deviance Statistic

The easiest way of understanding the deviance statistic is to return to the principles of maximum likelihood estimation. As described in section 3.4, we obtain ML estimates by maximizing numerically the log-likelihood function, the logarithm of the joint likelihood of observing all the sample data actually observed. The log-likelihood function, which depends on the hypothesized model and its assumptions, contains all the unknown parameters (the γ's and σ's) and the sample data. ML estimates are those values of the unknown parameters (the $\hat{\gamma}$'s and $\hat{\sigma}$'s) that maximize the log-likelihood.

As a by-product of ML estimation, the computer determines the magnitude of the log-likelihood function for this particular combination of observed data and parameter estimates. Statisticians call this number the *sample log-likelihood* statistic, often abbreviated as LL. Every program that uses ML methods outputs the LL statistic (or a transformation of it). In general, if you fit several competing models to the same data, the larger the LL statistic, the better the fit. This means that if the models you compare yield negative LL statistics, those that are *smaller* in absolute value—i.e., closer to 0—fit better. (We state this obvious point explicitly as there has been some confusion in the literature about this issue.)

The *deviance statistic* compares log-likelihood statistics for two models: (1) the *current* model, the model just fit; and (2) a *saturated* model, a more general model that fits the sample data perfectly. For reasons explained below, deviance is defined as this difference multiplied by −2:

$$\text{Deviance} = -2[LL_{\text{current model}} - LL_{\text{saturated model}}]. \tag{4.15}$$

For a given set of data, deviance quantifies *how much worse* the current model is in comparison to the best possible model. A model with a small deviance statistic is nearly as good as any you can fit; a model with a large deviance statistic is much worse. Although the deviance statistic may appear unfamiliar, you have used it many times in regression analysis, where it is identical to the residual sum of squares, $\left(\sum_{i=1}^{n} (Y_i - \hat{Y}_i)^2 \right)$.

To calculate a deviance statistic, you need the log-likelihood statistic for the saturated model. Fortunately, in the case of the multilevel model for change, this is easy because a saturated model contains as many parameters as necessary to achieve a perfect fit, reproducing every observed outcome value in the person-period data set. This means that the maximum of its likelihood function—the probability that it will perfectly reproduce the sample data—is 1. As the logarithm of 1 is 0, the log-likelihood statistic for the saturated model is 0. We can therefore drop the second term on the right-hand side of equation 4.15, defining the deviance statistic for the multilevel model for change as:

$$\text{Deviance} = -2LL_{\text{current model}}. \tag{4.16}$$

Because the deviance statistic is just −2 times the sample log-likelihood, many statisticians (and software packages) label it −2logL or −2LL. As befits its name, we prefer models with smaller values of deviance.

The multiplication by −2 invoked during the transition from log-likelihood to deviance is more than cosmetic. Under standard normal theory assumptions, the difference in deviance statistics between a pair of nested models fit to the identical set of data has a known distribution. This allows us to test hypotheses about differences in fit between competing models by comparing deviance statistics. The resultant *likelihood ratio tests* are so named because a difference of logarithms is equal to the logarithm of a ratio.

4.6.2 When and How Can You Compare Deviance Statistics?

Deviance statistics for the seven models fit to the alcohol use data appear in table 4.1. They range from a high of 670.16 for Model A to a low of

588.69 for Model D. We caution that you cannot directly interpret their magnitude (or sign). (Also notice that the deviance statistics for Models E, F, and G are identical. Centering one or more level-2 predictors has absolutely no effect on this statistic.)

To compare deviance statistics for two models, the models must meet certain criteria. At a minimum: (1) each must be estimated using the identical data; and (2) one must be *nested* within the other. The constancy of data criterion requires that you eliminate any record in the person-period data set that is missing for any variable in *either* model. A difference of even one record invalidates the comparison. The nesting criterion requires that you can specify one model by placing *constraints* on the parameters in the other. The most common constraint is to set one or more parameters to 0. A "reduced" model is nested within a "full" model if every parameter in the former also appears in the latter.

When comparing multilevel models for change, you must attend to a third issue before comparing deviance statistics. Because these models involve two types of parameters—fixed effects (the γ's) and variance components (the σ's)—there are three distinct ways in which full and reduced models can differ: in their fixed effects, in their variance components, or in some combination of each. Depending upon the method of estimation—full or restricted ML—only certain types of differences can be tested. This limitation stems from principles underlying the estimation methods. Under FML (and IGLS), we maximize the likelihood of the sample data; under RML (and RIGLS), we maximize the likelihood of the sample *residuals*. As a result, an FML deviance statistic describes the fit of the entire model (both fixed and random effects), but a RML deviance statistic describes the fit of only its stochastic portion of the model (because, during estimation, its fixed effects are assumed "known"). This means that if you have applied FML estimation, as we have here, you can use deviance statistics to test hypotheses about any combination of parameters, fixed effects, or variance components. But if you have used RML to fit the model, you can use deviance statistics to test hypotheses only about variance components. Because RML is the default method in some multilevel programs (e.g., SAS PROC MIXED), caution is advised. Before using deviance statistics to test hypotheses, be sure you are clear about which method of estimation you have used.

Having fit a pair of models that meets these conditions, conducting tests is easy. Under the null hypothesis that the specified constraints hold, the difference in deviance statistics between a full and reduced model (often called "delta deviance" or ΔD) is distributed asymptotically as a χ^2 distribution with degrees of freedom (*d.f.*) equal to the number of independent constraints imposed. If the models differ by one parameter, you have one degree of freedom for the test; if they differ by three parameters, you have three. As with any hypothesis test, you compare ΔD to a *critical value*, appropriate for that number of degrees of freedom, rejecting H_0 when the test statistic is large.[4]

4.6.3 Implementing Deviance-Based Hypothesis Tests

Because the models in table 4.1 were fit using Full IGLS, we can use deviance statistics to compare their goodness-of-fit, whether they differ by only fixed effects (as do Models B, C, D, and E, F, G) or both fixed effects and variance components (as does Model A in comparison to all others). Before comparing two models, you must: (1) ensure that the data set has remained the same across models (it does); (2) establish that the former is nested within the latter; and (3) compute the number of additional constraints imposed.

Begin with the two unconditional models. We obtain multilevel Model A from Model B by invoking three independent constraints: $\gamma_{10} = 0$, $\sigma_1^2 = 0$, and $\sigma_{01} = 0$. The difference in deviance statistics, $(670.16 - 636.61) = 33.55$, far exceeds 16.27, the .001 critical value of a χ^2 distribution on 3 *d.f.*, allowing us to reject the null hypothesis at the $p < .001$ level that all three parameters are simultaneously 0. We conclude that the unconditional growth model provides a better fit than the unconditional means model (a conclusion already suggested by the single parameter tests for *each* parameter).

Deviance-based tests are especially useful for comparing what happens when we simultaneously add one (or more) predictor(s) to each level-2 submodel. As we move from Model B to Model C, we add *COA* as a predictor of both initial status and rate of change. Noting that we can obtain the former by invoking two independent constraints on the latter (setting both γ_{01} and γ_{11} to 0) we compare the difference in deviance statistics of $(636.61 - 621.20) = 15.41$ to a χ^2 distribution on 2 *d.f.*. As this exceeds the .001 critical value (13.82), we reject the null hypothesis that both γ_{01} and γ_{11} are simultaneously 0. (We ultimately set γ_{11} to 0 because we are unable to reject its single parameter hypothesis test in Model D. Comparing Models D and E, which differ by only this term, we find a trivial difference in deviance of 0.01 on 1 *d.f.*.)

You can also use deviance-based tests to compare nested models with identical fixed effects and different random effects. Although the strategy is the same, we raise this topic explicitly for two reasons: (1) if you use restricted methods of estimation (RML or RIGLS), these are the only types of deviance comparisons you can make; and (2) they address an important question we have yet to consider: Must the complete set of random effects appear in every multilevel model?

In every model considered so far, the level-2 submodel for each individual growth parameter (π_{0i} and π_{1i}) has included a residual (ζ_{0i} or ζ_{1i}). This practice leads to the addition of *three* variance components: σ_0^2, σ_1^2, and σ_{01}. Must all three always appear? Might we sometimes prefer a more parsimonious model? We can address these questions by considering the consequences of removing a random effect. To concretize the discussion, consider the following extension of Model F, which eliminates the second level-2 residual, ζ_{1i}:

$$Y_{ij} = \pi_{0i} + \pi_{1i}TIME_{ij} + \varepsilon_{ij}$$
$$\pi_{0i} = \gamma_{00} + \gamma_{01}COA_i + \gamma_{02}CPEER_i + \zeta_{0i}$$
$$\pi_{1i} = \gamma_{10} + \gamma_{12}CPEER_i,$$

and $\varepsilon_{ij} \sim N(0, \sigma_\varepsilon^2)$ and $\zeta_{0i} \sim N(0, \sigma_0^2)$. In the parlance of multilevel modeling, we have "fixed" the individual growth rates, preventing them from varying randomly across individuals (although we allow them to be related to *CPEER*). Removing this one level-2 residual (remember, residuals are *not* parameters) eliminates *two* variance components (which *are* parameters): σ_1^2 and σ_{01}.

Because the fixed effects in this reduced model are identical to those in Model F, we can test the joint null hypothesis that both σ_1^2 and σ_{01} are 0 by comparing deviance statistics. When we fit the reduced model to data, we obtain a deviance statistic of 606.47 (not shown in table 4.1). Comparing this to 588.70 (the deviance for Model F) yields a difference of 18.77. As this exceeds the .001 critical value of a χ^2 distribution with 2 *d.f.* (13.82), we reject the null hypothesis. We conclude that there is residual variability in the annual rate of change in *ALCUSE* that could potentially be explained by other level-2 predictors and that we should retain the associated random effects in our model.

4.6.4 AIC and BIC Statistics: Comparing Nonnested Models Using Information Criteria

You can test many important hypotheses by comparing deviance statistics for pairs of nested models. But as you become a more proficient data analyst, you may occasionally want to compare pairs of models that are not nested. You are particularly likely to find yourself in this situation when you would like to select between alternative models that involve *different* sets of predictors.

Suppose you wanted to identify which subset of interrelated predictors best captures the effect of a single underlying construct. You might, for example, want to control statistically for the effects of parental socioeconomic status (*SES*) on a child outcome, yet you might be unsure which combination of many possible SES measures—education, occupation, or income (either maternal or paternal)—to use. Although you could use principal components analysis to construct summary measures, you might also want to compare the fit of alternative models with different subsets of predictors. One model might use only paternal measures; another might use only maternal measures; still another might be restricted only to income indicators, but for both parents. As these models would not be nested (you cannot recreate one by placing constraints on parameters in another), you cannot compare their fit using deviance statistics.

We now introduce two ad hoc criteria that you can use to compare the relative goodness-of-fit of such models: the Akaike Information Criterion (AIC; Akaike, 1973) and the Bayesian Information Criterion (BIC; Schwarz, 1978). Like the deviance statistic, each is based on the log-likelihood statistic. But instead of using the LL itself, each "penalizes" (i.e., decreases) the LL according to pre-specified criteria. The AIC penalty is based upon the number of model parameters. This is because adding parameters—even if they have no effect—will increase the LL statistic, thereby decreasing the deviance statistic. The BIC goes further. Its penalty is based not just upon the number of parameters, but also on the sample size. In larger samples, you will need a larger improvement before you prefer a more complex model to a simpler one. In each case, the result is multiplied by −2 so that the information criterion's scale is roughly equivalent to that of the deviance statistic. (Note that the number of parameters you consider in the calculations differs under full and restricted ML methods.) Under full ML, both fixed effects and variance components are relevant. Under restricted ML, as you would expect, only the variance component parameters are relevant.

Formally, we write:

$$\text{Information criterion} = -2[LL - (\text{scale factor})(\text{number of model parameters})]$$

$$= \text{Deviance} + 2(\text{scale factor})(\text{number of model parameters}).$$

For the AIC, the scale factor is 1; for the BIC, it is half the log of the sample size. This latter definition leaves room for some ambiguity, as it is not clear whether the sample size should be the number of individuals under study or the number of records in the person-period data set. In the face of this ambiguity, Raftery (1995) recommends the former formulation, which we adopt here.

AICs and BICs can be compared for any pair of models, regardless of whether one is nested within another, *as long as both are fit to the identical set of data.* The model with the smaller information criterion (either AIC or BIC) fits "better." As each successive model in table 4.1 is nested within a previous one, informal comparisons like these are unnecessary. But to illustrate how to use these criteria, let us compare Models B and C. Model B involves six parameters (two fixed effects and four variance components); Model C involves eight parameters (two additional fixed effects). In this sample of 82, we find that Model B has an AIC statistic of $636.6 + 2(1)(6) = 648.6$ and an BIC of $636.6 + 2(\ln(82)/2)(6) = 663.0$, while Model C has an AIC statistic of $621.2 + 2(1)(8) = 637.2$ and an BIC of $621.2 + 2(\ln(82)/2)(8) = 656.5$. Both criteria suggest that C is preferable to B, a conclusion we already reached via comparison of deviance statistics.

Comparison of AIC and BIC statistics is an "art based on science." Unlike the objective standard of the χ^2 distribution that we use to compare deviance statistics, there are few standards for comparing information criteria. While large differences suggest that the model with the smaller value is preferable, smaller differences are difficult to evaluate. Moreover, statisticians have yet to agree on what differences are "small" or "large." In his excellent review extolling the virtues of BIC, Raftery (1995) declares the evidence associated with a difference of 0–2 to be "weak," 2–6 to be "positive," 6–10 to be "strong," and over 10 to be "very strong." But before concluding that information criteria provide a panacea for model selection, consider that Gelman and Rubin (1995) declared these statistics to be "off-target and only by serendipity manage to hit the target in special circumstances" (p. 165). We therefore offer a cautious recommendation to examine information criteria and to use them for model comparison only when more traditional methods cannot be applied.

4.7 Using Wald Statistics to Test Composite Hypotheses About Fixed Effects

Deviance-based comparisons are not the only method of testing composite hypotheses. We now introduce the Wald statistic, a generalization of the "parameter estimate divided by its standard error" strategy for testing hypotheses. The major advantage of the Wald statistic is its generality: you can test composite hypotheses about multiple effects regardless of the method of estimation used. This means that if you use restricted methods of estimation, which prevent you from using deviance-based tests to compare models with different fixed effects, you still have a means of testing composite hypotheses about sets of fixed effects.

Suppose, for example, you wanted to test whether the entire true change trajectory for a particular type of adolescent—say, a child of non-alchoholic parents with an average value of *PEER*—differs from a "null" trajectory (one with zero intercept and zero slope). This is tantamount to asking whether the average child of non-alchoholic parents drinks no alcohol at age 14 and remains abstinent over time.

To test this composite hypothesis, you must first figure out the entire set of parameters involved. This is easier if you start with a model's composite representation, such as Model F: $Y_{ij} = \gamma_{00} + \gamma_{01}COA_i + \gamma_{02}CPEER_i + \gamma_{10}TIME_{ij} + \gamma_{12}CPEER_i \times TIME_{ij} + [\zeta_{0i} + \zeta_{1i}TIME_{ij} + \varepsilon_{ij}]$. To identify parameters, simply derive the true change trajectory for the focal group, here children of non-alchoholic parents with an average value of *CPEER*. Substituting $COA = 0$ and $CPEER = 0$ we have: $E[Y_j \mid COA = 0, CPEER = 0] = \gamma_{00} + \gamma_{01}(0) + \gamma_{02}(0) + \gamma_{10}TIME_{ij} + \gamma_{12}(0) \times TIME_{ij} = \gamma_{00} + \gamma_{10}TIME_{ij}$, where the expectation notation, $E[\ldots]$, indicates that this is the *average population trajectory* for the entire $COA = 0$, $CPEER = 0$ subgroup. Taking expectations eliminates the level-1 and level-2 residuals, because—like all residuals—they average to zero. To test whether this trajectory differs from the null trajectory in the population, we formulate the composite null hypothesis:

$$H_0: \gamma_{00} = 0 \text{ and } \gamma_{10} = 0. \tag{4.17}$$

This joint hypothesis is a composite statement about an entire population trajectory, not a series of separate independent statements about each parameter.

We now restate the null hypothesis in a generic form known as a *general linear hypothesis*. In this representation, each of the model's fixed effects is multiplied by a judiciously chosen constant (an integer, a decimal, a fraction, or zero) and then the sum of these products is equated to another constant, usually zero. This "weighted linear combination" of parameters and constants is called a *linear contrast*. Because Model F includes five fixed effects—even though only two are under scrutiny here—we restate equation 4.17 as the following general linear hypothesis:

$$H_0: 1\gamma_{00} + 0\gamma_{01} + 0\gamma_{02} + 0\gamma_{10} + 0\gamma_{12} = 0$$
$$0\gamma_{00} + 0\gamma_{01} + 0\gamma_{02} + 1\gamma_{10} + 0\gamma_{12} = 0. \tag{4.18}$$

Although each equation includes all five fixed effects, the carefully chosen multiplying constants (the *weights*) guarantee that only the two focal parameters, γ_{00} and γ_{10}, remain viable in the statement. While this may seem like little more than an excessively parameterized reshuffling of symbols, its structure allows us to invoke a widely used testing strategy.

Most software programs require you to express a general linear hypotheses in matrix notation. This allows decomposition of the hypothesis into two distinct parts: (1) a matrix of multiplying constants (e.g., the 0's and 1's in equation 4.18); and (2) a vector of parameters (e.g., the γ's). To construct the matrix of multiplying constants, commonly labeled a *constraints* or *contrast matrix*, C, simply lift the numbers in the general linear hypothesis equation en bloc and array them in the same order. From equation 4.18 we have:

$$C = \begin{bmatrix} 1 & 0 & 0 & 0 & 0 \\ 0 & 0 & 0 & 1 & 0 \end{bmatrix}.$$

To form the vector of fixed effects, commonly labeled the *parameter vector*, or γ, lift the parameters in the general linear hypothesis en bloc and array them in the same order as well:

$$\gamma = [\gamma_{00} \quad \gamma_{01} \quad \gamma_{02} \quad \gamma_{10} \quad \gamma_{12}].$$

The general linear hypothesis is formed from the product of the C matrix and the transposed γ vector:

$$H_0: \begin{bmatrix} 1 & 0 & 0 & 0 & 0 \\ 0 & 0 & 0 & 1 & 0 \end{bmatrix} \begin{bmatrix} \gamma_{00} \\ \gamma_{01} \\ \gamma_{02} \\ \gamma_{10} \\ \gamma_{12} \end{bmatrix} = \begin{bmatrix} 0 \\ 0 \end{bmatrix},$$

which can be written generically as: $H_0: C\gamma' = 0$. For a given model, the elements of C will change from hypothesis to hypothesis but the elements of γ will remain the same.

Any general linear hypothesis that can be written in this $C\gamma' = 0$ form can be tested using a Wald statistic. Instead of comparing a parameter estimate to its standard error, the Wald statistic compares the *square* of the weighted linear combination of parameters to its estimated variance. As the variance of an estimate is the square of its standard error, the Wald statistic then resembles a squared z-statistic. (Indeed, if you use a Wald statistic to test a null hypothesis about a single fixed effect, W reduces to the square of the usual z-statistic.) Under the null hypothesis and usual normal theory assumptions, W has a χ^2 distribution with degrees of freedom equal to the number of rows in the C matrix (because the number of rows determines the number of independent constraints the null hypothesis invokes). For this hypothesis, we obtain a Wald statistic of 51.01 on 2 *d.f.*, allowing us to reject the composite null hypothesis in equation 4.18 at the .001 level.

General linear hypotheses can address even more complex questions about change over time. For example, when we examined the OLS estimated change trajectories in figure 4.2, we noticed that among children of non-alchoholic parents, those with low values of *CPEER* tended to have a lower initial status and steeper slopes than those with high values of *CPEER*. We might therefore ask whether the former group "catches up" to the latter. This is a question about the "vertical" separation between these two groups" true change trajectories at some later age, say 16.

To conduct such a test, you must once again first figure out the specific parameters under scrutiny. As before, we do so by substituting appropriate predictor values into the fitted model. Setting *COA* to 0 (for the children of non-alchoholic parents) and now selecting −.363 and +.363 as the low and high values of *CPEER* (because they correspond to .5 standard deviations on either side of the centered variable's mean of 0) we have:

$$E[Y_j | COA = 0, CPEER = low] = \gamma_{00} + \gamma_{01}(0) + \gamma_{02}(-.363) + \gamma_{10}TIME_{ij}$$
$$+ \gamma_{12}(-.363) \times TIME_{ij}$$
$$= (\gamma_{00} - .363\gamma_{02}) + (\gamma_{10} - .363\gamma_{12})TIME_{ij}$$
$$E[Y_j | COA = 0, CPEER = high] = \gamma_{00} + \gamma_{01}(0) + \gamma_{02}(.363) + \gamma_{10}TIME_{ij}$$
$$+ \gamma_{12}(.363) \times TIME_{ij}$$
$$= (\gamma_{00} + .363\gamma_{02}) + (\gamma_{10} + .363\gamma_{12})TIME_{ij}.$$

The predicted *ALCUSE* levels at age 16 are found by substituting *TIME* = (16 − 14) = 2 into these equations:

$$E[Y_j | COA = 0, CPEER = low] = \gamma_{00} - .363\gamma_{02} + 2\gamma_{10} - 2(.363)\gamma_{12}$$
$$E[Y_j | COA = 0, CPEER = high] = \gamma_{00} + .363\gamma_{02} + 2\gamma_{10} + 2(.363)\gamma_{12}.$$

How do we express the "catching up" hypothesis? If the low *CPEER* group "catches up," the expected values of the two groups should be identical at age 16. We therefore derive the composite null hypothesis by equating their expected values:

$$\gamma_{00} - .363\gamma_{02} + 2\gamma_{10} - 2(.363)\gamma_{12} = \gamma_{00} + .363\gamma_{02} + 2\gamma_{10} + 2(.363)\gamma_{12}.$$

Simplifying yields the following constraint $\gamma_{02} + 2\gamma_{12} = 0$, which we can be re-expressed as:

$$H_0 : 0\gamma_{00} + 0\gamma_{01} + 1\gamma_{02} + 0\gamma_{10} + 2\gamma_{12} = 0. \tag{4.19}$$

Notice that unlike the composite null hypothesis in equation 4.18, which required two equations, this composite null hypothesis requires just one.

This is a result of a reduction in the number of independent constraints. Because the first hypothesis simultaneously tested *two* independent statements—one about γ_{00} and the other about γ_{10}—it required two separate equations. Because this hypothesis is just a *single* statement—albeit about two parameters, γ_{02} and γ_{12}—it requires just one. This reduction reduces the dimensions of the contrast matrix, C.

We next express the composite null hypothesis in matrix form. The parameter vector, γ, remains unchanged from equation 4.18 because the model has not changed. But because the null hypothesis has changed, the constraint matrix must change as well. Stripping off the numerical constants in equation 4.19 we have $C = [0 \quad 0 \quad 1 \quad 0 \quad 2]$.

As expected, C is just a single row reflecting its single constraint. The composite null hypothesis is:

$$H_0 : [0 \quad 0 \quad 1 \quad 0 \quad 2] \begin{bmatrix} \gamma_{00} \\ \gamma_{01} \\ \gamma_{02} \\ \gamma_{10} \\ \gamma_{12} \end{bmatrix} = [0],$$

which has the requisite $C\gamma' = 0$ algebraic form. Conducting this test we find that we can reject the null hypothesis at the usual level of statistical significance ($\chi^2 = 6.23$, $p = .013$). We conclude that these average true change trajectories do not converge by age 16. In other words, the alcohol consumption of children of non-alchoholic parents with *low CPEER* does not catch up to the alcohol consumption of children of non-alchoholic parents with *high CPEER*.

Because many research questions can be stated in this form, general linear hypothesis testing is a powerful and flexible technique. It is particularly useful for conducting *omnibus tests* of several level-2 predictors so that you can assess whether sets of predictors make a difference *as a group*. If we represent a nominal or ordinal predictor using a set of indicator variables, we could use this approach to test their overall effect and evaluate pair-wise comparisons among subgroups.

Although Wald statistics can be used to test hypotheses about variance components, we suggest that you do not do so. The small-sample distribution theory necessary for these tests is poorly developed. It is only in very large samples—that is, *asymptotically*—that the distribution of a W statistic involving variance components *converges* on a χ^2 distribution as your sample size tends to infinity. We therefore do not recommend the use of Wald statistics for composite null hypotheses about variance components.

4.8 Evaluating the Tenability of a Model's Assumptions

Whenever you fit a statistical model, you invoke assumptions. When you use ML methods to fit a linear regression model, for example, you assume that the errors are independent and normally distributed with constant variance. Assumptions allow you to move forward, estimate parameters, interpret results, and test hypotheses. But the validity of your conclusions rests on your assumptions' tenability. Fitting a model with untenable assumptions is as senseless as fitting a model to data that are knowingly flawed. Violations lead to biased estimates, incorrect standard errors, and erroneous inferences.

When you fit a multilevel model for change, you also invoke assumptions. And because the model is more complex, its assumptions are more complex as well, involving both structural and stochastic features at each level. The structural specification embodies assumptions about the true functional form of the relationship between outcome and predictors. At level-1, you specify the shape of the hypothesized individual change trajectory, declaring it to be linear (as we have assumed so far) or nonlinear (as we assume in chapter 6). At level-2, you specify the relationship between each individual growth parameter and time-invariant predictors. And, as in regular regression analysis, you can specify that the level-2 relationship is linear (as we have so far) or more complex (nonlinear, discontinuous, or potentially interactive). The stochastic specification embodies assumptions about that level's outcome (either Y_{ij} at level-1 or π_{0i} and π_{1i} at level-2) that remains unexplained by the model's predictors. Because you know neither their nature nor value, you make assumptions about these error distributions, typically assuming univariate normality at level-1 and bivariate normality at level-2.

No analysis is complete until you examine the tenability of your assumptions. Of course, you can never be completely certain about the tenability of assumptions because you lack the very data you need to evaluate their tenability: information about the population from which your sample was drawn. Assumptions describe *true* individual change trajectories, population relationships between *true* individual growth parameters and level-2 predictors, and true errors for each person. All you can examine are the *observed* properties of *sample* quantities—*fitted* individual change trajectories, *estimated* individual growth parameters, and *sample* residuals.

Must you check the assumptions underlying every statistical model you fit? As much as we would like to say yes, reality dictates that we say no. Repetitive model checking is neither efficient nor plausible. We suggest instead that you examine the assumptions of several initial models and then again in any model you cite or interpret explicitly.

We offer simple multilevel model checking strategies in the three sections below. Section 4.8.1 reviews methods for assessing functional form; although we introduced the basic ideas earlier, we reiterate them here for completeness. We then extend familiar strategies from regression analysis to comparable issues in the multilevel context: assessing normality (section 4.8.2) and homoscedasticity (section 4.8.3). Table 4.3 summarizes what you should look for at each stage of this work.

4.8.1 Checking Functional Form

The most *direct* way of examining the functional form assumptions in the multilevel model for change is to inspect "outcome versus predictors" plots at each level.

- *At level-1.* For each individual, examine empirical growth plots and superimpose an OLS-estimated individual change trajectory. Inspection should confirm the suitability of its hypothesized shape.
- *At level-2.* Plot OLS estimates of the individual growth parameters against each level-2 predictor. Inspection should confirm the suitability of the hypothesized level-2 relationships.

For the eight adolescents in figure 4.1, for example, the hypothesis of linear individual change seems reasonable for subjects 23, 32, 56 and 65, but less so for subjects 04, 14, 41, and 82. But it is hard to argue for systematic deviations from linearity for these four cases given that the departures observed might be attributable to measurement error. Inspection of empirical growth plots for the remaining adolescents leads to similar conclusions.

Examination of the level-2 assumptions is facilitated by figure 4.4, which plots OLS-estimated individual growth parameters against the two substantive predictors. In the left pair of plots, for *COA*, there is nothing to assess because a linear model is de facto acceptable for dichotomous predictors. In the right pair of plots for *PEER*, the level-2 relationships do appear to be linear (with only a few exceptions).

4.8.2 Checking Normality

Most multilevel modeling packages can output estimates of the level-1 and level-2 errors, ε_{ij}, ζ_{0i} and ζ_{1i}. We label these estimates, $\hat{\varepsilon}_{ij}$, $\hat{\zeta}_{0i}$ and $\hat{\zeta}_{1i}$, "raw residuals." As in regular regression, you can examine their behavior using exploratory analyses. Although you can also conduct formal tests for normality (using Wilks-Shapiro and Kolmogorov-Smirnov statistics, say), we prefer visual inspection of the residual distributions.

Table 4.3: Strategies for checking assumptions in the multilevel model for change, illustrated using Model F of tables 4.1 and 4.2 for the alcohol use data

Assumption and what to expect if the assumption is tenable	What we find in the alcohol use data		
	level-1 residual, $\hat{\varepsilon}_{ij}$	level-2 residual, $\hat{\zeta}_{0i}$	level-2 residual, $\hat{\zeta}_{1i}$
Shape. Linear individual change trajectories and linear relationships between individual growth parameters and level-2 predictors.	Empirical growth plots suggest that most adolescents experience linear change with age. For others, the small number of waves of data (3) makes it difficult to declare curvilinearity making the linear trajectory a reasonable approximation.	Because *COA* is dichotomous, there is no linearity assumption for $\hat{\pi}_{0i}$. With the exception of two extreme data points, the plot of $\hat{\pi}_{0i}$ vs. *PEER* suggests a strong linear relationship.	Because *COA* is dichotomous, there is no linearity assumption for $\hat{\pi}_{1i}$. Plot of $\hat{\pi}_{1i}$ vs. *PEER* suggests a weak linear relationship.
Normality. All residuals, at both level-1 and level-2, will be normally distributed.	A plot of $\hat{\varepsilon}_{ij}$ vs. normal scores suggests normality. We find further support for normality in a plot of standardized $\hat{\varepsilon}_{ij}$ vs *ID*, which reveals no unusual data points.	A plot of $\hat{\zeta}_{0i}$ vs. normal scores suggests normality. So does a plot of standardized $\hat{\zeta}_{0i}$ vs. *ID*, which reveals no unusual data points. There is slight evidence of a floor effect in the outcome.	A plot of $\hat{\zeta}_{1i}$ vs. normal scores suggests normality, at least in the upper tail. The lower tail seems compressed. We find further support for this claim when we find no unusual data points in a plot of standardized $\hat{\zeta}_{1i}$ vs. *ID*. There is also evidence of a floor effect in the outcome.
Homoscedasticity. Equal variances of the level-1 and level-2 residuals at each level of every predictor.	A plot of $\hat{\varepsilon}_{ij}$ vs. AGE suggests approximately equal variability at ages 14, 15, and 16.	A plot of $\hat{\zeta}_{0i}$ vs. COA suggests homoscedasticity at both values of *COA*. So does a plot vs. *PEER*, at least for values up to, and including, 2. Beyond this, there are too few cases to judge.	A plot of $\hat{\zeta}_{1i}$ vs. COA suggests homoscedasticity at both values of *COA*. So does a plot vs. *PEER* at least for values up to, and including, 2. Beyond this, there are too few cases to judge.

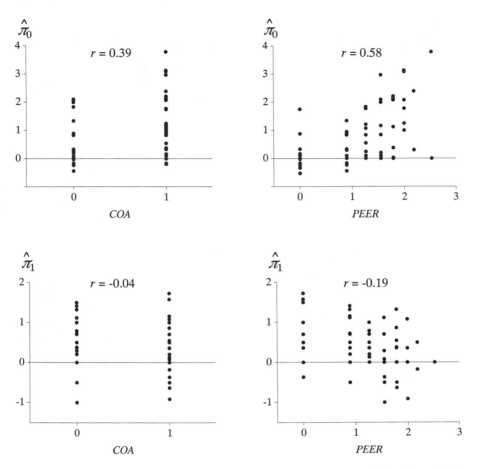

Figure 4.4. Examining the level-2 linearity assumption in the multilevel model for change. OLS estimated individual growth parameters (for the intercept and slope) plotted vs. selected predictors. Left panel is for the predictor *COA*; right panel is for the predictor *PEER*.

For each raw residual—the one at level-1 and the two at level-2— examine a *normal probability plot*, a plot of their values against their associated *normal scores*. If the distribution is normal, the points will form a line. Any departure from linearity indicates a departure from normality. As shown in the left column of figure 4.5, the normal probability plots for Model F for the alcohol use data appear linear for the level-1 residual, $\hat{\varepsilon}_{ij}$, and the first level-2 residual, $\hat{\zeta}_{0i}$. The plot for second level-2 residual, $\hat{\zeta}_{2i}$, is crooked, however, with a foreshortened lower tail falling closer to the center than anticipated. As the second level-2 residual describes unpredicted inter-individual variation in rates of change, we conclude that variability in this distribution's lower tail may be limited. This may

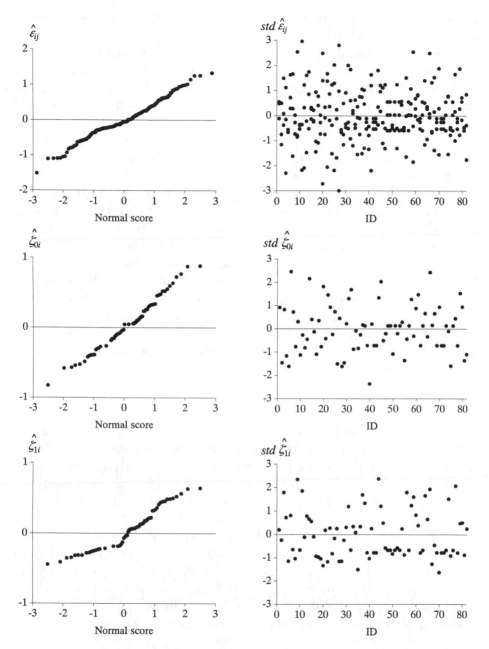

Figure 4.5. Examining normality assumptions in the multilevel model for change. Left panel presents normal probability plots for the raw residuals at level-1 and level-2. Right panel presents plots of standardized residuals at level-1 and level-2 vs. *ID* numbers.

be due to the bounded nature of *ALCUSE*, whose "floor" of zero imposes a limit on the possible rates of change.

Plots of *standardized* residuals—either univariate plots or bivariate plots against predictors—can also provide insight into the tenability of normality assumptions. If the raw residuals are normally distributed, approximately 95% of the standardized residuals will fall within ±2 standard deviations of their center (i.e., only 5% will be greater than 2). Use caution when applying this simple rule of thumb, however, because there are other distributions that are *not* normal in which about 5% of the observations also fall in these tails.

You can also plot the standardized residuals by *ID* to identify extreme individuals (as in the right panel of figure 4.5). In the top plot, the standardized level-1 residuals appear to conform to normal theory assumptions—a large majority fall within 2 standard deviations of center, with relatively few between 2 and 3, and none beyond. Plots of standardized level-2 residuals suggest that the negative residuals tend to be smaller in magnitude, "pulled in" toward the center of both plots. This feature is most evident for the second level-2 residual, $\hat{\zeta}_{1i}$, in the lower plot, but there is also evidence of its presence in the plot for $\hat{\zeta}_{0i}$. Again, compression of the lower tail may result from the fact that the outcome, *ALCUSE*, has a "floor" of zero.

4.8.3 Checking Homoscedasticity

You can evaluate the homoscedasticity assumption by plotting raw residuals against predictors: the level-1 residuals against the level-1 predictor, the level-2 residuals against the level-2 predictor(s). If the assumption holds, residual variability will be approximately equal at every predictor value. Figure 4.6 presents these plots for Model F of the alcohol use data.

The level-1 residuals, $\hat{\varepsilon}_{ij}$, have approximately equal range and variability at all ages; so, too, do the level-2 residuals plotted against *COA*. The plots of the level-2 residuals against *PEER* reveal a precipitous drop in variability at the highest predictor values (*PEER* > 2.5), suggesting potential heteroscedasticity in this region. But the small sample size (only 82 individuals) makes it difficult to reach a definitive conclusion, so we satisfy ourselves that the model's basic assumptions are met.

4.9 Model-Based (Empirical Bayes) Estimates of the Individual Growth Parameters

One advantage of the multilevel model for change is that it improves the precision with which we can estimate individual growth parameters. Yet

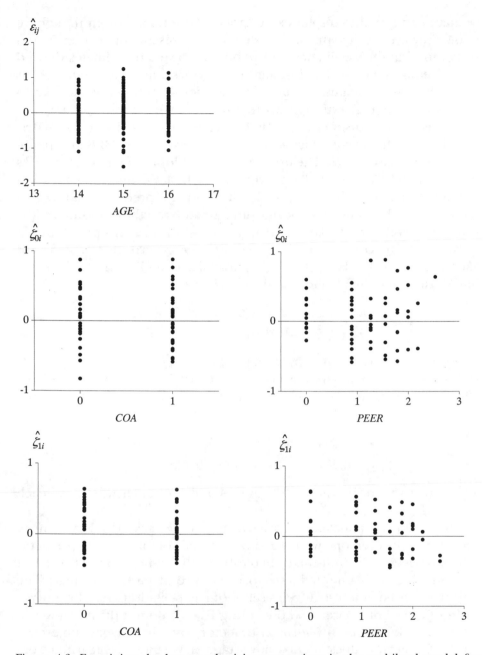

Figure 4.6. Examining the homoscedasticity assumptions in the multilevel model for change. Top panel presents raw level-1 residuals vs. the level-1 predictor *AGE*. Remaining panels present raw level-2 residuals vs. the two level-2 predictors, *COA* and *PEER*.

we have continued to display exploratory OLS estimates even though we know they are inefficient. In this section, we present superior estimates by combining OLS estimates with population average estimates derived from the fitted model. The resultant trajectories, known as *model-based* or *empirical Bayes* estimates, are usually your best bet if you would like to display individual growth trajectories for particular sample members.

There are two distinct methods for deriving model-based estimates. One is to explicitly construct a weighted average of the OLS and population average estimates. The other, which we adopt here, has closer links to the model's conceptual underpinnings: first we obtain population average trajectories based upon an individual's predictor values and second we add individual-specific information to these estimates (by using the level-2 residuals).

We begin by computing a population average growth trajectory for each person in the data set using a particular model's estimates. Adopting Model F for the alcohol use data, we have:

$$\hat{\pi}_{0i} = 0.394 + 0.571COA_i + 0.695CPEER_i$$
$$\hat{\pi}_{1i} = 0.271 - 0.151CPEER_i.$$

Substituting each person's observed predictor values into these equations yields his or her population average trajectory. For example, for subject 23, a child of an alchoholic parent whose friends at age 14 did not drink (resulting in a value of -1.018 for *CPEER*) we have:

$$\hat{\pi}_{0,23} = 0.394 + 0.571(1) + 0.695(-1.018) = 0.257$$
$$\hat{\pi}_{1,23} = 0.271 - 0.151(-1.018) = 0.425,$$

(4.20)

a trajectory that begins at 0.257 at age 14 and rises linearly by 0.425 each year.

This intuitively appealing approach has a drawback: it yields identical trajectories for everyone with the same specific combination of predictor values. Indeed, it is indistinguishable from the same approach used in Section 4.5.3 to obtain fitted trajectories for prototypical individuals. The trajectory in equation 4.20 represents our expectations for the *average* child of alchoholic parents whose young friends do not drink. However, what we seek here is an *individual* trajectory for this person, subject 23. His OLS trajectory does not take advantage of what we have learned from model fitting. Yet his population average trajectory does not capitalize on a key feature of the model: its explicit allowance for interindividual variation in initial status and rates of change.

The level-2 residuals, $\hat{\zeta}_{0i}$ and $\hat{\zeta}_{1i}$, which distinguish each person's growth parameters from his or her population average trajectory, provide the missing link. Because each person has his or her own set of residuals, we can add them to the model's fitted values:

$$\tilde{\pi}_{0i} = \hat{\pi}_{0i} + \hat{\zeta}_{0i}$$
$$\tilde{\pi}_{1i} = \hat{\pi}_{1i} + \hat{\zeta}_{1i},$$

(4.21)

where we place a ~ over the model-based estimates to distinguish them from the population average trajectories. Adding residuals to the population averages distinguishes each person from his or her peer group (defined by his or her predictor values). Most multilevel modeling software programs routinely provide these residuals (or the model-based estimates themselves). For subject 23, for example, the child of alchoholic parents whose peers did not drink, his level-2 residuals of 0.331 and 0.075 yield the following model-based estimates of his individual growth trajectory:

$$\tilde{\pi}_{0,23} = 0.257 + 0.331 = 0.588$$
$$\tilde{\pi}_{1,23} = 0.425 + 0.075 = 0.500.$$

Notice that both of these estimates are larger than the population average values obtained above.

Figure 4.7 displays the observed data for the eight individuals depicted in figure 4.1 and adds three types of fitted trajectories: (1) OLS-estimated trajectories (dashed lines); (2) population average trajectories (faint lines); and (3) model-based individual trajectories (bold lines). First, notice that across the plots, the population average trajectories (the faint lines) are the most stable, varying the least from person to person. We expect greater stability because these are *average* trajectories for groups of individuals who share particular predictor values. People who share identical predictor values will have identical average trajectories, even though their observed outcome data may differ. Population average trajectories do not reflect the behavior of individuals and hence are likely to be the least variable.

Next examine the model-based and OLS estimates (the bold and dashed lines), each designed to provide the individual information we seek. For three adolescents, the difference between estimates is small (subjects 23, 41, and 65), but for four others (subjects 4, 14, 56, and 82) it is pronounced and for subject 32, it is profound. We expect discrepancies like these because we estimate each trajectory using a different method and they depend upon the data in different ways. This does not mean that one of them is "right" and the other "wrong." Each has a set of statistical properties for which it is valued. OLS estimates are unbiased but inefficient; model-based estimates are biased, but more precise.

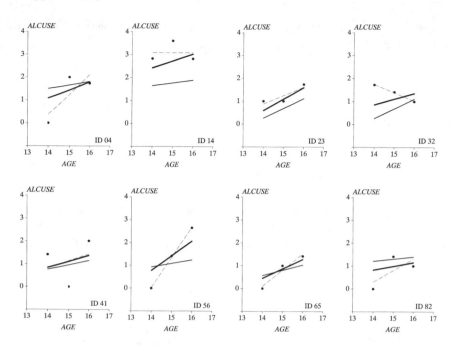

Figure 4.7. Model-based (empirical Bayes) estimates of the individual growth trajectories. Each plot presents the observed *ALCUSE* measurements (as data points), OLS fitted trajectories (dashed lines), population average trajectories (faint lines), and model-based empirical Bayes trajectories (bold lines).

Now notice how each model-based trajectory (in bold) falls between its OLS and population average trajectories (the dashed and faint lines). This is a hallmark of the model-based procedure to which we alluded earlier. Numerically, the model-based estimates are weighted averages of the OLS and population average trajectories. When OLS estimates are precise, they have greater weight; when OLS estimates are imprecise, the population average trajectories have greater weight. Because OLS trajectories differ markedly from person to person, the model-based trajectories differ as well, but their discrepancies are smaller because the population average trajectories are more stable. Statisticians use the term "borrowing strength" to describe procedures like this in which individual estimates are enhanced by incorporating information from others with whom he or she shares attributes. In this case, the model-based trajectories are *shrunk* toward the average trajectory of that person's peer group (those with the same predictor values). This combination yields a superior, more precise, estimate.

Model-based estimates are also more precise because they require estimation of fewer parameters. In positing the multilevel model for change, we assume that everyone shares the same level-1 residual variance, σ_ε^2. When we fit OLS trajectories, we estimate a separate level-1 variance for each individual in the sample. Fewer parameters in the multilevel model for change mean greater precision.

In choosing between OLS- and model-based trajectories, you must decide which criterion you value most, *unbiasedness* or *precision*. Statisticians recommend precision—indeed, increased precision is a fundamental motivation for fitting the multilevel model. But as we extol the virtues of model-based estimates, we conclude with a word of caution. Their quality depends heavily on the quality of the model fit. If the model is flawed, particularly if its level-2 components are specified incorrectly, then the model-based estimates will be flawed as well.

How might you use model-based estimates like these in practice? Stage (2001) provides a simple illustration of the power of this approach in his evaluation of the relationship between first-grade reading fluency and changes in oral reading proficiency in second-graders. He began by fitting a multilevel model for change to four waves of second-grade data, demonstrating that while first-grade performance was a strong predictor of initial status it was not a statistically significant predictor of rate of change. Stage went on to compute empirical Bayes estimates of the number of words each child was able to read by the end of second grade and he compared these estimates to: (1) the number of words each child was observed to have read at the end of second grade; and (2) the number of words each child was predicted to have read on the basis of simple OLS regression analyses within child. As Stage suggests, administrators might be better off assigning children to summer school programs (for remedial reading) not on the basis of observed or OLS-predicted end-of-year scores but rather on the basis of the empirical Bayes estimates, which yield more precise estimates of the child's status at the end of the year.

5

Treating *TIME* More Flexibly

Change is a measure of time
        —Edwin Way Teale

All the illustrative longitudinal data sets in previous chapters share two structural features that simplify analysis. Each is: (1) balanced—everyone is assessed on the identical number of occasions; and (2) time-structured—each set of occasions is identical across individuals. Our analyses have also been limited in that we have used only: (1) time-invariant predictors that describe immutable characteristics of individuals or their environment (except for *TIME* itself); and (2) a representation of *TIME* that forces the level-1 individual growth parameters to represent "initial status" and "rate of change."

The multilevel model for change is far more flexible than these examples suggest. With little or no adjustment, you can use the same strategies to analyze more complex data sets. Not only can the waves of data be irregularly spaced, their number and spacing can vary across participants. Each individual can have his or her own data collection schedule and the number of waves can vary without limit from person to person. So, too, predictors of change can be time-invariant or time-varying, and the level-1 submodel can be parameterized in a variety of interesting ways.

In this chapter, we demonstrate how you can fit the multilevel model for change under these new conditions. We begin, in section 5.1, by illustrating what to do when the number of waves is constant but their spacing is irregular. In section 5.2, we illustrate what to do when the number of waves per person differs as well; we also discuss the problem of missing data, the most common source of imbalance in longitudinal work. In section 5.3, we demonstrate how to include time-varying predictors in your data analysis. We conclude, in section 5.4, by discussing why and how you can adopt alternative representations for the main effect of *TIME*.

5.1 Variably Spaced Measurement Occasions

Many researchers design their studies with the goal of assessing each individual on an identical set of occasions. In the tolerance data introduced in chapter 2, each participant was assessed five times, at ages 11, 12, 13, 14, and 15. In the early intervention data introduced in chapter 3 and the alcohol use data introduced in chapter 4, each participant was assessed three times: at ages 12, 24, and 36 months or ages 14, 15, and 16 years. The person-period data sets from these time-structured designs are elegantly balanced, with a temporal variable that has an identical cadence for everyone under study (like *AGE* in tables 2.1, and 3.1).

Yet sometimes, despite a valiant attempt to collect time-structured data, actual measurement occasions will differ. Variation often results from the realities of fieldwork and data collection. When investigating the psychological consequences of unemployment, for example, Ginexi, Howe, and Caplan (2000) designed a time-structured study with interviews scheduled at 1, 5, and 11 months after job loss. Once in the field, however, the interview times varied considerably around these targets, with increasing variability as the study went on. Although interview 1 was conducted between 2 and 61 days after job loss, interview 2 was conducted between 111 and 220 days, and interview 3 was conducted between 319 and 458 days. Ginexi and colleagues could have associated the respondents' outcomes with the *target* interview times, but they argue convincingly that the number of days since job loss is a better metric for the measurement of time. Each individual in their study, therefore, has a *unique* data collection schedule: 31, 150, and 356 days for person 1; 23, 162, and 401 days for person 2; and so on.

So, too, many researchers design their studies knowing full well that the measurement occasions may differ across participants. This is certainly true, for example, of those who use an *accelerated cohort* design in which an age-heterogeneous cohort of individuals is followed for a constant period of time. Because respondents initially vary in age, and *age*, not *wave*, is usually the appropriate metric for analysis (see the discussion of time metrics in section 1.3.2), observed measurement occasions will differ across individuals. This is actually what happened in the larger alcohol-use study from which the small data set in chapter 4 was excerpted. Not only were those 14-year-olds re-interviewed at ages 15 and 16, concurrent samples of 15- and 16-year-olds were re-interviewed at ages 16 and 17 and ages 17 and 18, respectively. The advantage of an accelerated cohort design is that you can model change over a longer temporal period (here, the five years between ages 14 and 18) using fewer waves of data. Unfortunately, under the usual conditions, the data sets are then sparser at the earliest and latest ages, which can complicate the specification of the level-1 submodel.

In this section, we show how you can use the methods of previous chapters to analyze data sets with variably spaced measurement occasions. All you need to deal with are some minor coding issues for the temporal predictor in the person-period data set; model specification, parameter estimation, and substantive interpretation proceeds as before. To illustrate just how simple the analysis can be, we begin by discussing data sets in which the *number* of waves is constant but their *spacing* varies. We discuss data sets in which the *number* of waves varies as well in section 5.2.

5.1.1 The Structure of Variably Spaced Data Sets

We illustrate how to analyze data sets with variably spaced measurement occasions using a small sample extracted from the Children of the National Longitudinal Study of Youth (CNLSY). The data set, comprising children's scores on the reading subtest of the Peabody Individual Achievement Test (PIAT), includes three waves of data for 89 African-American children. Each child was 6 years old in 1986, the first year of data collection. During the second wave of data collection, in 1988, these children were to be 8; during the third wave, in 1990, they were to be 10. We focus here on an unconditional growth model, not the inclusion of level-2 predictors, because this second aspect of analysis remains unchanged.

Table 5.1 presents excerpts from the person-period data set. Notice that its structure is virtually identical to all person-period data sets shown so far. The only difference is that it contains *three* temporal variables denoting the passage of time: *WAVE*, *AGE*, and *AGEGRP*. Although we will include only one of these in any given model, a distinctive feature of time-unstructured data sets is the possibility of multiple metrics for clocking time (often called metameters).

WAVE is the simplest but least analytically useful of the three. Although its values—1, 2, and 3—reflect the study's design, they have little substantive meaning when it comes to addressing the research question. Because *WAVE* does not identify the child's age at each occasion, nor does it capture the chronological distance between occasions, it cannot contribute to a meaningful level-1 submodel. We mention this issue explicitly because empirical researchers sometimes postulate individual growth models using design variables like *WAVE* (or year of data collection) even though other temporal predictors are generally more compelling.

AGE is a better predictor because it specifies the child's actual age (to the nearest month) on the day each test was administered. A child like

Table 5.1: Excerpts from the person-period data set for the reading study

ID	WAVE	AGEGRP	AGE	PIAT
04	1	6.5	6.00	18
04	2	8.5	8.50	31
04	3	10.5	10.67	50
27	1	6.5	6.25	19
27	2	8.5	9.17	36
27	3	10.5	10.92	57
31	1	6.5	6.33	18
31	2	8.5	8.83	31
31	3	10.5	10.92	51
33	1	6.5	6.33	18
33	2	8.5	8.92	34
33	3	10.5	10.75	29
41	1	6.5	6.33	18
41	2	8.5	8.75	28
41	3	10.5	10.83	36
49	1	6.5	6.50	19
49	2	8.5	8.75	32
49	3	10.5	10.67	48
69	1	6.5	6.67	26
69	2	8.5	9.17	47
69	3	10.5	11.33	45
77	1	6.5	6.83	17
77	2	8.5	8.08	19
77	3	10.5	10.00	28
87	1	6.5	6.92	22
87	2	8.5	9.42	49
87	3	10.5	11.50	64
. . .	. . .	. . .	. . .	. . .

Note that *TIME* is clocked using three distinct variables: *WAVE*, *AGEGRP*, and *AGE*.

ID 04, who had just turned 6 at wave 1, has an *AGE* of 6.00 for that record; a child like *ID* 87, who would soon turn 7, has an *AGE* of 6.92. The average child is 6.5 years old at wave 1, as we would expect if births and testing occasions were randomly distributed. If data collection had proceeded according to plan, the average child would have been 8.5 and 10.5 years old at the next two waves. Not surprisingly, actual ages varied around these targets. By wave 2, the youngest child had just turned 8 while the oldest was well over 9. By wave 3, the youngest child had just turned 10 while the oldest was nearly 12. Like many longitudinal studies, the CNLSY suffers from "occasion creep"—over time, the temporal separation of occasions widens as the actual ages exceed design projections. In this data set, the average child is 8.9 years in wave 2 and nearly 11 years in wave 3.

The third temporal variable, *AGEGRP*, is a time-structured predictor that is more substantively meaningful than the design variable *WAVE*. Its values indicate the child's "expected age" on each measurement occasion (6.5, 8.5, and 10.5). This time-structured predictor clocks time on a scale that is comparable numerically to the irregularly spaced predictor *AGE*. Adding *AGEGRP* to the person-period data set allows us to demonstrate that the characterization of a data set as time-structured or irregular can depend on nothing more than the *cadence* of the temporal predictor used to postulate a model. If we postulate our model using *AGEGRP*, the data set is time-structured; if we postulate a comparable model using *AGE*, it is not.

The multilevel model for change does not care if the individual-specific cadence of the level-1 predictor is identical for everyone or if it varies from case to case. Because we fit the model using the actual numeric values of the temporal predictor, spacing is irrelevant. We can postulate and fit a comparable model regardless of the variable's cadence. Of far greater importance is the choice of the functional form for the level-1 submodel. Should it represent linear change or a more complex shape for the individual growth trajectory? Might this decision depend upon the specific temporal predictor chosen for model building?

To address these questions, figure 5.1 presents empirical change plots with superimposed OLS linear change trajectories for 9 children. Each panel plots each child's *PIAT* scores twice, once for each temporal predictor. We use •'s and a dashed line when plotting by *AGE*; we use +'s and a solid line when plotting by *AGEGRP*. With just three waves of data—whichever temporal predictor we use—it is difficult to argue for anything but a linear change individual growth model.

If we can postulate a linear change individual growth model using either temporal predictor, which one should we use? As argued above, we prefer *AGE* because it provides more precise information about the child at the moment of testing. Why set this information aside just to use the equally spaced, but inevitably less accurate, *AGEGRP*. Yet this is what many researchers do when analyzing longitudinal data—indeed, it is what *we* did in chapters 3 and 4. There, instead of using the participant's precise ages, we used integers: 12, 18, and 24 months for the children in chapter 3; 14, 15, and 16 for the teenagers in chapter 4. Although the loss of precision may be small, as suggested by the close correspondence between the pairs of fitted OLS trajectories in each panel of figure 5.1, there are children for whom the differential is much larger. To investigate this question empirically, we fit two multilevel models for change to these data: one using *AGEGRP*, another using *AGE* as the

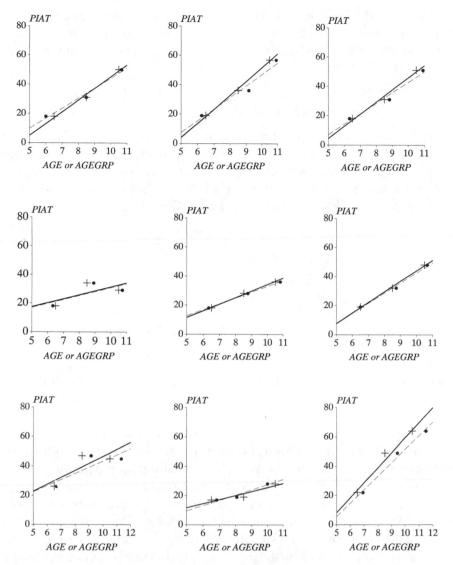

Figure 5.1. Comparing time-structured and time-unstructured representations of the effect of *TIME*. Empirical change plots with superimposed OLS trajectories for 9 participants in the reading study. The +'s and solid lines are for *TIME* clocked using the child's *target* age at data collection; the •'s and dashed lines are for *TIME* clocked using each child's observed age.

temporal predictor at level-1. Doing so allows us to demonstrate how to analyze irregularly spaced data sets *and* to illustrate the importance of assessing the merits of alternative metrics for time empirically.

5.1.2 Postulating and Fitting Multilevel Models with Variably Spaced Waves of Data

Regardless of which temporal representation we use, we postulate, fit, and interpret the multilevel model for change using the same strategies. Adapting the general specification of an unconditional growth model in equations 4.9a and 4.9b, let Y_{ij} be child i's *PIAT* score on occasion j and $TIME_{ij}$ represent either temporal variable:

$$
\begin{aligned}
Y_{ij} &= \pi_{0i} + \pi_{1i} TIME_{ij} + \varepsilon_{ij} \\
\pi_{0i} &= \gamma_{00} + \zeta_{0i} \\
\pi_{1i} &= \gamma_{10} + \zeta_{1i},
\end{aligned}
\tag{5.1a}
$$

where

$$
\varepsilon_{ij} \sim N(0, \sigma_\varepsilon^2) \text{ and } \begin{bmatrix} \zeta_{0i} \\ \zeta_{1i} \end{bmatrix} \sim N\left(\begin{bmatrix} 0 \\ 0 \end{bmatrix}, \begin{bmatrix} \sigma_0^2 & \sigma_{01} \\ \sigma_{10} & \sigma_1^2 \end{bmatrix} \right).
\tag{5.1b}
$$

If we center both *AGE* and *AGEGRP* on age 6.5—the average child's age at wave 1—the parameters have the usual interpretations. In the population from which this sample was drawn, γ_{00} represents the average child's true initial status (at age 6.5); γ_{10} represents the average child's annual rate of true change between ages 6 and 11; σ_ε^2 summarizes the within-child scatter around his or her own true change trajectory; and σ_0^2 and σ_1^2 summarize the between-child variability in initial status and annual rates of change.

Use of a generic representation $TIME_{ij}$ in the level-1 growth model (instead of a specific representation like $AGE - 6.5$ or $AGEGRP - 6.5$) yields these interpretations. We can postulate the same model for either predictor because $TIME_{ij}$ includes subscripts that are both person-specific (i) and time-specific (j). If *TIME* represents $AGEGRP - 6.5$, the data set is time structured; if we use $AGE - 6.5$, it is not. From a data-analytic perspective, you just specify the relevant temporal representation to your statistical software. From an interpretive perspective, the distinction is moot.

Table 5.2 presents the results of fitting these two unconditional growth models to these data: the first uses $AGEGRP - 6.5$; the second uses $AGE - 6.5$. Each was fit using full ML in SAS PROC MIXED. The parameter estimates for initial status, $\hat{\gamma}_{00}$, are virtually identical—21.16 and 21.06—as are those for the within-child variance, σ_ε^2: 27.04 and 27.45. But the similarities stop there. For the slope parameter, γ_{10}, the estimated growth

Table 5.2: Results of using alternative representations for the main effect of *TIME* ($n = 89$) when fitting an unconditional growth model to the CNLSY reading data

			Predictor representing *TIME*	
		Parameter	*AGEGRP* – 6.5	*AGE* – 6.5
Fixed Effects				
Initial status, π_{0i}	Intercept	γ_{00}	21.1629***	21.0608***
			(0.6143)	(0.5593)
Rate of Change, π_{1i}	Intercept	γ_{10}	5.0309***	4.5400***
			(0.2956)	(0.2606)
Variance Components				
Level-1:	within-person	σ_ε^2	27.04***	27.45***
Level-2:	In initial status	σ_0^2	11.05*	5.11
	In rate of change	σ_1^2	4.40***	3.30***
Goodness-of-fit				
	Deviance		1819.8	1803.9
	AIC		1831.9	1815.9
	BIC		1846.9	1830.8

$\sim p < .10$; $* p < .05$; $** p < .01$; $*** p < .001$.
The first model treats the data set as time-structured by using the predictor (*AGEGRP* – 6.5); the second model treats the data set as time-unstructured by using each child's actual age at each assessment, (*AGE* – 6.5).

Note: SAS Proc Mixed, Full ML. Also note that the covariance component, σ_{01}, is estimated, but not displayed.

rate is half a point larger in a model with *AGEGRP* – 6.5 (5.03 vs. 4.54). This cumulates to a two-point differential in PIAT scores over the four years under study. So, too, the two level-2 variance components are much larger for a model with *AGEGRP* – 6.5.

Why are these estimates larger when we treat the data set as time-structured, using *AGEGRP* – 6.5 as our level-1 predictor, than when we treat it as irregular, using *AGE* – 6.5? We obtain a larger fixed effect for linear growth because *AGEGRP* associates the data for waves 2 and 3 with earlier ages (8.5 and 10.5) than observed. If we amortize the same gain over a shorter time period, the slope must be steeper. We obtain larger estimated variance components because the model with the time-structured predictor fits less well—there is more unexplained variation in initial status and growth rates—than when we associate each child's data with his or her age at testing. In other words, treating this unstructured data set as though it is time-structured introduces error into the analysis—error that we can reduce by using the child's age at testing as the temporal predictor.

We conclude that the model with *AGEGRP* as the level-1 temporal predictor fits less well than the model with *AGE*. With the former representation, the slope is inappropriately larger—inaccurately implying more rapid gains—and there is more unexplained variation in initial status and rates of change. The superiority of the model with AGE as the temporal predictor is supported by its smaller AIC and BIC statistics. The bottom line: never "force" an unstructured data set to be structured. If you have several metrics for tracking time—and you often will—investigate the possibility of alternative temporal specifications. Your first choice, especially if tied to design, not substance, may not always be the best.

5.2 Varying Numbers of Measurement Occasions

Once you allow the spacing of waves to vary across individuals, it is a small leap to allow their *number* to vary as well. Statisticians say that such data sets are *unbalanced*. As you would expect, balance facilitates analysis: models can be parameterized more easily, random effects can be estimated more precisely, and computer algorithms will converge more rapidly.

Yet a major advantage of the multilevel model for change is that it is easily fit to unbalanced data. Unlike approaches such as repeated measures analysis of variance, with the multilevel modeling of change it is straightforward to analyze data sets with varying numbers of waves of data. To illustrate the general approach, we begin, in section 5.2.1, by introducing a new data set in which the number of waves per person varies widely, from 1 to 13. We extend this discussion in section 5.2.2, by discussing implementation and estimation problems that can arise when data are unbalanced. We conclude, in section 5.2.3, by discussing potential causes of imbalance—especially missing data—and how they can affect statistical analysis.

5.2.1 Analyzing Data Sets in Which the Number of
Waves per Person Varies

Murnane, Boudett, and Willett (1999) used data from the National Longitudinal Survey of Youth (NLSY) to track the labor-market experiences of male high school dropouts. Like many large panel studies, the NLSY poses a variety of design complications: (1) at the first wave of data collection, the men varied in age from 14 to 17; (2) some subsequent waves were separated by one year, others by two; (3) each wave's interviews were conducted at different times during the calendar year; and (4) respondents could describe more than one job at each interview. Person-specific schooling and employment patterns posed further problems. Not only could respondents drop out of school at different times and enter the

Table 5.3: Excerpts from the person-period data set for the high school dropout wage study

ID	EXPER	LNW	BLACK	HGC	UERATE
206	1.874	2.028	0	10	9.200
206	2.814	2.297	0	10	11.000
206	4.314	2.482	0	10	6.295
332	0.125	1.630	0	8	7.100
332	1.625	1.476	0	8	9.600
332	2.413	1.804	0	8	7.200
332	3.393	1.439	0	8	6.195
332	4.470	1.748	0	8	5.595
332	5.178	1.526	0	8	4.595
332	6.082	2.044	0	8	4.295
332	7.043	2.179	0	8	3.395
332	8.197	2.186	0	8	4.395
332	9.092	4.035	0	8	6.695
1028	0.004	0.872	1	8	9.300
1028	0.035	0.903	1	8	7.400
1028	0.515	1.389	1	8	7.300
1028	1.483	2.324	1	8	7.400
1028	2.141	1.484	1	8	6.295
1028	3.161	1.705	1	8	5.895
1028	4.103	2.343	1	8	6.900

labor force at different times, they also changed jobs at different times. To track wages on a common temporal scale, Murnane and colleagues decided to clock time from each respondent's first day of work. This allows each hourly wage to be associated with a temporally appropriate point in the respondent's labor force history. The resulting data set has an unusual temporal schedule, varying not only in spacing but length.

Table 5.3 presents excerpts from the person-period data set. To adjust for inflation, each hourly wage is expressed in constant 1990 dollars. To address the skewness commonly found in wage data and to linearize the individual wage trajectories, we analyze the natural logarithm of wages, *LNW*. Then, to express this outcome on its original scale, we take antilogs (e.g., $e^{(2.028)} = \$7.60$ per hour).

The temporal variable *EXPER* identifies the specific moment—to the nearest day—in each man's labor force history associated with each observed value of *LNW*. Notice the variability in the number and spacing of waves. Dropout 206 has three waves, for jobs held at 1.874, 2.814 and 4.314 years of experience after labor force entry. Dropout 332 has 10 waves, the first for a job held immediately after entering the labor force, the others for jobs held approximately every subsequent year. Dropout

1028 has 7 waves; the first three describe the first six months of work (at 0.004, 0.035, and 0.515 years). Across the full sample, 77 men have 1 or 2 waves of data, 82 have 3 or 4, 166 have 5 or 6, 226 have 7 or 8, 240 have 9 or 10, and 97 have more than 10. The earliest wave describes someone's first day of work; the latest describes a job held 13 years later.

This is the first data set we have presented in which the number of waves of data varies across individuals. Some men even have fewer than three waves—less than the minimum articulated in previous chapters. A major advantage of the multilevel model for change is that everyone can participate in the estimation, regardless of how many waves he contributes to the data set. Even the 38 men with just 1 wave of data and the 39 with just 2 waves are included in the estimation. Although they provide less, or no, information about within-person variation—and hence do not contribute to variance component estimation—they can still contribute to the estimation of fixed effects where appropriate. Ultimately, each person's fitted trajectory is based on a combination of his: (1) observed trajectory, and (2) a model-based trajectory determined by the values of the predictors.

You need no special procedures to fit a multilevel model for change to unbalanced data. All you need do is specify the model appropriately to your statistical software. As long as the person-period data set includes enough people with enough waves of data for the numeric algorithms to converge, you will encounter no difficulties. If the data set is severely unbalanced, or if too many people have too few waves for the complexity of your hypothesized model, problems may arise in the estimation. For now, we continue with this data set, which includes so many people with so many waves that estimation is straightforward. We discuss strategies for identifying and resolving estimation problems in section 5.2.2.

Table 5.4 presents the results of fitting three multilevel models for change to the wage data, using full ML in SAS PROC MIXED. First examine the results for Model A, the unconditional growth model. The positive and statistically significant fixed effect for *EXPER* indicates that inflation-adjusted wages rise over time. Because the outcome, *LNW*, is expressed on a logarithmic scale, its parameter estimate, $\hat{\gamma}_{10}$, is not a linear growth rate. As in regular regression, however, transformation facilitates interpretation. If an outcome in a linear relationship, Y, is expressed as a natural logarithm and $\hat{\gamma}_{10}$ is the regression coefficient for a predictor X, then $100(e^{(\hat{\gamma}_{10})} - 1)$ is the *percentage change* in Y per unit difference in X. Because *EXPER* is calibrated in years, this transformation yields an annual percentage growth rate in wages. Computing $100(e^{(0.0457)} - 1) = 4.7$, we estimate that the average high school dropout's inflation-adjusted hourly wages rise by 4.7% with each year of labor force participation.

Table 5.4: Results of fitting a taxonomy of multilevel models for change to the high school dropout wage data ($n = 888$)

		Parameter	Model A	Model B	Model C
Fixed Effects					
Initial status, π_{0i}	Intercept	γ_{00}	1.7156***	1.7171***	1.7215***
			(0.0108)	(0.0125)	(0.0107)
	$(HGC - 9)$	γ_{01}		0.0349***	0.0384***
				(0.0079)	(0.0064)
	$BLACK$	γ_{02}		0.0154	
				(0.0239)	
Rate of change, π_{1i}	Intercept	γ_{10}	0.0457***	0.0493***	0.0489***
			(0.0023)	(0.0026)	(0.0025)
	$(HGC - 9)$	γ_{11}		0.0013	
				(0.0017)	
	$BLACK$	γ_{12}		−0.0182**	−0.0161***
				(0.0055)	(0.0045)
Variance Components					
Level-1:	within-person	σ_ε^2	0.0951***	0.0952***	0.0952***
Level-2:	In initial status	σ_0^2	0.0543***	0.0518***	0.0518***
	In rate of change	σ_1^2	0.0017***	0.0016***	0.0016***
Goodness-of-fit					
	Deviance		4921.4	4873.8	4874.7
	AIC		4933.4	4893.8	4890.7
	BIC		4962.1	4941.7	4929.0

$\sim p < .10; \ * p < .05; \ ** p < .01; \ *** p < .001.$
Model A is an unconditional growth model; Model B includes the effects of highest grade completed $(HGC - 9)$ and race $(BLACK)$ on both initial status and rate of change; Model C is a reduced model in which $(HGC - 9)$ predicts only initial status and $BLACK$ predicts only rate of change.

Note: SAS Proc Mixed, Full ML. Also note that the covariance component, σ_{01}, is estimated, but not displayed.

After specifying a suitable individual growth model, you add level-2 predictors in the usual way. The statistically significant variance components in Model A, for both initial status and rate of change, suggest the wisdom of this action. Models B and C examine the effects of two predictors: (1) the race/ethnicity of the dropout; and (b) the highest grade he completed before dropping out. Although the sample includes 438 Whites, 246 African Americans, and 204 Latinos, analyses not shown here suggest that we cannot distinguish statistically between the trajectories of Latino and White dropouts. For this reason, these models include just one race/ethnicity predictor $(BLACK)$. Highest grade completed, *HGC*, is a continuous variable that ranges from 6th through 12th grade, with an average of 8.8 and a standard deviation of 1.4. To facilitate

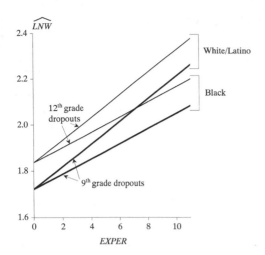

Figure 5.2. Displaying the results of a fitted multilevel model for change. Log wage trajectories from Model C of table 5.4 for four prototypical dropouts: Blacks and Whites/Latinos who dropped out in 9th and 12th grades.

interpretation, our analyses use a rescaled version, $HGC - 9$, which centers HGC around this substantively meaningful value near the sample mean (see section 4.5.4 for a discussion of centering).

Model B of Table 5.4 associates each predictor with initial status and rate of change. The estimated fixed effects suggest that $HGC - 9$ is related only to initial status while $BLACK$ is related only to the rate of change. We therefore fit Model C, whose level-2 submodels reflect this observation. The fixed effect for $HGC - 9$ on initial status tells us that dropouts who stay in school longer earn higher wages on labor force entry ($\hat{\gamma}_{01} = 0.0384$, $p < .001$), as we might expect because they are likely to have more skills than peers who left school earlier. The fixed effect for $BLACK$ on rate of change tells us that, in contrast to Whites and Latinos, the wages of Black males increase less rapidly with labor force experience ($\hat{\gamma}_{12} = -0.0161$, $p < .001$). The statistically significant level-2 variance components indicate the presence of additional unpredicted interindividual variation in both initial status and rate of change. In sections 5.3.3 and 6.1.2, we add other predictors that explain some of this remaining variation.

Figure 5.2 summarizes the effects in Model C by displaying wage trajectories for four prototypical dropouts: Blacks and Whites/Latinos who dropped out in 9th and 12th grades. We obtained these trajectories using the same two-stage process presented in section 4.5.3. We first substituted the two values of $BLACK$ (0 and 1) into Model C and then substituted in two prototypical values of $HGC - 9$ (0 and 3, to correspond to 9 and 12 years of education). The plots document the large and statistically significant effects of education and race on the wage trajectories. The longer a prospective dropout stays in school, the higher his wages on labor force entry. But race plays an important role, not on initial wages but on the rate of change. Although the average Black dropout initially earns an hourly wage indistinguishable from the average White or Latino dropout, his annual percentage increase is lower. Controlling for highest grade completed, the average annual percentage increase is $100(e^{(0.0489)} - 1) = 5.0\%$ for Whites and Latinos in comparison to $100(e^{(0.0328)} - 1) = 3.3\%$ for Blacks. Over time, this race differential overwhelms the initial advantage of remaining in school. Beyond 7 years of labor force participation, a Black male who left school in 12th grade earns a lower hourly wage than a White or Latino male who left in 9th.

5.2.2 Practical Problems That May Arise When Analyzing Unbalanced Data Sets

We encountered no problems when fitting models to the unbalanced data in section 5.2.1. The most complex model (C) converged in just three iterations and we could estimate every parameter in the model. But if your data set is severely unbalanced, or if too few people have enough waves of data, computer iterative algorithms may not converge and you may be unable to estimate one or more variance components.

Why does imbalance affect the estimation of variance components but not fixed effects? No matter how unbalanced the person-period data set, the estimation of fixed effects is generally no more difficult than the estimation of regression coefficients in a regular linear model. To demonstrate why, let us begin with a multilevel model—for simplicity, an unconditional growth model—expressed in composite form:

$$Y_{ij} = [\gamma_{00} + \gamma_{10}TIME_{ij}] + [\zeta_{0i} + \zeta_{1i}TIME_{ij} + \varepsilon_{ij}]. \tag{5.2a}$$

If we re-express the composite error term in the second set of brackets as: $\varepsilon_{ij}^* = [\zeta_{0i} + \zeta_{1i}TIME_{ij} + \varepsilon_{ij}]$, we obtain an equivalent representation of equation 5.2a:

$$Y_{ij} = \gamma_{00} + \gamma_{10}TIME_{ij} + \varepsilon_{ij}^*. \tag{5.2b}$$

Equation 5.2b resembles a standard regression model, with γ's instead of β's and ε_{ij}^* instead of ε_{ij}. The difference is that we do not assume that the composite residuals ε_{ij}^* are independent and normally distributed with mean 0 and variance $\sigma_{\varepsilon^*}^2$. Instead we assume that their constituents—ζ_{0i}, ζ_{1i}, and ε_{ij}—follow the assumptions:

$$\varepsilon_{ij} \sim N(0, \sigma_\varepsilon^2) \text{ and } \begin{bmatrix} \zeta_{0i} \\ \zeta_{1i} \end{bmatrix} \sim N \left(\begin{bmatrix} 0 \\ 0 \end{bmatrix}, \begin{bmatrix} \sigma_0^2 & \sigma_{01} \\ \sigma_{10} & \sigma_1^2 \end{bmatrix} \right).$$

It is these complex assumptions—about the variance components—that complicate estimation.

Now consider the following thought experiment. Suppose we are willing to make a simplifying assumption about the composite residuals, declaring them to be independent and normally distributed: $\varepsilon_{ij}^* \sim N(0, \sigma_{\varepsilon^*}^2)$. This is tantamount to assuming that both level-2 residuals, ζ_{0i} and ζ_{1i}, are always 0, as would be their associated variance components (i.e., both σ_0^2 and σ_1^2 are also 0). In the language of multilevel modeling, we would be *fixing* the intercept and rate of change, making them constant across individuals. Whether each person contributed one wave or many, estimation of the two fixed effects and the one variance component would then become a standard regression problem. All you would need are a sufficient number of distinct values of $TIME_{ij}$ in the person-period data set—enough distinct points in a plot of Y_{ij} vs. $TIME_{ij}$—to identify the level-1 submodel's functional form. In a time-structured data set, this plot would be composed of vertical stripes, one for each measurement occasion. This is why you would need at least three waves of data—the stripes would lie at just those three occasions. In an unstructured data set, the variable spacing of waves makes it easier to estimate fixed effects because the data points are more separated "horizontally." This allows you to relax the data minimum per person—allowing some people to have fewer than three waves—as long as you have enough distinct values of $TIME_{ij}$ to estimate the fixed effects.

If we are unwilling to make these simplifying assumptions—and we generally are—estimation of variance components can be difficult if too many people have too few waves. Variability in the spacing of waves helps, but may not resolve the problem. Estimation of variance components requires that enough people have sufficient data to allow quantification of within-person residual variation—variation in the residuals over and above the fixed effects. If too many people have too little data, you will be unable to quantify this residual variability.

When does the numeric task become so difficult that the variance components cannot be estimated? We offer no rules because so many issues are involved, including the degree of imbalance, the complexity of the model, the number of people with few vs. many waves, and the inclusion of time-varying predictors (discussed in section 5.3). Suffice it to say that when imbalance is severe enough, numeric computer algorithms can produce theoretically impossible values or fail to converge. Each statistical software program has its own way of informing the user of a problem; once discovered, we recommend that you be proactive and not automatically accept the default "solution" your program offers. Below, we discuss each of the two major estimation problems.

Boundary Constraints

Many population parameters have *boundary constraints*—limits beyond which they cannot theoretically lie. Like variances and correlation coefficients, the variance/covariance components in the multilevel model have clear boundaries: (1) a variance component cannot be negative; and (2) a covariance component, expressed in correlation form, must lie between −1 and +1. Because of the complexity of the estimation task—especially with unbalanced data—as well as the iterative nature of the computational algorithms, multilevel modeling programs occasionally generate parameter estimates that reach, or lie outside, these limits. When this happens, the program may output the implausible estimate or its boundary value (e.g., it might set a variance component to 0).

How will you know if you have encountered a boundary constraint? The warning signs differ across programs. If you use SAS PROC MIXED, the program log will note that "the G matrix [the variance-covariance matrix for the variance components] is not positive definite." By default, SAS sets the offending estimate to its boundary value. MLwiN does not provide a note; instead, it sets the offending estimate, and all associated estimates, to boundary values. If your output indicates that an estimate is exactly 0, you have likely encountered a boundary constraint. HLM will provide you with a warning message and modify its computational algorithm to avoid the problem. With all software, one clue that you may be approaching a "boundary" is if you find you need an excessive number of iterations to reach convergence.

We recommend that you never let a computer program arbitrarily make important decisions like these. Regardless of which program you use, you should be proactive about boundary constraints. Overspecification of the model's stochastic portion is the usual cause; model simplification is generally the cure. A practical solution is to compare alternative models that remove one, or more, offending random effects systematically until the model can be fit. This strategy, known as *fixing* a predictor's effect, usually resolves the problems.

We illustrate this approach using a small data set purposefully selected from the larger wage data set just analyzed. We constructed this sample for pedagogic purposes, hoping to create such extreme imbalance that boundary constraints would arise. This new data set is composed of the 124 men who had three or fewer waves of wage data: 47 men have three waves, 39 have two, and 38 have only one. The earliest value of *EXPER* is 0.002; the latest is 7.768. This data set is *not* a random sample of the original group.

Table 5.5 presents the results of fitting three models to this smaller data set; each is based upon Model C, the "final" model of table 5.4. As before, each was fit using ML in SAS PROC MIXED. In the first model, which is identical to Model C, the estimated variance component for linear growth, $\hat{\sigma}_1^2$, is exactly 0. This is a standard sign of a boundary

Table 5.5: Comparison of three alternative approaches to fitting Model C of table 5.4 to a severely unbalanced subset of the high school dropout wage data ($n = 124$)

		Parameter	A Default method	B Removing boundary constraints	C Fixing rates of change
Fixed Effects					
Initial status, π_{0i}	Intercept	γ_{00}	1.7373***	—	1.7373***
			(0.0476)		(0.0483)
	$(HGC - 9)$	γ_{01}	0.0462~	—	0.0458~
			(0.0245)		(0.0245)
Rate of	Intercept	γ_{10}	0.0516*	—	0.0518*
change, π_{1i}			(0.0211)		(0.0209)
	BLACK	γ_{12}	−0.0596~	—	−0.0601~
			(0.0348)		(0.0346)
Variance Components					
Level-1:	Within-person	σ_ε^2	0.1150***	0.1374***	0.1148***
Level-2:	In initial status	σ_0^2	0.0818**	0.0267	0.0842***
	In rate of change	σ_1^2	0.0000	−0.0072	—
Goodness-of-fit					
	Deviance		283.9	—	283.9
	AIC		297.9	—	295.9
	BIC		317.6	—	312.8

~$p < .10$; *$p < .05$; **$p < .01$; ***$p < .001$.
Model A uses the default option in SAS PROC MIXED; Model B removes boundary constraints for the variance components; Model C removes the level-2 residual for rate of change, eliminating the associated variance component (as well as the associated covariance component).

Note. SAS Proc Mixed, Full ML. Also note that the covariance component, σ_{01}, is estimated where appropriate, but not displayed.

problem, used by both SAS PROC MIXED and MLwiN. Estimates of 0 are always suspicious; here they indicate that the algorithm has encountered a boundary constraint. (Note that SAS allows the associated covariance component to be non-zero, whereas MLwiN would also set that term to 0.)

Model B in table 5.5 represents our dogged attempt to fit the specified model to data. To do so, we invoke a software option that relaxes the default boundary constraint permitting us to obtain a negative variance component. When analyzing severely unbalanced data, eliminating automatic fix-ups can help identify problems with boundary constraints. Unfortunately, in this case, the iterative algorithm does not converge (a different problem that we will soon discuss). Nevertheless, notice that the estimated variance component for rate of change at the last iteration is negative—a logical impossibility. This, too, is another sign suggesting the need for model simplification.

Model C in table 5.5 constrains the variance component for the linear growth rate, and its associated covariance component, to be 0. Notice that the deviance statistic for this model is identical to that of the first, suggesting the wisdom of fixing this parameter. This model fits no worse and involves fewer parameters (as reflected by the superior AIC and BIC statistics). This means that with this data set—which is *not* a random sample from the original—we cannot confirm the existence of any systematic residual variation in the slopes of the wage trajectories beyond the modest effect of *BLACK* shown in the final column of table 5.5.

Nonconvergence

As discussed in section 4.3, all multilevel modeling programs implement iterative numeric algorithms for model fitting. These algorithms compare fit criteria (such as the log-likelihood statistic) across successive iterations and declare convergence when the change in the fit criterion is sufficiently "small." Although the user can determine how small is "small enough," all programs have a default criterion, generally an arbitrarily small proportional change. When the criterion is met, the algorithm *converges* (i.e., stops iterating). If the criterion cannot be met in a large number of iterations, estimates should be treated with suspicion.

How many iterations are needed to achieve convergence? If your data set is highly structured and your model simple, convergence takes just a few iterations, well within the default values set by most programs. With unbalanced data sets and complex models, convergence can take hundreds or thousands of iterations although the algorithms in specialized packages (e.g., HLM and MLwiN) usually converge more rapidly than those in multipurpose programs (e.g., SAS PROC MIXED).

For every model you fit—but especially for models fit to unbalanced data—be sure to check that the algorithm has converged. In complex problems, the program's default limits on the maximum number of iterations may be too low to reach convergence. All packages allow you to increase this limit. If the algorithm still does not converge, sequentially increase the limit until it does. Some programs allow you to facilitate this search by providing "starting values" for the variance and covariance components.

No matter how many iterations you permit and no matter how much prior information you provide, there will be times when the algorithm will not converge. Nonconvergence can result from many factors, but two common causes are poorly specified models and insufficient data; their combination can be deadly. If you need an extremely large number of iterations to fit a model to data, closely examine the variance components and determine whether you have sufficient information to warrant allowing level-2 residuals for both initial status and rates of change. (If you are fitting nonlinear models using the methods of chapter 6, scrutinize other variance components as well.) Remember that any given data set contains a finite amount of information. You can postulate a complex model, but it is not always possible to fit that model to the available data.

We conclude by noting that other problems besides boundary constraints can cause nonconvergence. One problem, easily remedied, is a variable's scale. If an outcome's values are too small, the variance components will be smaller still; this can cause nonconvergence via rounding error issues. Simple multiplication of the outcome by 100, 1000, or another factor of 10 can usually ameliorate this difficulty. Predictor scaling can also cause problems but usually you want to adjust its metric in the *opposite* direction. For a temporal predictor, for example, you might move from a briefer time unit to a longer one (from days to months or months to years) so as to increase the growth rate's magnitude. These kinds of transformations have only cosmetic effects on your essential findings. (They will change the value of the log likelihood and associated statistics, but leave the results of tests unaffected.)

5.2.3 Distinguishing among Different Types of Missingness

No discussion of imbalance is complete without a complementary discussion of its underlying source. Although some researchers build imbalance into their design, most imbalance is unplanned, owing to scheduling problems, missed appointments, attrition, and data processing errors. Further imbalance accrues if individuals who miss a wave of data collection subsequently return to the sample. For example, although the NLSY has a low annual attrition rate—less than 5% of the original sample initially leave in each of the first 13 years—many participants miss one or two waves. In their exhaustive study of NSLY attrition, MaCurdy, Mroz, and Gritz (1998) find many differences among persisters, dropouts, and returnees. Of relevance for the wage analyses just presented are the findings that attrition is higher for both the unemployed and men who once earned high wages.

Unplanned imbalance, especially when it stems from attrition or other potentially systematic sources, may invalidate your inferences. The issue is not the technical ability to fit a model but rather a substantive question about credible generalization. To probe the issues, statisticians frame the problem, not in terms of imbalance, but rather in terms of *missing data*. When you fit a multilevel model for change, you implicitly assume that each person's observed records are a random sample of data from his or her underlying true growth trajectory. If your design is sound and has no built-in bias, and everyone is assessed on every planned occasion, your observed data will meet this assumption. If one or more individuals are not assessed on one or more occasions, your observed data may not meet this assumption. In this case, your parameter estimates may be biased and your generalizations incorrect.

Notice that we use the word "may," not "will," throughout the previous paragraph. This is because missingness, in and of itself, is not necessarily problematic. It all depends upon what statisticians call the *type of missingness*. In seminal work on this topic, Little (1995), refining earlier work with Rubin (Little & Rubin, 1987), distinguishes among three types of missingness: (1) *missing completely at random* (MCAR); (2) covariate-dependent dropout (CDD); and (3) *missing at random* (MAR) (see also Schafer, 1997). As Laird (1988) demonstrates, we can validly generalize the results of fitting a multilevel model for change under all three of these missingness conditions, which she groups together under rubric *ignorable nonresponse*.

When we say that data are MCAR, we argue that the observed values are a random sample of all the values that could have been observed (according to plan), had there been no missing data. Because time-invariant predictors are usually measured when a study begins, their values are rarely missing. As a result, when a multilevel model includes no time-varying predictors, the only predictor that can be missing is *TIME* itself (when a planned measurement occasion is missed). This means that longitudinal data are MCAR if the probability of assessment on any occasion is independent of: (1) the particular time; (2) the values of the substantive predictors; and (3) the values of the outcome (which are, by definition, unobserved). For the NLSY wage data just analyzed, we can make a case for the MCAR assumption if the probability of providing wage data at any point in time is independent of the particular moment in that individual's labor force history, all other predictors, and the unobserved wage. There cannot be particular moments when a man would be unlikely to grant an interview, as would be the case if men were unwilling to do so on specific days (which seems unlikely). But missingness must also not vary systematically by an individual's wage or other potentially unobserved characteristics. MaCurdy and colleagues (1998) convincingly demonstrate that these latter two conditions are implausible for the NLSY.

The conclusion that the MCAR assumption is untenable for the NLSY data is unsurprising as this assumption is especially restrictive—wonderful when met, but rarely so. Covariate dependent dropout (CDD) is a less restrictive assumption that permits associations between the probability of missingness and observed predictor values ("covariates"). Data can be CDD even if the probability of missingness is systematically related to either *TIME* or observed substantive predictors. For the NLSY wage data, we can argue for the validity of the CDD assumption even if there are particular moments when men are unlikely to grant interviews. Missingness can also vary by either race or highest grade completed (our two observed predictors). By including these observed predictors in the multilevel model, we deflect the possibility of bias, allowing appropriate generalization of empirical results.

The major difficulty in establishing the tenability of the MCAR and CDD assumptions is the requirement of demonstrating that the probability of missingness at any point in time is unrelated to the contemporaneous value of the associated outcome. Because this outcome is unobserved, you cannot provide empirical support as you lack the very data you need. Only a substantive argument and thought experiment will do. Any potential relationship between the unobserved outcome and the probability of missingness invalidates these assumptions. For example, if men with particularly high or low wages are less likely to participate in an NLSY interview, we cannot support either assumption. As this hypothesis is both tenable and likely, we cannot defend either assumption for the NLSY wage data (nor for many other longitudinal data sets).

Fortunately, there is an even less restrictive type of missingness—more common in longitudinal research—that still permits valid generalization of the multilevel model for change: the MAR assumption. When data are MAR, the probability of missingness can depend upon *any* observed data, for either the predictors or any outcome values. It cannot, however, depend upon any *un*observed value of either any predictor or the outcome. So if we are willing to argue that the probability of missingness in the NLSY depends only upon observed predictor values (that is, *BLACK* and *HGC*) *and* wage data, we can make a case for the MAR assumption. The allowance for dependence upon observed outcome data can account for a multitude of sins, often supporting the credibility of the MAR assumption even when MCAR and CDD assumptions seem far-fetched.

As general as it seems, you should not accept the MAR assumption without scrutiny. Greenland and Finkle (1995) examine this assumption in cross-sectional research and suggest that even it can be difficult to meet. To illustrate their point, they argue that someone's unwillingness to answer a question about sexual preference (i.e., heterosexual vs. homosexual) is likely correlated with his or her true sexual preference. We agree, but believe that there are many times when an individual's outcome values will adequately reflect such concerns. Yet even this assertion can be untrue. For example, a recovering alcoholic's willingness to continue participating in a study about abstinence is likely related to his or her ability to stay sober on each occasion. Such a systematic pattern—even if impossible to prove—invalidates the MAR assumption.

In practice, the burden of evaluating the tenability of these missingness assumptions rests with you. Any type of ignorable missingness permits valid inference; you just need to determine which seems most credible for your project. We suggest that you act as your own harshest critic—better you than the reviewers! As MAR is the least restrictive assumption, it provides the acid test. The key question is whether it is safe to assume that the probability of missingness is unrelated to unobserved concurrent outcomes (conditional on all observed outcomes). For the NLSY wage data, we can invent two plausible scenarios that undermine this assumption: If men are less likely to be interviewed at a particular wave if, at that time, they are earning especially: (1) *high* wages—because they might be less willing to take the time off from work to participate; or (2) *low* wages—because they might be less willing to reveal these low values to an interviewer. Because current wages (even unobserved) are strongly correlated with past and future wages, however, these risks are likely minimal. We therefore conclude that they are unlikely to be a major source of missingness for these data, supporting the credibility of the MAR assumption.[1]

If you cannot invoke one of these three missingness assumptions, you will need to add corrections to the multilevel model for change. Two different strategies are currently used: *selection models* and *pattern mixture models*. Under the selection approach, you build one statistical model for the "complete" data and a second model for the selection process that gave rise to the missingness. Under the pattern mixture approach, you identify a small number of missingness patterns and then fit a multilevel model stratified by these patterns. For further information, we direct your attention to the excellent papers by Hedeker and Gibbons (1997), Little (1995), and Little and Yau (1998).

5.3 Time-Varying Predictors

A time-varying predictor is a variable whose *values* may differ over time. Unlike their time-invariant cousins, which record an individual's static status, time-varying predictors record an individual's potentially differing status on each associated measurement occasion. Some time-varying predictors have values that change naturally; others have values that change by design.

In their four-year study of how teen employment affects the amount of time adolescents spend with their families, Shanahan, Elder, Burchinal, and Conger (1996) examined the effects of three time-varying predictors: (1) the average number of hours worked per week; (2) the total amount of money earned per year; and (3) whether earnings were used for nonleisure activities (e.g., schoolbooks or savings). At age $12\frac{1}{2}$, the average adolescent spent 16.3 hours per week with his or her family; over time, this amount declined at an average annual rate of 1.2 hours per week. Teen employment had both positive and negative effects. Although teens who made more money experienced steeper declines than peers who made less, those who spent some earnings on nonleisure activities or who worked especially long hours spent *more* time, on average, with their families (although their rates of decline were no shallower). The authors conclude that: "adolescent work constitutes a potentially positive source of social development, although this depends on how its multiple dimensions—earnings, spending patterns, [and] hours . . .—fit with the adolescent's broader life course" (p. 2198).

In this section, we demonstrate how you can include time-varying predictors in the multilevel model for change. We begin, in section 5.3.1, by showing how to parameterize, interpret, and graphically display a model that includes a time-varying predictor's main effect. In section 5.3.2, we allow the *effect* of a time-varying predictor to vary over time. In section 5.3.3, we discuss how to recenter time-varying predictors so as to facilitate interpretation. We conclude, in section 5.3.4, with some words of caution. Having described the analytic opportunities that time-varying predictors afford, we raise complex conceptual issues that can compromise your ability to draw clear convincing conclusions.

5.3.1 Including the Main Effect of a Time-Varying Predictor

Conceptually, you need no special strategies to include the main effect of a time-varying predictor in a multilevel model for change. The key to understanding why this is so lies in the *structure* of the person-period data set. Because each predictor—whether time-invariant or time-varying—has its own value on *each* occasion, it matters little whether these values vary across each person's multiple records. A time-invariant predictor's values remain constant; a time-varying predictor's values vary. There is nothing more complex to it than that.

Table 5.6: Excerpts from the person-period data set for the unemployment study

ID	MONTHS	CES-D	UNEMP
7589	1.3142	36	1
7589	5.0924	40	1
7589	11.7947	39	1
55697	1.3471	7	1
55697	5.7823	4	1
65641	0.3285	32	1
65641	4.1068	9	0
65641	10.9405	10	0
65441	1.0842	27	1
65441	4.6982	15	1
65441	11.2690	7	0
53782	0.4271	22	1
53782	4.2382	15	0
53782	11.0719	21	1

We illustrate the general approach using data from Ginexi and colleagues' (2000) study of the effects of unemployment on depressive symptoms (mentioned briefly in section 5.1). By recruiting 254 participants from local unemployment offices, the researchers were able to interview individuals soon after job loss (within the first 2 months). Follow-up interviews were conducted between 3 and 8 months and 10 and 16 months after job loss. Each time, participants completed the Center for Epidemiologic Studies' Depression (CES-D) scale (Radloff, 1977), which asks them to rate, on a four-point scale, the frequency with which they experience each of 20 depressive symptoms. CES-D scores can vary from a low of 0 for someone with no symptoms to a high of 80 for someone in serious distress.

Just over half the sample ($n = 132$) was unemployed at every interview. Others had a variety of re-employment patterns: 62 were always working after the first interview; 41 were still unemployed at the second interview but working by the third; 19 were working by the second interview but unemployed at the third. We investigate the effect of unemployment using the time-varying predictor, *UNEMP*. As shown in the person-period data set in table 5.6, *UNEMP* represents individual i's unemployment status at each measurement occasion. Because subjects 7589 and 55697 were consistently unemployed, their values of *UNEMP* are consistently 1. Because the unemployment status of the remaining cases *changed*, their values of *UNEMP* change as well: subject 65641 was working at both the second and third interviews (pattern 1-0-0); subject 65441 was working by the third (pattern 1-1-0); and subject 53782 was working at the second interview but unemployed again by the third (pattern 1-0-1). For any individual, *UNEMP* can be either 0 or 1 at each measurement occasion except the first (because, by design, everyone was initially unemployed).

We begin, as usual, with an unconditional growth model without substantive predictors:

$$Y_{ij} = \pi_{0i} + \pi_{1i}TIME_{ij} + \varepsilon_{ij}$$
$$\pi_{0i} = \gamma_{00} + \zeta_{0i} \tag{5.3a}$$
$$\pi_{1i} = \gamma_{10} + \zeta_{1i},$$

where

$$\varepsilon_{ij} \sim N(0,\sigma_\varepsilon^2) \text{ and } \begin{bmatrix} \zeta_{0i} \\ \zeta_{1i} \end{bmatrix} \sim N\left(\begin{bmatrix} 0 \\ 0 \end{bmatrix}, \begin{bmatrix} \sigma_0^2 & \sigma_{01} \\ \sigma_{10} & \sigma_1^2 \end{bmatrix} \right). \tag{5.3b}$$

Model A of table 5.7 presents the results of fitting this model to data, where $TIME_{ij}$ indicates the number of months (to the nearest day) between the date of interview j for person i and his date of initial unemployment. On the first day of job loss ($TIME_{ij} = 0$), we estimate that the average person has a non-zero CES-D score of 17.67 ($p<.01$); over time, this level declines linearly at a rate of 0.42 per month ($p<.001$). The variance components for both initial status and rates of change are statistically significant, suggesting the wisdom of exploring the effects of person-specific predictors.

Using a Composite Specification

Because many respondents eventually find work, the unconditional growth model likely tells an incomplete story. If employment alleviates depressive symptoms, might the reemployment of half the sample explain the observed decline? If you exclusively use level-1/level-2 representations, you may have difficulty postulating a model that addresses this question. In particular, it may not be clear where—in which model—the time-varying predictor should appear. So far, person-specific variables have appeared in level-2 submodels as predictors of level-1 growth parameters. Although you might therefore conclude that substantive predictors must always appear at level-2, this conclusion would be incorrect!

The easiest way of understanding how to include a time-varying predictor is to use the composite specification of the multilevel model. It is not that we cannot include a time-varying predictor in a model written using a level-1/level-2 specification (we will soon show how to do so), but rather that it is easier to learn how these predictors' effects operate and what types of models you might fit, if you start here.

We begin with the composite specification for the unconditional

Table 5.7: Results of fitting a taxonomy of multilevel models for change to the unemployment data ($n = 254$)

		Parameter	Model A	Model B	Model C	Model D
Fixed Effects						
Composite model	Intercept (initial status)	γ_{00}	17.6694** (0.7756)	12.6656*** (1.2421)	9.6167*** (1.8893)	11.2666*** (0.7690)
	TIME (rate of change)	γ_{10}	−0.4220*** (0.0830)	−0.2020* (0.0933)	0.1620 (0.1937)	
	UNEMP	γ_{20}		5.1113*** (0.9888)	8.5291*** (1.8779)	6.8795*** (0.9133)
	UNEMP by *TIME*	γ_{30}			−0.4652* (0.2172)	−0.3254** (0.1105)
Variance Components						
Level-1:	Within-person	σ_ε^2	68.85***	62.39***	62.03***	62.43***
Level-2:	In intercept	σ_0^2	86.85***	93.52***	93.71***	41.52***
	In rate of change	σ_1^2	0.36*	0.46**	0.45**	—
	In *UNEMP*	σ_2^2	—	—	—	40.45*
	In *UNEMP* by *TIME*	σ_3^2	—	—	—	0.71**
Goodness-of-fit						
	Deviance		5133.1	5107.6	5103.0	5093.6
	AIC		5145.1	5121.6	5119.7	5113.6
	BIC		5166.3	5146.4	5147.3	5148.9

$\sim p < .10$; $* p < .05$; $** p < .01$; $*** p < .001$.
These models predict depression scores (on the *CES-D*) in the months following unemployment as a function of the time-varying predictor *UNEMP*. Model A is an unconditional growth model (see equation 5.4). Model B adds the main effect of *UNEMP* as a fixed effect (see equation 5.5); Model C also adds the interaction between *UNEMP* and linear *TIME* (see equation 5.7). Model D allows *UNEMP* to have both fixed and random effects (see equation 5.10). Notice that we have changed the order in which the fixed effects appear to correspond to the composite specification of the model.

Note: Full ML, SAS Proc Mixed. Also note the models include all associated covariance parameters, which we do not display to conserve space.

growth model, formed by substituting the second and third equations in equation 5.3a into the first:

$$Y_{ij} = [\gamma_{00} + \gamma_{10} TIME_{ij}] + [\zeta_{0i} + \zeta_{1i} TIME_{ij} + \varepsilon_{ij}]. \qquad (5.4)$$

As in chapter 4, we use brackets to distinguish the model's fixed and stochastic portions. Because the fixed portion in the first bracket resembles a standard regression model, we can add the main effect of the time-varying predictor, *UNEMP*, by writing:

$$Y_{ij} = [\gamma_{00} + \gamma_{10}TIME_{ij} + \gamma_{20}UNEMP_{ij}] + [\zeta_{0i} + \zeta_{1i}TIME_{ij} + \varepsilon_{ij}]. \tag{5.5}$$

The two subscripts on *UNEMP* signify its time-varying nature. In writing equation 5.5, we assume that individual i's value of Y at time j depends upon: (1) the number of months of since job loss (*TIME*); (2) his or her contemporaneous value of *UNEMP*; and (3) three person-specific residuals, ζ_{0i}, ζ_{1i}, and ε_{ij}.

What does this model imply about the time-varying predictor's main effect? Because the fixed effects, the γ's, are essentially regression parameters, we can interpret them using standard conventions:

- γ_{10} is the population average monthly rate of change in CES-D scores, controlling for unemployment status.
- γ_{20} is the population average difference, over time, in CES-D scores between the unemployed and employed.

The intercept, γ_{00}, refers to a logical impossibility: someone who is employed (*UNEMP* = 0) on the first day of job loss (*TIME* = 0). As in regular regression, an intercept can fall outside the range of the data (or theoretical possibility) without undermining the validity of the remaining parameters.

We can delve further into the model's assumptions by examining figure 5.3, which presents four average population trajectories implied by the model. As in figure 3.4, we obtained these trajectories by substituting in specific values for the substantive predictor(s). But because *UNEMP* is time-varying, we substitute in *time-varying patterns* not constant values. Since everyone was initially unemployed, *UNEMP* can take on one of four distinct patterns: (1) 1 1 1, for someone consistently unemployed; (2) 1 0 0, for someone who soon finds a job and remains employed; (3) 1 1 0, for someone who remains unemployed for a while but eventually finds a job; and (4) 1 0 1, for someone who soon finds a job only to lose it. Each pattern yields a different population trajectory, as shown in figure 5.3.

The unbroken trajectory in the upper left panel represents the predicted change in depressive symptoms for people who remain unemployed during the study. Because their values of *UNEMP* do not change, their implied average trajectory is linear. In displaying this single line, we do not mean to suggest that everyone who is consistently unemployed follows this line. The person-specific residuals, ζ_{0i} and ζ_{1i}, allow different individuals to have unique intercepts and slopes. But every true trajectory for someone who is consistently unemployed is linear, regardless of its level or slope.

The remaining trajectories in figure 5.3 reflect different patterns of temporal variation in *UNEMP*. Unlike the population trajectories in pre-

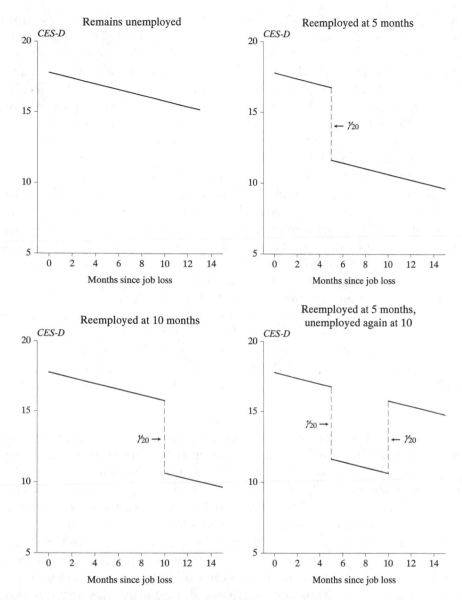

Figure 5.3. Identifying a suitable level-1 model for a time-varying predictor. Four average population trajectories implied by equation 5.5 for the effects of time-varying unemployment (*UNEMP*) on CES-D scores. In each panel, the magnitude of the effect of unemployment remains constant (at γ_{20}), but because *UNEMP* is time-varying, the model implies different population average trajectories corresponding to alternative patterns of unemployment and reemployment.

vious chapters, these are *discontinuous*. Discontinuity is a direct conse-quence of *UNEMP*'s dichotomous time-varying nature. The upper right panel, for the 1 0 0 pattern, presents a hypothesized population trajec-tory for someone who finds a job at 5 months and remain employed. The lower left panel, for the 1 1 0 pattern, presents a hypothesized trajectory for someone who finds a job at 10 months and remains employed. The lower right panel, for the 1 0 1 pattern, presents a hypothesized trajec-tory for someone who finds a job at 5 months only to lose it at 10.

In offering these hypothetical trajectories, we must mention two caveats. First, although we link the upper and lower segments in each panel using dashed lines, our model implies only the solid portions. We use the dashed lines to emphasize that a change in unemployment status is associated with a switch in trajectory. Second, these few trajectories are not the only ones implied by the model. As in the first panel, person-specific residuals—ζ_{0i} and ζ_{1i}—suggest the existence of many other discon-tinuous trajectories, each with its own intercept and slope. But because the model constrains the effect of *UNEMP* to be constant, the *gap* between trajectories—for any individual—will be identical, at γ_{20}, the parameter associated with *UNEMP*. (We relax this assumption in section 5.3.2.)

Model B of table 5.7 presents the results of fitting this model to data. The parameter estimate for TIME, $\hat{\gamma}_{10}$, suggests that the monthly rate of decline in CES-D, while still statistically significant, has been cut in half (to 0.20 from 0.42 in Model A). This suggests that reemployment explains some of the observed decline in CES-D scores. This conclusion is reinforced by: (1) the large statistically significant effect of *UNEMP*—the average CES-D score is 5.11 points higher ($p < .001$) among the unem-ployed; and (2) the poorer fit of Model A in comparison to Model B—the difference in deviance statistics is 25.5 on the addition of one parameter ($p < .001$) and the AIC and BIC statistics are much lower as well. (We discuss the variance components later in this section.)

The left panel of figure 5.4 displays prototypical trajectories for Model B. Rather than present many different discontinuous trajectories reflect-ing the wide variety of transition times for *UNEMP*, we present just two con-tinuous trajectories: the upper one for someone consistently unemployed; the lower one for someone consistently employed after 3.5 months. Displaying only two trajectories reduces clutter and highlights the most extreme contrasts possible. Because of this study's design, we start the fitted trajectory for *UNEMP* = 0 at 3.5 months, the earliest time when a participant could be interviewed while working. To illustrate what would happen were we to extrapolate this trajectory back to *TIME* = 0, we include the dashed line. Because the model includes only the main effect of *UNEMP*, the two fitted trajectories are constrained to be parallel.

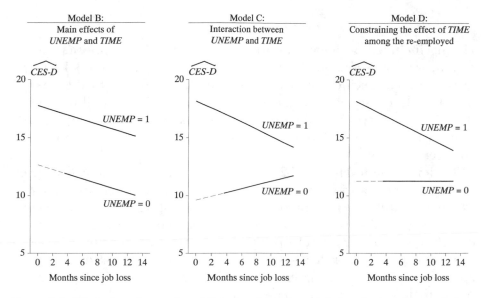

Figure 5.4. Displaying the results of fitted multilevel models for change that include a time-varying predictor. Prototypical trajectories from three models presented in table 5.7: Model B—the main effect of *UNEMP* and *TIME*, Model C—the interaction between *UNEMP* and *TIME*, and Model D—which constrains the effect of *TIME* to be 0 among the reemployed.

How do these two fitted trajectories display the main effect of unemployment status in Model B? Had the study followed just two static groups—the consistently unemployed and the consistently employed—these two trajectories would be the only ones implied by the model. But because *UNEMP* is time-varying, Model B implies the existence of many more depression trajectories, one for each possible *pattern* of unemployment/employment. Where are these additional trajectories? We find it helpful to think of the extremes shown as a conceptual *envelope* encompassing all discontinuous trajectories implied by the model. If *UNEMP* remains constant, an individual stays on one depression trajectory; if *UNEMP* changes, an individual shifts trajectories. As everyone in this study is unemployed at the first interview, everyone begins on the top trajectory. Those who find new jobs drop to the lower trajectory. Those who remain employed stay there. Those who lose their new jobs return to the upper trajectory. Conceptually, envision many dashed vertical lines running from the upper trajectory to the bottom (and back again) for individuals who change employment status. The set of these trajectories, which fall within the envelope shown, represent the complete set of prototypes implied by the model.

Using a Level-1/Level-2 Specification

Having included a time-varying predictor under the composite specification, we now show how you can specify the identical model using a
level-1/level-2 specification. This representation provides further insight
into how time-varying predictors' effects operate; it also allows you to
include time-varying predictors using software packages (e.g., HLM)
that require a level-1/level-2 specification of the multilevel model for
change.

To derive the level-1/level-2 specification that corresponds to a given
composite specification, you proceed backwards. In other words, just as
we can substitute level-2 submodels into a level-1 submodel to form a composite specification, so, too, can we *decompose* a composite model into its
constituent level-1 and level-2 parts. Because the time-specific subscript j
can appear only in a level-1 model, all time-varying predictors must
appear in at level-1. We therefore write the level-1 submodel for the composite main effects model in equation 5.5 as:

$$Y_{ij} = \pi_{0i} + \pi_{1i}TIME_{ij} + \pi_{2i}UNEMP_{ij} + \varepsilon_{ij}. \tag{5.6a}$$

Person-specific predictors that vary over time appear at level-1, not level-
2. If you have no time-invariant predictors, as here, the accompanying
level-2 models are brief:

$$\begin{aligned}
\pi_{0i} &= \gamma_{00} + \zeta_{0i} \\
\pi_{1i} &= \gamma_{10} + \zeta_{1i} \\
\pi_{2i} &= \gamma_{20}.
\end{aligned} \tag{5.6b}$$

You can verify that substituting these level-2 models into the level-1 model
in equation 5.6a yields the composite specification in equation 5.5. To
add the effects of time-invariant predictors, you include them, as usual,
in the level-2 submodels.

Notice that the third equation in equation 5.6b, for π_{2i}, the parameter
for *UNEMP*, includes no level-2 residual. All the multilevel models fit so
far have invoked a similar constraint—that the effect of a person-specific
predictor is constant across population members. Time-invariant predictors require this assumption because they have no *within*-person variation
to allow for a level-2 residual. But for time-varying predictors we could
easily modify the last model in equation 5.6b to be:

$$\pi_{2i} = \gamma_{20} + \zeta_{2i}. \tag{5.6c}$$

This allows the effect of *UNEMP* to vary randomly across individuals
in the population. Adding this residual relaxes the assumption that

the gap between postulated trajectories in figure 5.3 is constant. To fit the new model to data, we revise the distributional assumptions for the residuals as presented in equation 5.3b. Commonly, we expand the assumption of multivariate normality to include all three level-2 residuals:

$$
\varepsilon_{ij} \sim N\left(0, \sigma_\varepsilon^2\right) \text{ and } \begin{bmatrix} \zeta_{0i} \\ \zeta_{1i} \\ \zeta_{2i} \end{bmatrix} \sim N\left(\begin{bmatrix} 0 \\ 0 \\ 0 \end{bmatrix}, \begin{bmatrix} \sigma_0^2 & \sigma_{01} & \sigma_{02} \\ \sigma_{10} & \sigma_1^2 & \sigma_{12} \\ \sigma_{20} & \sigma_{21} & \sigma_2^2 \end{bmatrix} \right). \tag{5.6d}
$$

Notice that in adding one extra residual, ζ_{2i}, we add three extra variance components: σ_2^2, σ_{20} and σ_{21}.

Just because we *can* add these terms to our model does not mean that we should. Before doing so, we must decide whether the additional parameters are: (1) necessary; and (2) estimable using the available data. To address the first issue, consider whether the effect of employment on CES-D scores, controlling for time, *should* vary randomly across individuals. Before answering yes, remember that we are talking about *random* variation. If we expect the effect of unemployment to vary *systematically* across people, we can add substantive predictors that reflect this hypothesis. The question here is whether we should go further and add a residual that allows the effect of *UNEMP* to vary randomly. To be sure, much of our caution stems from concerns about the second point—the ability to estimate the additional parameters. With three (and sometimes fewer) measurement occasions per person, we often lack sufficient data to estimate additional variance components. Indeed, if we attempt to fit this more elaborate model, we encounter boundary constraints (as described in section 5.2.2). We therefore suggest that you resist the temptation to automatically allow the effects of time-varying predictors to vary at level-2 unless you have good reason, and sufficient data, to do so. (We will soon do so in section 5.3.2.)

As your models become more complex, we offer some practical advice (born of the consequences of the failure to follow it). When including time-varying predictors, we suggest that you write out the entire model before specifying your choice to a computer package. We suggest this extra step because it is not always obvious which random effects to include. In equation 5.6b, for example, the level-2 submodels require the first two parameters to be random and the third to be fixed. In other words, to fit this model you must use what appears to be an *inconsistent* set of level-2 submodels. As in many aspects of longitudinal analysis, the default or "standard" specifications may not yield the model *you* want to fit.

Time-Varying Predictors and Variance Components

In section 4.5.2, we discussed how the magnitude of variance components generally change on the inclusion of time-invariant predictors: (1) the level-1 variance component, σ_ε^2, remains relatively stable because time-invariant predictors cannot explain much within-person variation; but (2) the level-2 variance components, σ_0^2 and σ_1^2, will decline if the time-invariant predictors "explain" some of the between-person variation in initial status or rates of change, respectively. Time-varying predictors, in contrast, can affect all three variance components because they vary both within- *and* between-persons. And although you can interpret a decrease in the magnitude of the level-1 variance component, changes in level-2 variance components may not be meaningful, as we now show.

The general principles can be illustrated simply using Models A and B in table 5.7. Adding *UNEMP* to the unconditional growth model (Model A) reduces the magnitude of the within-person variance component, σ_ε^2, by 9.4% (from 68.85 to 62.39). Using strategies from section 4.4.3, equation 4.13, we conclude that time-varying unemployment status explains just over 9% of the variation in CES-D scores. This interpretation is straightforward because the time-varying predictor is added to the level-1 model, reducing the magnitude of the level-1 residual, ε_{ij}.

But ascribing meaning to observed changes in the level-2 variance components σ_0^2 and σ_1^2 can be nearly impossible. As we move from Model A to B both estimates *increase*! Although we alluded to this possibility in section 4.4.3, this is first example in which we observe such a pattern. The explanation for this seeming paradox—that changes in level-2 variance components do not assess the effects of time-varying predictors—lies in the associated level-1 submodel. When you add a time-varying predictor, as either a main effect or an interaction, you *change* the meaning of the individual growth parameters because:

- The intercept parameter, π_{0i}, now refers to the value of the outcome when *all* level-1 predictors, not only *TIME* but also the time-varying predictor, are zero.
- The slope parameter, π_{1i}, is now a *conditional* rate of change, controlling for the effects of the time-varying predictor.

Altering the population quantity that each parameter represents alters the meaning of the associated level-2 variance component. Hence, it makes no sense to compare the magnitude of these variance components across successive models.

This means that you must rely on changes in the time-varying predictors fixed effects, and associated goodness-of-fit statistics, when deciding whether to retain a time-varying predictor in your model. As tempting as it is to compute the percentage reduction in a variance component associated with the inclusion of a time-varying predictor, there is no consistently meaningful way of doing so.

5.3.2 Allowing the Effect of a Time-Varying Predictor to Vary over Time

Might unemployment status also affect the trajectory's slope? In previous chapters, we initially associated predictors with both initial status *and* rates of change. Yet because Model B includes only the *main* effects of *TIME* and *UNEMP*, the trajectories are constrained to be parallel.

There are many ways to specify a model in which the trajectories' slopes vary by unemployment status. The easiest approach, and the one we suggest you begin with, is to add the cross-product—here, between *UNEMP* and *TIME*—to the main effects model:

$$Y_{ij} = [\gamma_{00} + \gamma_{10}TIME_{ij} + \gamma_{20}UNEMP_{ij} + \gamma_{30}UNEMP_{ij} \times TIME_{ij}]$$
$$+ [\zeta_{0i} + \zeta_{1i}TIME_{ij} + \varepsilon_{ij}]. \tag{5.7}$$

Notice the close resemblance between this and the composite model that includes an interaction between a time-*invariant* predictor and *TIME* (shown in equation 4.3). The differences between the two are purely cosmetic: (1) the substantive predictor (here *UNEMP* and there *COA*) has an additional subscript *j* to indicate that it is time-varying; and (2) different subscripts reference the relevant fixed effects (the γ's).

Model C of table 5.7 presents the results of fitting this model to data. The interaction between *TIME* and *UNEMP* is statistically significant ($\hat{\gamma}_{30}$ = −0.46, $p < .05$). As with all interactions, we can interpret this effect in two ways: (1) the effect of unemployment status on CES-D scores varies over time; and (2) the rate of change in CES-D scores over time differs by unemployment status. Rather than delve into these interpretations, we draw your attention to the prototypical trajectories for this model displayed in the middle panel of figure 5.4. Here we find an unexpected pattern: while CES-D scores decline among the unemployed, the *reverse* is found among the re-employed—their CES-D scores appear to increase! The parameter estimate for the main effect of *TIME*, $\hat{\gamma}_{10} = 0.16$, suggests why we observe this anomaly—it is not statistically significant (it is even smaller than its standard error, 0.19). Although we estimate a non-zero rate of change among the re-employed, we might have obtained this estimate even if the true rate of change in the population was zero.

This suggests that it might be wise to constrain the trajectory among the re-employed to be flat, with a slope of 0, while allowing the trajectory among the unemployed to decline over time. Were we fitting a standard regression model, we might achieve this goal by removing the main effect of *TIME*:

$$Y_{ij} = [\gamma_{00} + \gamma_{20}UNEMP_{ij} + \gamma_{30}UNEMP_{ij} \times TIME_{ij}] + [\zeta_{0i} + \zeta_{1i}TIME_{ij} + \varepsilon_{ij}].$$

(5.8)

Had we fit this model to data and obtained fitted trajectories by unemployment status we would find: when $UNEMP = 0$, $\hat{Y}_{ij} = \hat{\gamma}_{00}$, when $UNEMP = 1$, $\hat{Y}_{ij} = (\hat{\gamma}_{00} + \hat{\gamma}_{20}) + \hat{\gamma}_{30}TIME_{ij}$.

This model's structural portion yields trajectories with the desired properties: (1) for the employed, we would have a flat line at level $\hat{\gamma}_{00}$; and (2) for the unemployed, we would have a slanted line, with intercept $\hat{\gamma}_{00} + \hat{\gamma}_{20}$ and slope $\hat{\gamma}_{30}$.

We do not fit this model, however, because of the lack of congruence between its structural and stochastic portions. Comparing the elements in the two sets of brackets in equation 5.8, notice that the model includes: (1) a random effect for *TIME*, ζ_{1i}, but no corresponding main effect (we removed γ_{10} from the model when we removed the main effect of *TIME*); and (2) a fixed effect for the *UNEMP* by *TIME* interaction (γ_{30}) and no corresponding random effect. We therefore postulate an alternative model in which the fixed and random effects are better aligned:

$$Y_{ij} = [\gamma_{00} + \gamma_{20}UNEMP_{ij} + \gamma_{30}UNEMP_{ij} \times TIME_{ij}]$$
$$+ [\zeta_{0i} + \zeta_{3i}UNEMP_{ij} \times TIME_{ij} + \varepsilon_{ij}].$$

(5.9)

Notice that the interaction term, *UNEMP* by *TIME*, appears as both a fixed and a random effect. But when we attempt to fit the model in equation 5.9 to data, we find that its AIC and BIC statistics are larger (worse) than that of Model C (we cannot conduct a formal test because this model is not fully nested within the other, nor do we present the results in table 5.7).

It might appear, then, that Model C is preferable. But before reaching this conclusion, we revisit a question raised in the previous section: Should the effect of *UNEMP* be constant across the population? When we previously attempted to allow this effect to vary randomly (by augmenting Model B, which included the main effect of *TIME*) we could not fit the model to data. But having constrained the model's structural portion so that the trajectory among the re-employed is flat, we notice an inconsistency in equation 5.8: it allows the intercept among the employed, γ_{00}, to vary randomly (through the inclusion of the residual, ζ_{0i}) but not the increment to this intercept associated with unemployment, γ_{20} (there is no corresponding residual, ζ_{2i}). Why should we allow the flat level of the trajectory among the re-employed to vary and *not* allow the increment to this flat level (which yields the the intercept among the unemployed) to vary randomly as well? Perhaps the fit of the model in equation 5.9 is poorer than Model C because of this unrealistically stringent constraint on the random effects.

We address this supposition by fitting Model D:

$$Y_{ij} = [\gamma_{00} + \gamma_{20}UNEMP_{ij} + \gamma_{30}UNEMP_{ij} \times TIME_{ij}]$$
$$+ [\zeta_{0i} + \zeta_{2i}UNEMP_{ij} + \zeta_{3i}UNEMP_{ij} \times TIME_{ij} + \varepsilon_{ij}], \quad (5.10)$$

which allows each fixed effect to have an associated random effect. The results of fitting this model are shown in the final column of table 5.7 and are graphed in the right panel of figure 5.4. Immediately upon layoff, the average unemployed person in the population has a CES-D score of 18.15 (=11.27 + 6.88). Over time, as they acclimate to their new status, the average unemployed person's CES-D scores decline at a rate of −0.33 per month ($p < .01$). CES-D scores among those who find a job are lower (by as much as 6.88 if the job is found immediately after layoff or as little as 2.97 if 12 months later (14.24 − 11.27). Once a formerly unemployed individual finds a job and keeps it, we find no evidence of systematic change in CES-D scores over time. We believe that this model provides a more realistic representation of the patterns of change in CES-D scores over time than Model C. Not only is it substantively compelling, its AIC statistic is superior (and its BIC nearly equivalent) even though it includes several additional parameters (the extra variance components shown in table 5.7 as well as the extra covariance components not shown).

We hope that this example illustrates how you can test important hypotheses about time-varying predictors' effects and investigate even more ways in which outcomes might change over time (here, how CES-D scores change not just with time but also re-employment). As we will show in chapter 6, the ability to include time-varying predictors opens up a world of analytic opportunities. Not only can level-1 individual growth models be smooth and linear, they can also be discontinuous and curvilinear. This allows us to postulate and fit level-1 submodels that better reflect our hypotheses about the population processes that give rise to sample data and assess the tenability of such hypotheses with data. But to adequately build a foundation for pursuing those types of analyses, we must consider other issues that arise when working with time-varying predictors, and we do so by beginning with issues of centering.

5.3.3 Recentering Time-Varying Predictors

In chapter 4, when discussing interpretation of parameters associated with time-invariant predictors, we introduced the practice of recentering:

subtracting a constant from a predictor's values to alter its parameter's meaning. In some analyses, we subtracted a predictor's overall sample mean (known as *grand-mean centering*); in others, we subtracted a substantively interesting value (such as 9 for *highest grade completed*). We now describe similar strategies you can use with time-varying predictors.

To concretize the discussion, let us return to the wage data for high school dropouts summarized in table 5.4. We can express Model C in composite form by writing: $Y_{ij} = [\gamma_{00} + \gamma_{10}TIME_{ij} + \gamma_{01}(HGC_i - 9) + \gamma_{12}BLACK_i \times TIME_{ij}] + [\zeta_{0i} + \zeta_{1i}TIME_{ij} + \varepsilon_{ij}]$. As did the original researchers, we now introduce the possibility that wages might be affected by a time-varying predictor, *UERATE*, the unemployment rate in the local geographic area:

$$Y_{ij} = [\gamma_{00} + \gamma_{10}TIME_{ij} + \gamma_{01}(HGC_i - 9) + \gamma_{12}BLACK_i \times TIME_{ij} + \gamma_{20}UERATE_{ij}]$$
$$+ [\zeta_{0i} + \zeta_{1i}TIME_{ij} + \varepsilon_{ij}]. \tag{5.11}$$

We restrict attention to the main effect of *UERATE* because extensive analysis suggests that its effect on log wages does not vary over time.

Adapting recentering strategies outlined in section 4.5.4 for time-invariant predictors, we could include *UERATE* in several different ways, each using one of the following:

- Its raw values
- Deviations around its grand mean *in the person-period data set* (7.73)
- Deviations from another meaningful constant (say, 6, 7 or 8, common unemployment rates during the time period under study)

Each strategy would lead to virtually identical conclusions. Were we to fit the model in equation 5.11 using each, we would find identical parameter estimates, standard errors, and goodness-of-fit statistics *with just one exception*: for the intercept, γ_{00}. Inspecting equation 5.11 clarifies why this is so. As in regression, adding a main effect does not alter the meaning of the model's remaining parameters. If *UERATE* is expressed on its raw scale, γ_{00} estimates the average log wage on the first day of work (*EXPER* = 0) for a black male who dropped out in ninth grade (*HGC* – 9 = 0) and who lives in an area with *no* unemployment (*UERATE* = 0). If *UERATE* is grand-mean centered, γ_{00} estimates the average log-wage for a comparable male who lives in an area with an "*average*" unemployment rate. But because this "average" would be computed in the person-period data set, in which both the measurement occasions and number of waves vary across people, it may not be particularly meaningful.

Table 5.8: Results of adding three alternative representations of the time-varying predictor for local area unemployment rate (*UERATE*) to Model C of table 5.4 for the high school dropout wage data (*n* = 888)

	Parameter	Model A: centered at 7	Model B: within-person centering	Model C: time-1 centered
Fixed Effects				
Initial status, π_{0i}	Intercept	γ_{00}	1.7490*** (0.0114)	1.8743*** (0.0295)	1.8693*** (0.0260)
	(*HGC* − 9)	γ_{01}	0.0400*** (0.0064)	0.0402*** (0.0064)	0.0399*** (0.0064)
	UERATE	γ_{20}	−0.0120*** (0.0018)	−0.0177*** (0.0035)	−0.0162*** (0.0027)
	Deviation of *UERATE* from centering value	γ_{30}		−0.0099*** (0.0021)	−0.0103*** (0.0019)
Rate of change, π_{1i}	Intercept	γ_{10}	0.0441*** (0.0026)	0.0451* (0.0027)	0.0448*** (0.0026)
	BLACK	γ_{12}	−0.0182*** (0.0045)	−0.0189*** (0.0045)	−0.0183*** (0.0045)
Variance Components					
Level-1:	within-person	σ_ε^2	0.0948***	0.0948***	0.0948***
Level-2:	In initial status	σ_0^2	0.0506***	0.0510***	0.0503***
	In rate of change	σ_1^2	0.0016***	0.0016***	0.0016***
Goodness-of-fit					
	Deviance		4830.5	4827.0	4825.8
	AIC		4848.5	4847.0	4845.8
	BIC		4891.6	4894.9	4893.7

~*p* < .10; *p < .05; **p < .01; ***p < .001.
Model A adds (*UERATE* − 7); Model B centers *UERATE* at each person's mean; Model C centers *UERATE* around each person's value of *UERATE* at his first measurement occasion.

Note: SAS Proc Mixed, Full ML. Also note that the covariance component, σ_{01}, is estimated, but not displayed.

We therefore often prefer recentering time-varying predictors *not* around the grand-mean but rather around a substantively meaningful constant—here, say 7. This allows γ_{00} to describe the average log-wage for someone whose local area has a 7% unemployment rate. The results of fitting this last model appear in the first column of table 5.8. As in section 5.2.1, we can interpret this parameter estimate by computing $100(e^{(-0.0120)} - 1) = -1.2$. We conclude that each one-percentage point difference in local area unemployment rate is associated with wages that are 1.2 percent lower.

Given that centering has so little effect on model interpretation, you may wonder why we raise this issue. We do so for three reasons: (1) the topic receives much attention in the multilevel literature (see, e.g., Kreft et al., 1995; Hofmann & Gavin, 1998); (2) some computer programs tempt analysts into recentering their predictors through the availability of simple toggle switches on an interactive menu; and (3) there are still other meaningful ways of recentering. Not only can you recenter around a *single* constant, you can recenter around *multiple constants*, one per person. It is this approach, also known as within-context or group-mean centering, to which we now turn.

The general idea behind within-context centering is simple: instead of representing a time-varying predictor using a single variable, decompose the predictor into multiple constituent variables, which, taken together, separately identify specific sources of variation in the outcome. Of the many ways of decomposing a time-varying predictor, two deserve special mention:

- *Within-person centering*: include the *average* unemployment rate for individual i, $\overline{UERATE_{i0}}$, as well as the deviation of each period's rate from this average, ($UERATE_{ij} - \overline{UERATE_{i0}}$).
- *Time-1 centering*: include *time-1*'s unemployment rate for individual i, $UERATE_{i1}$, as well as the deviation of each subsequent rate from this original value, ($UERATE_{ij} - UERATE_{i1}$).

Within-context centering provides *multiple* ways of representing a time varying predictor. Under within-person centering, you include a time-invariant *average* value and deviations from that average; under time-1 centering, you include the time-invariant *initial* value and deviations from that starting point. In both cases, as well as in the many other possible versions of within-context centering, the goal is to represent the predictor in a way that provides greater insight into its effects. (Of course, within-person centering raises interpretive problems of endogeneity, discussed in the following section.)

The last two columns of table 5.8 present the results of fitting the multilevel model for change with *UERATE* centered within-person (Model B) and around time-1 (Model C). Each contributes a particular insight into the negative effect of local unemployment on dropouts' wages. Model B reveals an association between wages and two aspects of the unemployment: (1) its average over time—the lower the average rate, the lower the wage; and (2) its relative magnitude, at each point in time, in comparison to this average. Model C demonstrates that wages are also associated with two other aspects of the time-varying unemployment rates: (1) their *initial* value, when the dropout first enters the labor force; and (2) the *increment* or *decrement*, at each subsequent point in time, from that initial value. Is either of these centered options clearly superior to the raw variable representation? Given that we cannot compare deviance statistics (because no model is nested within any other), comparison of AIC and BIC statistics suggests that all three are roughly comparable, with BIC giving the nod to Model A and AIC the nod to Model C.

These strategies for representing the effect of a time-varying predictor are hardly the only options. We offer them primarily in the hope that they will stimulate your thinking about substantively interesting ways of representing predictors' effects. We find routine recommendations to always, or never, center unconstructive. We prefer instead to recommend that you think carefully about which representations might provide the greatest insight into the phenomenon you are studying.

5.3.4 An Important Caveat: The Problem of Reciprocal Causation

Most researchers get very excited by the possibility that a statistical model could represent the relationship between changing characteristics of individuals and their environments, on the one hand, and individual outcomes on the other. We now dampen this enthusiasm by highlighting interpretive difficulties that time-varying predictors can present. The problem, known generally as *reciprocal causation* or *endogeneity*, is the familiar "chicken and egg" cliché: if X is correlated with Y, can you conclude that X *causes* Y or is it possible that Y *causes* X?

Many, but not all, time-varying predictors are subject to these problems. To help identify which are most susceptible, we classify time-varying predictors into four groups: *defined, ancillary, contextual,* and *internal*[2]. In the context of individual growth modeling, classification is based on the degree to which a predictor's values at time t_{ij} are: (1) assignable *a priori*; and (2) potentially influenced by the study participant's contemporaneous outcome. The more "control" a study participant has over his or her predictor values, the more clouded your inferences.

A time-varying predictor is *defined* if, in advance of data collection, its values are predetermined for everyone under study. Defined predictors are impervious to issues of reciprocal causation because no one—not the study participants nor the researchers—can alter their values. Most defined predictors are themselves functions of time. All representations of *TIME* are defined because their values depend solely on a record's time-period. Time-varying predictors that reflect other periodic aspects of time—such as season (fall, summer, etc) or anniversary (anniversary month, nonanniversary month)—are defined because once the metric for time is chosen, so, too, are their values. Predictors whose values are set by an external schedule are also defined. If Ginexi and colleagues (2000) added a variable representing each person's time-varying unemployment benefits, its values would be defined because payments reflect a uniform schedule. Similarly, when comparing the efficacy of time-varying drug tapering regimens in a randomized trial, an individual's dosage is defined if the researcher determines the entire dosing schedule *a priori*. Different people may take different doses at different times, but if the schedule is predetermined, the predictor is defined.

A time-varying predictor is *ancillary* if its values cannot be influenced by study participants because they are determined by a stochastic process totally external to them. We use the term "stochastic process" to emphasize that, unlike a defined predictor, an ancillary predictor can behave erratically over time. Ancillary predictors are impervious to issues of reciprocal causation because no one involved in the study directly affects their values. Most ancillary predictors assess potentially changing characteristics of the physical or social environment in which respondents live. In his study of marital dissolution, for example, South (1995) divided the United States into 382 local marriage markets and used census data to create a time-varying predictor assessing the availability of spousal alternatives in each market. His *availability index* contrasted the number of unmarried persons "locally available" to the respondent with the number of unmarried persons "locally available" to the respondent's spouse. As no respondent could be part of the local marriage market (because all were married), this predictor is ancillary. If some *were* part of the local market (as they would be in a study of marital *initiation*), this predictor would be approximately ancillary because: (1) the contribution of any individual to the index would be negligible (given that the smallest marriage market included over a half million people); and (2) few individuals move to a particular area because of the availability of spousal alternatives. Following this logic, the local area unemployment rate just used in the high school dropout wage analysis is approximately ancillary. Other ancillary predictors include weather (Young, Meaden, Fogg, Cherin, & Eastman, 1997) and treatment, if randomly assigned.

A *contextual* time-varying predictor also describes an "external" stochastic process, but the connection between units is closer—between husbands and wives, parents and children, teachers and students, employers and employees. Because of this proximity, contextual predictors can be influenced by an individual's contemporaneous outcome values; if so, they are susceptible to issues of reciprocal causation. To assess whether reciprocal causation is a problem, you must analyze the particular situation. For example, in their 30-year study of the effects of parental divorce on mental health, Cherlin, Chase-Lansdale and McRae (1998) included time-varying predictors denoting whether children had experienced a parental divorce during four developmental phases: 7–10, 11–15, 16–22, and 23–33. These contextual time-varying predictors are unlikely to create interpretive problems because it is doubtful that someone's level of emotional problems would influence either the occurrence or the timing of a parental divorce. But in their three-year study of the link between the quality of childcare centers and children's early cognitive and language development, Burchinal et al. (2000) face a thornier problem. Because parents may *choose* particular childcare centers precisely because they emphasize particular skills, observed links between center quality and child development may be due to a link between development and quality, not quality and development. If such criticisms seem reasonable, we suggest that you treat a contextual time-varying predictor as if it were internal, and address issues of reciprocal causation in ways we now describe.

Internal time-varying predictors describe an individual's potentially changeable status over time. Some describe *psychological* states (mood or satisfaction), while others describe *physical* states (respiratory function, blood levels), *social* states (married/unmarried, working/unemployed), or other personal attributes. In their four-year study of adolescent smoking, for example, Killen, Robinson, Haydel, et al. (1997) annually assessed dozens of internal predictors ranging from counts of the number of friends who smoke and the frequency of drinking to the adolescent's height and weight. And in their four-year study of conduct disorder in boys, Lahey, McBurnett, Loeber, & Hart (1995) collected annual data on receipt of various kinds of psychological treatment, both in-patient and out-patient, medication and talk therapy.

Internal time-varying predictors raise serious interpretive dilemmas. Isn't it reasonable to argue, for example, that as teens start smoking, they increase the number of friends who smoke, increase their frequency of drinking, and lose weight? So, too, isn't it possible that as a child's behavior worsens a parent may be more likely to initiate psychotherapy? Although the causal link may be from predictor to outcome, it may also run the opposite way. Some readers may believe that longitudinal data— and the associated statistical models—should resolve such concerns. But resolution of the directional arrow is more difficult. As long as a model links *contemporaneous* information about time-varying predictors and outcomes, we effectively convert a longitudinal problem into a cross-sectional one, fully burdened by questions of reciprocal causation.

Given the conceptual appeal of internal and contextual time-varying predictors, what should you do? We have two concrete recommendations.

First, use theory as a guide, play your own harshest critic, and determine whether your inferences are clouded by reciprocal causation. Second, if your data allow, consider coding time-varying predictors so that their values in each record in the person-period data set refer to a *previous* point in chronological time. After all, there is nothing about the multilevel model for change that requires contemporaneous data coding. Most researchers use contemporaneous values by default. Yet it is often more logical to link *prior* status on a predictor with current status on an outcome.

For example, in their study of conduct disorder (CD) in boys, Lahey and colleagues (1995) carefully describe three ways they coded the effect of time-varying predictors representing treatment:

> In each case, the treatment was considered to be present in a given year if that form of treatment had been provided during all or part of the *previous 12 months* (emphasis added). . . . In addition, the analyses of treatment were repeated using the cumulative number of years that the treatment had been received as the time-varying covariate to determine whether the accumulated number of years of treatment influenced the number of CD symptoms in each year. Finally, a 1-year time-lagged analysis was conducted to look at the effect of treatment on the number of CD symptoms in the following year. (p. 90)

By linking each year's outcomes to prior treatment data, the researchers diminish the possibility that their findings are clouded by reciprocal causation. So, too, by carefully describing several alternative coding strategies, each of which describes a predictor constructed from the prior year's data, the researchers appear more credible and thoughtful in their work.

How might we respond to questions about reciprocal causation in Ginexi and colleagues' (2000) study of the link between unemployment and depression? A critic might argue that individuals whose CES-D scores decline over time are more likely to find jobs than peers whose levels remain stable or perhaps increase. If so, the observed link between re-employment and CES-D scores might result from the effects of CES-D on employment, not employment on CES-D. To rebut this criticism, we emphasize that the re-employment predictor indicates whether the person is *currently* employed at each subsequent interview. As a result, the moment of re-employment is temporally prior to the collection of CES-D scores. This design feature helps ameliorate the possibility that the observed relationship between unemployment and depression is a result of reciprocal causation. Had the CES-D and re-employment data been collected simultaneously, it would have been more difficult to marshal this argument.

Our message is simple: just because you can establish a link between a time-varying predictor and a time-varying outcome does not guarantee that the link is causal. While longitudinal data can help resolve issues of temporal ordering, the inclusion of a time-varying predictor can muddy the very issues the longitudinal models were intended to address. Moreover, as we will show in the second half of this book, issues of reciprocal causation can be even thornier when studying event occurrence because the links between outcomes and predictors are often more subtle than the examples just presented suggest. This is not to say you should not include time-varying predictors in your models. Rather, it is to say that you must recognize the issues that such predictors raise and not naively assume that longitudinal data alone will resolve the problem of reciprocal causation.

5.4 Recentering the Effect of *TIME*

TIME is the fundamental time-varying predictor. It therefore makes sense that if recentering a substantive time-varying predictor can produce interpretive advantages, so, too, should recentering *TIME*. In this section, we discuss an array of alternative recentering strategies, each yielding a different set of level-1 individual growth parameters designed to address related, but slightly different, research questions.

So far, we have tended to recenter *TIME* so that the level-1 intercept, π_{0i}, represents individual i's true *initial status*. Of course, the moment corresponding to someone's "initial status" is context specific—it might be a particular chronological age in one study (e.g., age 3, 6.5, or 13) or the occurrence of a precipitating event in another (e.g., entry into or exit from the labor force). In selecting a sensible starting point, we seek an early moment, ideally during the period of data collection, inherently meaningful for the process under study. This strategy yields level-2 submodels in which all parameters are directly and intrinsically interpretable, and it ensures that the value of *TIME* associated with the intercept, π_{0i}, falls within *TIME*'s observed range. Not coincidentally, this approach also yields a level-1 submodel that reflects everyday intuition about intercepts as a trajectory's conceptual "starting point."

Although compelling, this approach is hardly sacrosanct. Once you are comfortable with model specification and parameter interpretation, a world of alternatives opens up. We illustrate some options using data from Tomarken, Shelton, Elkins, and Anderson's (1997) randomized trial evaluating the effectiveness of supplemental antidepressant medication for individuals with major depression. The study began with an overnight

Table 5.9: Alternative coding strategies for *TIME* in the antidepressant trial

WAVE	DAY	READING	TIME OF DAY	TIME	(TIME − 3.33)	(TIME − 6.67)
1	0	8 A.M.	0.00	0.00	−3.33	−6.67
2	0	3 P.M.	0.33	0.33	−3.00	−6.33
3	0	10 P.M.	0.67	0.67	−2.67	−6.00
4	1	8 A.M.	0.00	1.00	−2.33	−5.67
5	1	3 P.M.	0.33	1.33	−2.00	−5.33
6	1	10 P.M.	0.67	1.67	−1.67	−5.00
...						
11	3	3 P.M.	0.33	3.33	0.00	−3.33
...						
16	5	8 A.M.	0.00	5.00	1.67	−1.67
17	5	3 P.M.	0.33	5.33	2.00	−1.33
18	5	10 P.M.	0.67	5.67	2.33	−1.00
19	6	8 A.M.	0.00	6.00	2.67	−0.67
20	6	3 P.M.	0.33	6.33	3.00	−0.33
21	6	10 P.M.	0.67	6.67	3.33	0.00

hospital stay for 73 men and women who were already being treated with a nonpharmacological therapy that included bouts of sleep deprivation. During the pre-intervention night, the researchers prevented each participant from obtaining any sleep. The next day, each person was sent home with a week's worth of pills (placebo or treatment), a package of mood diaries (which use a five-point scale to assess positive and negative moods), and an electronic pager. Three times a day—at 8 A.M., 3 P.M., and 10 P.M.—during the next month, respondents were electronically paged and reminded to fill out a mood diary. Here we analyze the first week's data, focusing on the participants' positive moods. With full compliance, each person would have 21 assessments. Although two people were recalcitrant (producing only 2 and 12 readings), everyone else was compliant, filling out at least 16 forms.

Table 5.9 presents seven variables that represent related, but distinct, ways of clocking time. The simplest, *WAVE*, counts from 1 to 21; although great for data processing, its cadence has little intuitive meaning because few of us divide our weeks into 21 conceptual components. *DAY*, although coarse, has great intuitive appeal, but it does not distinguish among morning, afternoon, and evening readings. One way to capture this finer information is to add a second temporal variable, such as *READING* or *TIME OF DAY*. Although the metric of the former makes it difficult to analyze, the metric of the latter is easily understood: 0 for morning readings; 0.33 for afternoon readings; 0.67 for evening readings. (We could also use a 24-hour clock and assign values that were not equidistant.) Another way to distinguish within-day readings is to create a single variable that combines both aspects of time. The next three variables, *TIME*, *TIME* − 3.33, and *TIME* − 6.67, achieve this goal. The first, *TIME*, operates like our previous temporal variables—it is centered on initial status. The others are linear transformations of *TIME*: one centered on 3.33, the study's *midpoint*, and the other centered on 6.67, the study's *final* wave.

Having created these alternative variables, we could now specify a separate set of models for each. Instead of proceeding in this tedious fashion, let us write a general model that uses a generic temporal variable (T) whose values are centered around a generic constant (c):

$$Y_{ij} = \pi_{0i} + \pi_{1i}(T_{ij} - c) + \varepsilon_{ij}. \tag{5.12a}$$

We can then write companion level-2 models for the effect of treatment:

$$\begin{aligned} \pi_{0i} &= \gamma_{00} + \gamma_{01}TREAT_i + \zeta_{0i} \\ \pi_{1i} &= \gamma_{10} + \gamma_{11}TREAT_i + \zeta_{1i} \end{aligned} \tag{5.12b}$$

and invoke standard normal theory assumptions for the residuals. This same model can be used for most of the temporal variables in table 5.9 (except those that distinguish only between within-day readings).

Table 5.10 presents the results of fitting this general model using the three different temporal variables, *TIME*, *TIME* − 3.33, and *TIME* − 6.67. Begin with the initial status representation of *TIME*. Because we cannot reject null hypotheses for either linear change or treatment, we conclude that: (1) on average, there is no linear trend in positive moods over time in the placebo group ($\hat{\gamma}_{10} = -2.42$, *n.s.*); and (2) when the study began, the groups were indistinguishable ($\hat{\gamma}_{01} = -3.11$, *n.s.*) as randomization would have us expect. The statistically significant coefficient for the effect of *TREAT* on linear change ($\hat{\gamma}_{11} = 5.54$, $p < 0.05$) indicates that the trajectories' slopes differ. The prototypical trajectories in figure 5.5 illustrate these findings. On average, the two groups are indistinguishable initially, but over time, the positive mood scores of the treatment group increase while those of the control group decline. The statistically significant variance components for the intercept ($\hat{\sigma}_0^2 = 2111.33$, $p < .001$) and linear change ($\sigma_1^2 = 63.74$, $p < .001$) indicate that that substantial variation in these parameters has yet to be explained.

What happens as we move the centering constant from 0 (initial status), to 3.33 (the study's midpoint), to 6.67 (the study's endpoint)? As expected, some estimates remain identical, while others change. The general principle is simple: parameters related to the *slope* remain stable while those related to the *intercept* differ. On the stable side, we obtain

Table 5.10: Results of using alternative representations for the main effect of *TIME* when evaluating the effect of treatment on the positive mood scores in the antidepressant trial ($n = 73$)

		Parameter	Temporal predictor in level-1 model		
			TIME	(*TIME* – 3.33)	(*TIME* – 6.67)
Fixed Effects					
Level-1 intercept, π_{0i}	Intercept	γ_{00}	167.46*** (9.33)	159.40*** (8.76)	151.34*** (11.54)
	TREAT	γ_{01}	−3.11 (12.33)	15.35 (11.54)	33.80* (15.16)
Rate of change, π_{1i}	Intercept	γ_{10}	−2.42 (1.73)	−2.42 (1.73)	−2.42 (1.73)
	TREAT	γ_{11}	5.54* (2.28)	5.54* (2.28)	5.54* (2.28)
Variance Components					
Level-1:	within-person	σ_{ε}^2	1229.93***	1229.93***	1229.93***
Level-2:	In level-1 intercept	σ_0^2	2111.33***	2008.72***	3322.45***
	In rate of change	σ_1^2	63.74***	63.74***	63.74***
	Covariance	σ_{01}	−121.62*	90.83	303.28***
Goodness-of-fit					
	Deviance		12680.5	12680.5	12680.5
	AIC		12696.5	12696.5	12696.5
	BIC		12714.8	12714.8	12714.8

~$p < .10$; * $p < .05$; ** $p < .01$; *** $p < .001$.
TIME is centered around initial status, middle status, and final status.

Note: Full ML, SAS PROC MIXED.

identical estimates for the linear rate of change in the placebo group ($\hat{\gamma}_{10}$ = −2.42, *n.s.*) and the effect of treatment on that rate ($\hat{\gamma}_{11}$ = 5.54, $p < 0.05$). So, too, we obtain identical estimates for the residual variance in the rate of change ($\hat{\sigma}_1^2 = 63.74$, $p < .001$) and the within-person residual variance ($\hat{\sigma}_{\varepsilon}^2 = 1229.93$). And, most important, the deviance, AIC and BIC statistics remain unchanged because these models *are* structurally identical.

Where these models differ is in the location of their trajectories' *anchors*, around their starting point, midpoint, or endpoint. Because the intercepts refer to these anchors, each model tests a different set of hypotheses about them. If we change c, we change the anchors, which changes the estimates and their interpretations. In terms of the general model in equations 5.12a and 5.12b, γ_{00} assesses the elevation of the population average change trajectory at time c; γ_{01} assesses the differential elevation of this trajectory at time c between groups; σ_0^2 assesses the population variance in true status at time c; and σ_{01} assesses the population

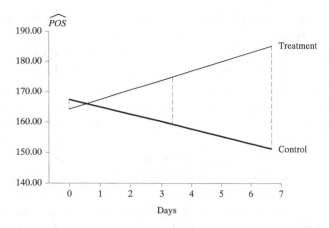

Figure 5.5. Understanding the consequences of rescaling the effect of *TIME*. Prototypical trajectories for individuals by *TREATMENT* status in the antidepressant experiment. The dashed vertical lines reflect the magnitude of the effect of *TREATMENT* if time is centered at the study's beginning (0), midpoint (3.33), and endpoint (6.67).

covariance between true status at time c and the per-unit rate of change in Y.

Although general statements like these are awkward, choice of a suitable centering constant can create simple, even elegant, interpretations. If we choose c to be 3.33, this study's midpoint, the intercept parameters assess effects at midweek. Because the treatment is still nonsignificant ($\hat{\gamma}_{01}$ = 15.35, *n.s.*), we conclude that the average elevation of the two trajectories remains indistinguishable at this time. If we choose c to be 6.67, this study's endpoint, the intercept parameters assess effects at week's end. Doing so yields an important finding: Instead of reinforcing the expected nonsignificant early differences between groups, we now find a statistically significant treatment effect ($\hat{\gamma}_{01}$ = 33.80, $p < .05$). After a week of antidepressant therapy, the positive mood score for the average member of the treatment group differs from that of the average member of the control group.

How can changing the centering constant for *TIME* have such a profound impact, especially since the fundamental model is unchanged? The dashed vertical lines in the prototypical plots in figure 5.5 provide an explanation. In adopting a particular centering constant, we cause the resultant estimates to describe the trajectories' behavior at that specific point in time. Changing the trajectory's anchor changes the location of the focal comparison. Of course, you could conduct *post hoc* tests of these contrasts (using methods of section 4.7) and obtain identical results. But when doing data analysis, it is sometimes easier to establish level-1 parameters that automatically yield readymade tests for hypotheses of greatest interest. We urge you to identify a scale for *TIME* that creates a level-1 submodel with directly interpretable parameters. Initial status often works well, but there are alternatives. The midpoint option is especially useful when *total study duration* has intrinsic meaning; the endpoint option is especially useful when *final status* is of special concern.

Statistical considerations can also suggest the need to recenter *TIME*. As shown in table 5.10, a change in center can change the interpretation, and hence values, of selected random effects. Of particular note is the effect that a recentering can have on σ_{01}, the covariance between a level-1 model's intercept and slope. Not only can a recentering affect this parameter's magnitude, it can also affect its sign. In these data, the covariance between intercept and slope parameters moves from -121.62 to 90.83 to 303.28 as the centering constant changes. These covariances (and their associated variances) imply correlation coefficients of -0.33, 0.25, and 0.66, respectively. As you might imagine, were we to choose an even larger centering constant, outside the range of the data, it would be possible to find oneself specifying a model in which the correlation between parameters is close to 1.00. As Rogosa and Willett (1985) demonstrate, you can always alter the correlation between the level-1 growth parameters simply by changing the centering constant.

Understanding that the correlation between level-1 individual growth parameters can change through a change of centering constants has important analytic consequences. Recall that in section 5.2.2, we alluded to the possibility that you might encounter boundary constraints if you attempted to fit a model in which the correlation between intercept and slope is so high that iterative algorithms may not converge and you cannot find stable estimates. We now introduce the possibility that the correlation between true intercept and true slope can be so high as to preclude model fitting. When this happens, recentering *TIME* can sometimes ameliorate your problem.

There is yet another reason you might recenter time: it can sometimes lead to a simpler level-1 model. For this to work, you must ask yourself: Is there a centering constant that might totally eliminate the need for an explicit intercept parameter? If so, you could decrease the number of parameters needed to effectively characterize the process under study. This is precisely what happened in the work of Huttenlocher, Haight, Bryk, Seltzer, and Lyons (1991). Using a sample of 22 infants and toddlers, the researchers had data on the size of children's vocabularies at up to six measurement occasions between 12 and 26 months. Reasoning that there must be an age at which we expect children to have *no* words, the researchers centered *TIME* on several early values, such as 9, 10, 11, and 12 months. In their analyses, they found that centering around age 12 months allowed them to eliminate the intercept parameter in their level-1 submodel, thereby dramatically simplifying their analyses.

We conclude by noting that there are other scales for *TIME* that alter not only a level-1 submodel's intercept but also its slope. It is possible, for example, to specify a model that uses neither a traditional intercept nor slope, but rather parameters representing initial and final status. To do so, you need to create two new temporal predictors, one to register each feature, and eliminate the stand-alone intercept term.

To fit a multilevel model for change in which the level-1 individual growth parameters refer to initial and final status, we write:

$$Y_{ij} = \pi_{0i}\left(\frac{max\ time - TIME_{ij}}{max\ time - min\ time}\right) + \pi_{1i}\left(\frac{TIME_{ij} - min\ time}{max\ time - min\ time}\right) + \varepsilon_{ij}. \qquad (5.13a)$$

In the context of the antidepressant medication trial, in which the earliest measurement is at time 0 and the latest at time 6.67, we have:

$$Y_{ij} = \pi_{0i}\left(\frac{6.67 - TIME_{ij}}{6.67}\right) + \pi_{1i}\left(\frac{TIME_{ij}}{6.67}\right) + \varepsilon_{ij}.$$

Although it may not appear so, this model is identical to the other linear growth models; it is just that its parameters have new interpretations. This is true despite the fact that equation 5.13a contains no classical "intercept" term and *TIME* appears twice in two different predictors.

To see how the individual growth parameters in this model represent individual i's initial and final status, substitute the minimum and maximum values for *TIME* (0 and 6.67) and simplify. When *TIME* = 0, we are describing someone's initial status. At this moment, the second term of equation 5.13a falls out and the first term becomes π_{0i} so that individual i's initial status is $\pi_{0i} + \varepsilon_{ij}$. Similarly, when *TIME* = 6.67, we are describing someone's final status. At this moment, the first term of equation 5.13a falls out and the second term becomes π_{1i} so that individual i's final status is $\pi_{1i} + \varepsilon_{ij}$.

We can then specify standard level-2 submodels—for example:

$$\begin{aligned} \pi_{0i} &= \gamma_{00} + \gamma_{01}TREAT_i + \zeta_{0i} \\ \pi_{1i} &= \gamma_{10} + \gamma_{11}TREAT_i + \zeta_{1i} \end{aligned} \qquad (5.13b)$$

and invoke standard normal theory assumptions about the residuals. When we fit this model to data, we find the same deviance statistic we found before—12,680.5—reinforcing the observation that this model is identical to the three linear models in table 5.10. And when it comes to the parameter estimates, notice the similarity between these and selected results in table 5.10:

$$\hat{\pi}_{0i} = 167.46 - 3.11 TREAT_i$$
$$\hat{\pi}_{1i} = 151.34 + 33.80 TREAT_i.$$

The first model provides estimates of initial status in the control group (167.46) and the differential in initial status in the treatment group (−3.11). The second model provides estimates of final status in the control group (151.34) and the differential in final status in the treatment group (33.80).

This unusual parameterization allows you to address questions about initial and final status simultaneously. Simultaneous investigation of these questions is superior to a piecemeal approach based on separate analyses of the first and last wave. Not only do you save considerable time and effort, you increase statistical power by using all the longitudinal data, even those collected at intermediate points in time.

6

Modeling Discontinuous and Nonlinear Change

Things have changed.

—Bob Dylan

All the multilevel models for change presented so far assume that individual growth is smooth and linear. Yet individual change can also be discontinuous or nonlinear. Patients' perceptions of their psychological well-being may abruptly shift when psychiatrists intervene and change their medications. Initial decreases in employee self-efficacy may gradually abate as new hires develop confidence with experience on the job.

This is not the first time we have confronted such possibilities. In the early intervention study of chapter 3, the trajectory of the child's cognitive development was nonlinear between infancy and age 12. To move forward and fit a model to these data, we focused on a narrower temporal period—the year of life between 12 and 24 months—in which the linearity assumption was tenable. In chapter 4, when changes in adolescent alcohol use seemed nonlinear, we transformed the outcome (and one of the predictors). Although the researchers used a nine-point scale to assess alcohol consumption, we analyzed the *square root* of scores on this scale, which yielded approximately linear change trajectories.

In this chapter, we introduce strategies for fitting models in which individual change is explicitly discontinuous or nonlinear. Rather than view these patterns as inconveniences, we treat them as substantively compelling opportunities. In doing so, we broaden our questions about the nature of change beyond the basic concepts of initial status and rate of change to a consideration of acceleration, deceleration, turning points, shifts, and asymptotes. The strategies that we use fall into two broad classes. *Empirical* strategies that let the "data speak for themselves." Under this approach, you inspect observed growth records systematically and identify a transformation of the outcome, or of *TIME*, that linearizes the individual change trajectory. Unfortunately, this approach can lead to interpretive difficulties, especially if it involves esoteric transformations or higher order polynomials. Under *rational* strategies, on the other hand, you use theory to hypothesize a substantively meaningful functional form for the individual change trajectory. Although rational strategies generally yield clearer interpretations, their dependence on good theory makes them somewhat more difficult to develop and apply.

We begin, in section 6.1, by describing ways of incorporating abrupt discontinuities into the individual change trajectory. This approach is especially useful when discrete shocks or time-limited treatments affect the life course. In section 6.2, we show how transformation of either the outcome or *TIME* can lead to a multilevel model for *linear* change based on the transformed variable(s). In section 6.3, we extend this basic idea by specifying trajectories that are polynomial functions of *TIME*. While admittedly atheoretical, we show that you can obtain curvilinear trajectories of almost any level of complexity just by adding higher order terms to a polynomial growth function. We conclude, in section 6.4, by surveying several models for individual change in which the outcome is an explicit nonlinear function of the growth parameters. This includes well-known trajectories such as the logistic and negative exponential growth curves, and others whose origins lie in theoretical work on expected patterns of change in the social, biological, and physical sciences.

6.1 Discontinuous Individual Change

Not all individual change trajectories are continuous functions of time. When analyzing the wage data for high school dropouts introduced in chapter 5, Murnane and colleagues (1999) asked whether the (log) wage trajectories might fail to remain the smooth functions of work experience postulated in section 5.2. In particular, they hypothesized that dropouts who obtain a GED (a General Education Development diploma, an alternative certificate awarded those who pass a high school equivalency examination) might command higher salaries. If so, their wage trajectories could exhibit a discontinuity—a shift in elevation and/or slope—upon GED receipt.

If you have reason to believe that individual change trajectories might shift in elevation and/or slope, your level-1 model should reflect this hypothesis. Doing so allows you to test ideas about how the trajectory's shape might change over time: here, how a dropout's wage trajectory might change not just with work experience but also upon GED receipt. This simple notion—that individual trajectories might suddenly shift in elevation or slope for identifiable reasons—has many applications. Psychologists following preschoolers into an early intervention program might test whether service provision produces a discontinuity in development. Organizational researchers following employees assigned to different work groups might assess whether contextual change produces a discontinuity in productivity.

To postulate a discontinuous individual change trajectory, you need to know not just *why* the shift might occur but also *when*. This is because your level-1 individual growth model must include one (or more) time-varying predictor(s) that specify whether and, if so, when each person experiences the hypothesized shift. In some studies, the precipitating event occurs at the same exact moment for everyone. When tracking students' test scores between adjacent grades, for example, all students take a summer break during the same months. If test scores decline when students are out of school, we would expect a discontinuity during vacation periods. In other studies, the precipitating event occurs at different times for different people and some participants may not experience the event at all. When tracking adolescent girls through menarche, for example, some will start their periods before data collection, others will do so during data collection, and still others may not do so for years to come. This suggests a model with a person-specific discontinuity, one that represents the time-varying menarcheal status of each sample member.

In this section, we discuss how to conceptualize, parameterize, and select among discontinuous individual change trajectories. We begin, in section 6.1.1, by outlining an array of options, each displaying a different discontinuity. In section 6.1.2, we offer strategies for choosing among them. We conclude, in section 6.1.3, by extending these ideas to a wider set of alternative trajectories. In our discussion, we emphasize the general case of person-specific discontinuities. At the end of the section, we apply these ideas to data sets in which the discontinuity occurs at a common point in time.

6.1.1 Alternative Discontinuous Level-1 Models for Change

To postulate a discontinuous level-1 individual growth model, you must first decide on its functional form. Although you can begin empirically, we prefer to focus on substance and the longitudinal process that gave rise to the data. What kind of discontinuity might the precipitating event create? What would a plausible level-1 trajectory look like? Before parameterizing models and constructing variables, we suggest that you: (1) take pen and paper and sketch some options; and (2) articulate—in

Table 6.1: Excerpts from the person-period data set for the high school dropout wage study

ID	LNW	EXPER	GED	POSTEXP	GED by EXPER
206	2.028	1.874	0	0	0
206	2.297	2.814	0	0	0
206	2.482	4.314	0	0	0
2365	1.782	0.660	0	0	0
2365	1.763	1.679	0	0	0
2365	1.710	2.737	0	0	0
2365	1.736	3.679	0	0	0
2365	2.192	4.679	1	0	4.679
2365	2.042	5.718	1	1.038	5.718
2365	2.320	6.718	1	2.038	6.718
2365	2.665	7.872	1	3.192	7.872
2365	2.418	9.083	1	4.404	9.083
2365	2.389	10.045	1	5.365	10.045
2365	2.485	11.122	1	6.442	11.122
2365	2.445	12.045	1	7.365	12.045
4384	2.859	0.096	0	0	0
4384	1.532	1.039	0	0	0
4384	1.590	1.726	1	0	1.726
4384	1.969	3.128	1	1.402	3.128
4384	1.684	4.282	1	2.556	4.282
4384	2.625	5.724	1	3.998	5.724
4384	2.583	6.024	1	4.298	6.024

words, not equations—the rationale for each. We recommend these steps because, as we demonstrate, the easiest models to specify may not display the type of discontinuity you expect to find.

We illustrate this approach using the high school dropout wage data of sections 5.2.1 and 5.3.3. There, our level-1 individual growth model expressed the natural logarithm of individual i's wages at time j (LNW_{ij} or, more generally, Y_{ij}) as a linear function of work experience since labor force entry ($EXPER_{ij}$):

$$Y_{ij} = \pi_{0i} + \pi_{1i}EXPER_{ij} + \varepsilon_{ij}, \tag{6.1a}$$

where $\varepsilon_{ij} \sim N(0, \sigma_\varepsilon^2)$. We also identified three additional predictors deserving inclusion: highest grade completed ($HGC - 9$), race ($BLACK$), and local area unemployment rate ($UERATE - 7$). To focus on specification of a discontinuous level-1 model, we have temporarily set these latter predictors aside. When we fit the postulated alternatives to data (in section 6.1.2), we will quickly reintroduce them.

Table 6.1 updates the person-period data set excerpted in table 5.3.

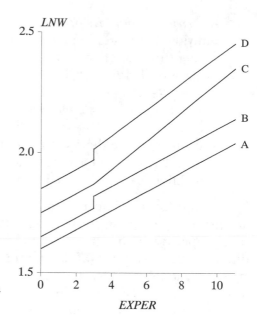

Figure 6.1. Comparing a linear change trajectory with three potential discontinuous change trajectories. Model A is a linear change trajectory, Model B postulates a shift in level but not slope, Model C postulates a shift in slope but not level, Model D postulates a shift in both level and slope.

Dropout 206 appears in the earlier table; dropouts 2365 and 4384 are new. The time-varying predictor, GED_{ij}, indicates whether the record for individual i at time j is "pre" or "post" GED receipt. Because pursuit of a GED is a personal decision, many high school dropouts do not receive it; those who do, earn it at different times. In this sample, 581 dropouts did not receive a GED; among the remaining 307, the *timing* of GED attainment varies. Dropout 206 did not earn a GED during data collection; his three values of this predictor remain at 0. Dropout 2365 received a GED 4.679 years after labor-force entry; his values of *GED* begin at 0 (for *EXPER* of 0.660, 1.679, 2.737, and 3.679) and change to 1 thereafter. Dropout 4384 received his GED 1.726 years after labor-force entry; his first two values of *GED* are 0 and his later ones are 1. We will soon describe the remaining predictors.

How might GED receipt affect individual i's wage trajectory? Figure 6.1 offers four plausible alternatives. The simplest answer—not at all—leads to the linear trajectory (A) with no discontinuity. If GED has an effect, it may take different forms. Upon receipt of a GED, we might find:

- *An immediate shift in* elevation, *but no shift in* slope. In trajectory B, individual i's wages increase abruptly upon GED receipt, but his subsequent rate of change is unaffected. This means that the *elevation* of his level-1 trajectory jumps, but its *slope* in the pre- and post-GED epochs remain the same.

- *An immediate shift in* slope, *but no shift in* elevation: In trajectory C, individual *i*'s wages remain stable upon GED receipt, but his subsequent rate of change increases. This means that the elevation of his level-1 trajectory is no higher at GED receipt, but its slope in the pre- and post-GED epochs differs.
- *Immediate shifts in both* elevation *and* slope: In trajectory D, individual *i*'s wages change in two ways as a result of GED receipt: they abruptly rise *and* their subsequent rate of change increases. This means that both the elevation and the slope of the level-1 trajectory differ pre- and post-GED receipt.

Even if we assume linear segments pre- and post-GED, these options are just the beginning. In sketching these trajectories, for example, we do not specify whether the magnitude of any shift (in elevation or slope) differs by the *timing* of GED attainment. If timing is irrelevant, we would specify a model in which the magnitude of any shift is constant regardless of when the GED was earned. But might the effect of GED attainment decline over time, if employers use work experience, not education, to signal background competence? This would lead to a growth model in which the magnitude of any GED shift diminishes over time.

As this discussion suggests, the array of possible discontinuous trajectories is vast. We do not attempt to catalogue them all, focusing instead on the major options in figure 6.1. By carefully walking through these alternatives, we hope to illustrate the general principles with sufficient clarity so that you can apply these ideas—with appropriate customization—to your own research.

Including a Discontinuity in Elevation, not Slope

We begin with the simplest type of discontinuity—one that immediately affects a trajectory's elevation but not its slope. In trajectory B, GED receipt immediately "bumps up" individual *i*'s wage trajectory, but has no effect on his subsequent rate of change. We can postulate a level-1 individual growth model of this type by adding the time-varying predictor GED_{ij} to the level-1 linear-change model in equation 6.1:

$$Y_{ij} = \pi_{0i} + \pi_{1i}EXPER_{ij} + \pi_{2i}GED_{ij} + \varepsilon_{ij}. \tag{6.2}$$

Because GED_{ij} distinguishes the pre- and post-GED epochs *for* individual *i*, it permits the elevation of his trajectory to differ upon GED receipt. The individual growth parameter π_{2i} captures the magnitude of this shift. Because GED takes on only two values—0 and 1—this magnitude is identical regardless of when the certificate is earned.

To verify that this growth model demonstrates the postulated discontinuity, substitute the two values of GED_{ij}. Before individual i passes the exam, $GED_{ij} = 0$, yielding the trajectory's pre-GED portion: $Y_{ij} = \pi_{0i} + \pi_{1i} EXPER_{ij} + \varepsilon_{ij}$. If, and when, individual i passes the exam, GED_{ij} becomes 1, yielding its post-GED portion:

$$Y_{ij} = \pi_{0i} + \pi_{1i} EXPER_{ij} + \pi_{2i}(1) + \varepsilon_{ij}$$
$$= (\pi_{0i} + \pi_{2i}) + \pi_{1i} EXPER_{ij} + \varepsilon_{ij}$$

We have two line segments with identical slopes, π_{1i}, but different intercepts: π_{0i}, pre-GED; $(\pi_{0i} + \pi_{2i})$, post-GED. This confirms that the individual growth parameter associated with GED_{ij}, π_{2i}, describes the magnitude of the hypothesized shift in elevation at GED receipt.

The top left panel of figure 6.2 plots a hypothetical true level-1 trajectory for a dropout from this population who received his GED three years after labor-force entry. To facilitate interpretation, the plot also includes a dashed line segment that continues the pre-GED portion of the trajectory into the post-GED era. Known as a *counterfactual*, this line segment illustrates what his wage trajectory *would have been* had he not earned his GED in year 3. Comparison of his post-GED trajectory and his counterfactual highlights the model's discontinuity. Although counterfactuals may appear unnecessary in simple cases like this, we will soon demonstrate their importance when working with more complex models.

Including a Discontinuity in Slope, not Elevation

To specify a level-1 individual growth model that includes a discontinuity in slope, not elevation, you need a different time-varying predictor. Unlike GED, this predictor must clock the passage of time (like *EXPER*). But unlike *EXPER*, it must do so within only one of the two epochs (pre- or post-GED receipt). Adding a second temporal predictor allows each individual change trajectory to have two distinct slopes: one before the hypothesized discontinuity and another after.

Construction of a suitable time-varying predictor to register the desired discontinuity is often the hardest part of model specification. To create a trajectory that differs only in slope, not elevation, we use $POSTEXP_{ij}$, which clocks labor force participation from the day of GED attainment (see table 6.1). Before individual i earns a GED, *POSTEXP* is 0. On the day a GED is earned, *POSTEXP* remains at 0. On the very next day, its values begin to climb, traveling in concert with the primary temporal predictor, here *EXPER*.

To clarify how *POSTEXP* operates, and its crucial relationship to *EXPER*, examine the data for dropout 2365. His first five records reveal

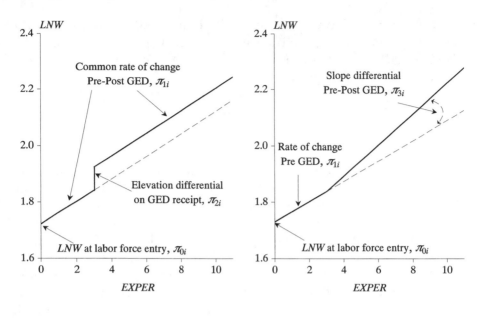

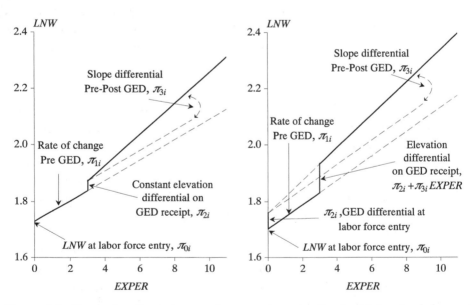

Figure 6.2. Alternative discontinuous change trajectories for the high school dropout wage data.

that he earned his GED 4.679 years after labor-force entry (the value of *EXPER* in the record when *GED* switches from 0 to 1). In the following record, after 5.718 years, *POSTEXP* = 1.038, the length of time since GED receipt. A fundamental feature of *POSTEXP*—indeed, any temporal predictor designed to register a shift in slope—is that the difference between each non-zero pair of consecutive values must be numerically identical to the difference between the corresponding pair of values for the basic temporal predictor (here, *EXPER*). To verify this behavior, examine the remaining person-period records for dropout 2365: the first pair differs by 1, the following by 1.154; the final by 0.923. This identical cadence ensures that these predictors move in lockstep, a feature we soon exploit. Notice, too, that because the timing of GED attainment is person-specific, the cadence of $POSTEXP_{ij}$ is also person-specific. For dropout 4384, for example, *POSTEXP* remains at 0 until his fourth record, when *EXPER* is 3.138 and *POSTEXP* is 1.402 (a difference of 1.726, the time when he earned his GED). If someone does not earn a GED, such as dropout 206, $POSTEXP_{ij}$ remains at 0 for every person-period record.

To postulate a level-1 individual growth trajectory with a discontinuity in slope but not level, we add this second temporal predictor to the basic model in equation 6.1:

$$Y_{ij} = \pi_{0i} + \pi_{1i}EXPER_{ij} + \pi_{3i}POSTEXP_{ij} + \varepsilon_{ij}. \tag{6.3}$$

To verify that this growth model displays the postulated discontinuity, we again divide the trajectory into its two components. Before GED receipt, $POSTEXP_{ij} = 0$, which yields the familiar linear-change trajectory— Pre-GED: $Y_{ij} = \pi_{0i} + \pi_{1i}EXPER_{ij} + \varepsilon_{ij}$. After GED receipt, we cannot eliminate $POSTEXP_{ij}$ because it takes on many different values. Instead, we find another equation with the same intercept as the pre-GED segment but two "slopes"—Post-GED: $Y_{ij} = \pi_{0i} + \pi_{1i}EXPER_{ij} + \pi_{3i}POSTEXP_{ij} + \varepsilon_{ij}$. Each slope assesses the effect of work experience, but it does so from a different origin: (1) π_{1i} captures the effects of *total* work experience (measured from labor force entry); and (2) π_{3i} captures the *added* effect of post-GED work experience (measured from GED receipt).

Key to understanding how these two slopes reflect the postulated discontinuity is the recognition that once someone earns a GED, his values of both *EXPER* and *POSTEXP* *increase at the same exact rate* (even though their values differ). A one-unit increase in one predictor parallels a one-unit increase in the other. This behavior, a consequence of our variable construction strategy, simplifies interpretation of the associated parameters. Before GED receipt, the trajectory's slope—which assesses the difference in log wages for a one-unit difference in time—is π_{1i}. After GED receipt, we can add these slopes together because a one-unit increase in *EXPER* is accompanied by a one-unit increase in *POSTEXP*. As a result, the post-GED trajectory has a slope of $(\pi_{1i} + \pi_{3i})$.

The important role of π_{3i} is highlighted in the top right panel of figure 6.2, which presents a hypothetical true level-1 change trajectory for a dropout who earned his GED three years after labor-force entry. As before, the dashed line represents the counterfactual—his postulated wage trajectory had he not earned a GED. π_{3i} does not represent his slope after GED attainment but rather the *increment* (or decrement) to what his slope would have been had he not earned his GED. If π_{3i} is 0, the slopes are the same; if π_{3i} is non-zero, the slopes differ.

Including Discontinuities in Both Elevation and Slope

We now postulate a level-1 individual growth model with discontinuities in both elevation and slope. But instead of offering one approach, we offer two. We begin by including all three predictors: *EXPER*, *GED*, and *POSTEXP*. We then specify a second growth model that includes *EXPER*, *GED*, and their statistical interaction (based upon ideas introduced in section 5.3.2). Although it may not be obvious, these two approaches, which appear similar, are not equivalent!

Let us begin by adding *GED* and *POSTEXP* to the basic level-1 individual growth model:

$$Y_{ij} = \pi_{0i} + \pi_{1i}EXPER_{ij} + \pi_{2i}GED_{ij} + \pi_{3i}POSTEXP_{ij} + \varepsilon_{ij}. \tag{6.4}$$

Once again, interpretation is clearer if we parse the model into its pre- and post-GED components. Before GED receipt, both *GED* and *POSTEXP* are 0, and we have the familiar simple linear-change trajectory:

$$\begin{aligned} Y_{ij} &= \pi_{0i} + \pi_{1i}EXPER_{ij} + \pi_{2i}(0) + \pi_{3i}(0) + \varepsilon_{ij} \\ &= \pi_{0i} + \pi_{1i}EXPER_{ij} + \varepsilon_{ij}. \end{aligned}$$

After degree attainment, *GED* becomes 1 and *POSTEXP* begins its steady climb in lockstep with *EXPER*. This yields a post-GED trajectory with a different intercept and two "slopes":

$$\begin{aligned} Y_{ij} &= \pi_{0i} + \pi_{1i}EXPER_{ij} + \pi_{2i}(1) + \pi_{3i}POSTEXP_{ij} + \varepsilon_{ij} \\ &= (\pi_{0i} + \pi_{2i}) + \pi_{1i}EXPER_{ij} + \pi_{3i}POSTEXP_{ij} + \varepsilon_{ij}. \end{aligned}$$

So unlike equation 6.3, which includes *EXPER* and *POSTEXP* and not *GED*, equation 6.4 describes a population in which the components of individual i's wage trajectory differ in both intercept and slope. As before, π_{1i} captures the effect of total work experience (*EXPER*) and π_{3i} captures the incremental effect of post-GED work experience (*POSTEXP*). With regard to the intercepts, however, π_{0i} now assesses individual i's log wages on his first day of labor force entry and π_{2i} assesses how much higher (or lower) his wages immediately rise on GED receipt.

The bottom left panel of figure 6.2 presents an illustrative change trajectory for a hypothetical dropout who earned his GED three years after labor-force entry. Before then, both *GED* and *POSTEXP* are 0, yielding a line with slope π_{1i}. Once he earns his GED, two things happen: *GED* changes from 0 to 1 *and POSTEXP* starts its climb in lockstep with *EXPER.* The change in *GED* creates the vertical shift in year 3; the additional annual change in *POSTEXP* creates the differential in slope. We include two counterfactuals in this display. The bottom dashed line describes what his post-GED trajectory would look like if GED receipt had no effect. The upper dashed line describes what his post-GED trajectory would look like if GED receipt affected his intercept but not slope. (Although there is a third counterfactual—what his post-GED trajectory would look like if GED receipt affected only his slope—we do not display this option.)

We now present an alternative approach for postulating a similar—but fundamentally different—individual growth model. Underlying this approach is the observation that some people might view *GED* and *POSTEXP* as nothing more than ordinary time-varying predictors. Although we understand this perspective, we believe that time-varying predictors like these, which may fundamentally alter the *shape* of the level-1 change trajectory, are different and deserve special attention.[1]

Nevertheless, how would we include *GED* in our level-1 individual growth model if it were just an "ordinary" time-varying predictor? In addition to the main effects in equation 6.2, we might explore interactions among the level-1 predictors (as in section 5.3.2). After all, including an interaction between *GED* and *TIME* would generate change trajectories with varying intercepts and slopes—the very properties we seek. We therefore offer the following level-1 individual growth model:

$$Y_{ij} = \pi_{0i} + \pi_{1i}EXPER_{ij} + \pi_{2i}GED_{ij} + \pi_{3i}(GED_{ij} \times EXPER_{ij}) + \varepsilon_{ij}. \qquad (6.5)$$

To understand how this model differs from its cousin in equation 6.4, let us substitute in the two values of *GED* and compute its separate segments. Pre-GED we have:

$$Y_{ij} = \pi_{0i} + \pi_{1i}EXPER_{ij} + \pi_{2i}(0) + \pi_{3i}(0 \times EXPER_{ij}) + \varepsilon_{ij}$$
$$= \pi_{0i} + \pi_{1i}EXPER_{ij} + \varepsilon_{ij}.$$

Post-GED we have:

$$Y_{ij} = \pi_{0i} + \pi_{1i}EXPER_{ij} + \pi_{2i}(1) + \pi_{3i}(1 \times EXPER_{ij}) + \varepsilon_{ij}$$
$$= (\pi_{0i} + \pi_{2i}) + (\pi_{1i} + \pi_{3i})EXPER_{ij} + \varepsilon_{ij}.$$

The intercepts of these two line segments differ by π_{2i}, the parameter associated with *GED*. The slopes differ by π_{3i}, the parameter associated with the *GED* by *EXPER* interaction.

The bottom right panel of figure 6.2 presents an illustrative trajectory for a hypothetical dropout who earned his GED three years after labor-force entry. Comparing the two bottom panels of figure 6.2 reveals some similarities between models. In both, π_{0i} assesses individual i's log wages at labor-force entry and π_{1i} assesses his annual growth in log wages before GED attainment.

But when it comes to the other parameters, there are some striking differences. First examine π_{3i}, the parameter associated with *POSTEXP* in equation 6.4 and the *GED* by *EXPER* interaction in equation 6.5. Oddly, even though this parameter is associated with *different* predictors, its interpretation is the same: it represents the increment (or decrement) to the slope in the post-GED epoch. In other words, π_{3i} consistently measures the effect of GED attainment on individual i's post-GED slope (even though it is associated with a different predictor in each model).

Next examine π_{2i}, the parameter associated with *GED*. Here we observe the *opposite* behavior. Even though this parameter is associated with the *same* predictor in each model, it does not represent the same quantity! In equation 6.4, π_{2i} assesses the magnitude of the instantaneous increment (or decrement) associated with GED attainment; in equation 6.5, π_{2i} assesses the magnitude of the increment (or decrement) associated with GED attainment at a particular—and not particularly meaningful—moment: the day of labor force entry!

These interpretative differences cannot be resolved by rescaling the temporal predictor, *EXPER*. This is because these growth models differ in fundamental ways. The unique features of the interaction model are that it:

- *Allows the magnitude of the instantaneous* GED *effect to vary over time.* Although the illustration in figure 6.2 depicts the elevation differential only for someone who earns his GED in year 3, the model allows this differential to vary with experience (over time). As shown in the figure, the general form of the differential is: Elevation differential on GED receipt: $\pi_{2i} + \pi_{3i}EXPER_{ij}$. In some situations, you may find a "jig" with a variable magnitude appealing; in others, you may not.
- *Does not include a single explicit parameter that assesses the instantaneous effect of* GED *attainment.* Although it is easy to estimate the GED effect at different values of total work experience (using the equation above), the model does not focus on this issue. This

realization has important implications when specifying level-2 submodels because π_{2i} does not assess the effect of GED attainment for individual i. Instead it assesses what the effect of GED attainment would be were individual i to have earned his GED on the day of labor-force entry.

These differences are more than cosmetic. Although both models include discontinuities in elevation and slope, they reflect fundamentally different assumptions about the behavior of the wage trajectories.

Which model better represents discontinuities in both elevation and slope? Not surprisingly, this question has no universal answer. As usual, theory should be your foremost guide. Should the instantaneous effect of GED receipt be constant or time-varying? Even when competing theories support each approach—as they do here—we still focus on substantive considerations. As we need empirical evidence to move the discussion forward, let us now postulate the associated level-2 submodels and fit these discontinuous trajectories to our wage data.

6.1.2 Selecting among the Alternative Discontinuous Models

Which of these hypothesized discontinuous change trajectories best suits our wage data? To answer this question, we fit a taxonomy of multilevel models for change. Because we have previously uncovered important effects of level-1 predictors *EXPER* and (*UERATE* − 7) and level-2 predictors (*HGC* − 9) and *BLACK* in table 5.8 and section 5.3.3, we begin with a more elaborate model for change than is usual.

Our "baseline" level-1 individual growth model contains the main effects of predictors *EXPER* and (*UERATE* − 7) at level-1:

$$Y_{ij} = \pi_{0i} + \pi_{1i}EXPER_{ij} + \pi_{2i}(UERATE_{ij} - 7) + \varepsilon_{ij}, \qquad (6.6a)$$

and the effects of (*HGC* − 9) and *BLACK* on initial status and rate of change at level-2:

$$\pi_{0i} = \gamma_{00} + \gamma_{01}(HGC_i - 9) + \zeta_{0i}$$
$$\pi_{1i} = \gamma_{10} + \gamma_{12}BLACK_i + \zeta_{1i} \qquad (6.6b)$$
$$\pi_{2i} = \gamma_{20},$$

where:

$$\varepsilon_{ij} \sim N(0, \sigma_\varepsilon^2) \text{ and } \begin{bmatrix} \zeta_{0i} \\ \zeta_{1i} \end{bmatrix} \sim N\left(\begin{bmatrix} 0 \\ 0 \end{bmatrix}, \begin{bmatrix} \sigma_0^2 & \sigma_{01} \\ \sigma_{10} & \sigma_1^2 \end{bmatrix} \right). \qquad (6.6c)$$

Notice that as in section 5.3.3, we fix the effect of (*UERATE* − 7) at level-2.

Table 6.2 briefly summarizes a taxonomy of discontinuous multilevel models for change fitted to the wage data. Each row presents a different multilevel model (labeled alphabetically in column 1). The table lists the fixed and random effects present in each model (columns 2 and 3), counts the number of fixed effects and variance components included (columns 4 and 5), lists the model deviance statistic (column 6), and provides the difference in deviance between the current model and a comparison model, along with the degrees of freedom associated with the test of difference in fit (column 7). Because we are comparing multilevel models for change that differ in both fixed effects and variance components, we use full ML for model fitting. The first model listed, Model A, is the "baseline." As you can see from both equation 6.6 and the table, the model contains five fixed effects: (1) the intercept, γ_{00}, (2) the effect of $(HGC - 9)$ on initial status, γ_{01}, (3) the main effect of $EXPER$, γ_{10}, (4) the effect of $BLACK$ on rate of change, γ_{12} (which shows up in the table in its "composite model" formulation as an interaction between $BLACK$ and $EXPER$), and (5) the main effect of $(UERATE - 7)$, γ_{20}. Model A also contains the four variance components listed in equation 6.6c: (1) the level-1 residual variance, σ_{ε}^2, (2) the level-2 variance of initial status, σ_0^2, (3) the level-2 variance of rate of change, σ_1^2, and (4) the level-2 covariance of initial status and rate of change, σ_{01}. Model A's deviance statistic is 4830.5 with a total of nine parameters.

Each subsequent model in the table builds systematically upon this baseline. Model B adds a discontinuity in elevation (but not slope) by including fixed and random effects for GED. Comparing its deviance statistic to that of Model A (last column) reveals a difference of 25.0 ($p < .001$). This suggests that the level-1 log wage trajectory does indeed display a discontinuity in elevation upon GED receipt. To determine whether the magnitude of this discontinuity varies across individuals, Model C excludes the three variance/covariance components associated with GED in Model B. Because its fit is significantly worse ($p < 0.05$), we retain those terms.

Models D and E explore a discontinuity in slope, not elevation. Although Model D, which includes both fixed and random effects for $POSTEXP$, represents a significant improvement over Model A ($p < .01$), Model E, which eliminates the associated variance component, fits nearly as well ($p > .25$). This suggests we might easily remove the variance component for $POSTEXP$ from the model. Before doing so, however, let us fit models with discontinuities in both elevation and slope because we will soon find that once the main effect of GED is included, the variance component associated with $POSTEXP$ differs.

Table 6.2: Comparison of fitting alternative discontinuous change trajectories to the high school dropout wage data ($n = 888$)

Model	Fixed effects	Variance components (in addition to σ^2_ε)	n parameters for... Fixed effects	n parameters for... Variance components	Deviance	Comparison model: ΔDeviance (df)
A	Intercept, *EXPER, HGC − 9, BLACK × EXPER, UERATE − 7*	Intercept, *EXPER*	5	4	4830.5	—
B	Model A + *GED*	Intercept, *EXPER, GED*	6	7	4805.5	A: 25.0*** (4)
C	Model B	Model B w/o *GED*	6	4	4818.3	B: 12.8** (3)
D	Model A + *POSTEXP*	Intercept, *EXPER, POSTEXP*	6	7	4817.4	A: 13.1** (4)
E	Model D	Model D w/o *POSTEXP*	6	4	4820.7	D: 3.3 (ns) (3)
F	Model A + *GED* and *POSTEXP*	Intercept, *EXPER, GED, POSTEXP*	7	11	4789.4	B: 16.2** (5) / D: 28.1*** (5)
G	Model F	Model F w/o *POSTEXP*	7	7	4802.7	F: 13.3** (4)
H	Model F	Model F w/o *GED*	7	7	4812.6	F: 23.3*** (4)
I	Model A + *GED* and *GED × EXPER*	Intercept, *EXPER, GED, GED × EXPER*	7	11	4787.0	B: 18.5*** (5)
J	Model I	Model I w/o *GED × EXPER*	7	7	4804.6	I: 17.6** (4)

~$p < .10$; *$p < .05$; **$p < .01$; ***$p < .001$.

Model A is identical to linear trajectory shown as Model A of table 5.8. Models B through J add discontinuities in elevation (B and C), rate of change (D and E), or both (F through J) at the moment of GED attainment. Parameter estimates and standard errors for Model F are tabulated in table 6.3.

Note: Full ML, SAS Proc Mixed.

Model F includes discontinuities in both elevation (through *GED*) and slope (through *POSTEXP*), each entered as fixed effects and variance components. To evaluate whether each predictor—on its own—merits inclusion, we make two comparisons, with: (1) Model B, which includes *GED*, so we can evaluate the effect of *POSTEXP*; and (2) Model D, which includes *POSTEXP*, so we can evaluate the effect of *GED*. In each case, we find evidence to support the predictor's inclusion, suggesting the need for a level-1 trajectory that includes both types of discontinuity.

Models G and H investigate whether we can simplify Model F by eliminating one or both variance components for *POSTEXP*. Given that Model E suggested that the variance component for *POSTEXP* might be unnecessary, these comparisons are especially important. Because each test rejects (at the .01 level for *POSTEXP* and the .001 level for *GED*), we continue with Model F, which includes variance components for both *GED* and *POSTEXP*.

Models I and J also include discontinuities in elevation and slope but they allow the magnitude of the GED differential to vary over time. Comparison with the relevant main effects model, B, confirms the need for the interaction. Comparison with Model J, which removes the interaction term's random effect, confirms the need for the associated variance components. Can we use the deviance statistic for this model, which is slightly smaller than Model F (4787.0 vs. 4789.3), to conclude that its fit is superior? However tempting this may be, we cannot do so as one model is not nested in the other. Although the associated AIC and BIC statistics would give a trivial nod to Model I (as the number of parameters remains unchanged), we share the original researchers' preference for Model F. This decision rests upon two considerations: (1) there is no reason to expect that the elevation differential associated with GED attainment should vary over time; and (2) this specification leads to level-2 models with greater substantive interest.

Before examining Model F in depth, we note that we fit six other multilevel models for change not shown in table 6.2. Each explored the possibility that one of the discontinuities—in either elevation or slope—was impacted by the substantive predictors (*HGC* − 9), *BLACK*, or (*UERATE* − 7). No effects were found. Although this may suggest an equitable society, it also indicates that GED attainment does not allow dropouts who leave school early, live in areas of high unemployment, or are Black to catch up with their peers.

Detailed results for Model F appear in table 6.3. To conserve space, we present estimates of only the fixed effects and variance components (not the associated covariance components). Figure 6.3 presents trajectories for four prototypical dropouts who obtained the GED after three years:

Table 6.3: Results of fitting Model F of table 6.2, a trajectory with discontinuities in elevation and slope to the high school dropout wage data ($n = 888$)

		Parameter	Estimate
Fixed Effects			
Composite model	Intercept	γ_{00}	1.7386***
			(0.0119)
	$(HGC - 9)$ on initial status	γ_{01}	0.0390***
			(0.0062)
	$(UERATE - 7)$	γ_{20}	−0.0117***
			(0.0018)
	$EXPER$ (rate of change)	γ_{10}	0.0415***
			(0.0028)
	$BLACK$ on rate of change	γ_{12}	−0.0196***
			(0.0045)
	GED	γ_{30}	0.0409~
			(0.0220)
	$POSTEXP$	γ_{40}	0.0094~
			(0.0055)
Variance Components			
Level-1:	within-person	σ_ε^2	0.0939***
Level-2:	In initial status	σ_0^2	0.0413***
	In rate of change	σ_1^2	0.0014***
	In GED discontinuity	σ_3^2	0.0163***
	In $POSTEXP$	σ_4^2	0.0034**
Goodness-of-fit			
	Deviance		4789.4
	AIC		4825.5
	BIC		4911.6

$\sim p < .10$; $* p < .05$; $** p < .01$; $*** p < .001$.

Note: Full ML, SAS Proc Mixed. Also note that all relevant covariance components were estimated (even though they are not displayed to conserve space).

Blacks and Whites/Latinos, who dropped out in 9th grade or 12th grade, and who live in communities where local area unemployment rates remain stable at 7%. On labor-force entry, a White male who dropped out in 9th grade and who lives in a community with an unemployment rate of 7% is expected to earn an hourly log wage of 1.7386 ($5.69 in constant 1990 dollars). Before GED attainment, log wages rise annually by 0.0415 (4.2% in raw wages). Upon GED receipt, log wages rise immediately by 0.0409 (4.2%) and then annually by 0.0415 + 0.0094 = 0.0509 (5.2%). Because the *GED* and *POSTEXP* fixed effects are not significantly different from 0 at conventional levels, the GED increment and the slope pre- and post-GED receipt may be no different for the *average* individual.

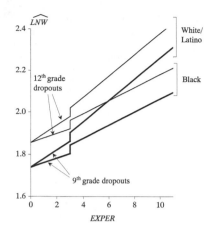

Figure 6.3. Displaying the results of fitting a discontinuous change trajectory to the high school dropout wage data. Log wage trajectories from table 6.3 for four prototypical dropouts—Blacks and Whites/Latinos who dropped out in 9th and 12th grades—each of whom earned a GED after three years in the labor force.

We retain both terms, however, because their associated random effects *are* statistically significant, indicating that for at least some people, GED receipt has either (or both) an immediate impact and a subsequent effect on wage growth. The effects of the three substantive predictors—local area unemployment rate, race, and highest grade completed—remain similar to those found in chapter 5.

6.1.3 Further Extensions of the Discontinuous Growth Model

It is easy to generalize these strategies to models with other discontinuities. Quality of theory and richness of data usually present greater obstacles to model development than do computer algorithms and algebraic constraints. You can parse each person's trajectory into discrete epochs by adding multiple discontinuities. You can include a discontinuity that occurs at a common point in time for everyone under study. Rather than catalogue the many possibilities, here we describe two simple extensions.

Dividing TIME into Multiple Phases

You can divide *TIME* into *multiple* epochs, allowing the trajectories to differ in elevation (and perhaps slope) during each. Suppose, for example, that some GED recipients subsequently graduated from community college. If we hypothesized a constant effect of work experience over time, we would add another time-varying predictor, *CC*, to the level-1 individual growth model: $Y_{ij} = \pi_{0i} + \pi_{1i}EXPER_{ij} + \pi_{2i}GED_{ij} + \pi_{3i}CC_{ij} + \varepsilon_{ij}$. Taken together, *GED* and *CC* create three distinct phases: one pre-GED and pre-CC, one post-GED but pre-CC, and a third post-GED and post-CC. π_{2i} assesses the immediate shift associated with GED receipt; π_{3i} assesses the immediate shift upon community college graduation.

You can also allow temporal slopes to differ across epochs. Researchers who conduct phased randomized experiments following individuals into, and out of, treatments can use such models. Consider a three-phase longitudinal study: during the *baseline* period, patients take their usual medication; during the *experimental* phase, they take an alternative; during *follow-up*, they return to the original drug. This "regression discontinuity" design leads to a level-1 individual growth trajectory with multiple discontinuities. If symptoms are linear with *TIME*, but elevation and slope differ with medication change, you might postulate that:

$$
\begin{aligned}
Y_{ij} = \pi_{0i} + \pi_{1i}TIME_{ij} + \pi_{2i}PHASE1_{ij} + \pi_{3i}TIMEP1_{ij} \\
+ \pi_{4i}PHASE2_{ij} + \pi_{5i}TIMEP2_{ij} + \varepsilon_{ij}.
\end{aligned}
\tag{6.7}
$$

The two epoch dummies, *PHASE1* and *PHASE2*, distinguish the two experimental phases from the baseline; the two additional temporal predictors, *TIMEP1* and *TIMEP2*, measure the additional passage of time during subsequent phases. These "phase" and "temporal" predictors function like *GED* and *POSTEXP* in the high school dropout data: *PHASE1* and *TIMEP1* allow for discontinuities in elevation and slope during phase 1; *PHASE2* and *TIMEP2* allow for additional discontinuities in phase 2. By specifying level-2 models and estimating fixed effects, you can address questions about the treatment's immediate (π_{2i}, π_{3i}) and long-term (π_{4i}, π_{5i}) effects.

Phased models like equation 6.7 are useful in naturalistic studies if you expect people to change in a step- or stage-like fashion. Psychologists often hypothesize such patterns when studying cognitive, affective, and moral development. With little adjustment, equation 6.7 could represent development across three contiguous cognitive or moral stages. Precipitating events can also lead you to divide time into discrete periods: transitions across grades and schools, entry or exit into prison, the deaths of parents or spouses. Although the nature of the discontinuity will be domain specific, the strategies for model specification are straightforward extensions of the principles above.

Discontinuities at Common Points in Time

In some data sets, the timing of the discontinuity will not be person-specific; instead, everyone will experience the hypothesized transition at a common point in time. You can hypothesize a similar discontinuous change trajectory for such data sets by applying the strategies outlined above.

Suppose, for example, we assessed students three times a year (fall, winter, and spring) for each of three grades (third, fourth, and fifth). Rather than postulating a linear trajectory, we might hypothesize a discontinuous alternative. If we thought, for example, that students made general progress through grades, but that within a grade, there might be even steeper progress, we might postulate that: $Y_{ij} = \pi_{0i} + \pi_{1i} (GRADE_{ij} - 4) + \pi_{2i}SEASON_{ij} + \varepsilon_{ij}$, where both $(GRADE_{ij}-4)$ and $SEASON_{ij}$ take on the values −1, 0, and 1. In this model, π_{0i} represents individual i's true test score in the middle of fourth grade, π_{1i} represents his true rate of linear growth across grades, and π_{2i} represents any additional linear growth that occurs during the academic year. This model would yield a zigzag trajectory. Alternatively, if we thought that growth would generally be linear but that fall readings might be low because test scores may drop during the summer when children are out of school, we could postulate that: $Y_{ij} = \pi_{0i} + \pi_{1i} (\text{ASSESSMENT\#}_{ij} - 1) + \pi_{2i}FALL_{ij} + \varepsilon_{ij}$. In this model, π_{0i} represents individual i's true initial test score at the beginning of third grade (controlling for the fact that it is a fall assessment), π_{1i} represents his true rate of linear growth across assessments, and π_{2i} represents the potential decrement in test scores associated with the fall assessment. This model yields an underlying linear trajectory punctuated by fall-specific drops.

Please treat these examples not as explicit directives but as inspirational starting points. If you have reason to hypothesize a particular type of discontinuity, you should develop a customized model that reflects your hypothesis and not adopt an "off-the-shelf" parameterization that may not. Once you move away from the standard linear change trajectory, your options grow as does your burden of proof. Good theory and a compelling rationale should always be your guides.

6.2 Using Transformations to Model Nonlinear Individual Change

We now consider smooth, but nonlinear, individual change trajectories. Certainly the easiest strategy for fitting such models is to transform either the outcome, or *TIME*, in the level-1 submodel so that a growth model that specifies linear change in the transformed outcome or predictor will suffice. When confronted by obviously nonlinear trajectories, we usually begin with the transformation approach for two reasons. First, a straight line—even on a transformed scale—is a simple mathematical form whose two parameters have clear interpretations. Second, because the metrics of many variables are ad hoc to begin with, transformation to another ad hoc scale may sacrifice little. If the original scale lacks well-accepted intu-

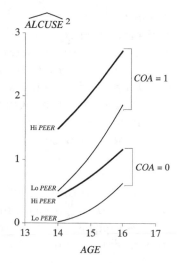

Figure 6.4. Re-expressing the prototypical trajectories in figure 4.3 for *ALCUSE* on the outcome's original scale. These prototypical trajectories are identical to those in figure 4.3 except that here we have squared the model's predicted values to reverse the effect of taking square roots before statistical analysis.

itive anchors, you lose nothing by using a transformed alternative. It matters not whether you conduct analyses in one arbitrary world (the original metric) or another (e.g., the "square root" metric). Either metric allows you to track individuals over time and to identify predictors associated with their differential patterns of change.

To support these assertions, reconsider the alcohol use data of chapter 4 and ask: What would those findings look like if we "de-transformed" the outcome back to its original nine-point scale? Figure 6.4 displays de-transformed trajectories based on the fitted trajectories in the final panel of figure 4.3. We obtained these trajectories by squaring the predicted values from the linear model fit to the square root data (thereby reversing the transformation, as "squaring" is the inverse of "square rooting"). As in figure 4.3, we display prototypical trajectories for children of alcoholics and nonalcoholics at low and high values of peer alcohol use. Reversing the transformation returns us to the original nine-point metric. Because this changes the scale of the vertical axis, the once linear change trajectories are now curved.

Despite this transition between metrics, the findings remain: children of alcoholics initially drink more but are no more likely to increase their drinking over time. But the "slopes" of the detransformed trajectories defy a "single number" summary. In the square root metric we originally analyzed, "annual rate of change" was meaningful because the trajectory was linear in the transformed world. But once we detransform back into the original measurement metric, the trajectory is curved and the rate of change is no longer constant over time: alcohol use increases more rapidly as time passes. Although you might think there is a conflict between these representations, each interpretation is correct *in its own world*. In the transformed metric, change in alcohol use is linear—its rate of change is *constant* over time. In the original metric of measurement, which we enter by detransformation, change in alcohol use is nonlinear— it *accelerates* over time. Our current formulation of the multilevel model for change assumes a level-1 linear change model. If change is not linear over time, we can seek an alternative metric for the outcome, or for time, in which this assumption holds. In the transformed world, the methods we have developed work well and we violate no assumptions. We then display the findings back in the detransformed metric of the original outcome to simplify communication of the findings.

This suggests a simple general strategy for modeling nonlinear change that capitalizes on the best of both worlds. Transform the outcome (or the level-1 *TIME* predictor) so that individual change becomes linear. Fit the multilevel model for change and test hypotheses in the transformed world, then detransform and present findings back in the original metric. The key to the success of this strategy is selection of a suitable transformation, a topic to which we now turn.

6.2.1 The Ladder of Transformations and the Rule of the Bulge

You can identify a suitable transformation for "correcting" nonlinearity in longitudinal data using the same methods you use for "correcting" nonlinearity in cross-sectional data. Rather than examining a single "outcome *vs.* predictor" plot, however, you examine multiple empirical growth plots, one for each sample member, seeking a transformation that works decently for most everyone under study.

A useful aid in this process is Mosteller and Tukey's (1977) ordered list of transformations known as the ladder of powers. On the left side of figure 6.5, we present our version of their ladder for transforming a generic variable "V," which appears on the center rung. Transformations in the upper half of the ladder, above V, are positive powers greater than 1, including the *square*, the *cube*, and the *fourth power*. Transformations in the lower half, below V, include the *logarithm*, *fractional powers* (representing the *square root*, the *cube root*, etc.), and *negative powers* (*inverses*). When we use a transformation in the upper half of the ladder (e.g., V^2, V^3), we say we move "up" in V; when we use a transformation in the lower half (e.g., $LOG(V)$, $1/V$), we say we move "down" in V.

To identify a suitable transformation, inspect the collection of empirical growth plots and apply what Mosteller and Tukey call the *rule of the*

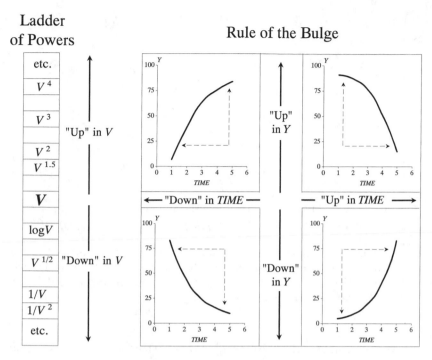

Figure 6.5. The ladder of transformations and the rule of the bulge. Guidelines for linearizing individual growth trajectories through judicious use of transformation.

bulge. We reprint their guidelines on the right side of figure 6.5. The idea is to match the general shape of the plots (discounting the effect of measurement error) to one of the four exemplars shown. You find linearizing transformations by moving "up" or "down" the ladder in the same direction(s) as the direction of the "bulge" in the exemplar. The arrows in figure 6.5 indicate the directions for each exemplar. In the upper left corner, the arrows point "up" in *Y* and "down" in *TIME*, suggesting that a curve with this shape can be linearized by moving "up" in *Y* (e.g., taking Y^2, Y^3 etc.) or "down" in *TIME* (e.g., taking *LOG(TIME)*, 1/*TIME* etc). In the bottom right corner, the arrows point "down" in *Y* and "up" in *TIME* suggesting that a curve with this shape can be linearized by moving "down" in *Y* or "up" in *TIME*. The further a transformation is from the ladder's center, the more dramatic its impact.

We suggest you experiment with several transformations before selecting one for analysis. The search process is hardly an exact science. Even for a single person, one transformation may not be equally successful at all points in time. As you eventually need to use the *same* transformation for everyone in your sample, selection involves some compromise so that,

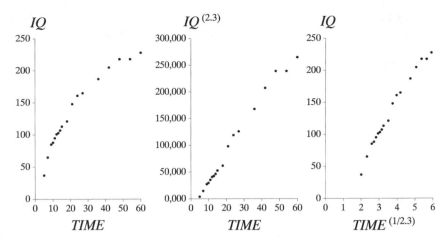

Figure 6.6. Comparing empirical growth plots for a single child in the Berkeley growth study. The left panel presents raw data; the middle panel presents the same data with the outcome, *IQ*, raised to the 2.3 power; the right panel presents the same data with the predictor, *TIME*, expressed as the 2.3th root of *AGE*.

overall, you can argue that the resultant transformed shape is linear for most everyone.

We illustrate this process in figure 6.6, which presents 20 waves of data for a single girl from the Berkeley Growth Study (Bayley, 1935). The left panel displays the child's cognitive trajectory on its original scale. Its curvilinear shape suggests that as she develops, her mental ability increases less rapidly—in other words, the curve decelerates. This matches the exemplar in the upper left corner of figure 6.5. To linearize this trajectory, we can either move "up" in Y (e.g., take Y^2, Y^3, etc.) or "down" in *TIME* (e.g., take $LOG(TIME)$, $1/TIME$, etc.) After trying several alternatives, we found that raising IQ to the 2.3rd power was a good compromise. The transformed trajectory appears in the middle panel. Notice its dogleg at about 20 months—a shift apparent in the original trajectory as well—which may be due to changes in the measurement method at this age. Transformation does not eliminate this discontinuity, but provides a reasonably linear change trajectory for both halves.

You can transform *either* the outcome *or TIME*, often using the inverse of the transformation that is best for the other. But applying the inverse of the "outcome" transformation to the predictor, or vice versa, will not produce the *identical* reduction in nonlinearity owing to differences in the range and scale of the variables and the presence of an intercept in the model. Such differences make it worth examining the effect of both types of transformation. For these data, taking the 2.3th *root* of age (shown in the right panel of figure 6.6) is not as successful in linearizing the trajectory as raising the outcome to the 2.3rd power. If both transformations are equally successful, the choice is yours. If one variable is measured on an easily understood or widely accepted scale—as *TIME* usually is—we recommend that you preserve its metric by transforming its partner. Here, a transformation of the cognitive outcome is more successful in removing nonlinearity and also preserves the metric of *TIME*.

We conclude by reiterating a caution mentioned in section 2.3.1. Notice that we examine *empirical growth plots* for each sample member (or a random subset), not the aggregate trajectory formed by joining within-occasion sample averages. However tempting it is to draw inferences about the shape of individual trajectories from the shape of the aggregate, their forms may not be identical. The forms are identical when change is linear with time but they may not be when change is nonlinear. Because you do not know the shape of the true individual trajectory—if you did, you wouldn't need to do this detective work—avoid this pitfall by always using *individual* plots to identify the shape of *individual* change. (We expand upon this point in section 6.4, when we introduce truly nonlinear trajectories.)

6.3 Representing Individual Change Using a Polynomial Function of *TIME*

We can also model curvilinear change by including several level-1 predictors that *collectively* represent a polynomial function of time. Although the resulting *polynomial growth model* can be cumbersome, it can capture an even wider array of complex patterns of change over time.

Table 6.4 presents an ordered series of polynomial growth models. Each relates the observed value of an outcome, *Y*, to *TIME* for individual *i* on occasion *j*. The first column labels the trajectory; the second presents the associated level-1 model; the last illustrates the trajectory's shape for the arbitrarily selected values of *Y*, *TIME*, and the individual growth parameters shown in the third column. As we add higher order functions of *TIME*, the true change trajectory becomes more complex. Below, we describe how to interpret results (section 6.3.1) and select among the alternatives (sections 6.3.2 and 6.3.3).

6.3.1 The Shapes of Polynomial Individual Change Trajectories

The "no change" and "linear change" models are familiar; the remaining models, which contain quadratic and cubic functions of *TIME*, are new. For completeness, we comment on them all.

Table 6.4: A taxonomy of polynomial individual change trajectories

		Illustrative example	
Shape	Level-1 model	Parameter values	Plot of the true change trajectory
No change	$Y_{ij} = \pi_{0i} + \varepsilon_{ij}$	$\pi_{0i} = 71$	
Linear change	$Y_{ij} = \pi_{0i} + \pi_{1i}TIME_{ij} + \varepsilon_{ij}$	$\pi_{0i} = 71$ $\pi_{1i} = 1.2$	
Quadratic change	$Y_{ij} = \pi_{0i} + \pi_{1i}TIME_{ij}$ $\quad + \pi_{2i}TIME_{ij}^2 + \varepsilon_{ij}$	$\pi_{0i} = 50$ $\pi_{1i} = 3.8$ $\pi_{2i} = -0.03$	
Cubic change	$Y_{ij} = \pi_{0i} + \pi_{1i}TIME_{ij}$ $\quad + \pi_{2i}TIME_{ij}^2 + \pi_{3i}TIME_{ij}^3$ $\quad + \varepsilon_{ij}$	$\pi_{0i} = 30$ $\pi_{1i} = 10$ $\pi_{2i} = -.2$ $\pi_{3i} = .0012$	
$\vdots$	$\vdots$	$\vdots$	$\vdots$

"No Change" Trajectory

The "no change" trajectory is known as a polynomial function of "zero order" because *TIME* raised to the 0th power is 1 (i.e., $TIME^0 = 1$). This model is tantamount to including a constant predictor, 1, in the level-1 model, as a multiplier of the sole individual growth parameter, the intercept, π_{0i}. The intercept represents the vertical elevation of the "no-change" trajectory at every point in time (71 in the example). Even though each trajectory is flat, different individuals can have different intercepts and so a collection of true "no change" trajectories is a set of vertically scattered horizontal lines. The "no change" trajectory is the level-1 submodel of the "unconditional means model" that we introduced in section 4.4.1. Here, we use the "no change" label to highlight its relationship with other polynomial trajectories.

"Linear Change" Trajectory

The "linear change" trajectory is known as a "first order" polynomial in time because *TIME* raised to the 1st power equals *TIME* itself (i.e., $TIME^1 = TIME$). Linear *TIME* is the sole predictor and the two individual growth parameters have the usual interpretations. This model allows each individual to possess a unique intercept and slope parameter that yield a collection of crisscrossing trajectories for a group of people. Associated level-2 models can link person-specific characteristics to interindividual heterogeneity in both intercept and slope.

"Quadratic Change" Trajectory

Adding $TIME^2$ to a level-1 individual growth model that already includes linear *TIME* yields a *second order* polynomial for *quadratic* change. Unlike a level-1 model that includes *only* $TIME^2$, a second order polynomial change trajectory includes two *TIME* predictors and three growth parameters (π_{0i}, π_{1i} and π_{2i}). The first two parameters have interpretations that are *similar*, but not identical, to those in the linear change trajectory; the third is new.

In the quadratic change model, π_{0i} still represents the trajectory's *intercept*, the value of *Y* when both predictors, here *TIME* and $TIME^2$, are 0. But π_{1i}, the parameter associated with *TIME*, does not represent a constant rate of change. Instead, it represents the *instantaneous rate of change* at one specific moment, when $TIME = 0.$[2] Although most people still use the "slope parameter" nomenclature, a quadratic change trajectory has no constant common slope. The rate of change changes smoothly over time. π_{2i}, the *curvature* parameter associated with level-1 predictor $TIME^2$, describes this changing rate of change. Hypothesizing a quadratic individual change trajectory allows you to formulate level-2 questions about interindividual differences in *intercept, instantaneous rate of change,* and *curvature.*

To develop your intuition about quadratic change, examine the sample trajectory in table 6.4. It has an instantaneous rate of change of 3.8 at *TIME* 0 and a curvature of −0.03. Because π_{1i} is positive, the trajectory initially rises, with true status having the intention of increasing by 3.8 in the first unit of time. But because π_{2i} is negative, this increase does not persist. With each passing unit of time, the magnitude of the outcome's rising value diminishes. In essence, π_{1i} and π_{2i} compete to determine the value of *Y*. The quadratic term will eventually win because, for numeric reasons alone, $TIME^2$ increases more rapidly than *TIME*. So, in this example, even though the linear term suggests that *Y* increases over time, the eventual domination of the quadratic term removes more than the linear term adds and causes the trajectory to peak and then decline.

Quadratic trajectories with a single "peak" are said to be *concave* to the time axis. The peak is called the "stationary point" because the slope momentarily goes to zero before reversing direction. Quadratic curves have one stationary point. If the curvature parameter is positive, the trajectory is *convex* to the time axis, with a single "trough." Whether positive or negative, the larger the magnitude of π_{2i}, the more dramatic its effect, rendering the curvature more extreme. The moment when the quadratic trajectory curve flips over, at either a peak or a trough, is $(-\pi_{1i}/2\pi_{2i})$, which in our example is a time of $-3.8/(2(-.03)) = 63.33$.

Higher Order Change Trajectories

Adding higher powers of *TIME* increases the complexity of the polynomial trajectory. The fourth row of table 6.4 presents a third-order polynomial that includes level-1 predictors *TIME*, $TIME^2$ and $TIME^3$. A third-order polynomial has two stationary points; here, one peak and one trough. A quartic polynomial, which adds $TIME^4$ to the cubic model, has three stationary points—either two peaks and one trough or two troughs and one peak depending on the parameters' signs. A fifth-order polynomial has four stationary points; a sixth order has five. By using higher order polynomials to represent individual change, you can represent trajectories of almost any level of complexity.

Interpretation of the individual growth parameters is more complex for higher order polynomials. Even the cubic model's parameters do not represent "initial status," "instantaneous growth rate," and "curvature" as they do in a quadratic. In general, we prefer the simpler representations; we use higher order polynomials only when other approaches fail.

In the next section, we describe strategies for selecting among competing polynomial forms. Before doing so, we inject a note of reality concerning the data collection demands that these models pose. The more complex the polynomial, the more waves of data you need to collect to be able to fit the trajectory to data. In a time-structured data set, you need at least one more wave of data per person than there are individual growth parameters in the level-1 individual growth model. A level-1 linear change trajectory requires at least three waves of data. A quadratic level-1 individual growth model requires at least four; a cubic at least five. And these are only the *minimum* requirements. Greater precision and power requires more waves. In analysis as in life, nothing comes without a cost.[3]

6.3.2 Selecting a Suitable Level-1 Polynomial Trajectory for Change

We illustrate strategies for selecting a level-1 polynomial change trajectory using data on 45 children tracked from first through sixth grade as part of a larger study reported by Keiley, Bates, Dodge, and Pettit (2000). Near the end of every school year, teachers rated each child's level of externalizing behavior using Achenbach's (1991) Child Behavior Checklist. The checklist uses a three-point scale (0 = rarely/never, 1 = sometimes, 2 = often) to quantify the frequency with which the child displays 34 aggressive, disruptive, or delinquent behaviors. The outcome, *EXTERNAL*, which ranges from 0 to 68, is the sum of these 34 scores.

Figure 6.7 presents empirical growth plots for 8 children. (For now, ignore the fitted trajectories and focus on the data points.) As a group, these cases span the wide array of individual change patterns in the data. Child D displays little change over time. Child C appears to decline linearly with age (at least through fourth grade). Children A, B, and G display some type of quadratic change, but their curvatures differ. For A, the curvature parameter appears negative; for B and G, positive. Child E may have two stationary points—a trough in second grade and a peak in fifth—suggesting a cubic trajectory. Children F and H may have three stationary points—a peak, a trough, and another peak—although with only six waves, it is difficult to distinguish true quartic change from occasion-specific measurement error.

When faced with this many different patterns, which polynomial trajectory makes sense? If there is no obvious winner, we suggest that you first adopt an exploratory approach and fit separate person-specific OLS models to each person's data. This process is relatively easy, albeit tedious:

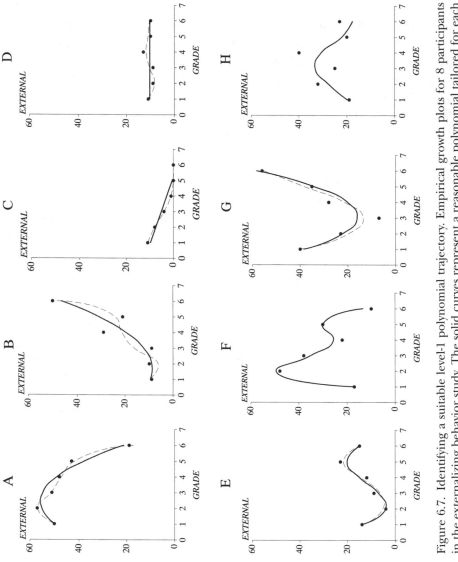

Figure 6.7. Identifying a suitable level-1 polynomial trajectory. Empirical growth plots for 8 participants in the externalizing behavior study. The solid curves represent a reasonable polynomial tailored for each child: a flat line for D; a linear trajectory for C; a quadratic trajectory for A, B, and G; a cubic for E; and a quartic for F and H. The dashed lines represent the highest order polynomial necessary—a quartic.

all you need do is create a set of temporal predictors that capture the requisite polynomial shapes—e.g., *TIME*, *TIME*^{2}, *TIME*^{3}, and so on—and then, for each child, use the set of predictors needed to represent the trajectory desired. The solid curves in figure 6.7 represent the choices just articulated: (1) "no change" for D; (2) linear for C; (3) quadratic for A, B and G; (4) cubic for E; and (5) quartic for F and H.

Comparison of observed and fitted values of *EXTERNAL* demonstrates the utility of this approach. Most of the fitted trajectories reasonably summarize each child's data record. But are ad hoc decisions like these optimal? Closer comparison of observed and fitted values suggests cause for concern. For child C, the small differences appear systematic. Perhaps he really has a quadratic trajectory with a wide flat trough that begins in sixth grade. Child H raises a different issue. Here, we fitted a quartic model (because of the two peaks and one trough), but the fitted trajectory seems quadratic (it has just one peak). Inspection of his regression results reveals that the cubic and quartic parameters are small and indistinguishable from zero. The imagined peaks and troughs may be measurement error. This suggests a need for parsimony when specifying polynomial trajectories, a recommendation we adopt more vigorously in coming sections.

But for now, we move in the opposite direction: fitting exploratory trajectories using a common more general shape. We do so for two reasons: (1) the decision-making process needed to fit custom trajectories to entire data sets can be tedious and counterproductive; and (2) we cannot easily specify a level-1 individual growth model unless we use a *common* shape for the trajectory across people. Instead of selecting a unique polynomial form for each child, we select the *highest order* polynomial needed to summarize individual change for *any child.* For the eight children in figure 6.7, we select a quartic because no child appears to need a higher order polynomial. While hardly parsimonious, a quartic can be fit easily to each child's data; the *data* will then demand the contribution of higher order terms as needed. If we use a quartic for Child A, for example, the estimated growth parameters for the cubic and quartic terms may be close, or equal, to 0.

The dashed curves in figure 6.7 display the results of fitting a quartic to each child's data. (Note that the solid lines already represent a quartic for children F and H.) This common trajectory simplifies implementation, but clearly overfits. While children E and G have fitted trajectories virtually identical to those specified using the case-specific approach, the others reveal more complex forms. Is this complexity necessary? The fitted trajectories for children A and B, which previously seemed quadratic, now have an extra "bump." And the fitted trajectory for child D, which seemed flat, now looks like a scaled down quartic!

Should you be parsimonious and potentially underestimate the trajectory's complexity or should you be cautious and potentially overfit? An answer to this question is clouded by the use of *sample* data to draw conclusions about *population* trajectories. When you inspect empirical plots, you try to account for measurement error, but this is easier said than done. For example, you may have been prepared to conclude that the true trajectories for A and B were quadratic, but now you might see them as more complex. So, too, the quartic hypothesized for child H may be unnecessarily complex. Fortunately, when exploratory analyses lead to conflicting conclusions, you can resolve this dilemma by comparing goodness-of-fit statistics across a series of models, as we now do.

6.3.3 Testing Higher Order Terms in a Polynomial Level-1 Model

Table 6.5 presents the results of fitting four models of increasing polynomial complexity at level-1 to the externalizing behavior data. Each was fit using Full IGLS in MLwiN. For simplicity, we include no other substantive predictors at either level-1 or level-2. Note, however, that as we increase the complexity of the fixed portion of the level-1 model we do add the associated random effects. We describe the rationale for this decision below as we describe the empirical results.

Let us begin with Model A, the "no-change" trajectory. The estimated grand mean is 12.96 ($p < .001$), which suggests that between first and sixth grades, the average child has a non-zero level of externalizing behavior. Examining the variance components, we find statistically significant variability both within-child (70.20, $p < .001$) and between-children (87.42, $p < .001$). We conclude that externalizing behavior varies from occasion to occasion and that children differ from each other.

Is this "no-change" trajectory adequate or should we add a linear *TIME* predictor to the level-1 individual growth model? We address this question by comparing Model A to the standard linear-change model (B). To facilitate interpretation, we express *TIME* as *GRADE*-1 so that the intercept (π_{0i}) refers to the level of externalizing behavior in first grade. We find that while the average child has a non-zero level of externalizing behavior in first grade (13.29, $p < .001$) this level does not change linearly over time on average (−0.13, *n.s.*). The statistically significant variance components ($\hat{\sigma}_0^2 = 123.52$, p < .001; $\hat{\sigma}_1^2 = 4.69$, p < .01) suggest, however, that children differ substantially from these averages. In other words, the *average* trajectory may be flat but many of the *individual* trajectories are not.

To determine whether Model B is preferable to Model A, we test the

Table 6.5: Comparison of fitting alternative polynomial change trajectories to the externalizing behavior data ($n = 48$)

	Parameter	Model A No change	Model B Linear change	Model C Quadratic change	Model D Cubic change
Fixed Effects					
Composite model					
Intercept (1st grade status)	γ_{00}	12.96***	13.29***	13.97***	13.79***
TIME (linear term)	γ_{10}		-0.13	-1.15	-0.35
$TIME^2$ (quadratic term)	γ_{20}			0.20	-0.23
$TIME^3$ (cubic term)	γ_{30}				0.06
Variance Components					
Level-1: Within-person	σ_{ε}^2	70.20***	53.72***	41.98***	40.10***
Level-2: In 1st grade status	σ_{0}^2	87.42***	123.52***	107.08***	126.09***
Linear term					
variance	σ_{1}^2		4.69**	24.60*	88.71
covar with 1st grade status	σ_{01}		-12.54*	-3.69	-51.73
Quadratic term					
variance	σ_{2}^2			1.22*	11.35
covar with 1st grade status	σ_{02}			-1.36	22.83~
covar with linear term	σ_{12}			-4.96*	-31.62
Cubic term					
variance	σ_{3}^2				0.08
covar with 1st grade status	σ_{03}				-3.06~
covar with linear term	σ_{13}				2.85
covar with quadratic term	σ_{23}				-0.97
Goodness-of-fit					
Deviance statistic		2010.3	1991.8	1975.8	1967.0
AIC		2016.3	2003.8	1995.8	1997.0
BIC		2021.9	2015.0	2014.5	2025.1

$\sim p < .10$; $* p < .05$; $** p < .01$; $*** p < .001$.
Model A is the "no change" trajectory; Model B is the linear change trajectory; Model C is the quadratic change trajectory; Model D is the cubic change trajectory.

Note: Full IGLS, MlwiN.

compound null hypothesis about the *set* of differences between models (in the linear growth rate, its associated variance component, and the extra covariance parameter, σ_{01}): H_0: $\gamma_{10} = 0$, $\sigma_1^2 = 0$, *and* $\sigma_{01} = 0$. As the difference in deviance statistics (18.5) far exceeds the 0.05 critical value of a χ^2 distribution on three *d.f.*, we reject H_0 and abandon the "no change" model.

You can use this same testing strategy to evaluate the impact of adding polynomial terms to the level-1 growth model, by comparing Model C to B and D to C, and so on. Before doing so, however, we draw your attention to a common dilemma that arises in model fitting. In Model B, although the variance component for linear growth (σ_1^2) is statistically significant, its associated fixed effect (γ_{10}) is not. This is not an inconsistency, but it requires interpretive care. The test for the variance component tells us that there is statistically significant variation in linear rates of change across children. The test for the fixed effect tells us that the average value of these rates is indistinguishable from 0. Yet we retain the fixed effect because the non-zero variance component suggests that we may be able to predict some of this variation with a level-2 predictor. We might find, for example, that the average slope for boys is positive and that the average slope for girls is negative. This would be of tremendous interest, even if the rates average to zero when boys and girls are pooled. We remind you that when selecting a functional form for level-1 model, you are as interested in the level-2 variance components as you are in the level-1 fixed effects.

We now compare the quadratic Model C to the linear Model B. Because we seek a level-1 individual growth model that describes the fundamental structure of these data, we include not just the additional fixed effect (for $TIME^2$) but also the required additional variance components: the population variance for curvature , σ_2^2, as well as its covariances with first grade status, σ_{02}, and linear growth, σ_{12}. To do otherwise would constrain the curvature parameter to be identical across individuals, a constraint that seems antithetical to the model-building exercise in which we are engaged. We find that the deviance statistic declines by 16.0, which exceeds the .01 critical value of a χ^2 distribution on four *d.f.* (13.27). We therefore reject the null hypothesis that all four parameters are simultaneously zero and conclude that there is potentially predictable variation in curvature across children.

Do we need to go further and adopt a cubic model? Comparison of Models D and C suggest that the answer is no. Addition of a cubic term adds one fixed effect and four random effects (σ_3^2, σ_{03}, σ_{13}, and σ_{23}), but the deviance statistic declines by only 8.8 (1975.8–1967.0), which is less than the associated .05 critical value of 11.07 (*d.f.* = 5).

We conclude that we should treat individual change in externalizing behavior as though it follows a quadratic trajectory. This conclusion, reinforced by the AIC and BIC statistics, is a realistic compromise that respects the many kinds of variation present in the data. This does not mean that *no child* follows a cubic trajectory, but rather that, overall, when individual change is hypothesized to be cubic, sufficient children end up with cubic parameters that are close enough to zero that there is too little systematic variation in this parameter to worry about. The reverse is true for the quadratic parameter: even though its average value is indistinguishable from 0, it displays sufficient variation to warrant inclusion.

Having selected a suitable polynomial individual change trajectory, model building proceeds as before, although there are extra individual growth parameters to explore. Each level-1 parameter has its own level-2 submodel. Level-1 quadratic change provides for three level-2 submodels; a level-1 cubic provides for four. To examine the effect of *FEMALE*, we would begin by postulating a level-2 association with each level-1 parameter:

$$Y_{ij} = \pi_{0i} + \pi_{1i}(GRADE_{ij} - 1) + \pi_{2i}(GRADE_{ij} - 1)^2 + \varepsilon_{ij}$$
$$\pi_{0i} = \gamma_{00} + \gamma_{01}FEMALE_i + \zeta_{0i}$$
$$\pi_{1i} = \gamma_{10} + \gamma_{11}FEMALE_i + \zeta_{1i}$$
$$\pi_{2i} = \gamma_{20} + \gamma_{21}FEMALE_i + \zeta_{2i}$$

where

$$\varepsilon_{ij} \sim N(0, \sigma_\varepsilon^2) \text{ and } \begin{bmatrix} \zeta_{0i} \\ \zeta_{1i} \\ \zeta_{2i} \end{bmatrix} \sim N\left(\begin{bmatrix} 0 \\ 0 \\ 0 \end{bmatrix}, \begin{bmatrix} \sigma_0^2 & \sigma_{01} & \sigma_{02} \\ \sigma_{10} & \sigma_1^2 & \sigma_{12} \\ \sigma_{20} & \sigma_{21} & \sigma_2^2 \end{bmatrix} \right).$$

Having explored a variety of gender differentials in these data, however, it turns out that none is statistically significant. *FEMALE* has no effect on first-grade status, instantaneous rate of change in first grade, or curvature.

6.4 Truly Nonlinear Trajectories

All the individual growth models described so far—including the curvilinear ones presented in this chapter—share an important mathematical property: they are *linear in the individual growth parameters*. Why do we use the label "linear" to describe trajectories that are blatantly nonlinear? The explanation for this apparent paradox is that this mathematical property depends not on the *shape* of the underlying growth trajectory but rather *where*—in which portion of the model—the nonlinearity arises. In all previous models, nonlinearity (or discontinuity) stems from the representation of the *predictors*. To allow the hypothesized trajectory to deviate from a straight line, *TIME* is either transformed or expressed using higher order polynomial terms. In the truly nonlinear models we now discuss, nonlinearity arises in a different way—through the *parameters*.

In this section, we consider models that are not linear in the parameters. We begin, in section 6.4.1, by introducing the notion of *dynamic consistency*, a key concept for understanding the distinction between our previous models and the truly nonlinear ones we discuss here. In section 6.4.2, we illustrate a general approach for fitting truly nonlinear models by analyzing a data set in which the level-1 individual growth trajectory is hypothesized to follow a logistic curve. In section 6.4.3, we expand this approach, surveying a range of other truly nonlinear growth models including the hyperbolic, inverse polynomial, and exponential trajectories. We conclude, in section 6.4.3, by describing how researchers have historically translated substantive theories about nonlinear growth into mathematical representations that can be fit to data.

6.4.1 What Do We Mean by Truly Nonlinear Models?

To highlight the distinction between models that are linear in the parameters and those that are not, consider the following simple quadratic level-1 trajectory: $Y_{ij} = \pi_{0i} + \pi_{1i}TIME_{ij} + \pi_{2i}TIME_{ij}^2 + \varepsilon_{ij}$. We compute individual *i*'s value of *Y* at time *j*—say at *TIME* = 2—by substituting in this value: $Y_{i2} = \pi_{0i}(1) + \pi_{1i}(2) + \pi_{2i}(2)^2 + \varepsilon_{i2}$. For reasons that will soon become apparent, we add the implicit multiplier 1 next to the intercept, π_{0i}. This addition is not essential to our explanation, but it simplifies the argument.

Ignoring ε_{i2} for a moment, individual *i*'s hypothesized true value of *Y* at *TIME* = 2 is the *sum* of three quantities: $\pi_{0i}(1)$, $\pi_{1i}(2)$, and and $\pi_{2i}(2)^2$. Each has a similar form: it is an *individual growth parameter* multiplied by a *number*: π_{0i} times 1, plus π_{1i} times 2, plus π_{2i} times 2^2. All true values of *Y*, for all values of *TIME*, share this property—they are the *sum* of several terms, each of which is the *product* of an *individual growth parameter* and a *numerical weight* whose value is either constant (such as the "1" multiplying π_{0i}) or dependent upon the measurement occasion (such as the 2 and 2^2 multiplying π_{1i} and π_{2i}). We say that this portion of the growth model is a "weighted linear composite of the individual growth parameters" or, more simply, that true individual change is "linear in the parameters."

Individual growth models that are *linear in the parameters* have important spatial properties to which we alluded in chapter 2. These properties are apparent only at the group level—that is, when you summarize everyone's changes using an "average trajectory." As described in section 2.3, we can derive this average trajectory in one of two ways, by computing: (1) the *curve of the averages*—estimate the average outcome on each measurement occasion and then plot a curve through these averages; or (2) *the average of the curves*—estimate the growth parameters for each individual trajectory, average these values, and then plot the result. If an individual growth model is linear in the parameters, it will not matter which approach you use because the "curve of the averages" and the "average of the curves" will be identical. In addition, the average trajectory possesses the same *functional form* (i.e., the same general *shape*) as the constituent individual trajectories: the average of a heterogeneous group of straight lines will be a straight line, the average of a heterogeneous group of quadratics will be quadratic, and so on.[4]

These two properties—(1) the coincidence between the "curve of the averages" and the "average of the curves," and (2) the equivalence in functional form between individual and average trajectories—were labeled *dynamic consistency* by Keats (1983). Many common functions are dynamically consistent, including the straight line, the quadratic, and all polynomials. If a function is linear in the parameters, it will be dynamically consistent.

The concept of dynamic consistency has two important consequences for analysis. First, it reinforces why you should never draw conclusions about the shape of an individual change trajectory from the shape of an average trajectory drawn through occasion-specific means. If true change is not dynamically consistent, your conclusions about the model's functional form will be incorrect. Second, any level-1 model that is dynamically consistent—that is any polynomial, any model with a transformed outcome, or any discontinuous model—can be fit using standard software for multilevel modeling.

Trajectories that are not dynamically consistent are less tractable. Many important level-1 models that arise from substantive theories—such as the *logistic model* for individual change that we examine next—are not linear in the parameters, and as such, are *not* dynamically consistent. We have already alluded to this possibility in section 2.3.1, when we stated that the average of a set of logistic trajectories is not a logistic but a smoothed *step-function*. We now illustrate how to proceed when the logical level-1 individual growth model is a logistic curve.

6.4.2 The Logistic Individual Growth Curve

We introduce the fitting of truly nonlinear change trajectories using data on cognitive growth collected by our colleague, Terry Tivnan (1980).

During a three-week period, Tivnan repeatedly played a two-person checkerboard game, Fox n' Geese, with 17 first- and second-graders. The game begins with a "fox," represented by one black marker, in the back rank of one side of the board, and four "geese," represented by four white markers, in the back rank of the opposite side. Each player takes turns moving his or her pieces, one square at a time, as in checkers, around the light squares. The fox can move forward or backward, but geese can only move forward. Players have opposing goals: the geese try to trap the fox so that it cannot move; the fox tries to reach the other side of the board without being trapped.

Fox n' Geese is a useful tool for studying cognitive growth because: (1) there exists a strategy that will always lead to victory; (2) this strategy is not immediately obvious to someone who has never played the game; and (3) this strategy can be deduced, even by most young children, through successive rounds of play. In early games, children move their geese at random. As they deduce the winning strategy, move-making becomes directed and adept. Tivnan played up to 27 games with each child. He summarized each child's performance in each game as the number of moves completed before making a catastrophic error (*NMOVES*). The greater the number of moves made before error, the greater a child's skill.

Figure 6.8 presents data for eight children. Each panel plots *NMOVES* versus "game number," Tivnan's metric for time. Although *NMOVES* always lies between 1 and 20, change trajectories vary considerably both within and among children. Some children (ids 04, 07, 08, and 15) initially made fatal errors but soon found effective strategies and survived up to 20 moves. Others (ids 11 and 12) took longer to learn. And some (ids 01 and 06) never discovered the optimal strategy, making early fatal errors in every game.

It makes no sense to hypothesize that the true individual change trajectories that gave rise to these sample data are linear. Instead, knowledge of the game of Fox n' Geese and inspection of these plots suggests a nonlinear level-1 model with three features may be appropriate:

- A *lower asymptote.* Each child's trajectory rises from a lower asymptote ("floor") of one because all players, regardless of skill, must make at least one move.
- An *upper asymptote.* Each child's trajectory approaches an upper asymptote ("ceiling") because he or she can make only a finite number of moves before all four geese can no longer move. Based on figure 6.8, 20 appears to be a reasonable upper asymptote.
- A *smooth curve joining these asymptotes.* Learning theory suggests that each child's true trajectory will smoothly traverse the region

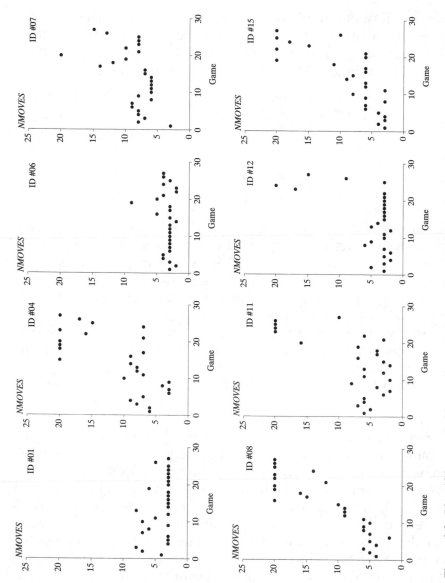

Figure 6.8. Identifying a suitable nonlinear trajectory. Empirical growth plots for 8 children in the Fox n' Geese study. Note that each resembles a classic logistic function.

between asymptotes. We further expect that each trajectory will accelerate away from the floor as the child initially deduces the winning strategy and decelerate toward the ceiling as the child finds it increasingly difficult to refine the strategy further.

Although you may not be aware of the fact, these three features define a *logistic* (or "S-shaped") *trajectory*. We therefore adopt the following logistic function as the hypothesized individual change trajectory for Tivnan's experiment:

$$Y_{ij} = 1 + \frac{19}{1 + \pi_{0i} e^{-(\pi_{1i} TIME_{ij})}} + \varepsilon_{ij}, \tag{6.8}$$

where Y_{ij} represents the number of moves child i makes prior to a fatal error in game j and $\varepsilon_{ij} \sim N(0, \sigma_\varepsilon^2)$. Like the level-1 linear-change growth model, the logistic model has two individual growth parameters, which we have also labeled π_{0i} and π_{1i}. But because of how and where these parameters appear in the model, they have interpretations that are somewhat different from what we usually anticipate.

To clarify what this level-1 logistic trajectory postulates about the relationship between *NMOVES* and *TIME* and how we interpret its parameters, figure 6.9 presents true trajectories for nine children with specific combinations of parameter values for π_{0i} and π_{1i}. We selected these trajectories so that the three children in each panel have a common value of π_{0i} (150, 15, or 1.5) but *differing* values of π_{1i} (0.1, 0.3, or 0.5). We computed each trajectory by substituting the various combinations of parameter values into the structural part of equation 6.8 and calculating the hypothesized value of *NMOVES* at various values of *TIME*.

Because the level-1 logistic model is not linear in *TIME*, its parameters do not have the usual interpretations. Instead, π_{0i} and π_{1i} take on roles that *relate* to, but differ from, their roles in a linear model. For example, π_{0i} is *not* the intercept but it is *related to* and determines the value of the intercept. This can be deduced from figure 6.9 because the three curves in each panel, which share a common value of π_{0i}, also have a common intercept. Substituting 0 for *TIME* into equation 6.8, we find an expression for the intercept to be $1 + \{19/(1 + \pi_{0i})\}$.

Similarly, the second individual growth parameter, π_{1i}, is not a slope per se, but it does determine the rapidity with which the trajectory approaches the upper asymptote. Comparison of the three curves in each panel illustrates this point. When π_{1i} is small (the lower curve in each panel), the logistic trajectory rises slowly; but even when the trajectory begins at a higher elevation (as in the right panel), it never actually reaches the upper asymptote. When π_{1i} is large (the upper curve in each

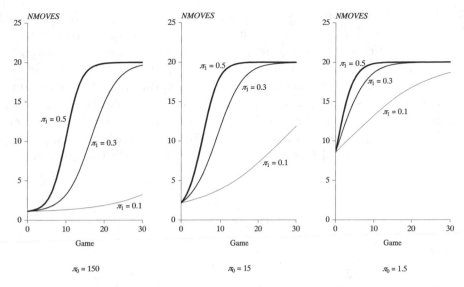

Figure 6.9. Understanding a logistic change trajectory. Hypothesized true logistic change trajectories for 9 children with varying individual growth parameters in equation 6.9.

panel), the logistic trajectory rises more rapidly. Even though these curvilinear trajectories have no single slope, the greater the value of π_{1i}, the more rapidly the curve approaches its upper asymptote. For ease of exposition, we will therefore refer to this second parameter as a "slope."

Where in the model do the asymptotes appear? Unlike other logistic trajectories with which you may be familiar, the model in equation 6.8 invokes two constraints: (1) all children have identical lower and upper asymptotes; and (2) these asymptotes are set to specific values—1 and 20. To verify these assertions, examine equation 6.8 and think about what would happen to Y as $TIME$—the number of games—tends to infinity, in either direction. (Although $TIME$ cannot strictly tend toward minus infinity, the curve in equation 6.8 can.) As $TIME$ tends to *minus* infinity, the denominator of the second term on the right side increases dramatically, driving Y toward 1. As $TIME$ tends to *plus* infinity, the denominator of the second term decreases toward 1, driving Y toward 20. If you have sufficient data, you can postulate a level-1 logistic growth trajectory with asymptotes that vary from child to child. But in this small data set with high and erratic outcome variability, we cannot estimate child-specific asymptotes with any precision. After considering the rules of the Fox n' Geese game and inspecting the data, we decided to pin the asymptotes at 1 and 20 for every child (1 and 20 are the minimum and maximum values of *NMOVES* in figure 6.8).

The logistic level-1 change trajectory in equation 6.8 is *not* linear in the individual growth parameters, π_{0i} and π_{1i}. No matter how hard you try to manipulate the equation algebraically, you cannot express the outcome as a weighted linear composite of π_{0i} and π_{1i}, because the parameters: (1) appear in the denominator; and (2) are exponentiated. But dynamic inconsistency does not prevent us from specifying a regular linear level-2 model for variation in the individual growth parameters across children. If children differ in their trajectories—as the plots in figure 6.8 suggest— a simple pair of initial level-2 submodels for inter-individual differences in change is:

$$\pi_{0i} = \gamma_{00} + \zeta_{0i}$$
$$\pi_{1i} = \gamma_{10} + \zeta_{1i}$$

(6.9a)

where

$$\begin{bmatrix} \zeta_{0i} \\ \zeta_{1i} \end{bmatrix} \sim N\left(\begin{bmatrix} 0 \\ 0 \end{bmatrix}, \begin{bmatrix} \sigma_0^2 & \sigma_{10} \\ \sigma_{10} & \sigma_1^2 \end{bmatrix} \right).$$

(6.9b)

This model stipulates that level-1 logistic individual growth parameters— π_{0i} and π_{1i}—differ across children around unknown population average values, γ_{00} and γ_{10}, with unknown population residual variances σ_0^2 and σ_1^2, and covariance σ_{10}.

If a level-1 individual growth model is not dynamically consistent, we cannot use software designed solely for fitting linear models to fit the new multilevel model for change. Many standard multilevel packages do not fit these models, although some feature additional procedures for specific kinds of nonlinear level-1 models. HLM, MLwiN, and STATA all have procedures for specifying and fitting a limited form of logistic change trajectory at level-1. SAS offers a particularly flexible routine, PROC NLMIXED, for fitting a level-1 trajectory of *any* kind, whether linear in the parameters or not. In addition, the random effects—ε_{ij}, ζ_{0i}, and ζ_{1i}— need not be drawn from a normal distribution because binomial, poisson, and user-defined distributions are also supported. Rather than delve into numeric details here, suffice it to say that the software uses an iterative maximum likelihood procedure for estimating fixed effects and variance components (see Pinheiro & Bates, 1995, 2000). In what follows, we use this procedure to fit a logistic multilevel model for change using standard normal theory assumptions codified in equation 6.9b.

Model A of table 6.6 presents the results of fitting the logistic multi-level model for change in equations 6.8 and 6.9a to the Fox n' Geese data. Not only are both fixed effects statistically significant ($p < .001$); there is also predictable inter-child variation in π_{1i} ($\sigma_1^2 = 0.0072$, p < .05).

Table 6.6: Results of fitting logistic change trajectories to the Fox n' Geese data ($n = 17$)

		Parameter	Model A	Model B
Fixed Effects				
"Intercept," π_{0i}	Intercept	γ_{00}	12.9551***	12.8840***
	$(READ - \overline{READ})$	γ_{01}		−0.3745
"Slope," π_{0i}	Intercept	γ_{10}	0.1227***	0.1223***
	$(READ - \overline{READ})$	γ_{11}		0.0405
Variance Components				
Level-1:	Within-person	σ_ε^2	13.4005***	13.4165***
Level-2:	In "intercept"	σ_0^2	0.6761	0.5610
	In "slope"	σ_1^2	0.0072*	0.0060~
	Covariance between "intercept" and "slope"	σ_{01}	−0.0586	−0.0469
Goodness-of-fit				
	Deviance		2479.7	2477.8
	AIC		2491.7	2493.8
	BIC		2496.7	2500.5

~$p < .10$; *$p < .05$; **$p < .01$; ***$p < .001$.
Model A is an unconditional logistic change model; Model B associates the level-2 predictor, *READ*, with both the "intercept" and "slope" of the logistic change trajectory.

Note: SAS PROC NLMIXED, Adaptive Gaussian Quadrature.

Because these parameters are not actually intercepts or slopes, we use a plot to facilitate interpretation. The left panel of figure 6.10 presents a prototypical trajectory for the average child. As hypothesized, this trajectory begins low and rises smoothly and nonlinearly over time. It does not near the potential asymptote of 20 because the game's difficulty prevents the average child from becoming proficient.

We next ask whether we can predict variation in the individual growth parameters. We illustrate this process by asking whether the level-1 individual growth parameters differ by the children's scores on a standardized reading test. Model B of table 6.6 postulates the following level-2 submodel:

$$\pi_{0i} = \gamma_{00} + \gamma_{01}\left(READ_i - \overline{READ}\right) + \zeta_{0i}$$
$$\pi_{1i} = \gamma_{10} + \gamma_{11}\left(READ_i - \overline{READ}\right) + \zeta_{1i}, \tag{6.10a}$$

where we: (1) center *READ* on the sample mean to preserve the comparability of level-2 parameters across models; and (2) invoke the same assumptions about the level-2 residuals articulated in equation 6.9b. Neither relationship with either individual growth parameter is

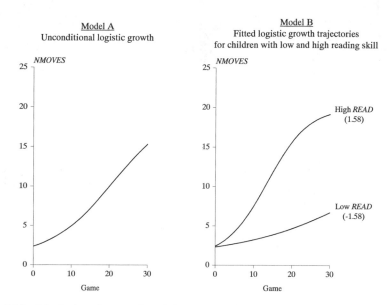

Figure 6.10. Displaying the results of fitting a logistic change trajectory to the Fox n' Geese data. Model A: Unconditional logistic growth model. Model B: Allows parameters in the logistic growth model to vary by the child's reading skill.

statistically significant, possibly because of the small sample size. Notice, however, that better readers approach the upper asymptote more rapidly ($\hat{\gamma}_{11} = 0.0405$, $p < .18$). This differential is apparent in the right panel of figure 6.10, which displays fitted trajectories for two prototypical children with reading scores two standard deviations above and below the sample mean. Although initially indistinguishable, better readers are more likely to near the upper asymptote of 20 after playing the full complement of 27 games.

6.4.3 A Survey of Truly Nonlinear Change Trajectories

By now you should realize that you can represent individual change using a virtually limitless number of mathematical functions. Even the "simple" class of polynomials in table 6.4 is infinite in size, one for each order of polynomial. If you multiply these by all the possible transformations—for both the outcome and temporal predictor—and all the possible discontinuities, the number of trajectories grows without bound. And these are only the dynamically consistent trajectories! There also exist an infinite number of trajectories that are not dynamically consistent. And in an infinite universe, everything that is possible, is possible *an infinite number of times.* In other words, for *any set* of longitudinal data, there exists an *unlimited* number of ways to select an *equally well-fitting* level-1 individual growth trajectory!

How then can you possibly specify a suitable model for your data and purposes? Clearly, you need more than empirical evidence. Among a group of well-fitting growth models, blind numeric comparison of descriptive statistics, goodness-of-fit, and regression diagnostics will rarely pick out the best one. As you might expect, we recommend that you blend theory and empirical evidence, articulating a rationale that you can translate into a statistical model. This recommendation underscores an important point that can often be overlooked in the heat of data analysis: *substance is paramount.* The best way to select an appropriate individual growth model is to work within an explicit theoretical framework. We suggest that you ask not "What is the best model for the job?" but, rather, "What model is most theoretically sound?"

As we highlight the importance of substance, we hasten to note that we are statisticians writing about methods, not substantive researchers writing about theories of change. In some ways, this makes the task we are about to embark upon—an overview of substantively motivated individual growth models—impossible. To be complete, we would need to survey all models for change within dozens of substantive areas—psychiatry, criminology, sociology, medicine, economics, and so on. This is not possible, nor is it our job. Instead, we offer a taste of available models by reviewing ideas within two substantive areas: biology and cognitive science. Through these discipline-based examples, we hope to scaffold your work in *your* substantive field.

Thoughtful work on trajectories of biological change goes back more than a century. The Count de Montbeillard conducted the first documented study of biological growth between 1759 and 1777, when he recorded annual data on the physical stature of his son (see Tanner, 1964). Since then, biologists have used an enormous variety of curvilinear models to represent change. Mead and Pike (1975) divided this large set into four classes: (1) polynomial, (2) hyperbolic, (3) inverse polynomial, and (4) exponential.

Having discussed the polynomial family in section 6.3, we spend little time on them here. Few biologists use polynomials because these trajectories do not flatten out asymptotically as the curve approaches an upper or lower limit. This feature is impractical when modeling physical constructs like height (but perhaps not weight!), as body shape and size typically level off with age. Because hypotheses about biological change often include a plateau, investigators usually choose growth models from the other families of curves, which do contain asymptotes.

Table 6.7: Selected curvilinear trajectories, and level-1 models, used for truly nonlinear change over time

Family	Specific curve	Level-1 model
Hyperbolic	Rectangular hyperbola	$Y_{ij} = \alpha_i - \dfrac{1}{\pi_{1i} TIME_{ij}} + \varepsilon_{ij}$
Inverse polynomial	Inverse quadratic	$Y_{ij} = \alpha_i - \dfrac{1}{\left(\pi_{1i} TIME_{ij} + \pi_{2i} TIME_{ij}^2\right)} + \varepsilon_{ij}$
Exponential	Simple exponential	$Y_{ij} = \pi_{0i} e^{\left(\pi_{1i} TIME_{ij}\right)} + \varepsilon_{ij}$
	Negative exponential	$Y_{ij} = \alpha_i - (\alpha_i - \pi_{0i}) e^{-\pi_{1i} TIME_{ij}} + \varepsilon_{ij}$
	Logistic	$Y_{ij} = \alpha_{1i} + \dfrac{(\alpha_{2i} - \alpha_{1i})}{\left(1 + \pi_{0i} e^{-\pi_{1i} TIME_{ij}}\right)} + \varepsilon_{ij}$

Table 6.7 presents examples of the most popular models from the three remaining families. The table contains one example from the hyperbolic and inverse polynomial classes, and three from the exponential class. Whenever a parameter's role resembles its role in a corresponding polynomial model, we have tried to use consistent labels. For example, although π_{0i} is not always *equal* to the trajectory's intercept, if we label it "π_{0i}," it is usually associated with, and in someway determines, the intercept. Similarly, π_{1i} is not always *equal* to the trajectory's slope—after all, curves do not *have* a single slope—but when we use this label, the parameter usually determines the rate at which the outcome grows or approaches an asymptote. In addition to the familiar "π" symbols, these models include several "α" parameters with labels like α_i, α_{1i}, and α_{2i}. Each represents a type of asymptote, again for the *ith* individual. Learning what each parameter represents in its trajectory is a critical part of using these models.

Hyperbolic Growth

The *rectangular hyperbola* is one of the simplest nonlinear models for individual change. *TIME* enters as a *reciprocal* in the denominator of the model's right side. This model possesses an important property for modeling biological and agricultural growth: over time, its outcome smoothly approaches—but never reaches—an asymptote. This behavior is apparent in the top left panel of figure 6.11, which plots rectangular hyperbolae for specific combinations of values of the individual growth parameters. In plotting these trajectories, we ignore the level-1 residuals and display only the true change trajectory. We therefore label the ordi-

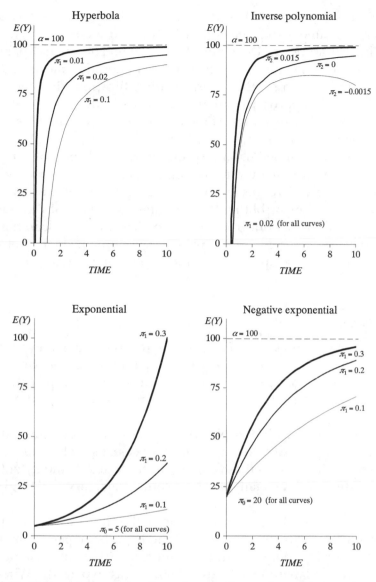

Figure 6.11. Understanding how alternative nonlinear change trajectories represent different patterns of change over time.

nate $E(Y)$, the "population expectation of Y." For clarity, we also drop the subscript i.

All three hyperbolae decelerate smoothly toward a ceiling, regardless of the parameters' specific values. What role does each parameter play in determining the trajectory's shape? To highlight these roles, we have selected the hyperbolic trajectories in figure 6.11 so that α_i is constant at

100 and π_{1i} is 0.01, 0.02, or 0.10. Notice that all three trajectories approach the same asymptote, 100, which is the value of α_i. The trajectories differ, however, in their *rate* of approach. The second parameter, π_{1i}, determines this rate: the *smaller* its value, the more rapid the approach. The reason for this *inverse* relationship (in which *higher* values reflect a *less rapid* approach) is that π_{1i} appears in the model's *denominator*. Notice, too, that seemingly small differences in π_{1i} yield large differences in the shape of the trajectory.[5]

Despite these compelling interpretations, the hyperbolic model is rarely an ideal representation for biological change. The problem is that as *TIME* nears zero, *Y* zooms off toward negative infinity (we avoided this calamity in our plots by blithely eliminating every part of the trajectory below $Y = 0$). This disconcerting behavior leads most researchers, who otherwise like this curve's simplicity and asymptotic properties, to reject it in favor of the inverse polynomial and exponential models that we now describe.

Inverse Polynomial Growth

The family of *inverse polynomials* extends the rectangular hyperbola by adding higher powers of *TIME* to the denominator of the quotient on the model's right side. Table 6.7 presents one example: the *inverse quadratic*. The inverse quadratic contains three parameters, the first two of which, α_i and π_{1i}, function like their identically named peers in the rectangular hyperbola. We display three true inverse quadratic trajectories in upper right panel of figure 6.11 at selected values of the third parameter, π_{2i}. In drawing these curves, we fix α_i at 100 and π_{1i} at 0.02, as we did for the middle rectangular hyperbolic trajectory in the upper left panel of figure 6.11.

Comparing these three inverse polynomial trajectories, notice that π_{2i} acts to disturb the "usual" curvature of the hyperbolic path to asymptote (just as π_{2i} does in a pure quadratic). When $\pi_{2i} = 0$, the inverse quadratic and the rectangular hyperbola are identical (as expected). When π_{2i} is positive, the inverse quadratic approaches its asymptote *more* rapidly than the corresponding hyperbola; when π_{2i} is negative, the inverse quadratic approaches its asymptote *less* rapidly. If π_{2i} is sufficiently large and negative, the trajectory can even turn away from its asymptote (as the lower trajectory illustrates). Unfortunately, like the rectangular hyperbola, the inverse quadratic trajectory departs dramatically for negative infinity at small values of *TIME*.

These specifications just touch the surface of the pool of possible hyperbolic and inverse polynomials. You can generate others by modifying the expressions in table 6.7, for example, by: (1) changing the negative signs into positive signs; (2) adding higher orders of *TIME* (with additional growth parameters) to the denominator of the inverse quadratic; or (3) combining either model with a standard polynomial to create a "mixture" model. Even within these families, you will be amazed at the complexity of shape that can be generated. But if you devise new models using our examples as springboards, remember that parameter interpretation changes with changing mathematical structure. We suggest that you always plot intuitive exemplars to help illustrate the model's behavior.

Exponential Growth

Exponential growth is probably the most widely used class of truly nonlinear models. This theoretically compelling group has been used for centuries to model biological, agricultural, and physical growth. This class includes a wide range of different functional forms, but all contain an *exponent of e*, the base of the natural logarithm. The last three rows of table 6.7 present three popular alternatives. The two bottom panels of figure 6.11 display underlying true trajectories for two of these: the simple exponential and the negative exponential.

The simple exponential growth model, also known as an *explosive* trajectory, is used by biologists to model the unbridled breeding of bacteria in the presence of unlimited nutrients. Those who say the earth's population is "increasing exponentially" are referring to this model. We find it helpful to view this model as an extension of a linear model whose outcome has been transformed logarithmically. This link to linearity allows us to interpret its parameters, π_{0i} and π_{1i}, in ways that are related to the intercept and slope of a straight line. The bottom left panel of figure 6.11 presents three true explosive growth trajectories in which the first parameter, π_{0i}, is set to 5. Notice that this is also the value of its intercept, and so trajectories with higher values of π_{0i} begin at higher elevations. The second growth parameter, π_{1i}, determines how rapidly the trajectory "explodes" upwards: the higher the value of π_{1i}, the more rapid the rise toward infinity.

Next consider the *negative exponential* growth model, displayed in the lower right panel of figure 6.11. This model dampens the unbridled growth of the simple exponential by including a ceiling beyond which the curve cannot rise. Epidemiologists use this model to track the growth of the number of new cases in epidemics, which necessarily level off when there is no one new left to infect. Agricultural researchers use this model to track crop yields and other outcomes that demand a single upper asymptote. The negative exponential model has three parameters, α_i, π_{0i}, and π_{1i}. As shown in the lower right panel of figure 6.11, π_{0i} and α_i represent the intercept and asymptote, and π_{1i} determines how rapidly the trajectory approaches its asymptote; the larger its value, the more rapid the approach.

The logistic trajectory is widely used in biology because it contains both a lower and an upper asymptote. We have already used one version when modeling children's performance in the game of Fox n' Geese. Additional parameters α_{1i} and α_{2i} represent the lower and upper asymptotes, respectively. Figure 6.9 already presented three sets of logistic growth trajectories whose values of α_{1i} and α_{2i} are set to 1 and 20, respectively. In these displays, we forced everyone to have these common asymptotes. With sufficient data, you can estimate the asymptotes as well. The remaining parameters, π_{0i} and π_{1i}, again determine, but do not equal, the trajectory's intercept and the rapidity with which it transits between asymptotes. This is also apparent in figure 6.9, in which the trajectories with the lower values of π_{0i} have lower intercepts and those with the higher values of π_{1i} approach the upper asymptote more rapidly.

6.4.4 From Substantive Theory to Mathematical Representations of Individual Growth

Having surveyed a broad array of nonlinear trajectories, we now take a different approach. Instead of proceeding from mathematical model to application, we proceed from substantive theory to mathematical model. In doing so, we hope to show how you can translate theories into mathematical representations that you can then examine empirically with data. To focus our discussion, we draw from the literature on human cognition (see, e.g., Guire and Kowalski, 1979; and Lewis, 1960).

We begin with Robertson's (1909) theory of human learning, developed from a well-known law of physical chemistry known as the *autocatalytic principle*. Some chemical reactions are accelerated by *catalysts*, an external substance that helps the reaction along. Others *autocatalyze*—that is, the products of the reaction itself accelerate the process. According to the autocatalytic principle, the rate at which a reaction proceeds is proportional to the product of: (1) the amount of catalyst currently available; and (2) the amount of chemical as yet unreacted.

Robertson hypothesized that learning could be modeled as an autocatalytic brain process that follows the same mathematical law. Expressing this principle in cognitive terms, we represent the rate at which learning occurs as:

$$\begin{pmatrix} \text{Rate at} \\ \text{which} \\ \text{learning} \\ \text{occurs} \end{pmatrix} \propto \begin{pmatrix} \text{Amount of} \\ \text{learning} \\ \text{that has} \\ \text{already} \\ \text{occurred} \end{pmatrix} \times \begin{pmatrix} \text{Amount of} \\ \text{learning} \\ \text{that has} \\ \text{yet to} \\ \text{occur} \end{pmatrix},$$

where the symbol "$\propto$" means "is proportional to." For readers familiar with differential calculus, we can express this equation in mathematical terms by writing:

$$\frac{dY}{dt} = kY(\alpha - Y),$$

where Y is the amount learned by time t, α is an upper limit (or asymptote) on the amount that can be learned, and k is a constant of proportionality. For readers unfamiliar with differential calculus, the expression on the left side of this equation is known as the *first derivative of Y with respect to time*. Fortunately, we know this quantity quite well by another name: the *rate of change in Y over time*.

Robertson's learning equation appears to differ dramatically from our other models, but any difference is more cosmetic than real. The model above is a *first-order differential equation* framed in terms of the *rate of change in Y*. Using the standard practices of calculus, we can re-express this differential equation in terms of Y by integrating this expression with respect to time. Using terms that reflect our usual notation, and adding subscripts i and j to represent persons and occasions, integration of Robertson's equation yields:

$$Y_{ij} = \frac{\alpha_i}{1 + \pi_{0i}e^{(-\pi_{1i}TIME_{ij})}},$$

where π_{0i} and π_{1i} are constants related to the original constant of proportionality k.[6] Examining the truly nonlinear models in table 6.7, we see that Robertson's autocatalytic hypothesis leads to a logistic trajectory for learning. Over time, the knowledge of individual i rises smoothly from a lower asymptote (of 0) toward an upper asymptote of α_i at a rate determined by π_{1i}. A plot of this curve would follow a logistic trajectory (such as those in figure 6.9).

As we demonstrate how to translate substantive theory into models, we must highlight a serious flaw in our argument: the omission of measurement errors. Our formulation of Robertson's learning curve includes no residuals, no errors, no ε's. You could argue that this omission is not problematic because this curve represents Robertson's hunch about *true* change—about the deep functioning of the human brain. But a representation without errors implies that we can measure brain functioning infallibly, which we cannot. The number of moves a child makes when playing Fox n' Geese does not tell us about the inner functioning of his or her brain. It tells us about the child's actions, which we hope reflect cognitive ability. To account for measurement error, we restate Robertson's hypothesis by adding a residual term:

$$Y_{ij} = \frac{\alpha_i}{1 + \pi_{0i}e^{(-\pi_{1i}TIME_{ij})}} + \varepsilon_{ij}.$$

When you fit this trajectory to data, you evaluate its goodness of fit and assess whether a logistic trajectory provides a reasonable description. But because measurement is not infallible, neither are your conclusions.

Robertson's elegant argument is compelling, in part, because of its scientific roots. By beginning with a hypothesis about the neurological mechanism that drives learning, he used mathematical symbols, algebra, and calculus to express his hypothesis as a statistical model. Psychologists who fit a logistic model to empirical data on learning—as we did for the Fox n' Geese data—implicitly test Robertson's hypotheses about brain function. Our ability to fit a logistic trajectory to our data provides support for his autocatalytic hypothesis. Of course, we do not argue that consistency between the data and model *proves* Robertson's hypothesis. Other hypotheses and other models may fit equally well, or better. But such is the nature of empirical research: the best we can do is fail to disprove Robertson's, or anyone's, hypothesis.

Robertson hypothesized other models for learning based upon his biochemical conceptions. For example, he hypothesized that the neurological autocatalytic process that drives learning could be retarded by a reverse chemical process that dampens growth. Addition of a dampening effect does not alter the logistic trajectory's fundamental shape, but it modifies its asymptote and the rate at which this asymptote is approached. In building this model, Robertson argued that each learner could keep only a *given* amount of material in his or her brain. This suggested that the rate at which learning took place was proportional not to $Y(\alpha - Y)$ as above, but rather to $Y\alpha$.

He applied these ideas to the study of rote memorization, by making subjects repeatedly read and recall lists of nonsense syllables. He thereby developed another individual growth trajectory that still enjoys wide application today:

$$Y_{ij} = \pi_{0i}e^{(\pi_{1i}TIME_{ij})} + \varepsilon_{ij},$$

where we use our usual notation system and add a residual term. In Robertson's study, Y_{ij} represents the number of syllables individual i could recall on occasion j and $TIME_{ij}$ is the number of repetitions required to achieve that level of memorization. Comparing this equation to those in table 6.7, we find that this dampened autocatalytic process is a simple exponential trajectory in which π_{0i} and π_{1i} play the familiar "intercept" and "rate" roles. Although Robertson (1908) fit this model to data, he found it imperfect because he believed that the curve should eventually "tip over" (as in a logistic function).

Clark Hull (1943, 1952) developed a variety of statistical models for human learning trajectories from theoretical conceptions about the acquisition and extinction of simple responses. His neurological hypotheses about how simple learned responses are evoked led him to postulate the following negative exponential trajectory for the development of what he called *habit strength*:

$$Y_{ij} = \alpha_i \left(1 - e^{(-\pi_{1i}TIME_{ij})}\right) + \varepsilon_{ij},$$

where we use our usual notation and add a residual term. Outcome Y_{ij} represents habit strength for individual i on occasion j and $TIME$ represents the corresponding number of trials. α_i is an upper asymptote parameter and π_{1i} is a "rate" parameter that determines how rapidly individual i approaches that asymptote. Examining table 6.7, you will see that this is just a negative exponential growth model with intercept zero (because Hull assumed that habit strength must be zero prior to the trials).

Hull's negative exponential model for individual change remains popular today. It is identical to an earlier learning curve proposed by the Russian physicist Schukarew (1907), which was based—like Robertson's—on the application of principles of physical chemistry to brain function. Throughout the last century, the negative exponential learning curve has arisen repeatedly in many domains, exciting cognitive theorists and empiricists alike. Estes and Burke used it to develop a statistical theory of learning (Estes, 1950; Estes and Burke, 1955), Grice (1942) applied it to the negotiation of mazes by hungry albino rats and Hicklin (1976) applied it to mastery learning and IQ growth in human beings.

Psychologist L. L. Thurstone broke theoretical and empirical ground on human learning in his doctoral thesis (1917) and elsewhere (1930). Stressing the importance of specifying "a rational equation for the learning function," he hypothesized that attainment (Y) could depend upon practice (a proxy for time) for individual i on occasion j, using the following model:

$$Y_{ij} = \pi_{0i} + \frac{(\alpha_i - \pi_{0i})TIME_{ij}}{\pi_{1i} + TIME_{ij}} + \varepsilon_{ij},$$

where, as usual, π_{0i} represents the intercept, α_i represents the upper asymptote, and π_{1i} determines how rapidly the curve approaches

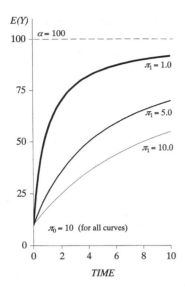

Figure 6.12. Understanding alternative nonlinear change trajectories implied by Thurstone's learning equation.

asymptote. We present illustrative true trajectories for Thurstone's learning trajectory in figure 6.12, for three values of the rate parameter, π_{1i} (1, 5, and 10). Notice that the curve resembles a hyperbola, with a finite intercept (instead of the infinite intercept of the rectangular hyperbola in figure 6.11). The larger the value of the rate parameter, the *less* rapidly the curve approaches the asymptote. As with the rectangular hyperbola, this inverse relationship arises because the rate parameter appears in the equation's denominator. Thurstone used this trajectory to model the learning of mazes by rats, and his equation was subsequently modified and extended by many colleagues, including Gulliksen (1934, 1953), to represent more complex kinds of learning.

7

Examining the Multilevel Model's Error Covariance Structure

> Change begets change. Nothing propagates so fast. . . . The mine which Time has slowly dug beneath familiar objects is spring in an instant, and what was rock before, becomes but sand and dust.
>
> —Charles Dickens

In previous chapters, we often emphasized the *fixed* effects in the multi-level model for change. Doing so made great sense because the fixed effects typically provide the most direct answers to our research questions. In this chapter, in contrast, we focus on the model's *random* effects as embodied in its error covariance structure. Doing so allows us to both describe the particular error covariance structure that the "standard" multilevel model for change invokes and it also allows us to broaden its representation to other—sometimes more tenable assumptions—about its behavior.

We begin, in section 7.1, by reviewing the "standard" multilevel model for change, expressed in composite form. In section 7.2, we closely examine this model's random effects, demonstrating that the composite error term is indeed both heteroscedastic and autocorrelated, as we would prefer for longitudinal data. But we also find that this error covariance structure may not be as general as we might like and, in some settings, alternatives may have greater appeal. This brings us to section 7.3, in which we compare several alternative error covariance structures and provide strategies for choosing among them.

7.1 The "Standard" Specification of the Multilevel Model for Change

Throughout this chapter, we use a small, time-structured data set first presented in Willett (1988). On each of four days, spaced exactly one week apart, 35 people completed an inventory that assesses their performance

Table 7.1: Ten cases from a person-level data set containing scores on an opposite's naming task across four occasions of measurement, obtained weekly, and a baseline measurement of *COG*, a measure of cognitive skill, obtained in the first week

ID	OPP1	OPP2	OPP3	OPP4	COG
01	205	217	268	302	137
02	219	243	279	302	123
03	142	212	250	289	129
04	206	230	248	273	125
05	190	220	229	220	81
06	165	205	207	263	110
07	170	182	214	268	99
08	96	131	159	213	113
09	138	156	197	200	104
10	216	252	274	298	96

on a timed cognitive task called "opposites naming." At wave 1, each person also completed a standardized instrument assessing general cognitive skill. Table 7.1 presents the first ten cases in the person-level data set (we use this format to conserve space), which includes values of: (a) *OPP1*, *OPP2*, *OPP3*, and *OPP4*—the individual's opposites-naming score on each occasion; and (b) *COG*, the baseline cognitive skill score. The full person-period data set has 140 records, 4 per person. In what follows, we assume that any skill improvement over time results from practice, not cognitive development. That said, research interest centers on determining whether opposites-naming skill increases more rapidly with practice among individuals with stronger cognitive skills.

We specify the "standard" multilevel model for change in the usual manner. For individual i on occasion j, we assume that opposites-naming score, Y_{ij}, is a linear function of *TIME*:

$$Y_{ij} = \pi_{0i} + \pi_{1i}TIME_j + \varepsilon_{ij}, \qquad (7.1a)$$

where subscript i has been omitted from predictor *TIME* because the data are time-structured and

$$\varepsilon_{ij} \overset{iid}{\sim} N(0, \sigma_\varepsilon^2). \qquad (7.1b)$$

To allow the individual growth parameters to take on their usual interpretations for person i—π_{0i} as the true initial level and π_{1i} as the true weekly rate of change—we scale *TIME* so that the first measurement occasion is labeled 0 and the others are labeled 1, 2, and 3. In the "standard" model we also assume that the random effects ε_{ij} are drawn from a univariate normal distribution with zero mean and unknown variance σ_ε^2. To further clarify the meaning of this assumption—our focus in this chapter—we add the notation "*iid*," which declares that the errors are mutually *independent*, across occasions and persons, and *identically distributed*. We discuss the implications of this assumption in detail further below.

To allow individual change trajectories to differ systematically across people, we posit a level-2 submodel in which the cognitive skills score (*COG*) is associated with both growth parameters:

$$\pi_{0i} = \gamma_{00} + \gamma_{01}\left(COG_i - \overline{COG}\right) + \zeta_{0i}$$
$$\pi_{1i} = \gamma_{10} + \gamma_{11}\left(COG_i - \overline{COG}\right) + \zeta_{1i} \, ,$$

(7.2a)

where

$$\begin{bmatrix} \zeta_{0i} \\ \zeta_{1i} \end{bmatrix} \overset{iid}{\sim} N\left(\begin{bmatrix} 0 \\ 0 \end{bmatrix}, \begin{bmatrix} \sigma_0^2 & \sigma_{01} \\ \sigma_{10} & \sigma_1^2 \end{bmatrix} \right).$$

(7.2b)

To facilitate interpretation, we center the continuous predictor *COG* on its sample mean. The level-2 fixed effects capture the effect of cognitive skill on the average trajectories of change; the level-2 random effects, ζ_{0i} and ζ_{1i}, represent those parts of the level-2 outcomes that remain "unexplained" by cognitive skill. In the "standard" multilevel model, we assume that these random effects have zero mean and that everyone draws them independently from a normal distribution. To allow for the possibility that even after accounting for cognitive skill, the unpredicted portions of a person's true intercept and true slope may be intertwined, we assume that each person draws both level-2 residuals simultaneously from a *bivariate* normal distribution with variances σ_0^2 and σ_1^2 *and* covariance σ_{01}.

Table 7.2 presents the results of fitting this "standard" multilevel model for change to the opposites-naming data. Because we focus in this chapter on the model's stochastic portion, we use restricted, not full, maximum likelihood (see section 4.3 for a comparison of methods). For an individual of average cognitive skill, initial level of opposites-naming skill is estimated to be 164.4 ($p < .001$); this average person's weekly rate of linear change is estimated to be 27.0 ($p < .001$). Individuals whose cognitive skills differ by one point have an initial opposites-naming score that is 0.11 lower (although this decrement is not statistically significant, $p = .82$); their average weekly rate of linear change is 0.43 higher ($p < .01$). Even after including cognitive skill as a predictor of both initial status and change, we detect statistically significant level-2 residual variation in

Table 7.2: Change in opposite's naming over a four-week period as a function of baseline IQ

		Parameter	Estimate
Fixed Effects			
Initial status, π_{0i}	Intercept	γ_{00}	164.37***
	$(COG - \overline{COG})$	γ_{01}	−0.11
Rate of change, π_{1i}	Intercept	γ_{10}	26.96***
	$(COG - \overline{COG})$	γ_{11}	0.43**
Variance Components			
Level-1:	Within-person variance	σ_ε^2	159.48***
Level-2:	Variance in ζ_{0i}	σ_0^2	1236.41***
	Variance in ζ_{1i}	σ_1^2	107.25***
	Covariance of ζ_{0i} and ζ_{1i}	σ_{01}	−178.23*
Goodness-of-fit			
	Deviance		1260.3
	AIC		1268.3
	BIC		1274.5

$\sim p < .10$; $* p < .05$; $** p < .01$; $*** p < .001$.
Parameter estimates, approximate p-values, and goodness-of-fit statistics from fitting a standard multilevel model for change ($n = 35$).

Note: SAS PROC MIXED, Restricted ML.

both initial status (1236.41, $p < .001$) and rate of change (107.25, $p < .001$). We also detect a statistically significant negative covariance (−178.2, $p < .05$) between the level-2 residuals, ζ_{0i} and ζ_{1i}, which suggests that, after controlling for cognitive skill, those with weaker initial opposites-naming skills improve at a faster rate, on average, than those with stronger initial skills. To interpret this estimate more easily, we compute the partial correlation between change and initial status to find:

$$\hat{\rho}_{\pi_0\pi_1|COG} = -178.25/\sqrt{1236.41 \times 107.25} = -0.49.$$

Finally, the estimated level-1 residual variance, $\hat{\sigma}_\varepsilon^2$, is 159.5.

7.2 Using the Composite Model to Understand Assumptions about the Error Covariance Matrix

To understand the error covariance structure in the "standard" multilevel model for change, we move to the composite representation obtained by collapsing the level-2 submodels in equation 7.2a into the level-1 submodel in equation 7.1a:

$$Y_{ij} = \{\gamma_{00} + \gamma_{01}(COG_i - \overline{COG}) + \zeta_{0i}\}$$
$$+ \{\gamma_{10} + \gamma_{11}(COG_i - \overline{COG}) + \zeta_{1i}\}TIME_j + \varepsilon_{ij}. \tag{7.3}$$

Multiplying out and rearranging terms yields:

$$Y_{ij} = \left[\gamma_{00} + \gamma_{10}TIME_j + \gamma_{01}(COG_i - \overline{COG}) + \gamma_{11}(COG_i - \overline{COG}) \times TIME_j\right]$$
$$+ \left[\varepsilon_{ij} + \zeta_{0i} + \zeta_{1i}TIME_j\right], \tag{7.4}$$

where random effects ε_{ij}, ζ_{0i}, and ζ_{1i} retain the distributional assumptions of equations 7.1b and 7.2b.

As in section 4.2, brackets distinguish the model's *structural* and *stochastic* portions. Its structural portion contains our hypotheses about the way that opposites-naming skill changes with time and depends on baseline cognitive skill. Its stochastic portion contains the *composite residual*, which we now label *r*, for convenience. The value of *r* for individual *i* on occasion *j* is:

$$r_{ij} = \left[\varepsilon_{ij} + \zeta_{0i} + \zeta_{1i}TIME_j\right], \tag{7.5}$$

which is a weighted linear combination of the original three random effects from the level-1/level-2 specification (ε_{ij}, ζ_{0i} and ζ_{1i}, with constants 1, 1, and $TIME_j$ acting as the weights). Our major focus in this chapter is on the statistical properties of r_{ij}.

But before examining these properties, let us simplify the composite model in equation 7.4 by substituting r_{ij} as defined in equation 7.5 into equation 7.4:

$$Y_{ij} = \left[\gamma_{00} + \gamma_{10}TIME_j + \gamma_{01}(COG_i - \overline{COG}) + \gamma_{11}(COG_i - \overline{COG}) \times TIME_j\right] + r_{ij}. \tag{7.6}$$

The composite model now looks like a regular multiple regression model, with the "usual" error term replaced by "*r*." This reinforces the notion, discussed in chapter 4, that you can conceptualize the multilevel analysis of change as a multiple regression analysis in the person-period data set, in which you regress the outcome on the main effects of *TIME*, a level-2 predictor (*COG*), and their statistical interaction.

Because of the special nature of "*r*," our standard practice is to fit the model in equation 7.6 by GLS regression analysis, not OLS, making specific assumptions about the distribution of the residuals. But before doing so, let's suppose for a hypothetical moment that we were willing to invoke the simpler OLS assumptions that all the r_{ij} are independent and normally distributed, with zero means and homoscedastic variance (σ_r^2, say).

We could codify these simple distributional assumptions for all the residuals simultaneously in one grand statement:

$$
\begin{bmatrix} r_{11} \\ r_{12} \\ r_{13} \\ r_{14} \\ r_{21} \\ r_{22} \\ r_{23} \\ r_{24} \\ \vdots \\ r_{n1} \\ r_{n2} \\ r_{n3} \\ r_{n4} \end{bmatrix} \sim N \left(\begin{bmatrix} 0 \\ 0 \\ 0 \\ 0 \\ 0 \\ 0 \\ 0 \\ 0 \\ \vdots \\ 0 \\ 0 \\ 0 \\ 0 \end{bmatrix}, \begin{bmatrix} \sigma_r^2 & 0 & 0 & 0 & 0 & 0 & 0 & 0 & \cdots & 0 & 0 & 0 & 0 \\ 0 & \sigma_r^2 & 0 & 0 & 0 & 0 & 0 & 0 & \cdots & 0 & 0 & 0 & 0 \\ 0 & 0 & \sigma_r^2 & 0 & 0 & 0 & 0 & 0 & \cdots & 0 & 0 & 0 & 0 \\ 0 & 0 & 0 & \sigma_r^2 & 0 & 0 & 0 & 0 & \cdots & 0 & 0 & 0 & 0 \\ 0 & 0 & 0 & 0 & \sigma_r^2 & 0 & 0 & 0 & \cdots & 0 & 0 & 0 & 0 \\ 0 & 0 & 0 & 0 & 0 & \sigma_r^2 & 0 & 0 & \cdots & 0 & 0 & 0 & 0 \\ 0 & 0 & 0 & 0 & 0 & 0 & \sigma_r^2 & 0 & \cdots & 0 & 0 & 0 & 0 \\ 0 & 0 & 0 & 0 & 0 & 0 & 0 & \sigma_r^2 & \cdots & 0 & 0 & 0 & 0 \\ \vdots & \vdots & \vdots & \vdots & \vdots & \vdots & \vdots & \vdots & \ddots & \vdots & \vdots & \vdots & \vdots \\ 0 & 0 & 0 & 0 & 0 & 0 & 0 & 0 & \cdots & \sigma_r^2 & 0 & 0 & 0 \\ 0 & 0 & 0 & 0 & 0 & 0 & 0 & 0 & \cdots & 0 & \sigma_r^2 & 0 & 0 \\ 0 & 0 & 0 & 0 & 0 & 0 & 0 & 0 & \cdots & 0 & 0 & \sigma_r^2 & 0 \\ 0 & 0 & 0 & 0 & 0 & 0 & 0 & 0 & \cdots & 0 & 0 & 0 & \sigma_r^2 \end{bmatrix} \right)
$$

$$(7.7)$$

where, because we have four waves of data per person, we have four residuals (one per occasion) for each of the n sample members. With a different number of waves of data, we would simply rescale the component vectors and matrices.

While equation 7.7 may appear needlessly complex for describing the behavior of the residuals in an OLS analysis, it provides a convenient and generalizable form for codifying assumptions on residuals that we will find useful, later. It says that the complete set of residuals in the analysis has a *multivariate normal distribution*. The statement has several important features:

- It contains a *vector of random variables* whose distribution is being specified. To the left of the "is distributed as" sign ("~"), a column contains all the random variables whose distribution is being specified. This *vector* contains all the model's residuals, which, for our data, run from the four residuals for person 1 (r_{11}, r_{12}, r_{13}, and r_{14}), to those for person 2 (r_{21}, r_{22}, r_{23}, r_{24}) and so on, through the four residuals for person n.
- It states the *distribution type*. Immediately after the ~, we stipulate that every element in the residual vector is normally distributed ("N"). Because the vector has many ("multi") entries, the residuals have a *multivariate* normal distribution.
- It contains a *vector of means*. Also to the right of the ~, inside the

parentheses and before the comma, is a vector of hypothesized means, one for each residual. All these elements are 0, reflecting our belief that the population mean of each residual is 0.

- It contains an *error* (or, residual) *covariance matrix*. The last entry in equation 7.7 is the error covariance matrix, which contains our hypotheses about the residual variances and covariances. Under classical OLS assumptions, this matrix is *diagonal*—all elements are zero, except those along the main diagonal. The off-diagonal zero values represent the *residual independence* assumption, which stipulates that that the residuals do not covary. Along the diagonal, all residuals have an identical population variance, σ_r^2. This is the *residual homoscedasticity* assumption.

The distributional statement in equation 7.7 is inappropriate for longitudinal data. Although we expect the composite residuals to be independent across people and normally distributed with zero means, *within people* we expect them to be *heteroscedastic* and *correlated over time*. We can write an error covariance matrix that reflects these new "longitudinal" assumptions as:

$$
\begin{bmatrix} r_{11} \\ r_{12} \\ r_{13} \\ r_{14} \\ r_{21} \\ r_{22} \\ r_{23} \\ r_{24} \\ \vdots \\ r_{n1} \\ r_{n2} \\ r_{n3} \\ r_{n4} \end{bmatrix} \sim N \left(\begin{bmatrix} 0 \\ 0 \\ 0 \\ 0 \\ 0 \\ 0 \\ 0 \\ 0 \\ \vdots \\ 0 \\ 0 \\ 0 \\ 0 \end{bmatrix}, \begin{bmatrix} \sigma_{r_1}^2 & \sigma_{r_1 r_2} & \sigma_{r_1 r_3} & \sigma_{r_1 r_4} & 0 & 0 & 0 & 0 & \cdots & 0 & 0 & 0 & 0 \\ \sigma_{r_2 r_1} & \sigma_{r_2}^2 & \sigma_{r_2 r_3} & \sigma_{r_2 r_4} & 0 & 0 & 0 & 0 & \cdots & 0 & 0 & 0 & 0 \\ \sigma_{r_3 r_1} & \sigma_{r_3 r_2} & \sigma_{r_3}^2 & \sigma_{r_3 r_4} & 0 & 0 & 0 & 0 & \cdots & 0 & 0 & 0 & 0 \\ \sigma_{r_4 r_1} & \sigma_{r_4 r_2} & \sigma_{r_4 r_3} & \sigma_{r_4}^2 & 0 & 0 & 0 & 0 & \cdots & 0 & 0 & 0 & 0 \\ 0 & 0 & 0 & 0 & \sigma_{r_1}^2 & \sigma_{r_1 r_2} & \sigma_{r_1 r_3} & \sigma_{r_1 r_4} & \cdots & 0 & 0 & 0 & 0 \\ 0 & 0 & 0 & 0 & \sigma_{r_2 r_1} & \sigma_{r_2}^2 & \sigma_{r_2 r_3} & \sigma_{r_2 r_4} & \cdots & 0 & 0 & 0 & 0 \\ 0 & 0 & 0 & 0 & \sigma_{r_3 r_1} & \sigma_{r_3 r_2} & \sigma_{r_3}^2 & \sigma_{r_3 r_4} & \cdots & 0 & 0 & 0 & 0 \\ 0 & 0 & 0 & 0 & \sigma_{r_4 r_1} & \sigma_{r_4 r_2} & \sigma_{r_4 r_3} & \sigma_{r_4}^2 & \cdots & 0 & 0 & 0 & 0 \\ \vdots & \vdots & \vdots & \vdots & \vdots & \vdots & \vdots & \vdots & \ddots & \vdots & \vdots & \vdots & \vdots \\ 0 & 0 & 0 & 0 & 0 & 0 & 0 & 0 & \cdots & \sigma_{r_1}^2 & \sigma_{r_1 r_2} & \sigma_{r_1 r_3} & \sigma_{r_1 r_4} \\ 0 & 0 & 0 & 0 & 0 & 0 & 0 & 0 & \cdots & \sigma_{r_2 r_1} & \sigma_{r_2}^2 & \sigma_{r_2 r_3} & \sigma_{r_2 r_4} \\ 0 & 0 & 0 & 0 & 0 & 0 & 0 & 0 & \cdots & \sigma_{r_3 r_1} & \sigma_{r_3 r_2} & \sigma_{r_3}^2 & \sigma_{r_3 r_4} \\ 0 & 0 & 0 & 0 & 0 & 0 & 0 & 0 & \cdots & \sigma_{r_4 r_1} & \sigma_{r_4 r_2} & \sigma_{r_4 r_3} & \sigma_{r_4}^2 \end{bmatrix} \right)
$$

(7.8)

where, again, the dimensions of the vectors and matrices reflect the design of the opposites-naming study.

The new distributional specification in equation 7.8 allows the residuals in the composite model to have a multivariate normal distribution with zero means and a *block diagonal*, not *diagonal*, error covariance structure. The term "block diagonal" means that all the matrix's elements are zero, except those within the "blocks" arrayed along the diagonal, one per person. The zero elements outside the blocks indicate that each person's residuals are independent of all others'—in other words, the residuals for person i have zero covariance with everyone else's residuals. But the non-zero covariance parameters *within* each block allow the residuals to covary *within person*. In addition, the multiple distinct parameters along each block's diagonal allow the variances of the within-person residuals to differ across occasions. These distinctions between the *diagonal* and the *block diagonal* error covariance matrices demark the fundamental difference between a cross-sectional and longitudinal design.[1]

Notice that the blocks of the error covariance matrix in equation 7.8 are identical across people. This *homogeneity assumption* says that, in an analysis of change, although the composite residuals may be heteroscedastic and dependent *within* people, the entire error structure is repeated identically across people—that is, everyone's residuals are identically heteroscedastic and autocorrelated. This assumption is not absolutely necessary, as it can be tested and relaxed in limited ways (provided you have sufficient data). Yet we typically invoke it for practical reasons, as it improves dramatically the parsimony with which we can specify the model's stochastic portion. Limiting the number of unique variance/covariance components in a hypothesized model improves the rapidity with which iterative model fitting converges. If we allowed each person in this study to possess a unique set of variance components, for example, we would be need to estimate $10n$ variance components—$6n$ more than the number of observations on the outcome in the person-period data set!

Adopting the homogeneity assumption allows us to express the distributional assumptions in equation 7.8 in more parsimonious terms by writing:

$$r \sim N\left(0, \begin{bmatrix} \Sigma_r & 0 & 0 & \cdots & 0 \\ 0 & \Sigma_r & 0 & \cdots & 0 \\ 0 & 0 & \Sigma_r & \cdots & 0 \\ \vdots & \vdots & \vdots & \ddots & 0 \\ 0 & 0 & 0 & 0 & \Sigma_r \end{bmatrix}\right). \quad (7.9)$$

Equation 7.9 says that the complete vector of residuals r has a multivariate normal distribution with mean vector 0 and a block-diagonal error covariance matrix constituted from submatrices, Σ_r and 0, where:

$$\Sigma_r = \begin{bmatrix} \sigma_{r_1}^2 & \sigma_{r_1 r_2} & \sigma_{r_1 r_3} & \sigma_{r_1 r_4} \\ \sigma_{r_2 r_1} & \sigma_{r_2}^2 & \sigma_{r_2 r_3} & \sigma_{r_2 r_4} \\ \sigma_{r_3 r_1} & \sigma_{r_3 r_2} & \sigma_{r_3}^2 & \sigma_{r_3 r_4} \\ \sigma_{r_4 r_1} & \sigma_{r_4 r_2} & \sigma_{r_4 r_3} & \sigma_{r_4}^2 \end{bmatrix}. \quad (7.10)$$

Again, the dimensions of Σ_r reflect the design of the opposite-naming study.

When you investigate random effects in the analysis of change, you anticipate that the composite residuals will have a multivariate distributional form like equation 7.8 or 7.9. As part of your analyses, you estimate the elements of this error covariance matrix, which means that—under the homogeneity assumption—you estimate the elements of the error covariance submatrix Σ_r, in equation 7.10.

This specification of the Σ_r error covariance submatrix—and hence the shape of the full error covariance matrix—is very general. It contains a set of error variance and covariance parameters (four of the former and six of the latter, for the opposites-naming data), each of which can take on an appropriate value. But when you specify a *particular* multilevel model for change, you invoke specific assumptions about these values. Most important for our purposes here is that the "standard" multilevel model for change invokes a specific mathematical structure for the r_{ij}. As we show below, this model constrains the error covariance structure much more than that specified in equations 7.9 and 7.10.

What does the error covariance submatrix Σ_r of the "standard" multilevel model for change look like? When we presented this model earlier in the book, we focused on its ability to represent hypotheses about fixed effects. Does it also provide a reasonable covariance structure for the composite residuals? Fortunately, most of its behavior is exactly what you would hope and expect. First, because a weighted linear combination of normally distributed variables is also normally distributed, for example, each composite residual in equation 7.5 is also normally distributed, as specified in equation 7.9. Second, because the mean of a weighted linear combination of random variables is equal to an identically weighted linear combination of the means of those variables, the mean of the composite residual in equation 7.5 must also be zero, as specified in equation 7.9. Third, the error covariance matrix of the composite residuals is indeed block diagonal, as specified in equation 7.9. But fourth, in the standard multilevel model for change, the elements of the Σ_r error covariance blocks in equations 7.9 and 7.10 possess a powerful dependence on time. As this is both the most interesting—and potentially troublesome—aspect of the standard model, we delve into this feature in some detail below.

7.2.1 Variance of the Composite Residual

We begin by examining what the "standard" multilevel model for change hypothesizes about the composite residual's variance. Straightforward algebraic manipulation of r_{ij} in equation 7.5 provides an equation for the diagonal elements of the error covariance submatrix Σ_r, in equation 7.10, for the standard multilevel model for change, in terms of *TIME* and the model's variance components. Under the standard multilevel model for change, the population variance of the composite residual at *TIME* t_j is:

$$\sigma^2_{r_j} = \text{Var}(\varepsilon_{ij} + \zeta_{0i} + \zeta_{1i}t_j) = \sigma^2_\varepsilon + \sigma^2_0 + 2\sigma_{01}t_j + \sigma^2_1 t^2_j. \qquad (7.11)$$

We can use this equation to obtain estimates of composite residual variance on each occasion for the opposites-naming data. Substituting the four associated values of *TIME* (0, 1, 2, and 3) and estimates of the variance components from table 7.2 into equation 7.11, we have:

$$\hat{\sigma}^2_{r_1} = \hat{\sigma}^2_\varepsilon + \hat{\sigma}^2_0 + 2\hat{\sigma}_{01}(0) + \hat{\sigma}^2_1(0^2)$$
$$= 159.5 + 1236.4 + 2 \times -178.2(0) + 107.3(0^2) = 1395.9$$
$$\hat{\sigma}^2_{r_2} = \hat{\sigma}^2_\varepsilon + \hat{\sigma}^2_0 + 2\hat{\sigma}_{01}(1) + \hat{\sigma}^2_1(1^2)$$
$$= 159.5 + 1236.4 + 2 \times -178.2(1) + 107.3(1^2) = 1146.8$$
$$\hat{\sigma}^2_{r_3} = \hat{\sigma}^2_\varepsilon + \hat{\sigma}^2_0 + 2\hat{\sigma}_{01}(2) + \hat{\sigma}^2_1(2^2)$$
$$= 159.5 + 1236.4 + 2 \times -178.2(2) + 107.3(2^2) = 1112.3$$
$$\hat{\sigma}^2_{r_4} = \hat{\sigma}^2_\varepsilon + \hat{\sigma}^2_0 + 2\hat{\sigma}_{01}(3) + \hat{\sigma}^2_1(3^2)$$
$$= 159.5 + 1236.4 + 2 \times -178.2(3) + 107.3(3^2) = 1294.4.$$

Rewriting the estimated error covariance sub-matrix $\hat{\Sigma}_r$ in equation 7.10 with its diagonal entries replaced by their estimates, we have:

$$\hat{\Sigma}_r = \begin{bmatrix} 1395.9 & \hat{\sigma}_{r_1 r_2} & \hat{\sigma}_{r_1 r_3} & \hat{\sigma}_{r_1 r_4} \\ \hat{\sigma}_{r_2 r_1} & 1146.8 & \hat{\sigma}_{r_2 r_3} & \hat{\sigma}_{r_2 r_4} \\ \hat{\sigma}_{r_3 r_1} & \hat{\sigma}_{r_3 r_2} & 1112.3 & \hat{\sigma}_{r_3 r_4} \\ \hat{\sigma}_{r_4 r_1} & \hat{\sigma}_{r_4 r_2} & \hat{\sigma}_{r_4 r_3} & 1294.4 \end{bmatrix}. \qquad (7.12)$$

So, under the standard multilevel model for change, composite residual variance for the opposites-naming data differs across occasions, revealing anticipated heteroscedasticity. For the opposites-naming data, composite residual variance is greatest at the beginning and end of data collection and smaller in between. And, while not outrageously heteroscedastic, this situation is clearly beyond the bland homoscedasticity that we routinely assume for residuals in cross-sectional data.

Based on the algebraic representation in equation 7.11, what can we say about the general temporal dependence of composite residual variance in the "standard" multilevel model for change? We can gain insight into this question by *completing the square* in equation 7.11:

$$\sigma_{r_j}^2 = \left(\sigma_\varepsilon^2 + \frac{\sigma_0^2 \sigma_1^2 - \sigma_{01}^2}{\sigma_1^2} \right) + \sigma_1^2 \left(t_j + \frac{\sigma_{01}}{\sigma_1^2} \right)^2. \qquad (7.13)$$

Because t_j appears in a term that is *squared,* equation 7.13 indicates that composite residual variance in the "standard" multilevel model for change has a *quadratic* dependence on time. It will be at its minimum at time $t = -(\sigma_{01}/\sigma_1^2)$ and will increase *parabolically and symmetrically* over time on either side of this minimum. For the opposites-naming data, we have:

$$-\frac{\hat{\sigma}_{01}}{\hat{\sigma}_1^2} = -\left(\frac{-178.23}{107.25} \right) = 1.66$$

$$\left(\hat{\sigma}_\varepsilon^2 + \frac{\hat{\sigma}_0^2 \hat{\sigma}_1^2 - \hat{\sigma}_{01}^2}{\hat{\sigma}_1^2} \right) = \left(159.48 + \frac{(1236.41)(107.25) - (-178.23)^2}{107.25} \right) = 1099.7,$$

which tells us that, under the standard multilevel model for change, the composite residual variance has an estimated minimum of almost 1100, occurring about two-thirds of the way between the second and third measurement occasions in the case of the opposites-naming data.

So, ask yourself! Does it make sense to assume, in real data—as the "standard" multilevel model for change does implicitly—that composite residual variance increases parabolically over time from a single minimum? For the "standard" model to make sense, and be applied in the real world, your answer must be yes. But are other patterns of heteroscedasticity possible (or likely)? In longitudinal data, might residual heteroscedasticity possess both a minimum *and* a maximum? Might there be even *multiple* minima and maxima? Might composite residual variance *decline* from a maximum, on either side of some fiducial time, rather than increasing from a minimum? Although compelling, none of these options is possible under the "standard" multilevel model for change.

Before concluding that the model we have spent so long developing is perhaps untenable because of the restriction it places on the error covariance matrix, let us quickly offer some observations that we hope will assuage your concerns. Although the "standard" multilevel model for change assumes that composite residual variance increases parabolically from a minimum with time, the temporal dependence of residual heteroscedasticity need not be *markedly* curved. The magnitude of the curvature depends intimately on the magnitude of the model's variance/covariance components. If all three level-2 components—σ_0^2, σ_1^2, and σ_{01}—are near zero, for example, the error covariance matrix is actually close to *homoscedastic*, with common variance σ_ε^2. Or, if level-2 residual slope variability, σ_1^2, and residual initial status/slope covariance, σ_{01}, are near zero, composite residual variance will still be homoscedastic, but with common variance $(\sigma_\varepsilon^2 + \sigma_0^2)$. In both cases, the "curvature" of the parabolic temporal dependence approaches zero and heteroscedasticity flattens.

In our own experience, these situations are common. The first occurs when the level-2 predictors "explain" most, or all, of the between-person variation in initial status and rate of change. The second occurs when the slopes of the change trajectories do not differ much across people—a common occurrence when study duration is short. Finally, as the sizes of the residual slope variance σ_1^2 and initial status/slope covariance, σ_{01}, differ relative to one another, the time at which minimum residual variance occurs can easily move beyond the temporal limits of the period of observation. When this happens, which is often, no minimum is evident within the period of observation, the composite residual variance appears to either increase or decrease monotonically over the time period under study. We conclude from these special cases and the general temporal dependence of the residual variance that, while the composite residual variance is indeed functionally constrained in the "standard" multilevel model for change, it is also capable of adapting itself relatively smoothly to many common empirical situations. Nonetheless, in any analysis of change, it makes great sense to check the hypothesized structure of the error covariance matrix—whether obtained implicitly, by adopting the standard model, or not—against data just as it is important to check the tenability of the hypothesized structure of the fixed effects. We illustrate the checking process in section 7.3.

7.2.2 Covariance of the Composite Residuals

We now examine the temporal dependence in the *covariance* of the composite residuals in the "standard" multilevel model for change. These covariances appear in the off-diagonal elements of the error covariance submatrix Σ_r, in equation 7.10. Again, mathematical manipulation of the composite residual in equation 7.5 provides the covariance between composite residuals at *TIMES* t_j and and $t_{j'}$:

$$\sigma_{r_j r_{j'}} = \sigma_0^2 + \sigma_{01}(t_j + t_{j'}) + \sigma_1^2 t_j t_{j'}, \tag{7.13}$$

where all terms have their usual meanings. For the opposites-naming data, substitution of appropriate values for time and estimates of the variance components from table 7.2 let us fill out the rest of $\hat{\Sigma}_r$ in equation 7.12 with numerical values:

$$\hat{\Sigma}_r = \begin{bmatrix} 1395.9 & 1058.2 & 880.0 & 701.7 \\ 1058.2 & 1146.8 & 916.2 & 845.2 \\ 880.0 & 916.2 & 1112.3 & 988.8 \\ 701.7 & 845.2 & 988.8 & 1294.4 \end{bmatrix}. \tag{7.14}$$

Notice the somewhat imperfect "band diagonal" structure, in which the overall magnitude of the residual covariances tends to decline in diagonal "bands" the further you get from the main diagonal. The magnitude of the residual covariance is around 900 to 1050 in the band immediately below the main diagonal, between 840 to 880 in the band beneath that, and about 700 in the band beneath that. We often anticipate a band diagonal structure in longitudinal studies because we expect the strength of the correlation between pairs of residuals to decline as they become more temporally remote, within person.

The expression for the covariance between composite residuals in equation 7.13 and the estimated error covariance matrix in equation 7.14 allow us to make some general comments about the temporal dependence of the composite residual covariance in the "standard" multilevel model for change. The dependence is powerful, principally because the covariance contains the *product* of pairs of times (the third term in equation 7.13). This product dramatically affects the magnitude of the error covariance when time values are large. Special cases are also evident—as in equation 7.14—the magnitude of the error covariance depends on the magnitudes of the three level-2 variance components. If all three level-2 components are close to zero, the composite residual covariances will also be near zero and the error covariance matrix in equations 7.9 and 7.10 becomes diagonal (in addition to being homoscedastic, as described in section 7.2.1). Regular OLS assumptions then apply, even for longitudinal data. Similarly, if only the level-2 residual slope variability, σ_1^2, and residual initial status/slope covariance, σ_{01}, are both vanishingly small, then the composite residual covariance takes on a constant value, σ_0^2. In this case, the error covariance matrix is *compound symmetric*, with the following structure:

$$\Sigma_r = \begin{bmatrix} \sigma_\varepsilon^2 + \sigma_0^2 & \sigma_0^2 & \sigma_0^2 & \sigma_0^2 \\ \sigma_0^2 & \sigma_\varepsilon^2 + \sigma_0^2 & \sigma_0^2 & \sigma_0^2 \\ \sigma_0^2 & \sigma_0^2 & \sigma_\varepsilon^2 + \sigma_0^2 & \sigma_0^2 \\ \sigma_0^2 & \sigma_0^2 & \sigma_0^2 & \sigma_\varepsilon^2 + \sigma_0^2 \end{bmatrix}. \tag{7.15}$$

Compound symmetric error covariance structures are particularly common in longitudinal data, especially if the slopes of the change trajectories do not differ much across people. Regardless of these special cases, however, the most sensible question to ask of your data is whether the error covariance structure that the "standard" multilevel model for change demands is realistic when applied to data in practice? The answer to this question will determine whether the standard model can be applied ubiquitously, a question we soon address in section 7.3.

7.2.3 Autocorrelation of the Composite Residuals

Finally, for descriptive purposes, we can also estimate the autocorrelations imposed among the composite residuals in the "standard" multilevel model for change. Applying the usual formula for computing a correlation coefficient from two variances and their covariance, we have:

$$\rho_{r_j r_{j'}} = \sigma_{r_j r_{j'}} / \sqrt{\sigma_{r_j}^2 \sigma_{r_{j'}}^2} \, ,$$

which yields a composite residual autocorrelation matrix of:

$$\begin{bmatrix} 1.00 & 0.84 & 0.71 & 0.52 \\ 0.84 & 1.00 & 0.81 & 0.69 \\ 0.71 & 0.81 & 1.00 & 0.82 \\ 0.52 & 0.69 & 0.82 & 1.00 \end{bmatrix}.$$

The approximate band-diagonal substructure of the error covariance matrix in the "standard" model is even more apparent in the error correlation matrix. For observations separated by one week, the residual autocorrelation is about 0.8; for observations separated by two weeks, the residual autocorrelation is about 0.70; for observations separated by three weeks, the residual autocorrelation is about 0.5. These magnitudes, regardless of temporal placement, are considerably larger than the *zero* autocorrelation anticipated among residuals in an OLS analysis.

7.3 Postulating an Alternative Error Covariance Structure

To postulate an appropriate multilevel model for change, any properties imposed on the model's composite residual—either implicitly by the assumptions of the model itself, or explicitly—must match those required by data. In specifying the model's stochastic portion, you should allow for heteroscedasticity and autocorrelation among the composite residuals. But what type of heteroscedasticity and autocorrelation makes the most sense? Is the composite residual *as specified by default* in the "standard" multilevel model for change, uniformly appropriate? Do its random effects always have the properties required of real-world residuals in the study of change? If you can answer yes to these questions, the "standard" multilevel model for change makes sense. But to determine whether you can safely answer yes, it is wise to evaluate the credibility of some plausible alternative error covariance structures, as we do now.

Fortunately, it is easy to specify alternative covariance structures for the composite residual and determine analytically which specification—the "standard" or an alternative—fits best. You already possess the analytic tools and skills needed for this work. After hypothesizing alternative models—as we describe below—you can use familiar goodness of fit statistics (deviance, AIC, and BIC) to compare their performance. Each model will have identical fixed effects but a different error covariance structure. The main difficulty you will encounter is not doing the analysis itself but rather identifying the error structures to investigate from among the dizzying array of options.

Table 7.3 presents six particular error covariance structures that we find to be the most useful in longitudinal work: *unstructured, compound symmetric, heterogeneous compound symmetric, autoregressive, heterogeneous autoregressive* and *Toeplitz.* The table also presents the results of fitting the multilevel model for change in equation 7.6 to the opposites-naming data, imposing each of the designated error structures. The table also presents selected output from these analyses: goodness-of-fit statistics; parameter estimates for the variance components and approximate p-values; and the fitted error covariance matrix of the composite residual, $\hat{\Sigma}_r$. As in table 7.2, we fit these models with SAS PROC MIXED and restricted ML. Because each has identical fixed effects, we could have used either full or restricted methods to compare models. We chose restricted methods because the obtained goodness-of-fit statistics then reflect only the fit of only model's stochastic portion, which is our focus here.

You compare these models in the usual way. A smaller deviance statistic indicates better fit, but because an improvement generally requires additional parameters, you must either formally test the hypotheses (if the models are nested) or use AIC and BIC statistics. Both penalize the log-likelihood of the fitted model for the number of parameters estimated, with the BIC exacting a higher penalty for increased complexity. The smaller the AIC and BIC statistics, the better the model fits.

7.3.1 Unstructured Error Covariance Matrix

An unstructured error covariance matrix is exactly what you would anticipate from its name: it has a general structure, in which each element of

Table 7.3: Selection of alternative error covariance matrices for use with the multilevel model for change in opposite naming, including goodness-of-fit statistics, variance component estimates, and fitted error covariance matrix

Description	Hypothesized error covariance structure, Σ_r	Goodness-of-fit			Variance components		Fitted error covariance Matrix, $\hat{\Sigma}_r$
		$-2LL$	AIC	BIC	Parameter	Estimate	
Unstructured	$\begin{bmatrix} \sigma_1^2 & \sigma_{12} & \sigma_{13} & \sigma_{14} \\ \sigma_{21} & \sigma_2^2 & \sigma_{23} & \sigma_{24} \\ \sigma_{31} & \sigma_{32} & \sigma_3^2 & \sigma_{34} \\ \sigma_{41} & \sigma_{42} & \sigma_{43} & \sigma_4^2 \end{bmatrix}$	1255.8	1275.8	1291.3	σ_1^2 σ_2^2 σ_3^2 σ_4^2 σ_{21} σ_{31} σ_{32} σ_{41} σ_{42} σ_{43}	1344.8*** 1150.3*** 1235.7*** 1205.9*** 1005.6*** 946.1*** 1028.4*** 583.1* 846.5*** 969.2***	$\begin{bmatrix} 1344.8 & 1005.6 & 946.1 & 583.1 \\ 1005.6 & 1150.3 & 1028.4 & 846.5 \\ 946.1 & 1028.4 & 1235.7 & 969.2 \\ 583.1 & 846.5 & 969.2 & 1205.9 \end{bmatrix}$
Compound symmetry	$\begin{bmatrix} \sigma^2+\sigma_1^2 & \sigma_1^2 & \sigma_1^2 & \sigma_1^2 \\ \sigma_1^2 & \sigma^2+\sigma_1^2 & \sigma_1^2 & \sigma_1^2 \\ \sigma_1^2 & \sigma_1^2 & \sigma^2+\sigma_1^2 & \sigma_1^2 \\ \sigma_1^2 & \sigma_1^2 & \sigma_1^2 & \sigma^2+\sigma_1^2 \end{bmatrix}$	1287.0	1291.0	1294.2	σ^2 σ_1^2	331.3*** 900.1***	$\begin{bmatrix} 1231.4 & 900.1 & 900.1 & 900.1 \\ 900.1 & 1231.4 & 900.1 & 900.1 \\ 900.1 & 900.1 & 1231.4 & 900.1 \\ 900.1 & 900.1 & 900.1 & 1231.4 \end{bmatrix}$
Heterogeneous compound symmetry	$\begin{bmatrix} \sigma_1^2 & \sigma_1\sigma_2\rho & \sigma_1\sigma_3\rho & \sigma_1\sigma_4\rho \\ \sigma_2\sigma_1\rho & \sigma_2^2 & \sigma_2\sigma_3\rho & \sigma_2\sigma_4\rho \\ \sigma_3\sigma_1\rho & \sigma_3\sigma_2\rho & \sigma_3^2 & \sigma_3\sigma_4\rho \\ \sigma_4\sigma_1\rho & \sigma_4\sigma_2\rho & \sigma_4\sigma_3\rho & \sigma_4^2 \end{bmatrix}$	1285.0	1295.0	1302.7	σ_1^2 σ_2^2 σ_3^2 σ_4^2 ρ	1438.0*** 1067.7*** 1147.9*** 1305.6*** 0.7367***	$\begin{bmatrix} 1438.0 & 912.9 & 946.5 & 1009.5 \\ 912.9 & 1067.7 & 815.6 & 869.8 \\ 946.5 & 815.6 & 1147.9 & 901.9 \\ 1009.5 & 869.8 & 901.9 & 1305.6 \end{bmatrix}$

Model	Covariance structure	Fit statistics			Parameters	Estimated covariance matrix			
Autoregressive	$\begin{bmatrix} \sigma^2 & \sigma^2\rho & \sigma^2\rho^2 & \sigma^2\rho^3 \\ \sigma^2\rho & \sigma^2 & \sigma^2\rho & \sigma^2\rho^2 \\ \sigma^2\rho^2 & \sigma^2\rho & \sigma^2 & \sigma^2\rho \\ \sigma^2\rho^3 & \sigma^2\rho^2 & \sigma^2\rho & \sigma^2 \end{bmatrix}$	1265.9	1269.9	1273.0	σ^2 = 1256.7*** ρ = 0.8253***	1256.7	1037.2	856.1	706.6
						1037.2	1256.7	1037.2	856.1
						856.1	1037.2	1256.7	1037.2
						706.6	856.1	1037.2	1256.7
Heterogeneous autoregressive	$\begin{bmatrix} \sigma_1^2 & \sigma_1\sigma_2\rho & \sigma_1\sigma_3\rho^2 & \sigma_1\sigma_4\rho^3 \\ \sigma_2\sigma_1\rho & \sigma_2^2 & \sigma_2\sigma_3\rho & \sigma_2\sigma_4\rho^2 \\ \sigma_3\sigma_1\rho^2 & \sigma_3\sigma_2\rho & \sigma_3^2 & \sigma_3\sigma_4\rho \\ \sigma_4\sigma_1\rho^3 & \sigma_4\sigma_2\rho^2 & \sigma_4\sigma_3\rho & \sigma_4^2 \end{bmatrix}$	1264.8	1274.8	1282.6	σ_1^2 = 1340.7*** σ_2^2 = 1111.1*** σ_3^2 = 1213.2*** σ_4^2 = 1233.9*** ρ = 0.8199***	1340.7	1000.7	857.3	708.9
						1000.7	1111.1	951.9	787.1
						857.3	951.9	1231.2	1003.1
						708.9	787.1	1003.1	1233.9
Toeplitz	$\begin{bmatrix} \sigma^2 & \sigma_1 & \sigma_2 & \sigma_3 \\ \sigma_1 & \sigma^2 & \sigma_1 & \sigma_2 \\ \sigma_2 & \sigma_1 & \sigma^2 & \sigma_1 \\ \sigma_3 & \sigma_2 & \sigma_1 & \sigma^2 \end{bmatrix}$	1258.1	1266.1	1272.3	σ^2 = 1246.9*** σ_1 = 1029.3*** σ_2 = 896.6*** σ_3 = 624.1**	1246.9	1029.3	896.6	624.1
						1029.3	1246.9	1029.3	896.6
						896.6	1029.3	1246.9	1029.3
						624.1	896.6	1029.3	1246.9

~p < .10; *p < .05; **p < .01; ***p < .001.

Note: SAS PROC MIXED, Restricted ML.

Σ_r takes on the value that the data demand. For the opposites-naming data, an unstructured error covariance matrix has 10 unknown parameters: 4 variances and 6 covariances. In table 7.3, we represent these parameters as σ_1^2, σ_2^2, σ_3^2, σ_4^2, σ_{21}, σ_{31}, σ_{32}, σ_{41}, σ_{42}, and σ_{43}. (Notice that in expressing the various error covariance matrices in table 7.3, we constantly reuse the same symbols—σ^2, σ_1^2, σ_2^2, σ_{21}, ρ, and so on. Use of the same symbol does not imply that we are estimating the same parameter. For example, we use the symbol σ_1^2 for two entirely different purposes in the unstructured and compound symmetric error structures and each of these differs from its use in the level-2 submodel in equation 7.2b)

The great appeal of an unstructured error covariance structure is that it places no restrictions on the structure of Σ_r. For a given set of fixed effects, its deviance statistic will always be the smallest of any error covariance structure. If you have just a few waves of data, this choice can be attractive. But if you have many waves, it can require an exorbitant number of parameters. For 20 waves, you would need 20 variance parameters and 190 covariance parameters—210 parameters in all—whereas the "standard" model requires only 3 variance components (σ_0^2, σ_1^2, and σ_ε^2) and one covariance component, σ_{01}.

In most analyses, a more parsimonious structure is desirable. Yet because the unstructured error covariance model always has the lowest deviance statistic, we usually begin exploratory comparisons here. For the opposites-naming data, we find a deviance statistic of 1255.8 for this model, about 4.5 points *less* than that for the "standard" model. But this modest improvement uses up 10 degrees of freedom (as opposed to the 4 in the "standard" model). It should come as no surprise then that the AIC and BIC statistics, which both penalize us for overuse of unknown parameters, are *larger* under this assumption than they are under the "standard" multilevel model (1275.8 vs. 1268.3 for AIC; 1291.3 vs. 1274.5 for BIC). So, of the two potential error structures, we prefer the "standard" to the unstructured. The excessive size of BIC, in particular (it is 16.8 points larger!), suggests that we are "wasting" considerable degrees of freedom in choosing an unstructured form for Σ_r.

7.3.2 Compound Symmetric Error Covariance Matrix

A *compound symmetric* error covariance matrix requires just two parameters, labeled σ^2 and σ_1^2 in table 7.3. Under compound symmetry, the diagonal elements of Σ_r are homoscedastic (with variance $\sigma^2 + \sigma_1^2$) on all occasions, and all pairs of residuals have a constant covariance, regardless of the times with which they are associated.

As we would expect, this model fits less well than the multilevel model with an unstructured Σ_r. But it also fits less well than the "standard" multilevel model. All three of its goodness-of-fit statistics are much larger: deviance is 26.7 points larger, AIC is 22.7 points larger, and BIC is 19.7 points larger. Interestingly, as specified in equation 7.15, a compound symmetric Σ_r is a special case of the "standard" model, when there is little or no residual variation (and hence no residual covariation) in the true slopes of the change trajectories across people. Since we know, from hypothesis tests in table 7.2, that the residual slope variability and covariability are *not* zero for these data, it comes as no surprise that compound symmetry is not an acceptable error covariance structure for these data. This form is most attractive, then, when you find little or no residual variance in slopes among the individual change trajectories.

7.3.3 Heterogeneous Compound Symmetric Error Covariance Matrix

The third error covariance matrix in table 7.3 is *heterogeneous compound symmetric*. In our example, this extension of the compound symmetric structure requires five parameters. Under heterogeneous compound symmetry, the diagonal elements of Σ_r are heteroscedastic (with variances σ_1^2, σ_2^2, σ_3^2, and σ_4^2 on each occasion for these data). In addition, all pairs of errors have their own covariance (you can see this most easily in the fitted error covariance matrix in table 7.3). Specifically, these covariances are the products of the corresponding error *standard deviations* and a constant error autocorrelation parameter, labeled ρ, whose magnitude is always less than or equal to unity.

Based on the deviance statistics alone, a model with a heterogeneously compound symmetric Σ_r fits the opposites-naming data better than a compound symmetric model (1285.0 vs. 1287.0), but still not as well as the "standard" (1285 vs. 1260.3). So, too, the AIC and BIC statistics penalize the heterogeneous compound symmetry model for its additional parameters (AIC = 1295.0; BIC = 1302.7), over both the compound symmetric and the "standard." We conclude that the heterogeneous compound symmetric model is probably less acceptable—for these data—than any other multilevel model fit so far.

7.3.4 Autoregressive Error Covariance Matrix

The fourth potential error covariance matrix in table 7.3 has an *autoregressive* (actually, *first-order* autoregressive) structure. Many researchers are drawn to an autoregressive error structure because its "band-diagonal" shape seems appropriate for growth processes. When Σ_r is first-order autoregressive, the elements on the main diagonal of Σ_r are homoscedastic (with variance σ^2). In addition, pairs of errors have identical covariances in bands parallel to the leading diagonal (again, examine the fitted error covariance matrix in table 7.3). These covariances are the product of the residual variance, σ^2, and an error autocorrelation parameter, labeled ρ, whose magnitude is again always less than, or equal to, unity. Error variance σ^2 is multiplied by ρ to provide the error covariances in the first band immediately below the leading diagonal, by ρ^2 in the band beneath that, by ρ^3 in the band beneath that, and so on. Thus, because the magnitude of ρ is always fractional, the error covariances in the bands of Σ_r decline, the further you go from the leading diagonal. Although an autoregressive Σ_r "saves" considerable degrees of freedom—it uses only two variance components—its elements are tightly constrained: the identical covariances in any band must be the same fraction of any entry in the previous band as are the entries in the following band of them.

Although the autoregressive model fits the opposites-naming data reasonably well, its constraints on the variance components' relative magnitudes prevent it from fitting as well as the "standard" multilevel model for change. Both the deviance statistic (1265.9) and AIC statistic (1269.9) are slightly larger than their peers in the "standard" multilevel model. On the other hand, the BIC statistic is slightly smaller in this model than it is in the "standard" (1273.0 for the former, 1274.5 for the latter), owing to the burden of additional parameters (4 vs. 2) in the "standard." Interestingly, although it cannot compete in a world of deviance, the autoregressive model is superior to the unstructured model according to AIC and BIC (as you might expect, given the number of the unknown parameters required by each model, 2 vs. 10).

7.3.5 Heterogeneous Autoregressive Error Covariance Matrix

The heterogeneous autoregressive error structure is a relaxed version of the strict autoregressive structure just described. Its main diagonal elements are heteroscedastic (with variances σ_1^2, σ_2^2, σ_3^2, and σ_4^2 for the four waves here). In addition, the bands of constant covariances between pairs of errors that appeared parallel to the main diagonal in the regular autoregressive model are free to differ in magnitude along the bands (again, examine the fitted error covariance matrix in table 7.3). This is achieved by multiplying the same error autocorrelation parameter, ρ, that appeared above by the product of the relevant error standard deviations. Thus, the band diagonal structure—with the magnitudes of the covariances declining across the bands, away from the main diagonal—is somewhat preserved, but loosened by the inclusion of additional variance components. A model with heterogeneous autoregressive Σ_r spends additional degrees of freedom, but benefits from additional flexibility over its simpler sibling.

As you might expect, the model with heterogeneous autoregressive error structure benefits in terms of the deviance statistic over the homogeneous autoregressive case, but can be penalized from the perspective of AIC and BIC. For these data, the model with heterogeneous autoregressive Σ_r fits less well than the "standard" multilevel model for change. Notice that, in the heterogeneous autoregressive model, the deviance statistic (1264.8), AIC (1274.8), and BIC (1282.6) are all larger than the equivalent statistics in the "standard." As with its homogeneous sibling, although it cannot compete in terms of deviance, the heterogeneous autoregressive model is superior to the unstructured model according to both the AIC and BIC statistics.

7.3.6 Toeplitz Error Covariance Matrix

For the opposites-naming data, the Toeplitz error covariance structure represents a far superior option. The Toeplitz structure has some of the characteristics of the autoregressive structure, in that it has bands of identical covariances arrayed parallel to the main diagonal. However, these elements are not forced to be an identical fraction of the elements in the prior band. Instead, their magnitudes within each band are determined by the data and are not constrained to stand in identical ratios to one another. For the opposites-naming data, we need four variance components to specify a Toeplitz structure (σ^2, σ_1, σ_2, and σ_3 in table 7.3) and so Σ_r is more flexible than the homogeneous autoregressive structure but more parsimonious that its heterogeneous sibling.

For these data, a Toeplitz error covariance structure fits better than the "standard" multilevel model for change *and* better than all other error covariance structures we have tested, regardless of which goodness-of-fit statistic you consult. The deviance statistic is 1258.1 (as opposed to 1260.3), AIC is 1266.1 (as opposed to 1268.3), and BIC is 1272.3 (as opposed to 1274.5). As we discuss below, however, these differences in goodness-of-fit are relatively small.

7.3.7 Does Choosing the "Correct" Error Covariance
Structure Really Matter?

The error covariance structures presented in table 7.3 are but the beginning. Even though the Toeplitz structure appears marginally more successful than the implicit error covariance structure of the "standard" multilevel model for change, it is entirely possible that there are other error structures that would be superior for these data. Such is the nature of all data analysis. In fitting these alternative models, we have refined our estimates of the variance components and have come to understand better the model's stochastic component. We would argue that, for these data, the "standard" multilevel model for change performs well—its deviance, AIC, and BIC statistics are only marginally worse than those of the Toeplitz model. The difference in BIC statistics—2.2 points—is so small that adopting Raftery's (1995) guidelines, we would conclude that there is only weak evidence that adoption of a Toeplitz error structure improves on the "standard" multilevel model.

If you focus exclusively on the deviance statistic, however, an unstructured error covariance matrix always leads to the best fit. This model will always fit better than the "standard" model, and than any other model that is constrained in some way. The question is: How much do we sacrifice if we choose the unstructured model over these others? For these data, it cost 10 degrees of freedom to achieve this best fit for the model's stochastic portion—five more degrees of freedom than any other error covariance structure considered here. Although some might argue that losing an additional handful of degrees of freedom is a small price to pay for optimal modeling of the error structure—a consequence of the fact that we have only four waves of data—in other settings, with larger panels of longitudinal data, few would reach this conclusion.

Perhaps most important, consider how choice of an error covariance structure affects our ability to address our research questions, especially given that it is the *fixed effects*—and not the *variance components*—that usually embody these questions. Some might say that refining the error covariance structure for the multilevel model for change is akin to rearranging the deck chairs on the *Titanic*—it rarely fundamentally changes our parameter estimates. Indeed, regardless of the error structure chosen, estimates of the fixed effects are unbiased and may not be affected much by choices made in the stochastic part of the model (providing that neither the data, nor the error structure, are idiosyncratic).

But refining our hypotheses about the error covariance structure *does* affect the *precision* of estimates of the fixed effects and will therefore impact hypothesis testing and confidence interval construction. You can see this happening in table 7.4, which displays estimates of the fixed effects and asymptotic standard errors for three multilevel models for change in opposites-naming: the "standard" model (from table 7.2) and models with a Toeplitz and unstructured error covariance matrix (from table 7.3). Notice that that the magnitudes of the estimated fixed effects

Table 7.4: Change in opposite's naming score over a four-week period, as a function of baseline IQ

		Parameter	Model with . . .		
			Standard error covariance structure	Toeplitz error covariance structure	Unstructured error covariance structure
Fixed Effects					
Initial status, π_{0i}	Intercept	γ_{00}	164.37***	165.10***	165.83***
			(6.206)	(5.923)	(5.952)
	$(COG - \overline{COG})$	γ_{01}	−0.11	−0.00	−0.07
			(0.504)	(0.481)	(0.483)
Rate of change, π_{1i}	Intercept	γ_{10}	26.96***	26.895***	26.58***
			(1.994)	(1.943)	(1.926)
	$(COG - \overline{COG})$	γ_{11}	0.43**	0.44**	0.46**
			(0.162)	(0.158)	(0.156)
Goodness-of-fit					
	Deviance		1260.3	1258.1	1255.8
	AIC		1268.3	1266.1	1275.8
	BIC		1274.5	1272.3	1291.3

$\sim p < .10$; $* p < .05$; $** p < .01$; $*** p < .001$.

Parameter estimates (standard errors), approximate p-values, and goodness-of-fit statistics after fitting a multilevel model for change with standard, Toeplitz and unstructured error covariance structures ($n = 35$).

Note: SAS PROC MIXED, Restricted ML.

are relatively similar (except, as you might expect, for γ_{01}, which is not statistically significant anyway). But also notice that the respective asymptotic standard errors decline as the error covariance structure is better represented. The standard errors are generally smaller in the Toeplitz and unstructured models than in the "standard," although differences between the Toeplitz and unstructured models are less unanimous. You should find it reassuring that—given the widespread application of the "standard" multilevel model for change—the differences in precision shown here are small and likely inconsequential. Of course, this conclusion is specific to these data; ensuing differences in precision may be greater in some data sets, depending on the design, the statistical model, the choices of error covariance structure, and the nature of the forces that bind the repeated observations together. To learn more about this topic, we refer interested readers to Van Leeuwen (1997); Goldstein, Healy, and Rasbash (1994); and Wolfinger (1993, 1996).

8

Modeling Change Using Covariance Structure Analysis

Change does not necessarily assure progress, but progress
implacably requires change.
— Henry S. Commager

Interestingly, the multilevel model for change can be mapped directly
onto the general mathematical framework provided by *covariance
structure analysis* (also known as *structural equation modeling*). The resulting analytic approach has become known as *latent growth modeling*. At its
core, a latent growth model is essentially a multilevel model for change.
But not only does the mapping of the multilevel model for change onto
the general covariance structure model provide an alternative approach
to model specification and estimation, its flexibility can dramatically
extend your analytic reach.

In this chapter, we describe how to conceptualize, postulate, fit, and
interpret a latent growth model. We begin, in section 8.1, by reviewing the
general covariance structure model. Here, as in the rest of the chapter, we
assume a basic level of familiarity with covariance structure analysis. If you
lack this foundation, we suggest that you review the method using one of
the excellent books that are available (e.g., Bollen, 1989). In section 8.2,
we map the multilevel model for change onto the general covariance
structure model. In section 8.3, we illustrate an invaluable extension of
this approach that allows you to investigate whether change in one construct is related to change in another. We conclude, in section 8.4, by
listing additional extensions of the basic approach briefly.

8.1 The General Covariance Structure Model

Covariance structure analysis (CSA) can be viewed as an extension of multivariate regression analysis and path analysis that contains elements of

both factor analysis and test theory. In a succinct historical overview, Bollen (1989, pp. 4–9) describes its roots in a variety of disciplines including statistics, sociology, economics, psychology, and psychometrics. With the advent of easily accessible computer software, CSA has become commonplace throughout the social sciences.

CSA was initially devised as a comprehensive technique for testing complex hypotheses about relationships among many quantities on a single occasion. Its utility derives not just from its generality but also from the compelling rationale underlying its core statistical model. The model was devised and refined to match researchers' intuitions about the way the world works. Perhaps what is most interesting for longitudinal analysis is that, although originally devised for cross-sectional data, we can manipulate the model so that it can be used to represent change over time.

To support the development of this mapping, in this section we review the basic CSA model. To concretize our presentation, we use Conger, Elder, Lorenz, & Simons (1994) study of the relationships among parental depression, marital conflict, and adolescent adjustment in a cross-sectional sample of 220 White middle-class families. For each family, their data set includes a variety of information about three members: the mother, the father, and a female seventh-grader.

Figure 8.1 presents a *path diagram* that succinctly represents our substantive hypotheses (based upon arguments presented in the original paper). Like all path diagrams, figure 8.1 makes explicit *two* important distinctions:

- Between underlying *constructs* and the *indicators* that measure them
- Between *outcomes* and *predictors*

These distinctions are embodied in the geometric shapes—the circles, rectangles, and arrows—of the path diagram. Before delving into the models, we explore the rationale and consequences of these distinctions.

The first distinction is between a *construct* and its *indicators*. In a path diagram, we represent the former by circles; the latter by squares. The circle in the upper left corner of figure 8.1 contains the construct "Paternal Depression." This construct has two indicators: "Depression Subscale of the SCL-90-R" and "Observer Rating of Depression." A single-headed arrow leads from the construct to each indicator. This representation couples our theoretical interest in the underlying level of the father's depressive mood (the circle) with the recognition that we cannot directly observe his true mood; instead, we measure the values of *indicators* (the squares). Paternal depression is a *construct*; the two ratings are observed *indicators* of its true level.

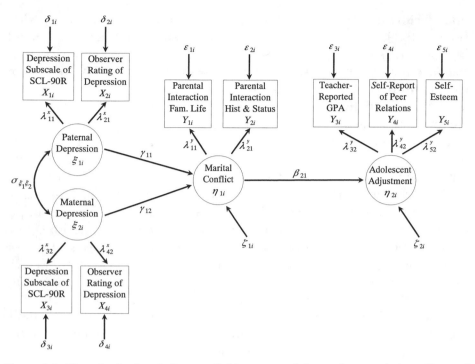

Figure 8.1. Hypothesized path diagram linking parental depression, marital conflict, and adolescent adjustment.

It seems natural to distinguish a construct from its indicators. Depression is something unseen within the father, something *latent* (hidden) that underpins his functioning and *drives* his observed indicators, each in its own metric. This belief—that a construct causes its indicators to take on specific values—is reflected in the direction of the single-headed arrow *from* the construct *to* the indicator. Similar collections of circles, boxes, and arrows link the remaining constructs to their respective indicators. Parallel indicators of "Maternal Depression" appear in the lower left corner. In the center, the construct "Marital Conflict" is measured by two indicators: parental interaction during a discussion of family life (e.g., parenting strategies and household chores) and parental interaction during a discussion of the history and status of the relationship. On the right, the construct "Adolescent Adjustment" is measured by three indicators: teacher-reported GPA, self-report of peer relations, and a standardized measure of self esteem.

Because of the realities of data collection—using real instruments with real people—indicators are *fallible* measures of a construct's true value. Some indicators, such as the standardized depression scale, will measure the construct well. Others, such as an observer's rating of depression, will be less precise. To account for this fallibility, we distinguish the construct's true value from its indicators' observed values and we also allow for the presence of measurement error. We show errors of measurement—one per indicator—using a short single-headed arrow pointing *into* the indicator from the side opposite the arrow denoting the contribution of the construct. As in classical test theory, the path diagram stipulates that an indicator's "observed" value derives from the "true" contribution of the construct plus the "error" introduced by measurement. In CSA, this distinction among "observed," "true," and "error" scores is captured in the specification of a *measurement model* (described below in sections 8.1.1 and 8.1.2).

The second distinction is between "outcomes" and "predictors." Use of these labels implies that you have a theory that makes the outcome a *consequence* of the predictor. Outcomes like marital conflict may be a consequence of predictors like maternal and paternal depression and, in its turn, adolescent adjustment may be affected by parental marital conflict. If parental depression is deeper, marital conflict may be greater. With higher levels of parental conflict at home, the adolescent may be less well adjusted. This is a *causal theory of action*; you expect less depressed parents to live more harmonious lives and their children to be better adjusted. Even though you may not be able to *test* this causal theory with cross-sectional data, you can still *hypothesize* that some effects are *causes* and others are *consequences*.

CSA allows you to make even subtler distinctions between "predictors" and "outcomes." As the path diagram in figure 8.1 suggests, *one predictor's outcome may be another outcome's predictor*. In the first half of the diagram, marital conflict is an outcome, being "caused" by parental depression. In the second half, marital conflict is a predictor of adolescent adjustment. To account for these multiple roles, CSA distinguishes between two conditions: *exogeneity* and *endogeneity*. A construct is *exogenous* if forces entirely outside the hypothesized system determine its values. In figure 8.1, paternal and maternal depression are exogenous; if they have causes, which they undoubtedly do, they do not appear in this system. An *endogenous* variable or construct, in contrast, is determined *within* the system. In this example, marital conflict and adolescent adjustment are endogenous.

We express the central hypotheses in CSA using statements about *relationships among constructs*. This recognizes that, although your data measure the values of indicators, your research questions demand that these indicators be purged of their measurement error when examining the underlying true relationships. You are not interested in the relationship between *observer ratings* of parental depression and marital conflict, but in the relationship between the true values that underlie these indicators, which themselves are related. Although one way of absolving measurement error is to disattenuate sample correlation and regression coefficients using external estimates of reliability, CSA instead uses each construct's multiple indicators to tease out the measurement error from the underlying true score.

A path diagram uses several devices to display simultaneous hypotheses about relationships among constructs. A single-headed arrow represents a hypothesized "causal" relationship between an exogenous and endogenous construct. The two arrows joining paternal and maternal depression (two exogenous constructs) to marital conflict (an endogenous construct) represent a hypothesized link between parents' moods and the health of their marriage. A single-headed arrow can also represent a hypothesized relationship between endogenous constructs, as shown in figure 8.1 for marital conflict and adolescent adjustment. A double-headed arrow allows exogenous constructs to *covary*, as in the extreme left of figure 8.1 which links maternal and paternal depression. This type of relationship is qualitatively different from that hypothesized between exogenous and endogenous constructs or among endogenous constructs. Single-headed arrows represent our causal theories of action—we believe the prior construct *causes* the latter. Double-headed arrows simply allow constructs to be *associated*; we have no theory about whether one causes the other. This is similar to what happens in regular regression, where you hypothesize that predictors "cause" the outcome, but the predictors themselves can be interrelated.

The general CSA model contains two types of submodels: *measurement models*, which distinguish a construct from its indicators; and a *structural model*, which represents the hypothesized nature and interrelationship of the constructs. Because there are two types of constructs—exogenous and endogenous—there are two measurement models. In the sections below, we describe all three: the *X*-measurement model (section 8.1.1), the *Y*-measurement model (section 8.1.2), and the structural model (section 8.1.3).

8.1.1 The *X*-Measurement Model

Our example contains two exogenous constructs: paternal and maternal depression. On the left side of figure 8.1, we provide notation for scores for adolescent i on: (1) the four observed indicators (X_{1i}, X_{2i}, X_{3i}, and X_{4i}); (2) the four corresponding measurement errors (δ_{1i}, δ_{2i}, δ_{3i}, and δ_{4i}); and (3) the two underlying constructs (ξ_{1i}, ξ_{2i}). An *X*-measurement model connects these quantities.

How should we represent this model? Classical test theory coupled with some thought about the nature of empirical measurement suggests one reasonable form. Begin by focusing on the first indicator, the father's score on the depression scale of the SCL-90: X_{1i} in figure 8.1. Classical test theory suggests that this "observed score" is the sum of a "true score" and an "error," which suggests the simple model $X_{1i} = \xi_{1i} + \delta_{1i}$. Although sensible, this specification ignores several issues that arise when we have multiple indicators of each underlying construct. We therefore use the more general representation:

$$X_{1i} = \tau_{x1} + \lambda_{11}^{x}\xi_{1i} + \delta_{1i}, \tag{8.1}$$

where the new symbols, τ_{x1} and λ_{11}^{x}, account for differences in the *centering* and *scaling* of each indicator.

The parameter τ_{x1} represents the mean of observed indicator X_{1i} across all adolescents in the population. We assume that, in the population of fathers, observed depression scores will be scattered around this unknown value, which we estimate using sample data. Including τ_{x1} allows observed scores on different indicators of the same construct to have different means. Conceptually, τ_{x1} "centers" the indicator's observed value on its population average. It allows the observed indicator for adolescent i to be expressed as the sum of: (1) the average across all adolescents; and (2) adolescent i's deviation from that average. This decomposition resembles the familiar decomposition of one-way analysis of variance. All exogenous indicators have their own mean parameters: τ_{x2}, τ_{x3}, and τ_{x4}. Although we do not display these parameters in figure 8.1 to avoid clutter, they are present in the model and estimated during analysis.

The measurement model multiplies the value of the construct for adolescent i, ξ_{1i}, by a scaling factor or "loading," λ_{11}^{x}. The construct ξ_{1i} is similar to the "true score" of classical test theory. A scaling factor allows the indicators and the construct to be measured on different scales. This permits a single latent construct to drive the values of several indicators, each measured in it own metric. The presence of different loadings allows the construct paternal depression (measured in some underlying "true" metric, as yet undetermined) to be simultaneously rescaled into both a depression score and an observer rating.

Finally, a comment on δ_{1i}, adolescent i's "measurement error." After we remove the population mean of X and rescale the underlying construct, what is left? If you subscribe to the tenets of classical test theory, you will respond "measurement error," because you believe that "observed" must equal "true" plus "error." And, so far, we have treated "δ" in the X-measurement model as the "error" that remains after the systematic part of the observed score is removed (or otherwise represented). We prefer, however, to think of the δ part of the measurement model as *that part of indicator X that does not depend on underlying construct ξ*. The δ may be measurement error, but it may also be nothing more than a portion of the observed score that is determined by constructs currently unspecified in the analysis.

We now generalize these ideas to the remaining indicators. The path diagram in figure 8.1 includes two exogenous constructs, each supporting a pair of indicators, each with its own means and loadings. An adolescent whose father has a "high" value on the underlying construct of depression will probably have high values on both associated indicators, the depression scale and the observer rating. Similarly, a mother's depressive mood construct will be reflected in her depression scale and observer rating. Using the format of equation 8.1, we write all four hypothesized measurement relationships as:

$$
\begin{aligned}
X_{1i} &= \tau_{x1} + \lambda^x_{11}\xi_{1i} + \delta_{1i} \\
X_{2i} &= \tau_{x2} + \lambda^x_{21}\xi_{1i} + \delta_{2i} \\
X_{3i} &= \tau_{x3} + \lambda^x_{32}\xi_{2i} + \delta_{3i} \\
X_{4i} &= \tau_{x4} + \lambda^x_{42}\xi_{2i} + \delta_{4i}.
\end{aligned}
\tag{8.2}
$$

Make sure, by reference to figure 8.1, that you understand what each equation represents. Each indicator has a unique mean parameter, loading, and error score, but there are only two underlying "true" scores, ξ_{1i} and ξ_{2i}, corresponding to the two exogenous constructs (paternal and maternal depression).

The four equations in 8.2 constitute the X-measurement model for our example. We can write these statements more parsimoniously in a matrix equation:

$$
\begin{bmatrix} X_{1i} \\ X_{2i} \\ X_{3i} \\ X_{4i} \end{bmatrix} = \begin{bmatrix} \tau_{x1} \\ \tau_{x2} \\ \tau_{x3} \\ \tau_{x4} \end{bmatrix} + \begin{bmatrix} \lambda^x_{11} & 0 \\ \lambda^x_{21} & 0 \\ 0 & \lambda^x_{32} \\ 0 & \lambda^x_{42} \end{bmatrix} \begin{bmatrix} \xi_{1i} \\ \xi_{2i} \end{bmatrix} + \begin{bmatrix} \delta_{1i} \\ \delta_{2i} \\ \delta_{3i} \\ \delta_{4i} \end{bmatrix}.
\tag{8.3}
$$

As you review this matrix representation, we draw your attention to three important features. First, the observed scores on the four exogenous indicators (X_{1i}, X_{2i}, X_{3i}, X_{4i}) appear as a four-element (4×1) *vector*, as do the four mean parameters (τ_{x1}, τ_{x2}, τ_{x3}, τ_{x4}) and the four errors (δ_{1i}, δ_{2i}, δ_{3i}, δ_{4i}). If we had additional indicators of each construct, we would generalize this representation by extending the length of each vector. Second, scores on the two latent constructs (ξ_{1i}, ξ_{2i}) are contained in a two-element (2×1) vector, because we hypothesize that two exogenous constructs generate scores on the four observed indicators. If we hypothesized the existence of additional exogenous constructs, this vector would grow in length. Third, all scaling factors $(\lambda^x_{11}, \lambda^x_{21}, \lambda^x_{32}, \lambda^x_{42})$ appear in a two-dimensional (4×2) *loading matrix* that allows each indicator to be an appropriately rescaled version of its corresponding construct. If you have additional indicators or constructs, you would enlarge the loading matrix accordingly. Notice that four elements in this matrix are set to zero to ensure that indicators X_{1i} and X_{2i} are *not* linked to construct ξ_{2i}, and that indicators X_{3i} and X_{4i} are *not* linked to construct ξ_{1i}. This "each indicator to its own construct" restriction is not required. An indicator can measure more than one construct; the model can easily handle this duplicity by freeing up one or more of the zeroes in the loading matrix. The restrictions in equation 8.3 are simply consequences of the hypotheses embodied in figure 8.1.

We now abbreviate the matrix representation of the X-measurement model in equation 8.3 using bold symbols: $\mathbf{X}$, ξ, and $\boldsymbol{\delta}$ to represent *vectors* of observed, true, and error scores, respectively; $\boldsymbol{\tau}_x$ to represent a *vector* of population means; and $\boldsymbol{\Lambda}_x$ to represent a *matrix* of loadings:

$$\mathbf{X} = \boldsymbol{\tau}_x + \boldsymbol{\Lambda}_x \xi + \boldsymbol{\delta} \tag{8.4}$$

where

$$\mathbf{X} = \begin{bmatrix} X_{1i} \\ X_{2i} \\ X_{3i} \\ X_{4i} \end{bmatrix}, \boldsymbol{\tau}_x = \begin{bmatrix} \tau_{x1} \\ \tau_{x2} \\ \tau_{x3} \\ \tau_{x4} \end{bmatrix}, \boldsymbol{\Lambda}_x = \begin{bmatrix} \lambda^x_{11} & 0 \\ \lambda^x_{21} & 0 \\ 0 & \lambda^x_{32} \\ 0 & \lambda^x_{42} \end{bmatrix}, \xi = \begin{bmatrix} \xi_{1i} \\ \xi_{2i} \end{bmatrix}, \text{and } \boldsymbol{\delta} = \begin{bmatrix} \delta_{1i} \\ \delta_{2i} \\ \delta_{3i} \\ \delta_{4i} \end{bmatrix}. \tag{8.5}$$

In what follows, we use equation 8.4 as "shorthand" for the X-measurement model.[1]

When modeling the behavior of observed indicators, constructs, and error scores, we must also account for variability across individuals. Not only will the unseen values of the latent construct differ across individuals, so will the unseen measurement errors. "True" and "error" variability is then pooled according to the measurement model and is revealed as the interindividual variability in indicators. To handle this eventuality, and to provide variance components that assess this variability, the X-measurement model hypothesizes that the construct and error score vectors are drawn from multivariate normal distributions with mean vectors and covariance matrices, as we now describe.

First, let's deal with the error scores. As with all residuals, we assume that the errors have a zero mean. To model their distribution, we need a covariance matrix that captures hypothesized variation in their values across people. In the X-measurement model, the covariance matrix Θ_δ contains the population variability in the δ's. Since there are four errors in our measurement model, their population covariance matrix is a symmetric four-by-four matrix:

$$\Theta_\delta = \text{Cov}\begin{bmatrix} \delta_{1i} \\ \delta_{2i} \\ \delta_{3i} \\ \delta_{4i} \end{bmatrix} = \begin{bmatrix} \sigma^2_{\delta_1} & \sigma_{\delta_1\delta_2} & \sigma_{\delta_1\delta_3} & \sigma_{\delta_1\delta_4} \\ \sigma_{\delta_2\delta_1} & \sigma^2_{\delta_2} & \sigma_{\delta_2\delta_3} & \sigma_{\delta_2\delta_4} \\ \sigma_{\delta_3\delta_1} & \sigma_{\delta_3\delta_2} & \sigma^2_{\delta_3} & \sigma_{\delta_3\delta_4} \\ \sigma_{\delta_4\delta_1} & \sigma_{\delta_4\delta_2} & \sigma_{\delta_4\delta_3} & \sigma^2_{\delta_4} \end{bmatrix} \tag{8.6}$$

where the main diagonal contains four variance parameters that represent interindividual variability *in* each error and the six covariance parameters capture bivariate associations *among* errors. Notice that the errors need not be *homoscedastic* and *independent*. It is entirely possible that, given the unique metrics of the indicators and their simultaneous measurement, these errors will be both heteroscedastic and intercorrelated.

In some analyses, you will assume that certain elements of this error covariance matrix are zero and that others are identical. For instance, if you think that error variances in depression scores should be identical for mothers and fathers, you would constrain the first and third diagonal elements to be identical. If you think that observer ratings of parental depression are homoscedastic across parents, you would force the second and fourth diagonal elements to be identical. If you think that none of the errors are correlated across indicators, you would set all off-diagonal elements to zero. Or you might selectively constrain a subset to zero to reflect hypotheses about the potential association of the errors of measurement within-indicator across-parent, but not across-indicator within-parent. A beauty of CSA is that you can entertain these various hypotheses just by specifying an appropriate structure for the error covariance matrix and asking whether the model fits your data. We return to this issue in section 8.2.

The underlying "true" latent construct scores also differ across individuals. And so, within the rubric of the X-measurement model, we must allow them to vary. We do this, as you might expect, by assuming that the exogenous constructs are drawn from a multivariate normal distribution, with mean vector κ and covariance matrix Φ. In our example, where the maternal and paternal depression are the two exogenous constructs, the means are contained in κ, a (2×1) vector:

$$\kappa = \text{Mean}\begin{bmatrix} \xi_{1i} \\ \xi_{2i} \end{bmatrix} = \begin{bmatrix} \mu_{\xi_1} \\ \mu_{\xi_2} \end{bmatrix} \tag{8.7}$$

and the variances and covariance in Φ, a (2×2) matrix:

$$\Phi = \mathrm{Cov}\begin{bmatrix} \xi_{1i} \\ \xi_{2i} \end{bmatrix} = \begin{bmatrix} \sigma^2_{\xi_1} & \sigma_{\xi_1\xi_2} \\ \sigma_{\xi_2\xi_1} & \sigma^2_{\xi_2} \end{bmatrix}. \tag{8.8}$$

The two elements in κ represent the population averages of the exogenous constructs of maternal and paternal depression; the elements of Φ represent their variation (two elements, on the diagonal) and covariation (one element, off the diagonal). Of particular interest in equation 8.8 is the parameter in the lower left corner, $\sigma_{\xi_2\xi_1}$. This parameter captures the magnitude and direction of the bivariate association between the two exogenous constructs on the left side of figure 8.1 represented by a double-headed arrow.

We conclude our presentation of the X-measurement model by highlighting a redundancy with important implications. Because the model includes two mean vectors, τ_x and κ, the population means of exogenous indicators are overspecified. Taking expectations throughout equation 8.4 for a generic exogenous indicator X (recalling that the mean of measurement error, δ, is zero), the mean of X is a weighted linear composite of τ_x and μ_ξ, $\mu_X = \tau_x + \lambda^x\mu_\xi$. This expression has two interesting special cases. First, if you force the population mean of the exogenous construct to zero (i.e., set $\mu_\xi = 0$) by constraining the appropriate element of the κ vector to zero, the mean of X must be represented entirely by τ_x, which will become μ_X. If you instead set τ_x to zero, the mean of X must be forced into the corresponding element of κ. In this case, the mean of X will be a rescaled version of the mean of the underlying construct, $\lambda^x\mu_\xi$. This redundancy in the modeling of the means allows for multiple equivalent specifications of the CSA model, as we demonstrate later in section 8.2.1.

8.1.2 The Y-Measurement Model

The Y-measurement model describes relationships among the endogenous constructs and their indicators. Its name derives from the convention of labeling endogenous indicators Y (to correspond to the notion of an "outcome"). The symbol η represents the associated construct and ε represents corresponding errors of measurement.

Figure 8.1 includes two endogenous constructs; their values for adolescent i are represented by η_{1i} (marital conflict) and η_{2i} (adolescent adjustment). Marital conflict supports two indicators, Y_{1i} and Y_{2i}, the two observer ratings of parental interaction. Adolescent adjustment supports three indicators: teacher-reported GPA (Y_{3i}), a self-report of peer relations (Y_{4i}), and a standardized measure of self-esteem (Y_{5i}). Each construct is linked to its corresponding indicators by single-headed arrows with associated scaling factors, λ_{11}^y, λ_{21}^y, λ_{32}^y, λ_{42}^y, and λ_{52}^y. Scores on each indicator have population means of τ_{y1}, τ_{y2}, τ_{y3}, τ_{y4}, and τ_{y5}, and are disturbed by errors of measurement, ε_{1i}, ε_{2i}, ε_{3i}, ε_{4i}, and ε_{5i}, respectively.

A *Y*-measurement model hypothesizes population relationships between endogenous constructs and indicators. As you would expect, it is structurally similar to the *X*-measurement model. There are five endogenous construct/indicator relationships in figure 8.1:

$$Y_{1i} = \tau_{y1} + \lambda_{11}^y \eta_{1i} + \varepsilon_{1i}$$

$$Y_{2i} = \tau_{y2} + \lambda_{21}^y \eta_{1i} + \varepsilon_{2i}$$

$$Y_{3i} = \tau_{y3} + \lambda_{32}^y \eta_{2i} + \varepsilon_{3i} \qquad (8.9)$$

$$Y_{4i} = \tau_{y4} + \lambda_{42}^y \eta_{2i} + \varepsilon_{4i}$$

$$Y_{5i} = \tau_{y5} + \lambda_{52}^y \eta_{2i} + \varepsilon_{5i}.$$

which can be written in matrix form:

$$\begin{bmatrix} Y_{1i} \\ Y_{2i} \\ Y_{3i} \\ Y_{4i} \\ Y_{5i} \end{bmatrix} = \begin{bmatrix} \tau_{y1} \\ \tau_{y2} \\ \tau_{y3} \\ \tau_{y4} \\ \tau_{y5} \end{bmatrix} + \begin{bmatrix} \lambda_{11}^y & 0 \\ \lambda_{21}^y & 0 \\ 0 & \lambda_{32}^y \\ 0 & \lambda_{42}^y \\ 0 & \lambda_{52}^y \end{bmatrix} \begin{bmatrix} \eta_{1i} \\ \eta_{2i} \end{bmatrix} + \begin{bmatrix} \varepsilon_{1i} \\ \varepsilon_{2i} \\ \varepsilon_{3i} \\ \varepsilon_{4i} \\ \varepsilon_{5i} \end{bmatrix}. \qquad (8.10)$$

Elements, vectors, and matrices in this new measurement model have their usual interpretation, and the entire model can be referred to in shorthand as:

$$\mathbf{Y} = \boldsymbol{\tau}_y + \boldsymbol{\Lambda}_y \boldsymbol{\eta} + \boldsymbol{\varepsilon} \qquad (8.11)$$

with

$$\mathbf{Y} = \begin{bmatrix} Y_{1i} \\ Y_{2i} \\ Y_{3i} \\ Y_{4i} \\ Y_{5i} \end{bmatrix}, \boldsymbol{\tau}_y = \begin{bmatrix} \tau_{y1} \\ \tau_{y2} \\ \tau_{y3} \\ \tau_{y4} \\ \tau_{y5} \end{bmatrix}, \boldsymbol{\Lambda}_y = \begin{bmatrix} \lambda_{11}^y & 0 \\ \lambda_{21}^y & 0 \\ 0 & \lambda_{32}^y \\ 0 & \lambda_{42}^y \\ 0 & \lambda_{52}^y \end{bmatrix}, \boldsymbol{\eta} = \begin{bmatrix} \eta_{1i} \\ \eta_{2i} \end{bmatrix}, \text{and } \boldsymbol{\varepsilon} = \begin{bmatrix} \varepsilon_{1i} \\ \varepsilon_{2i} \\ \varepsilon_{3i} \\ \varepsilon_{4i} \\ \varepsilon_{5i} \end{bmatrix} \qquad (8.12)$$

If appropriate, you can expand this representation to contain additional constructs and indicators simply by lengthening the respective vectors and enlarging the corresponding loading matrix.[2]

As before, we must also account for population variability across individuals in the observed indicator, construct, and error scores. As with the *X*-measurement model, the *Y*-measurement model hypothesizes that the error vector in equation 8.10 is drawn from a multivariate normal distribution with zero mean vector and covariance matrix, Θ_ε. Since there are five elements in the error vector, the population error covariance matrix is a symmetric five-by-five matrix:

$$
\Theta_\varepsilon = \text{Cov}
\begin{bmatrix}
\varepsilon_{1i} \\
\varepsilon_{2i} \\
\varepsilon_{3i} \\
\varepsilon_{4i} \\
\varepsilon_{5i}
\end{bmatrix}
=
\begin{bmatrix}
\sigma^2_{\varepsilon_1} & \sigma_{\varepsilon_1\varepsilon_2} & \sigma_{\varepsilon_1\varepsilon_3} & \sigma_{\varepsilon_1\varepsilon_4} & \sigma_{\varepsilon_1\varepsilon_5} \\
\sigma_{\varepsilon_2\varepsilon_1} & \sigma^2_{\varepsilon_2} & \sigma_{\varepsilon_2\varepsilon_3} & \sigma_{\varepsilon_2\varepsilon_4} & \sigma_{\varepsilon_2\varepsilon_5} \\
\sigma_{\varepsilon_3\varepsilon_1} & \sigma_{\varepsilon_3\varepsilon_2} & \sigma^2_{\varepsilon_3} & \sigma_{\varepsilon_3\varepsilon_4} & \sigma_{\varepsilon_3\varepsilon_5} \\
\sigma_{\varepsilon_4\varepsilon_1} & \sigma_{\varepsilon_4\varepsilon_2} & \sigma_{\varepsilon_4\varepsilon_3} & \sigma^2_{\varepsilon_4} & \sigma_{\varepsilon_4\varepsilon_5} \\
\sigma_{\varepsilon_5\varepsilon_1} & \sigma_{\varepsilon_5\varepsilon_2} & \sigma_{\varepsilon_5\varepsilon_3} & \sigma_{\varepsilon_5\varepsilon_4} & \sigma^2_{\varepsilon_5}
\end{bmatrix}
\tag{8.13}
$$

where the main diagonal contains five variance parameters assessing interindividual variability *in* each error and the ten covariance parameters capture bivariate associations *among* errors. As before, errors can be heteroscedastic and correlated or these conditions can be constrained and tested with data.

An interesting departure from the symmetry of the two measurement models is that we do not specify a mean vector and covariance matrix for the scores on the endogenous constructs, as we did for the exogenous constructs (using vector $\boldsymbol{\kappa}$ and matrix $\boldsymbol{\Phi}$). Their equivalents are not needed because variability in the endogenous constructs is ultimately modeled in the final part of the CSA model, the *structural model*, to which we now turn.

8.1.3 The Structural Model

The *structural model* codifies hypothesized relationships among the exogenous and endogenous constructs. The path diagram in figure 8.1 contains several such relationships. First, on the left, we hypothesize that the two exogenous constructs, paternal and maternal depression (ξ_{1i} and ξ_{2i}), predict the first endogenous construct, marital conflict (η_{1i}). We quantify the magnitude and direction of these relationships using a pair of structural "regression" parameters, γ_{11} and γ_{12}. Although these parameters describe relationships among constructs, they are similar in interpretation to ordinary regression coefficients: they represent the difference in the "outcome" construct per unit difference in the "predictor" construct. If positive, greater maternal and paternal depression are associated with higher levels of marital conflict. Second, the path diagram hypothesizes that the first endogenous construct, marital conflict (η_{1i}), predicts adolescent adjustment (η_{2i}). This relationship is captured in the structural regression parameter β_{21}. If negative, daughters of less conflicted parents are better adjusted in school. We purposefully use different symbols—γ and β—to distinguish the two types of relationship: (1) between endogenous and exogenous constructs; and (2) among endogenous constructs. This difference in symbolic representation reappears below when we codify the structural model algebraically.

As in regular regression, we must account for the reality that our predictions may not be "perfect." To represent the possibility that there may be a "residual" left over after we predict marital conflict using maternal and paternal depression, figure 8.1 includes a residual, ζ_{1i}, using a short arrow, pointing slightly backwards into the construct η_{1i}. If maternal and paternal depression predict marital conflict successfully, this "true" residual will be small. Each individual/family pairing has its own values of parental depression and marital conflict and its own true residual, hence the subscript i. Similarly, because marital conflict may not perfectly predict adolescent adjustment, a second "true" residual, ζ_{2i}, takes up the slack in η_{2i}. Although these residuals pertain to the regression of construct upon construct, they serve the same function as ordinary residuals: they represent the "unpredicted" portion of an outcome after accounting for predictors.

With these additions, we can now express the hypothesized relationships among constructs. In this example, we postulate a pair of simultaneous structural (construct-level) regression equations:

$$
\begin{aligned}
\eta_{1i} &= \alpha_1 + \gamma_{11}\xi_{1i} + \gamma_{12}\xi_{2i} + \zeta_{1i} \\
\eta_{2i} &= \alpha_2 + \beta_{21}\eta_{1i} + \zeta_{2i}.
\end{aligned}
\tag{8.14}
$$

The first equation postulates that marital conflict depends simultaneously on maternal and paternal depression. The second equation postulates that adolescent adjustment depends on marital conflict. The two new parameters—α_1 and α_2—provide the population means of the two endogenous constructs, which went unrepresented in the Y-measurement model. These parameters function as intercepts, representing the population value of the "outcome" construct when the values of the "predictor" constructs are zero. They can be interpreted using familiar strategies from regression.[3]

The structural model in equation 8.14 can be represented in matrix form using the construct score vectors defined in the measurement models, along with new parameter matrices that contain the structural regression parameters used to define the construct-level relationships in the path diagram:

$$
\begin{bmatrix} \eta_{1i} \\ \eta_{2i} \end{bmatrix} = \begin{bmatrix} \alpha_1 \\ \alpha_2 \end{bmatrix} + \begin{bmatrix} \gamma_{11} & \gamma_{12} \\ 0 & 0 \end{bmatrix} \begin{bmatrix} \xi_{1i} \\ \xi_{2i} \end{bmatrix} + \begin{bmatrix} 0 & 0 \\ \beta_{21} & 0 \end{bmatrix} \begin{bmatrix} \eta_{1i} \\ \eta_{2i} \end{bmatrix} + \begin{bmatrix} \zeta_{1i} \\ \zeta_{2i} \end{bmatrix}
\tag{8.15}
$$

which, as before, can be presented in "shorthand" version as

$$\eta = \alpha + \Gamma\xi + B\eta + \zeta \tag{8.16}$$

where score vectors ξ and η have been defined in equations (8.5) and (8.12), and

$$\alpha = \begin{bmatrix} \alpha_1 \\ \alpha_2 \end{bmatrix}, \Gamma = \begin{bmatrix} \gamma_{11} & \gamma_{12} \\ 0 & 0 \end{bmatrix}, B = \begin{bmatrix} 0 & 0 \\ \beta_{21} & 0 \end{bmatrix}, \text{ and } \zeta = \begin{bmatrix} \zeta_{1i} \\ \zeta_{2i} \end{bmatrix}. \tag{8.17}$$

You should check, by matrix multiplication, that equations 8.15 and 8.14 are identical. As before, expansion of the various score vectors and parameter matrices would allow us to include additional constructs and more complex relationships among constructs.

Notice that in writing the matrix representation we set five elements of Γ and B to zero. Two of these zeros, elements of Γ, which would have been labeled γ_{21} and γ_{22} had they been needed, represent the pair of paths between parents' depression and adolescent adjustment that have been omitted from figure 8.1 as a result of the researchers' theory that these exogenous constructs affect adolescent adjustment *indirectly* through marital conflict. A third zero, in the upper right corner of B, sits in place of coefficient β_{12}, a parameter that, if present, would permit adolescent adjustment to "reverse" predict marital conflict.[4] The remaining two zeros, on the diagonal of B, are never included in a structural model because they represent something nonsensible: the prediction of an endogenous construct by itself!

We conclude our specification of the structural model by accounting for variability across individuals in the endogenous constructs. Inspection of equations 8.15 and 8.16 indicates that the "total" variation in η is a composite of "true" variation (that part of the variability in η that can be predicted by the exogenous and other endogenous constructs), and residual variation, due to ζ. As with the measurement models, the structural model hypothesizes that the residual vector is drawn from a multivariate normal distribution with zero mean vector and a covariance matrix, Ψ. Since there are two elements in the true residual vector, the population covariance matrix of the residuals is a symmetric two-by-two matrix:

$$\Psi = \text{Cov}\begin{bmatrix} \zeta_{1i} \\ \zeta_{2i} \end{bmatrix} = \begin{bmatrix} \sigma^2_{\zeta_1} & \sigma_{\zeta_1\zeta_2} \\ \sigma_{\zeta_2\zeta_1} & \sigma^2_{\zeta_2} \end{bmatrix}, \tag{8.18}$$

where the main diagonal contains variance parameters representing interindividual variability *in* each true residual and the covariance parameter captures any hypothesized bivariate association *among* residuals. As before, these true residuals can be heteroscedastic and correlated or these conditions can be constrained and tested with data.

8.1.4 Fitting the CSA Model to Data

Having postulated an *X*-measurement model, a *Y*-measurement model, and a structural model, you can use software to fit these models simultaneously to data. The major task in CSA is to specify the correct shapes, sizes, and contents of the vectors and matrices that constitute your models. Some programs ask that you directly describe the matrices; others offer a point-and-click graphical interface that accepts (or constructs) a path diagram. In what follows, we use the LISREL software (Joreskog & Sorbom, 1996), but identical analyses can be implemented using programs such as EQS (Bentler, 1995), and MPLUS (Muthen, 2001). A variety of estimation methods are available, including GLS and ML. Whichever method you select, all programs provide: (1) parameter estimates, standard errors, *z*- (or *t*)-statistics, and *p*-values; (2) measures of model goodness-of-fit; and (3) some type of residual analysis. Because our earlier discussions of estimation and hypothesis testing apply directly, we now turn to extending the basic CSA model to longitudinal data.

8.2 The Basics of Latent Growth Modeling

For years, empirical researchers with longitudinal data conducted "between-wave" analyses using CSA that "chained" their outcome values over time: allowing status at time 1 to predict status at time 2, status at time 2 to predict status at time 3, and so on. Although not invalid, such analyses do not address questions about *change over time*. Instead, they assess the *stability of the rank order of individuals* on the outcome *over time*, addressing questions like: Does an individual who scores high at the beginning of time remain high on all future occasions?

More recently, methodologists have made it possible to use the CSA model to study individual change over time by mapping the multilevel model for change onto the general CSA model. This approach has become known as *latent growth modeling* (or *latent growth curve analysis*) and many people have contributed to its development. Meredith and Tisak (1984, 1990; see also Tisak & Meredith, 1990) showed how the CSA model provided a framework for representing interindividual differences in development. McArdle and colleagues extended their approach, demonstrating its flexibility to problems in psychology and the social sciences (McArdle, 1986a, 1986b, 1989, 1991; McArdle, Anderson & Aber, 1987;

McArdle & Epstein, 1987; McArdle, Hamagami, Elias, & Robbins, 1991). Muthén and colleagues explored and extended the model, and have done important work on time-unstructured data and missing data (Muthén, 1989, 1991, 1992; Muthén & Satorra, 1989).

In this section, we introduce this kind of analysis by showing how you can fit a latent growth model by mapping the multilevel model for change onto the general CSA model. The mapping is straightforward, with particular pieces of the CSA model acting as "containers" for different facets of the multilevel model for change:

- The Y-measurement model contains the level-1 individual change trajectory.
- The structural model contains the level-2 model for interindividual differences in change.
- The X-measurement model "loads" time-invariant predictors into the level-2 model.

To concretize our presentation, we use three waves of self-report data collected by Barnes, Farrell, and Banerjee (1994). At each of three points in time—the *beginning* of seventh grade, the *end* of seventh grade, and the *end* of eighth grade—1122 adolescents used a six-point scale to rate how frequently they had consumed beer, wine, and liquor during the previous month. The composite rating of alcohol use that we analyze is the average of these three items. The top panel of table 8.1 presents example data for five boys and five girls from the dataset. Column 1 contains an *ID*, columns 3 through 5 contain the three assessments of alcohol use (*ALC1, ALC2,* and *ALC3*) and column 2 indicates the adolescent's gender (*FEMALE*). We describe the remaining three variables in section 8.3. Our goal here is to determine whether the trajectories of change in alcohol use differ for boys and girls.

Notice that table 8.1 presents a *person-level* data set, not a *person-period* data set (see section 2.1 for a discussion of the distinction). To fit a latent growth model with CSA, your data must be organized at the person level, in a *multivariate* format. Values of the outcome for each occasion (*ALC1, ALC2,* and *ALC3*) appear in a separate column. Each person has a single row, with *multiple* (multi-) *variables* (-variate) containing the time-varying data. Unlike our previous representations of the multilevel model for change, CSA analyzes a *covariance structure.* As such, it requires a sample covariance matrix (and mean vector) that summarizes the associations among (and levels of) the variables, including the repeated measures of the outcome, as input. The sample covariance matrix and mean vector for the full sample appear in the bottom of table 8.1. Once you have specified a latent growth model, you (or more

Table 8.1: The alcohol use study: excerpts from the multivariate format data set and estimated means and variance/covariance matrix

Excerpts from the multivariate format data set

	FEMALE	ALC1	ALC2	ALC3	PEER1	PEER2	PEER3
0018	0	1.00	1.33	2.00	3	2	2
0021	0	1.00	2.00	1.67	1	1	1
0236	0	3.33	4.33	4.33	2	1	3
0335	0	1.00	1.33	1.67	1	2	1
0353	0	2.00	2.00	1.67	1	1	2
0555	1	2.67	2.33	1.67	2	3	1
0850	1	1.33	1.67	1.33	3	1	2
0883	1	3.00	2.67	3.33	4	5	1
0974	1	1.00	1.67	2.67	1	5	6
1012	1	1.00	1.67	2.33	1	2	4

Estimated means and variance/covariance matrix for transformed data

Variable	Mean	Covariances						
		FEMALE	ALC1	ALC2	ALC3	PEER1	PEER2	PEER3
FEMALE	0.612	0.238						
ALC1	0.225	−0.008	0.136					
ALC2	0.254	−0.013	0.078	0.155				
ALC3	0.288	−0.005	0.065	0.082	0.181			
PEER1	0.177	−0.009	0.066	0.045	0.040	0.174		
PEER2	0.290	−0.022	0.064	0.096	0.066	0.072	0.262	
PEER3	0.347	−0.024	0.060	0.074	0.132	0.071	0.112	0.289

Note that these sample statistics were computing after taking natural logarithms of both the alcohol use and peer pressure variables.

accurately, your software) compare the sample estimates to the mathematical implications of your models for the structure of the underlying population covariance matrix and mean vector. Because you compare sample and predicted covariance matrices (and mean vectors), the data must be formatted in a way that supports the estimation of covariance matrices (and mean vectors).

Notice that there is no dedicated column that records the values of time. This is unnecessary as each outcome has a unique but constant temporal value. *ALC1* was measured at the *start* of seventh grade; *ALC2* was measured at the *end* of seventh grade; *ALC3* was measured at the *end* of eighth grade. To identify the occasions of measurement, you specify the values of time when specifying your CSA model. For this reason, it is easiest to fit latent growth models when your data are *time-structured*—that is, when everyone is measured on the same (or similar sets of) occasions (see section 5.1 for discussion).

8.2.1 Mapping the Level-1 Model onto the Y-Measurement Model

We begin, as usual, by using empirical change plots to identify a suitable level-1 growth model for individual change over time. Visual inspection of the data in table 8.1 suggests heterogeneity in both initial status and rate of change. Graphical analyses in the full sample confirm this and suggest that we can posit a level-1 linear-change model if we analyze the natural logarithm of alcohol use. Letting Y_{ij} represent (log) alcohol use for adolescent i on occasion j we write:

$$Y_{ij} = \pi_{0i} + \pi_{1i}TIME_j + \varepsilon_{ij},\tag{8.19}$$

where $TIME_j = (GRADE_j\text{-}7)$. Because everyone is assessed on the same three occasions, *TIME* includes just a single subscript j. Since $GRADE_j$ takes on the value 7 at the "start of seventh grade," 7.75 at the "end of seventh grade," and 8.75 at the "end of eighth grade," $TIME_j$ takes on three values: 0, 0.75, and 1.75. To write the model more generally, we identify these three occasions as t_1, t_2, and t_3. The parameters of the level-1 model have the usual interpretations: π_{0i} represents individual i's true initial status at the beginning of seventh grade and π_{1i} represents individual i's true annual rate of linear change in log-alcohol use during the two-year period.

We now diverge slightly from our usual practice. Because each individual has three values of the observed outcome, we can use equation 8.19 to write down how each value—Y_{i1}, Y_{i2}, and Y_{i3}, for individual i—is related to the three values of *TIME* employed in the data collection design—t_1, t_2, and t_3—and the two individual growth parameters, π_{0i} and π_{1i}:

$$
\begin{aligned}
Y_{i1} &= \pi_{0i} + \pi_{1i}t_1 + \varepsilon_{i1}\\
Y_{i2} &= \pi_{0i} + \pi_{1i}t_2 + \varepsilon_{i2}\\
Y_{i3} &= \pi_{0i} + \pi_{1i}t_3 + \varepsilon_{i3}.
\end{aligned}\tag{8.20}
$$

With a little algebraic manipulation, we can write these equations in matrix form as:

$$\begin{bmatrix} Y_{i1} \\ Y_{i2} \\ Y_{i3} \end{bmatrix} = \begin{bmatrix} 0 \\ 0 \\ 0 \end{bmatrix} + \begin{bmatrix} 1 & t_1 \\ 1 & t_2 \\ 1 & t_3 \end{bmatrix} \begin{bmatrix} \pi_{0i} \\ \pi_{1i} \end{bmatrix} + \begin{bmatrix} \varepsilon_{i1} \\ \varepsilon_{i2} \\ \varepsilon_{i3} \end{bmatrix}. \tag{8.21}$$

While the level-1 individual growth model in equation 8.21 may appear dramatically different from the familiar representation in equation 8.19, it says the same exact thing: observed values of Y are related to $TIME$ (t_1, t_2, and t_3) and a pair of individual parameters (π_{0i} and π_{1i}); on each occasion, there is also a unique measurement error (ε_{i1}, ε_{i2}, and ε_{i3}). Equation 8.21 simply uses vectors and matrices to contain the various numerical values and parameters. (Don't be misled by the strange vector of zeros to the immediate right of the equals sign. This cosmetic addition simply facilitates our subsequent mapping onto the CSA model.)

Given this representation, we can map the hypothesized individual growth model in equation 8.21 onto the CSA Y-measurement model. Recall from equation 8.11 that the Y-measurement model is:

$$\mathbf{Y} = \boldsymbol{\tau}_y + \boldsymbol{\Lambda}_y \boldsymbol{\eta} + \boldsymbol{\varepsilon}. \tag{8.22}$$

So if we set the score vectors in equation 8.22 to

$$\mathbf{Y} = \begin{bmatrix} Y_{i1} \\ Y_{i2} \\ Y_{i3} \end{bmatrix}, \boldsymbol{\eta} = \begin{bmatrix} \pi_{0i} \\ \pi_{1i} \end{bmatrix}, \boldsymbol{\varepsilon} = \begin{bmatrix} \varepsilon_{i1} \\ \varepsilon_{i2} \\ \varepsilon_{i3} \end{bmatrix} \tag{8.23}$$

and the parameter matrices $\boldsymbol{\tau}_y$ and $\boldsymbol{\Lambda}_y$ to

$$\boldsymbol{\tau}_y = \begin{bmatrix} 0 \\ 0 \\ 0 \end{bmatrix}, \boldsymbol{\Lambda}_y = \begin{bmatrix} 1 & t_1 \\ 1 & t_2 \\ 1 & t_3 \end{bmatrix}, \tag{8.24}$$

we obtain the model in equation 8.21. This demonstrates that, as long as we specify the score vectors and parameter matrices correctly, the Y-measurement model in CSA can "contain" our level-1 individual growth trajectory in the multilevel modeling of change.

Equations 8.20 and 8.21 state that, for adolescent i, measurement error disturbs his or her true status by ε_{i1} on the first measurement occasion, ε_{i2} on the second, and ε_{i3} on the third. But we have not yet made any assumptions about the level-1 error covariance structure. Are the errors homoscedastic and independent? Heteroscedastic and autocorrelated? The flexibility of the Y-measurement model, which allows us to specify these distributions using the $\boldsymbol{\Theta}_\varepsilon$ matrix in equation 8.13, permits great flexibility. During analysis, you can compare the goodness of fit of models with alternative error structures and select an appropriate one. For these data, additional analyses not shown here suggest the tenability of the assumption that the level-1 errors are distributed independently and heteroscedastically over time within person. We therefore write the Θ_ε parameter matrix as:

$$\Theta_\varepsilon = \begin{bmatrix} \sigma^2_{\varepsilon_1} & 0 & 0 \\ 0 & \sigma^2_{\varepsilon_2} & 0 \\ 0 & 0 & \sigma^2_{\varepsilon_3} \end{bmatrix}. \tag{8.25}$$

Unlike what happens in a conventional CSA, the Λ_y loading matrix in equation 8.24 is a set of known times and constants, not a set of unknown parameters to be estimated. In this sense, the Y-measurement model "forces" the individual growth parameters, π_{0i} and π_{1i}, into the endogenous construct vector η, creating what we call the *latent growth vector*. This notion—that, in a latent growth model, the CSA η-vector can be forced to contain the individual growth parameters, π_{0i} and π_{1i}—is critical for level-2 analyses of interindividual differences in change, because it means that this variation can be modeled in a structural model, as we now show.

8.2.2 Mapping the Level-2 Model onto the Structural Model

As we have noted before, even if everyone in the population shares a common functional form for their individual change, their trajectories may differ because of interindividual variation in the growth parameters. Some adolescents may have different intercepts, others may have different slopes. In previous chapters, we used level-2 submodels to represent this variation, both *unconditional models* (that include no substantive predictors) and *conditional* models (that add time-invariant predictors at level-2). We now demonstrate how to specify models of both types by adopting a suitable form for the CSA structural model.

An Unconditional Latent Growth Model

An *unconditional* growth model (as in equations 4.9a and 4.9b) allows individual growth parameters to differ across people but does not relate their variation to predictors. This is equivalent to hypothesizing that the individual growth parameters are *distributed* across people in the population. To investigate heterogeneity in change trajectories, we assume, as before, that adolescents draw their intercepts and slopes from a multivariate normal distribution:

$$\begin{bmatrix} \pi_{0i} \\ \pi_{1i} \end{bmatrix} \sim N\left(\begin{bmatrix} \mu_{\pi_0} \\ \mu_{\pi_1} \end{bmatrix}, \begin{bmatrix} \sigma^2_{\pi_0} & \sigma_{\pi_0\pi_1} \\ \sigma_{\pi_1\pi_0} & \sigma^2_{\pi_1} \end{bmatrix} \right). \tag{8.26}$$

This equation is, in fact, an *unconditional level-2 model for interindividual differences in change* in that it allows different people to have different intercepts and slopes.

Each of the five unique level-2 parameters in equation 8.26 plays an important role. The two means, μ_{π_0} and μ_{π_1}, describe the average intercept and slope of the true change trajectory across everyone in the population. They address the question: What is the trajectory of true change in (log) alcohol use during grades seven and eight? The two variance parameters, $\sigma^2_{\pi_0}$ and $\sigma^2_{\pi_1}$, summarize population interindividual differences in true initial status and true rate of change. They answer the question: Do the change trajectories of true (log) alcohol use differ across adolescents? The off-diagonal covariance parameter, $\sigma_{\pi_0\pi_1}$, summarizes the strength and direction of the population relationship between true initial status and true rate of change. It answers the question: Is there an association between true initial status and true rate of change in (log) alcohol use among adolescents through grades seven and eight?

The unconditional distribution of the individual growth parameters in equation 8.26 can be represented in a CSA structural model because it allows the latent growth vector to differ across people. The requisite structural model is simple—it stipulates that the latent growth vector is the sum of an *average* and a *residual* (that is, a deviation from the average):

$$\begin{bmatrix} \pi_{0i} \\ \pi_{1i} \end{bmatrix} = \begin{bmatrix} \mu_{\pi_0} \\ \mu_{\pi_1} \end{bmatrix} + \begin{bmatrix} \zeta_{0i} \\ \zeta_{1i} \end{bmatrix}, \tag{8.27}$$

which has the form of the CSA structural model in equation 8.16:

$$\eta = \alpha + \Gamma\xi + B\eta + \zeta \tag{8.28}$$

with all of its parameter matrices except α set to zero:

$$\alpha = \begin{bmatrix} \mu_{\pi_0} \\ \mu_{\pi_1} \end{bmatrix}, \Gamma = \begin{bmatrix} 0 & 0 \\ 0 & 0 \end{bmatrix}, B = \begin{bmatrix} 0 & 0 \\ 0 & 0 \end{bmatrix} \tag{8.29}$$

and with the latent growth vector η defined as in equation 8.23.

In an unconditional growth model, the properties of the latent residual vector, ζ, are particularly interesting. Because the α vector explicitly represents the population averages of the intercepts and slopes, π_{0i} and π_{1i}, the elements of the latent residual vector contain deviations of π_{0i} and π_{1i} from their population means. As with all residuals, these deviations have a mean of zero. In addition, equation 8.18 reminds us that the latent residual vector ζ has a covariance matrix Ψ:

$$\Psi = \mathrm{Cov}[\zeta] = \begin{bmatrix} \sigma^2_{\pi_0} & \sigma_{\pi_0 \pi_1} \\ \sigma_{\pi_1 \pi_0} & \sigma^2_{\pi_1} \end{bmatrix}. \tag{8.30}$$

Because Ψ contains the level-2 variance and covariance parameters from equation 8.26, we are most interested in this matrix.

Together, equations 8.21 and 8.26 represent an unconditional multi-level model for change. Equations 8.22 through 8.25 and 8.27 through 8.30 re-express this model as a CSA model. By carefully specifying score and parameter matrices, the Y-measurement model becomes the level-1 model (including our assumptions about the distribution of the measurement errors) and the structural model becomes the level-2 model (including our assumptions about the distribution of the level-2 residuals). The top panel of figure 8.2 presents a path diagram that reflects this specification. Notice how the loadings link the observed measures of alcohol use (Y_{i1}, Y_{i2}, and Y_{i3}) to their respective latent constructs (π_{0i} and π_{1i}) and force the latter to become the true intercept and slope of the hypothesized individual change trajectory.

The most important implication of this mapping is that you can use standard CSA methods to fit the multilevel model for change. Doing so provides estimates of the parameters in the α vector, Θ_ε matrix and Ψ matrix, which address research questions about change. Model A in table 8.2 presents FML parameter estimates and goodness-of-fit statistics, along with approximate p-values, for the unconditional model just specified. Notice that this model fits particularly well ($\chi^2 = 0.05$, $d.f. = 1$, $p = .83$).

We interpret the fixed effects as usual. The estimated average true intercept is 0.226 ($p < .001$); the estimated average true slope is 0.036 ($p < .001$). Because alcohol use is expressed as a natural logarithm, we can use the methods of section 5.2.1 to interpret the slope in percentage terms. Computing $100(e^{(0.0360)} - 1) = 3.66\%$, we conclude that during grades seven and eight, the average adolescent increases his or her alcohol consumption by just under 4% a year. Substituting the intercept and slope estimates into the hypothesized change trajectory in equation 8.19 provides an algebraic expression for the average fitted change trajectory in (log) alcohol use:

$$\hat{Y} = 0.2257 + 0.0360(GRADE_j - 7).$$

As with other models that use a logarithmically transformed outcome, we can take antilogs (computing $e^{(0.2257+0.0360(GRADE-7))}$) and plot a fitted

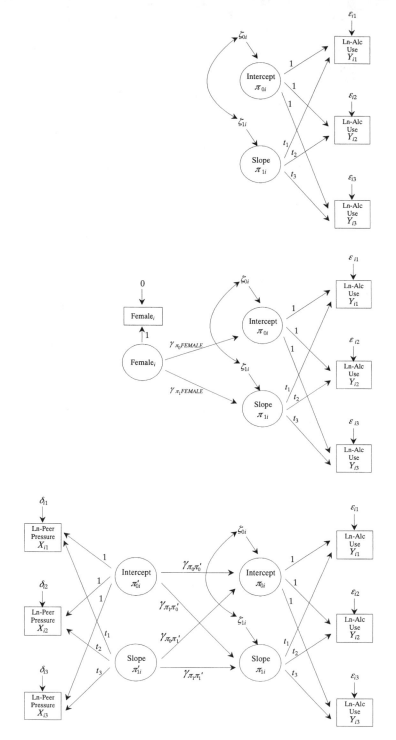

Figure 8.2. Three hypothesized path diagrams of a latent growth model for (log) alcohol use. Top panel: Unconditional model. Middle panel: Adding FEMALE as a time-invariant predictor. Bottom panel: Adding PEER PRESSURE as a time-varying predictor for a cross-domain analysis of change.

Table 8.2 Results of fitting selected latent growth models for the alcohol use data ($n = 1222$)

		Model A	Model B	Model C	Model D
Fixed Effects					
Latent	μ_{π_0}	0.2257***			
growth	μ_{π_1}	0.0360***			
model	α_0		0.2513***	0.2481***	0.0667***
	α_1		0.0312**	0.0360***	0.0083
	$\gamma_{\pi_0 FEMALE}$		−0.0419~	−0.0366~	
	$\gamma_{\pi_1 FEMALE}$		0.0079		
	$\gamma_{\pi_0 \pi_0'}$				0.7985***
	$\gamma_{\pi_0 \pi_1'}$				0.0805
	$\gamma_{\pi_1 \pi_0'}$				−0.1433~
	$\gamma_{\pi_1 \pi_1'}$				0.5767**
Variance Components					
Level-1	$\sigma^2_{\varepsilon_1}$	0.0485***	0.0489***	0.0488***	0.0481***
	$\sigma^2_{\varepsilon_2}$	0.0758***	0.0756***	0.0756***	0.0763***
	$\sigma^2_{\varepsilon_3}$	0.0768***	0.0771***	0.0772***	0.0763***
Level-2:	$\sigma^2_{\pi_0}$	0.0871***	0.0864***	0.0865***	0.0422***
	$\sigma^2_{\pi_1}$	0.0198***	0.0195***	0.0195***	0.0092~
	$\sigma_{\pi_0 \pi_1}$	−0.0125***	−0.0122***	−0.0122***	−0.0064
Distribution of the Exogenous Construct PEER PRESSURE					
	$\mu_{\pi_0'}$				0.1882***
	$\mu_{\pi_1'}$				0.0962***
	$\sigma^2_{\pi_0'}$				0.0698***
	$\sigma^2_{\pi_1'}$				0.0285**
	$\sigma_{\pi_0' \pi_1'}$				0.0012
Goodness-of-fit					
	χ^2	0.05	1.54	1.82	11.54
	df	1	2	3	4
	p	0.83	0.46	0.61	0.0211

~$p < .10$; *$p < .05$; **$p < .01$; ***$p < .001$.
Model A is an *unconditional growth model*; Models B and C are *conditional growth models* that include the effect of *FEMALE*; Model D is a *cross-domain model* that includes *PEER PRESSURE*.

Note: LISREL VII, Full ML.

trajectory in terms of raw alcohol use, as in the left panel of figure 8.3. Although this trajectory may appear linear, this is only because the estimated slope is small and the study duration limited making the underlying curvature almost imperceptible.

The level-1 variance components summarize the error variances at each occasion. Their values—0.049, 0.076, and 0.077—suggest some

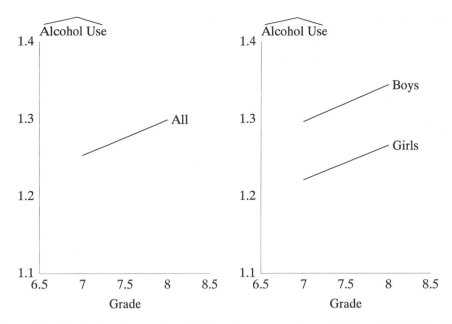

Figure 8.3. Displaying the results of fitting a latent growth model. Fitted trajectories for alcohol use from an unconditional model (left panel) and a conditional model including the main effect of FEMALE (right panel).

heteroscedasticity from t_1 to t_2 but little from t_2 to t_3. We can also use these components to estimate the reliability of the outcome measure—the proportion of observed variance in the outcome that is true variance—on each occasion. Computing $\{(0.136 - 0.0490)/0.136\}$, $\{(0.155 - 0.076)/0.155\}$, and $\{(0.181 - 0.077)/0.181\}$, we conclude that the reliability of the outcome is moderate—0.64, 0.51, and 0.58—on each occasion.

The level-2 variance components summarize population heterogeneity in true intercept $(0.087, p < .001)$ and true slope $(0.020, p < .001)$. Because both are non-zero, we conclude that adolescents vary in both their initial status and their rate of change. Finally, examine the estimated covariance between true intercept and true rate of change. As in the unconditional growth model in section 4.4.2, we can combine this with the estimated variances, to find that the estimated correlation between true intercept and true rate of change is −0.30. This moderate, but statistically significant $(p < .001)$, value suggests that adolescents who report lower seventh-grade alcohol use have more rapid rates of change as they age.

Including a Time-Invariant Predictor in the
X-Measurement Model

Given that the unconditional growth model suggests that there is predictable interindividual heterogeneity in the change trajectories, a natural next step is to ask whether we can actually predict some of this heterogeneity. To keep our illustration simple, we investigate whether heterogeneity in the alcohol use trajectories depends on the time-invariant dichotomy *FEMALE*. In a latent growth model, we address such questions by incorporating the predictor into the structural (level-2) model.

The key idea here is to take advantage of the unused *X*-measurement model to jimmy predictors into the structural model. In this example, this task is disarmingly simple because there is just one predictor, *FEMALE*, which we are prepared to assume is measured without error. As a result, and although it may appear a bit unusual, we will derive an *X*-measurement model from the following tautology: $FEMALE_i = 0 + 1(FEMALE_i) + 0$, which facilitates comparison with the standard *X*-measurement model:

$$\mathbf{X} = \boldsymbol{\tau}_x + \boldsymbol{\Lambda}_x \boldsymbol{\xi} + \boldsymbol{\delta} \qquad (8.31)$$

Examination of equation 8.31 in light of this tautology suggest that you can incorporate a single predictor into a CSA model by specifying an *X*-measurement model in which four conditions hold. First, the exogenous indicator vector $\mathbf{X}$ must contain a single element—here, the predictor *FEMALE*. Second, the measurement error vector $\boldsymbol{\delta}$ must contain a single element whose value is fixed at zero; this embodies our belief that gender is measured infallibly (i.e., with "zero" error). Third, the $\boldsymbol{\tau}_x$ vector must contain a single element whose value is also fixed at zero. We discuss this counterintuitive condition below (given that it is the mean of the exogenous indicator X that would naturally reside in $\boldsymbol{\tau}_x$ if the latter were free to accept it). Fourth, the matrix of loadings $\boldsymbol{\Lambda}_x$ must contain a single element whose value is fixed to 1; this recognizes that the scales of the exogenous construct for gender and its indicator are identical. As a result, by specifying an *X*-measurement model in which the score vectors are constrained to be:

$$\mathbf{X} = [FEMALE_i] \text{ and } \boldsymbol{\delta} = [0] \qquad (8.32)$$

and the parameter matrices are fixed at

$$\boldsymbol{\tau}_x = [0] \text{ and } \boldsymbol{\Lambda}_x = [1], \qquad (8.33)$$

we "force" *FEMALE* into the exogenous construct $\boldsymbol{\xi}$. And, since $\boldsymbol{\xi}$ will play the role of "predictor" in a forthcoming structural model, we have succeeded in "inserting" our chosen predictor into the level-2 submodel.

Why did we force the sole element of τ_x to be *zero* when we know full well that *FEMALE* does *not* have a zero mean? The answer to this paradox is rooted in a basic redundancy that is built into CSA via the X-measurement model (see the discussion at the end of section 8.1.1). The X-measurement model represents means using two parameter vectors: τ_x and κ. If we take expectations throughout for a single generic exogenous indicator (recalling that the population mean of measurement error δ is zero), the mean of the observed exogenous indicator X is a linear combination of the contents of τ_x and the population mean of the underlying exogenous construct, μ_ξ:

$$E\{X_i\} = E\{\tau_x + \lambda^x \xi_i + \delta_i\}$$
$$E\{X_i\} = \tau_x + \lambda^x E\{\xi_i\} + E\{\delta_i\}$$
$$\mu_X = \tau_x + \lambda^x \mu_\xi.$$

This suggests that, by being crafty about the way we specify τ_x and κ, we can *force* the mean of exogenous indicator X *either* entirely into τ_x, or entirely into the mean of the exogenous construct μ_ξ, or force it to be shared between them.

Of course, these multiple specifications are redundant. Your choice affects neither model fit nor substantive findings. It does, however, affect parameter interpretation and this usually suggests a particular specification. Consider, for example, what happens if we force the population mean of the exogenous *construct* to be zero (i.e., set $\mu_\xi = 0$) by fixing the appropriate element of the κ vector at zero. To accommodate the mean of the exogenous *indicator* X, we would have to free the corresponding element of τ_x, which would assume a value of μ_X. Permitting the population mean of X to be estimated in τ_x is equivalent to centering the observed indicator on its sample average, which translates into a centering of the predictor in the level-2 model. Based upon arguments about centering presented in section 4.5, we choose *not* to do this here, as we prefer to enter dichotomous predictors (like *FEMALE*) in their raw state. (With continuous predictors, we often prefer the other option.)

Now consider what happens if we force τ_x to zero. The mean of the exogenous construct ξ *cannot* be zero (*FEMALE*'s mean has to go *somewhere*) and so the corresponding element of κ must be freed up. The latent construct acquires a non-zero mean and the mean of exogenous indicator X becomes a rescaled version of the mean of its underlying construct, $\lambda^x \mu_\xi$. This is equivalent to passing *FEMALE* into the level-2 structural model in its raw state. As far as the exogenous indicators are concerned, the CSA model contains a redundancy because you do not need two independent places to model their means. This leaves you with two choices: (1) you can capture the mean directly in τ_x (the most usual practice), in which case the mean of the exogenous construct will be zero and the predictor will be centered on its average; or (2) you can force the mean into the exogenous construct, μ_ξ, by freeing up the appropriate element of κ and constraining the corresponding element of τ_x to zero, in which case the predictor will *not* be centered. The choice is yours and should be governed by considerations articulated in section 4.5.

The *X*-measurement model in equations 8.31 through 8.33 forces the exogenous construct ξ to become *FEMALE*. As a result, the population mean of *FEMALE* appears as the sole element of the κ vector

$$\kappa = \text{Mean}[\xi] = [\mu_{FEMALE}], \tag{8.34}$$

and the population variance of *FEMALE* appears as the sole element of covariance matrix Φ associated with the exogenous construct

$$\Phi = \text{Cov}[\xi] = [\sigma^2_{FEMALE}]. \tag{8.35}$$

We estimate both of these when the CSA model is ultimately fit.

Although we do not demonstrate it, you can easily allow the *X*-measurement model in equations 8.31 through 8.33 to accommodate additional time-invariant predictors and several indicators of each predictor construct. As Willett and Sayer (1994) show, you simply expand the exogenous indicator and construct score vectors to include sufficient elements to contain the new indicators and constructs, and you expand the parameter matrix Λ_x to include the requisite loadings (under the usual requirements for identification).

Once the *X*-measurement model is specified, you use the CSA structural model to specify hypothesized relationships between the true change trajectory's growth parameters and the predictor(s). The level-1 individual growth model remains unchanged. To do so, we modify the unconditional structural model in equations 8.27 through 8.30 so that the newly defined exogenous construct vector (here, *FEMALE*) becomes active in the equation's right side. We achieve this by using the latent regression-weight matrix Γ, which is present in the structural model solely for the purpose of modeling the relationship between the η and ξ vectors. To predict interindividual differences in change, we free those parameters in the Γ matrix that permit the regression of true intercept and slope on predictor:

$$\begin{bmatrix} \pi_{0i} \\ \pi_{1i} \end{bmatrix} = \begin{bmatrix} \alpha_0 \\ \alpha_1 \end{bmatrix} + \begin{bmatrix} \gamma_{\pi_0 FEMALE} \\ \gamma_{\pi_1 FEMALE} \end{bmatrix} [FEMALE_i] + \begin{bmatrix} 0 & 0 \\ 0 & 0 \end{bmatrix} \begin{bmatrix} \pi_{0i} \\ \pi_{1i} \end{bmatrix} + \begin{bmatrix} \zeta_{0i} \\ \zeta_{1i} \end{bmatrix}, \tag{8.36}$$

which is a CSA structural model with constituent parameter matrices:

$$\boldsymbol{\alpha} = \begin{bmatrix} \alpha_0 \\ \alpha_1 \end{bmatrix}, \boldsymbol{\Gamma} = \begin{bmatrix} \gamma_{\pi_0 FEMALE} \\ \gamma_{\pi_1 FEMALE} \end{bmatrix}, \mathbf{B} = \begin{bmatrix} 0 & 0 \\ 0 & 0 \end{bmatrix}. \tag{8.37}$$

In this model, α_0 and α_1 represent the population average intercept and slope of the (log) alcohol use trajectory for boys (when $FEMALE = 0$). Latent regression parameters $\gamma_{\pi_0 FEMALE}$ and $\gamma_{\pi_1 FEMALE}$ represent the increments to these parameters for girls. In the middle panel of figure 8.3, we extend the baseline path diagram in the top panel to include $FEMALE$ as a time-invariant predictor. Notice that $FEMALE$ is introduced as an exogenous indicator/construct pairing on the diagram's left side.

When we introduce a time-invariant predictor into the structural model, the elements of the latent residual vector $\boldsymbol{\zeta}$ in equation 8.36 represent those portions of the intercept and slope of the true trajectory that are not (linearly) related to the predictor. They are the "adjusted" values of true intercept and slope, controlling for the effect of $FEMALE$. The latent residual vector $\boldsymbol{\zeta}$ in equation 8.36 is distributed with zero mean vector, and covariance matrix $\boldsymbol{\Psi}$:

$$\boldsymbol{\Psi} = \mathrm{Cov}[\boldsymbol{\zeta}] = \begin{bmatrix} \sigma^2_{\pi_0 | FEMALE} & \sigma_{\pi_0 \pi_1 | FEMALE} \\ \sigma_{\pi_1 \pi_0 | FEMALE} & \sigma^2_{\pi_1 | FEMALE} \end{bmatrix}, \tag{8.38}$$

which now contains the partial variances and covariance of true intercept and slope, controlling for the linear effects of the predictor (hence the addition of the symbol "$|FEMALE$"—for "conditional on $FEMALE$" to each element in equation 8.38). If our predictions are successful, the partial variances on the diagonal of $\boldsymbol{\Psi}$ will be smaller than their unconditional cousins in equation 8.30. In other words, the *proportional decline in the variance of either true intercept or slope* on the diagonal of $\boldsymbol{\Psi}$ on inclusion of a predictor provide a pseudo-R^2 statistic that summarizes how much of the between-person variation in the change trajectory is associated with the predictor of change (see section 4.4.3 for a related discussion in the regular multilevel model for change).

The results of fitting this conditional model are shown in Model B of table 8.2. Notice that the model fits well ($\chi^2 = 1.54$, $d.f. = 2$, $p = .46$). We interpret the fixed effects using familiar strategies. The elements of $\hat{\boldsymbol{\alpha}}$ indicate that the average adolescent boy has an estimated true intercept of 0.2513 ($p < .001$) and an estimated true slope of 0.0312 ($p < .01$). The latent regression coefficients $\hat{\gamma}_{\pi_0 FEMALE}$ and $\hat{\gamma}_{\pi_1 FEMALE}$ assess gender differentials in these quantities. Because $\hat{\gamma}_{\pi_0 FEMALE}$ is -0.0419 ($p < .10$), we conclude that girls initially report a lower level of alcohol consumption. But $\hat{\gamma}_{\pi_1 FEMALE}$ is indistinguishable from 0 (0.0079, $p > .10$), we conclude that there is no discernible gender differential in rates of change.

Before moving to the variance components, notice that we do not compare goodness-of-fit statistics for Models A and B. However tempting it might be, you cannot compare the fit statistics of these models because *FEMALE* was not included in the sample covariance matrix for Model A. To test the effect of *FEMALE* on the *entire* change trajectory, you must assess the joint impact of $\gamma_{\pi_0 FEMALE}$ and $\gamma_{\pi_1 FEMALE}$ by comparing Model B to a baseline that is identical in all aspects except that it fixes these two parameters to zero. When we fit this baseline, we find that it, too, fits well ($\chi^2 = 5.36$, *d.f.* = 4, $p > .25$). Comparing it to Model B, we are unable to reject the joint hypothesis that $\gamma_{\pi_0 FEMALE}$ and $\gamma_{\pi_1 FEMALE}$ are jointly zero ($\Delta\chi^2 = 3.816$, $\Delta d.f. = 2$, $p < .15$).

Now examine the variance components for Model B. The level-1 variance components estimate the variances of the level-1 errors and describe the fallibility of outcome measurement on each occasion. They remain similar in magnitude and interpretation to those in Model A. The level-2 variance components are the estimated partial variances and covariance of true intercept and true slope, controlling for the linear effects of *FEMALE*. Comparing these partial variances with their unadjusted cousins, we see that inclusion of *FEMALE* as a level-2 predictor in the model reduces them only slightly, by 0.8% and 1.5%, respectively. This suggests that, despite being a statistically significant predictor of intercept ($p < .10$), *FEMALE* is of minor importance in explaining extant inter-individual variation in true change.

Given that *FEMALE* seems to have no impact on the true rate of change, we fit an additional model, Model C, which is identical to Model B except that $\gamma_{\pi_1 FEMALE}$ is fixed to zero. Using parameter estimates from Model C, we have the following two fitted change trajectories for prototypical boys and girls:

$$\textbf{Boys}: \quad \hat{Y} = 0.2481 + 0.0360(GRADE - 7)$$

$$\textbf{Girls}: \quad \hat{Y} = (0.2481 - 0.0366) + 0.0360(GRADE - 7)$$

By substituting a range of appropriate values for *GRADE* and antilogging the obtained fitted values, we constructed the prototypical change trajectories in the right panel of figure 8.3. Boys and girls enter seventh grade displaying differences in alcohol use, but their upwardly curved trajectories remain parallel.

8.3 Cross-Domain Analysis of Change

We now demonstrate an interesting way of incorporating a time-varying predictor into a latent growth model. The last three columns of table 8.1

present time-varying data on peer pressure to drink. On each occasion, each adolescent used a six-point scale to report the number of times during the past month a friend offered him or her alcoholic drink. These data allow us to determine whether adolescents' use of alcohol increases more rapidly if peer pressure also increases more rapidly. In statistical terms, this is a question about whether the rate of change in alcohol use can be predicted by the rate of change in peer pressure—a kind of "growth on growth" analysis.[5] For simplicity, in what follows, we set aside the time-invariant predictor *FEMALE*. Once you understand the general strategy, you can simultaneously include predictors of both types.

8.3.1 Modeling Individual Change in both the X- and Y-Measurement Models

To determine whether *change in an outcome* is associated with *change in a time-varying predictor*, we can simultaneously model individual change in *both* variables and investigate the relationship between the two sets of individual growth parameters. To achieve this, in the current data set, we must specify individual growth models for both alcohol use and peer pressure.

Individual change in the outcome is modeled as before, in the Y-measurement model. Equations 8.22 through 8.25 offer the usual specification. The individual growth parameters that describe true change in self-reported (log) alcohol use are forced, as usual, into latent growth vector η.

In the new context, however, we use the X-measurement model to represent individual *change* in the time-varying predictor. Exploratory analyses (not included here) suggest that the natural log of peer pressure for adolescent i on occasion j, X_{ij}, is a linear function of time, so we write the level-1 growth model as:

$$X_{ij} = \pi'_{0i} + \pi'_{1i}TIME_j + \delta_{ij}, \qquad (8.39)$$

where we add superscripts to the parameters to distinguish them from those representing change in (log) alcohol use. Even though equation 8.39 describes the behavior of a predictor, we interpret its parameters in the usual way: π'_{0i} represents individual i's true initial peer pressure at the beginning of seventh grade and π'_{1i} represents individual i's true rate of linear change in peer pressure during the two-year period.

The level-1 individual growth model for the exogenous peer pressure change trajectory in equation 8.39 becomes the X-measurement model:

$$\mathbf{X} = \boldsymbol{\tau}_x + \boldsymbol{\Lambda}_x \boldsymbol{\xi} + \boldsymbol{\delta}, \qquad (8.40)$$

with vectors that contain the empirical growth record, the individual growth parameters, and the errors of measurement:

$$\mathbf{X} = \begin{bmatrix} X_{i1} \\ X_{i2} \\ X_{i3} \end{bmatrix}, \, \boldsymbol{\xi} = \begin{bmatrix} \pi'_{0i} \\ \pi'_{1i} \end{bmatrix}, \, \boldsymbol{\delta} = \begin{bmatrix} \delta_{i1} \\ \delta_{i2} \\ \delta_{i3} \end{bmatrix} \tag{8.41}$$

and with parameter matrices that contain the usual known values and constants:

$$\boldsymbol{\tau}_x = \begin{bmatrix} 0 \\ 0 \\ 0 \end{bmatrix}, \, \boldsymbol{\Lambda}_x = \begin{bmatrix} 1 & t_1 \\ 1 & t_2 \\ 1 & t_3 \end{bmatrix}. \tag{8.42}$$

Notice that the $\boldsymbol{\xi}$ vector in equation 8.41 is a "latent growth vector" containing individual growth parameters that describe the true change trajectory in exogenous peer pressure. As in any X-measurement model, the population mean vector and covariance matrix of these level-1 parameters will appear in the $\boldsymbol{\kappa}$ vector and $\boldsymbol{\Phi}$ matrix:

$$\boldsymbol{\kappa} = \begin{bmatrix} \mu_{\pi'_0} \\ \mu_{\pi'_1} \end{bmatrix}, \, \boldsymbol{\Phi} = \begin{bmatrix} \sigma^2_{\pi'_0} & \sigma_{\pi'_0\pi'_1} \\ \sigma_{\pi'_1\pi'_0} & \sigma^2_{\pi'_1} \end{bmatrix}. \tag{8.43}$$

For these data, we assume that the level-1 error vector $\boldsymbol{\delta}$ has a covariance matrix $\boldsymbol{\Theta}_\delta$ similar to the heteroscedastic error covariance matrix specified for the Y-measurement model in equation 8.25:

$$\boldsymbol{\Phi}_\delta = \mathrm{Cov}[\boldsymbol{\delta}] = \begin{bmatrix} \sigma^2_{\delta_1} & 0 & 0 \\ 0 & \sigma^2_{\delta_2} & 0 \\ 0 & 0 & \sigma^2_{\delta_3} \end{bmatrix}. \tag{8.44}$$

When you use both Y- and X-measurement models, you specify these covariance matrices, $\boldsymbol{\Theta}_\delta$ and $\boldsymbol{\Theta}_\varepsilon$, as well as a matrix of their *covariances*, $\boldsymbol{\Theta}_{\delta\varepsilon}$, which accounts for any correlation between measurement errors in the exogenous and endogenous indicators. This feature has both substantive and psychometric advantages. In this study, this feature allows us to assume that the measurement errors in the outcome and predictor may covary across adolescents within-occasion:

$$\boldsymbol{\Phi}_{\delta\varepsilon} = \mathrm{Cov}(\boldsymbol{\delta\varepsilon}) = \begin{bmatrix} \sigma_{\delta_1\varepsilon_1} & 0 & 0 \\ 0 & \sigma_{\delta_2\varepsilon_2} & 0 \\ 0 & 0 & \sigma_{\delta_3\varepsilon_3} \end{bmatrix}. \tag{8.45}$$

8.3.2 Modeling the Relationship between Change Trajectories in the Structural Model

Having specified level-1 individual growth models for both the outcome and predictor, we now use the CSA structural model to represent the relationship between these trajectories. For these data, this means that we link interindividual variation in the growth parameters for (log) alcohol use and (log) peer pressure:

$$
\begin{bmatrix} \pi_{0i} \\ \pi_{1i} \end{bmatrix} = \begin{bmatrix} \alpha_0 \\ \alpha_1 \end{bmatrix} + \begin{bmatrix} \gamma_{\pi_0\pi_0'} & \gamma_{\pi_0\pi_1'} \\ \gamma_{\pi_1\pi_0'} & \gamma_{\pi_1\pi_1'} \end{bmatrix} \begin{bmatrix} \pi_{0i}' \\ \pi_{1i}' \end{bmatrix} + \begin{bmatrix} 0 & 0 \\ 0 & 0 \end{bmatrix} \begin{bmatrix} \pi_{0i}' \\ \pi_{1i}' \end{bmatrix} + \begin{bmatrix} \zeta_{0i} \\ \zeta_{1i} \end{bmatrix},
$$

which you will recognize as a CSA structural model:

$$
\eta = \alpha + \Gamma \xi + B\eta + \zeta \tag{8.46}
$$

with score vectors:

$$
\eta = \begin{bmatrix} \pi_{0i} \\ \pi_{1i} \end{bmatrix}, \, \xi = \begin{bmatrix} \pi_{0i}' \\ \pi_{1i}' \end{bmatrix}, \, \zeta = \begin{bmatrix} \zeta_{0i} \\ \zeta_{1i} \end{bmatrix} \tag{8.47}
$$

and parameter matrices:

$$
\alpha = \begin{bmatrix} \alpha_0 \\ \alpha_1 \end{bmatrix}, \, \Gamma = \begin{bmatrix} \gamma_{\pi_0\pi_0'} & \gamma_{\pi_1\pi_1'} \\ \gamma_{\pi_1\pi_0'} & \gamma_{\pi_1\pi_1'} \end{bmatrix}, \, B = \begin{bmatrix} 0 & 0 \\ 0 & 0 \end{bmatrix}. \tag{8.48}
$$

The latent regression-weight matrix Γ contains the level-2 regression parameters that capture the potential relationship between changes in (log) alcohol use and changes in (log) peer pressure.

The bottom panel of figure 8.2 extends the path diagram in the top panel to include time-varying peer pressure as an exogenous predictor of change. On the left side, the observed indicators of peer pressure, X_{i1}, X_{i2}, and X_{i3}, are the consequence of a pair of exogenous constructs representing the intercept and slope of the true trajectory of change in peer pressure, π_{0i}' and π_{1i}'. The individual growth parameters, in turn, predict the intercept and slope parameters of the endogenous true change trajectory in alcohol use. Our principal interest is in the γ coefficients in the center of the diagram.

Model D of table 8.2 presents the results of fitting this model to data; we omit estimates of the level-1 error structure to save space. Although the goodness-of-fit statistic is larger than ideal, the model fits reasonably well ($\chi^2 = 11.54$, $d.f. = 4$, $p = .0211$). For perspective, we compare Model D to a baseline (not shown) which constrains the four regression coefficients in Γ to zero ($\chi^2 = 342.34$, $d.f. = 8$, $p = .0000$). The huge decline in goodness-of-fit confirms that change in peer pressure is an important predictor of change in alcohol use ($\Delta\chi^2 = 330.8$, $\Delta d.f. = 4$).

Just two of the four fixed effects are statistically significant. Initial level of (log) peer pressure is positively related to initial level of (log) alcohol use ($\hat{\gamma}_{\pi_0\pi_0'} = 0.7985$, $p < .001$) and rate of change in (log) peer pressure is positively related to rate of change in (log) alcohol use ($\hat{\gamma}_{\pi_1\pi_1'} = 0.5767$, $p < .01$). At the beginning of seventh grade, adolescents drink more if their friends drink more, and those who experience more rapid growth in peer pressure experience more rapid growth in alcohol use. We cannot reject the null hypotheses for the other coefficients ($\gamma_{\pi_0\pi_1'}$ and $\gamma_{\pi_1\pi_0'}$).

The variance/covariance components summarize the outcome variation in (log) alcohol use that remains after accounting for changes in (log) peer pressure. We find predictable variation in both the intercept ($\hat{\sigma}^2_{\pi_0} = 0.0422$, $p < .001$) and slope ($\hat{\sigma}^2_{\pi_1} = 0.0092$, $p < .10$). When we compare these values to their cousins in the baseline (not shown), we find steep declines from their original values of 0.0762 and 0.0161. This suggests that about 47% and 43% of the variability in (log) alcohol use intercepts and slopes, respectively, is predicted by change in (log) peer pressure. Once we account for change in (log) peer pressure, the covariance is indistinguishable from 0, indicating no residual relationship between initial status and rate of change.

The remaining estimates in table 8.2 summarize change in (log) peer pressure. The estimated intercept and slope of the average trajectory indicate that (log) peer pressure is non-zero in seventh grade ($\hat{\mu}_{\pi_0'} = 0.1882$, $p < .001$) and climbs steadily over time ($\hat{\mu}_{\pi_1'} = 0.0962$, $p < .001$). Computing $100(e^{.0962}-1) = 10.1\%$, we estimate that peer pressure increases annually by about 10%. Finally, the statistically significant estimated variances for intercepts $\hat{\sigma}^2_{\pi_0'} = 0.0698$, $p < .001$) and slopes ($\hat{\sigma}^2_{\pi_1'} = 0.0285$, $p < .01$) suggest that the trajectories of change in (log) peer pressure differ across adolescents. But because the covariance is indistinguishable from 0, we find no relationship between initial status and rate of change.

8.4 Extensions of Latent Growth Modeling

As you might expect, the basic latent growth model presented in this chapter is just the beginning of what can be achieved with CSA and longitudinal data. The CSA framework is an invaluable tool for exploring complex relationships in longitudinal data, especially relationships among constructs that simultaneously change over time. In the section, we briefly describe some possible extensions. Our goal is not to be exhaustive, but rather to give you a flavor of the exciting possibilities.

We begin by noting that you can easily extend the general CSA model to handle longer longitudinal data records. Extra waves do not alter the basic *Y*-measurement model in equations 8.22 through 8.25; all you need do is extend the dimensions of the constituent score vectors and parameter matrices. If we had four waves of data, for example, the empirical growth record on the left side of equation 8.21 would contain an extra element for the additional wave, and an extra row would be tacked onto the τ_y vector, the Λ_y matrix and the ε vector:

$$\begin{bmatrix} Y_{i1} \\ Y_{i2} \\ Y_{i3} \\ Y_{i4} \end{bmatrix} = \begin{bmatrix} 0 \\ 0 \\ 0 \\ 0 \end{bmatrix} = \begin{bmatrix} 1 & t_1 \\ 1 & t_2 \\ 1 & t_3 \\ 1 & t_4 \end{bmatrix} \begin{bmatrix} \pi_{0i} \\ \pi_{1i} \end{bmatrix} + \begin{bmatrix} \varepsilon_{i1} \\ \varepsilon_{i2} \\ \varepsilon_{i3} \\ \varepsilon_{i4} \end{bmatrix}.$$

The same is true when you use an X-measurement model to represent exogenous change.

So, too, the occasions of measurement need not be equally spaced. You can collect panel data at irregular intervals either for convenience (e.g., at the beginning and end of the school year) or because you want to estimate features of the trajectory more precisely by clustering points around times of greater interest. Irregularly timed data collection is accommodated by specifying appropriate times within the Λ_y matrix of loadings in equation 8.24 (or in the Λ_x matrix in equation 8.42 for exogenous change). Although we did not emphasize it, the temporal spacing of the three waves of data in section 8.2 is irregular. Even if subgroups of respondents are assessed at different sets of irregularly spaced occasions, you can still fit a latent growth model by dividing the sample into subgroups with identical temporal designs and using multigroup analysis with suitable constraints on the relevant parameters across subgroups.[6]

It is easy to extend the basic level-1 individual growth model to include any number of functional forms. Not only can it be a polynomial of any order (provided sufficient data are available), you can also accommodate any curve in which individual status is linear in the parameters (see section 6.4). By comparing the goodness-of-fit of competing nested models, you can evaluate the need for the additional curvilinear terms, as before. To accommodate such modifications, you first add appropriate terms to the level-1 model in equation 8.19. This forces the additional individual growth parameters that represent the curvilinearity into the η latent growth vector in equations 8.21 and 8.23 and adds columns to the Λ_y loading matrix in equation 8.24. With four waves of data, for instance, we could postulate quadratic individual change by writing:

$$\begin{bmatrix} Y_{i1} \\ Y_{i2} \\ Y_{i3} \\ Y_{i4} \end{bmatrix} = \begin{bmatrix} 0 \\ 0 \\ 0 \\ 0 \end{bmatrix} = \begin{bmatrix} 1 & t_1 & t_1^2 \\ 1 & t_2 & t_2^2 \\ 1 & t_3 & t_3^2 \\ 1 & t_4 & t_4^2 \end{bmatrix} \begin{bmatrix} \pi_{0i} \\ \pi_{1i} \\ \pi_{2i} \end{bmatrix} + \begin{bmatrix} \varepsilon_{i1} \\ \varepsilon_{i2} \\ \varepsilon_{i3} \\ \varepsilon_{i4} \end{bmatrix}.$$

As suggested in section 8.2.1, you can also explicitly model the covariance structure of the level-1 measurement errors. The population error covariance matrix $\boldsymbol{\Theta}_\varepsilon$ in equation 8.25 can be specified very generally. You need not accept the independence and homoscedasticity assumptions of classical analysis, nor the band-diagonal configuration imposed by repeated-measures ANOVA. As in chapter 7, you can compare the fit of alternative error structures and determine the best structure analytically. You enjoy identical flexibility when modeling exogenous change in the X-measurement model.

Another extension is the ability to examine *intervening effects* in which an exogenous predictor does not act directly on endogenous change, but indirectly via the influence of intervening factors, each of which may be time-invariant or time-varying. To accomplish this, you would employ the currently unused latent regression parameter matrix **B** in the structural model, which permits endogenous constructs to predict each other.

But perhaps the most important extension of latent growth modeling is when you use it to model simultaneous change in several domains. Section 8.3 offers a simple example of this, but you can extend this strategy to change in multiple exogenous and endogenous domains. When studying students' progress through school, for example, you could simultaneously model endogenous change in math, science, and reading achievement. All you need do is extend the empirical growth record in equation 8.23 to include sufficient rows for each wave in each domain. With three waves of data in math, science, and reading, for example, the empirical growth record would have 9 rows, as would the $\boldsymbol{\tau}_y$ vector, the $\boldsymbol{\Lambda}_y$ parameter matrix and the $\boldsymbol{\varepsilon}$ error vector. In contrast, the latent growth vector $\boldsymbol{\eta}$, would expand only to accommodate the set of individual growth parameters required to represent the three kinds of change:

$$
\begin{bmatrix} Y_{i1}^m \\ Y_{i2}^m \\ Y_{i3}^m \\ Y_{i1}^s \\ Y_{i2}^s \\ Y_{i3}^s \\ Y_{i1}^r \\ Y_{i2}^r \\ Y_{i3}^r \end{bmatrix} = \begin{bmatrix} 0 \\ 0 \\ 0 \\ 0 \\ 0 \\ 0 \\ 0 \\ 0 \\ 0 \end{bmatrix} + \begin{bmatrix} 1 & t_1 & 0 & 0 & 0 & 0 \\ 1 & t_2 & 0 & 0 & 0 & 0 \\ 1 & t_3 & 0 & 0 & 0 & 0 \\ 0 & 0 & 1 & t_1 & 0 & 0 \\ 0 & 0 & 1 & t_2 & 0 & 0 \\ 0 & 0 & 1 & t_3 & 0 & 0 \\ 0 & 0 & 0 & 0 & 1 & t_1 \\ 0 & 0 & 0 & 0 & 1 & t_2 \\ 0 & 0 & 0 & 0 & 1 & t_3 \end{bmatrix} \begin{bmatrix} \pi_{0i}^m \\ \pi_{1i}^m \\ \pi_{0i}^s \\ \pi_{1i}^s \\ \pi_{0i}^r \\ \pi_{1i}^r \end{bmatrix} + \begin{bmatrix} \varepsilon_{i1}^m \\ \varepsilon_{i2}^m \\ \varepsilon_{i3}^m \\ \varepsilon_{i1}^s \\ \varepsilon_{i2}^s \\ \varepsilon_{i3}^s \\ \varepsilon_{i1}^r \\ \varepsilon_{i2}^r \\ \varepsilon_{i3}^r \end{bmatrix},
$$

where superscripts *m*, *s*, and *r* denote the three achievement domains. You can specify the *X*-measurement model similarly to represent simultaneous change in several exogenous domains. And, then, via latent regression coefficient matrix Γ in the structural model, you can investigate relationships among the several exogenous and endogenous changes simultaneously.

PART II

9

A Framework for Investigating
Event Occurrence

Tempora mutantur nos et mutamur in illis. (Times change, and we change with them.)
— Anonymous, quoted in Holinshed's Chronicles, 1578

Each year, on the third Thursday in November, thousands of U.S. smokers participate in the Great American Smoke Out, pledging to throw away their cigarettes and never smoke again. Some participants successfully remain abstinent for the rest of their lives, but many others relapse and start smoking again. When are people at greatest risk of relapse—immediately after quitting, when the physical symptoms of nicotine withdrawal are strongest, or several weeks later, when the social supports of the Smoke Out disappear? Who is at greatest risk of relapse—those who smoked the most or those who tried to quit on their own, without enrolling in a smoking cessation program?

Questions such as these—about the occurrence and timing of events—arise throughout the social and behavioral sciences. Criminologists ask questions about recidivism. Organizational researchers ask questions about employee turnover. Psychiatrists ask questions about the onset and recurrence of mental illness. Educators ask questions about graduation and dropping out. Psychologists ask questions about the attainment of developmental milestones. Yet despite their ubiquity, many general applied statistics books fail to discuss methods for addressing such questions, creating the misimpression that standard techniques, such as regression analysis and analysis of variance, suffice. Unfortunately, not only are these usually versatile methods ill-suited for modeling event occurrence, they may also conceal more than they reveal.

Researchers who want to study event occurrence must learn how to think about their data in new and unfamiliar ways. Even traditional methods for data description—the use of means and standard deviations—fail to serve researchers well. In this chapter, we introduce the essential features of event occurrence data, explaining how and why they create the need for new analytic methods. To provide a context, in section 9.1, we describe three studies of event occurrence, each conducted in a different discipline. In section 9.2, we identify the three major methodological features of each study: (1) a well-defined "event" whose occurrence is being explored; (2) a clearly identified "beginning of time"; and (3) a substantively meaningful metric for clocking time. We conclude, in section 9.3, by highlighting the hallmark feature of event occurrence data that makes the new statistical methods we soon discuss necessary—the problem known as censoring.

9.1 Should You Conduct a Survival Analysis? The "Whether" and "When" Test

Researchers use survival analysis in a variety of contexts that share a common characteristic: interest centers on describing *whether events occur* or *when events occur*. Data can be collected prospectively or retrospectively, over a short period of time or a long one, in an experiment or an observational study. Time can be measured in years, months, days, or seconds. The target event can occur once—e.g., graduation from high school or birth of a first child—or repeatedly over time—starting a job, leaving a job, buying a house, selling a house. Event occurrence can be beyond the individual's control—for example, having a miscarriage or getting fired—or well within his or her purview—for example, having an abortion or stopping breast-feeding. Because the earliest applications of survival analysis were in the study of human lifetimes, where the event of interest is death, the statistical terminology is shrouded in foreboding language. This leads many to assume incorrectly that the methods are appropriate only when studying negative events such as disease onset, recidivism, divorce, and drug use. Yet the methods care little about the valence of the event being studied. They lend themselves equally well to the study of positive events (e.g., marriage, childbirth, graduation) and neutral events (e.g., buying a car, entering day care).

To determine whether a research question calls for survival analysis, we find it helpful to apply a simple mnemonic we refer to as "the whether and when test." If your research questions include either word—whether or when—you probably need to use survival methods. To illustrate the range of research questions for which the methods are suitable, below we describe three studies that "pass" this test.

9.1.1 Time to Relapse among Recently Treated Alcoholics

Mental health professionals have long recognized that many people who participate in inpatient alcohol treatment programs start drinking again after program cessation. To decrease the risk of relapse, clinicians have developed after-care programs that support patients once they leave the hospital setting. Cooney, Kadden, Litt, and Getter (1991) evaluated the comparative efficacy of two such regimens—one that emphasized *coping skills* and another that emphasized *interaction skills*. The researchers randomly assigned 89 individuals, newly released from a 21-day alcohol treatment program, to the two treatments. Each person was followed for up to two years. The researchers hypothesized that treatment efficacy would vary as a function of the patients' level of psychosocial functioning at release. Specifically, they expected that interaction therapy would be more effective for people with lower levels of psychopathology, and coping skills therapy would be more effective for those with higher levels.

Research interest centered on whether and, if so, when the newly released patients relapsed to alcohol use. The researchers focused most closely on the subsequent occurrence of the first day of "heavy drinking," defined as consuming three or more ounces of alcohol (ethanol) in a 24-hour period. The outcome, time to first drink, was measured as the number of days between the day of release from the program and the first day of heavy drinking. During the two-year follow-up, 57 patients (64.0%) relapsed, 28 (31.5%) remained abstinent, and 4 (4.5%) who remained abstinent for a while ultimately disappeared. Among those who relapsed, some did so immediately after release—in as few as 5 or 10 days after leaving the hospital—while others relapsed much later—after 366, 370, and even 677 days.

Cooney and colleagues wanted to do more than describe these relapse patterns; they also wanted to predict time to relapse using two variables: level of psychopathology and type of after care. Initially, they explored the main effects of each predictor and found that neither had any effect. Then, they explored the statistical interaction between the predictors and found that certain after-care regimens were especially successful for certain types of individuals. As hypothesized, interaction therapy was more effective for those with low psychopathology scores while coping skills therapy was more effective for those with high scores. The researchers concluded that no single after-care program fits every patient's needs. To ensure better outcomes, programs and patients must be matched.

9.1.2 Length of Stay in Teaching

Since the 1990s, the demand for U.S. public school teachers has increased dramatically as student enrollments have grown and practicing teachers have retired. In most school districts, the greatest demand is in fields—such as special education—that traditionally have high turnover rates. To determine how long special educators stay in teaching and to identify factors associated with their stay or leave decisions, Singer (1993) examined the career histories of 3941 special educators newly hired in Michigan between 1972 and 1978. Using state administrative databases, she reconstructed the employment history of each teacher from his or her date of hire through 1985. All teachers were tracked for a minimum of seven years. Those in the earlier entry cohorts were tracked for longer periods of time, up to 13 years for those hired in 1972.

Research interest centered on whether and if so, when, the teachers first left teaching. The outcome—length of stay in teaching—was measured as the *number of years* between a teacher's dates of hire and departure from the Michigan public schools. Across the entire sample, 2507 teachers (63.6%) left teaching before 1985; the remaining 1434 (36.4%) were still teaching when data collection ended. Special educators were most likely to leave during the first five years of teaching. Approximately one-tenth of those still working at the beginning of each of these years left by the end of the year. Teachers who survived these initially "hazardous" years were much less likely to leave in later years.

After describing this pattern of risk, Singer investigated whether certain types of special educators were especially likely to leave. As researchers studying regular educators' careers have found, young women were particularly likely to depart, probably to begin a family. She also found that three predictors assessing the "opportunity costs" of continuing to teach in the public schools (in comparison to switching to another occupation) predicted teachers' stay-or-leave behavior. Special educators with the best job opportunities outside of teaching (those who provided support services or who taught students with speech, hearing, or vision disabilities) were at greatest risk of leaving as were those with higher test scores and those paid comparatively low salaries. Singer concluded that special educators, like their regular educator colleagues, assess the "costs" of continuing to teach, and those with better options elsewhere are more likely to leave.

9.1.3 Age at First Suicide Ideation

Suicide is a major cause of death among adolescents and the rate at which young Americans attempt suicide is increasing dramatically. Yet because the total number of suicides is small and people who take their own lives cannot be interviewed, researchers studying this problem generally use post-hoc informant interviews to contrast adolescents who attempt suicide with those who do not. Bolger, Downey, Walker, and Steininger (1989) approached this issue from a different perspective by focusing *not* on suicidal behavior but on suicidal *ideation*. After recruiting 391 undergraduates, ages 16 to 22, the researchers administered a 15-minute anonymous questionnaire that asked "Have you ever thought of committing suicide?" and if so, "At what age did the thought first occur to you?" Also included were questions assessing demographic risk factors (e.g., age, race, and gender) and potentially precipitating life experiences (e.g., death of a family member, parental absence during childhood).

Research interest centered on whether and, if so, when each undergraduate first thought of suicide. A total of 275 (70.3%) students reported having had suicidal thoughts; the remaining 116 (29.7%) did not. Of those who responded affirmatively, some reported initial thoughts as young as age 6, while others reported initial thoughts as late as age 21—an age *older* than some of the *younger* members of the sample. Adolescence was the time of greatest risk—149 (54.2%) respondents who had ever thought about suicide reported that their initial thoughts were manifested between ages 12 and 15.

Bolger and colleagues also investigated whether suicide ideation could be predicted using descriptors of an individual's demographic and precipitating life circumstances. They examined not only whether a predictor had a consistent effect across the life span but also whether a predictor's effect *differed* across developmental stages. Beyond a consistent gender differential (females were always at greater risk of suicidal ideation), the effects of two other predictors differed by developmental stage. In pre-adolescence, but not adolescence, parental absence was associated with increased risk of suicide ideation. In adolescence, however, a pronounced race differential appeared, with whites at greater risk than blacks. Because suicidal thoughts were so widespread and seemed to occur at developmentally vulnerable ages, the researchers concluded that suicidal ideation may be a normal part of development, and that at early ages, at least, may be precipitated by exposure to difficult life experiences.

9.2 Framing a Research Question about Event Occurrence

In introducing these studies, we emphasize their *substantive* features—their research questions, data collection plans, and findings. To determine whether a research question lends itself to survival analysis, you must also examine a study's *methodological* features. Even though these studies are set in different disciplines, they share three methodological features that make them a candidate for survival analysis. Specifically, each has a clearly defined:

- *Target event,* whose occurrence is being studied
- *Beginning of time,* an initial starting point when no one under study has yet experienced the target event
- *Metric for clocking time,* a meaningful scale in which event occurrence is recorded

Using this terminology, the alcohol relapse study is asking: In how many days, after release from the hospital, do recently treated alcoholics start drinking again? Similarly, the teacher turnover study is asking: In how many years, after initial hire, do special educators stop teaching? To demonstrate why these *methodological* features render a research question amenable to survival analysis, we now discuss each in turn.

9.2.1 Defining Event Occurrence

Event occurrence represents an individual's transition from one "state" to another "state." A recently treated ex-alcoholic is abstinent (state 1) until he or she starts drinking (state 2). A newly hired teacher is working (state 1) until he or she leaves the schools (state 2). An individual has never thought about suicide (state 1) until he or she does (state 2). We use the term "state" because it is so generic, applicable across many substantive disciplines. Some states are physical (living in a homeless shelter, living in a rented home); others are psychological (depressed or healthy) or social (married or divorced). The only requirement for survival analysis is that, in any particular research setting, the states be both mutually exclusive (nonoverlapping) and exhaustive (of all possible states).

In most applications, each person can occupy only two possible states: drinking or abstinent, employed or unemployed. In other applications, each individual can occupy three or more possible states. A researcher studying student careers, for example, might track high school freshmen as they transition from being in school (state 1) to one of two alternatives: dropping out (state 2) or graduation (state 3). A researcher studying marital duration might track newlyweds to see whether their marriages (state 1) end in divorce (state 2) or with death (state 3). For now, and for most of our presentation, we assume that all individuals can occupy only one of *two* states. In section 15.5, we extend these two-state methods, using an approach known as competing risks survival analysis, to situations in which individuals can occupy three or more states.

Some states can be occupied only once in a lifetime—first word, first step, puberty, high school graduation, and death, to name a few. Once a person enters these states, he or she can never reenter them. Many other states—including depression, incarceration, pregnancy, and marriage—can be occupied again and again. In their analysis of the course of psychopathology over time, Lavori et al. (1996) track individuals as they cycle into, and out of, up to seven depressive episodes. Methodologists use the term "spell" to refer to a single transition into (or out of) one of a series of repeatable states. Economists study employment spells, the length of time between hire and leaving a job; psychiatrists study illness spells, the length of time between onset and remission. Our presentation focuses exclusively on the most common and fundamental problem: the analysis of single spells.

States must be defined precisely, with clear guidelines indicating the behaviors, responses, or scores constituting each state. Empirical researchers often find this the most difficult requirement to meet. In some situations, you can invoke generally accepted definitions and criteria. Psychologists, for example, routinely use the DSM classification system. Other times, you may use simple face-valid definitions: a woman is pregnant or not, an adult is employed or not. But there will be studies in which state definition is difficult. When examining developmental milestones among infants, for example, who determines which word is the child's "first?" Is maternal report sufficient, or must the child be able to produce the word for an observer? How clearly articulated must the pronunciation be? These issues are not just academic because the definition of states provides the definition of outcomes. In their review of recidivism among sex offenders, for example, Furby, Weinrott, and Blackshaw (1989) show how different definitions of states—commission of the same crime, of any sex crime, of any crime—yield different outcomes that are associated with different predictors. As state definition is primarily a measurement issue, we simply raise it here. Prior to data collection, we urge you to give serious consideration to a wide range of alternative definitions so that the one you ultimately select provides the most meaningful basis for analysis.

9.2.2 Identifying the "Beginning of Time"

The "beginning of time" is a moment when *everyone* in the population occupies one, and only one, of the possible states. On the day they are released from the hospital, all recently treated alcoholics are abstinent; on the day they were hired, all new special educators are teaching. Over time, as individuals move from the original state to the next, they experience the target event. The timing of this transition—the distance from the "beginning of time" until event occurrence—is referred to as the *event time*.

To identify the "beginning of time" in a given study, imagine placing everyone in the population on a time-line, an axis with the "beginning of time" at one end and the last moment when event occurrence could be observed at the other. The goal is to "start the clock" when no one in the population has yet experienced the event but everyone is at least (theoretically) eligible to do so. In the language of survival analysis, you want to start the clock when everyone in the population is *at risk* of experiencing the event.

Because birth is often meaningful and convenient, it is a popular start time, especially in studies that track developmental sequences and milestones. Although it may seem awkward to report it this way, any study that uses *age* as the metric for time is actually using *birth* to denote time's beginning. Hence, in the suicide ideation study, which examined the age of first suicidal thought, the "beginning of time" is birth.

Another common way of setting the beginning of time is to tie it to the occurrence of a precipitating event—one that places all individuals in the population *at risk* of experiencing the target event. In the alcohol relapse study, the beginning of time is the date of hospital release because at that moment no one is drinking, yet everyone could theoretically start drinking again. In the teacher turnover study, the beginning of time is the date of hire because at that moment everyone is working, yet everyone could immediately quit. The choice of precipitating event varies widely across disciplines and research questions. Options have included: entry into high school (Roderick, 1994), graduation from college (Rayman & Brett, 1995), release from jail (Harris & Koepsell, 1996), marriage (South, 2001), divorce or separation (Wu, 1995), hospital admission (Strober, Freeman, Bower, & Rigali, 1996), hospital release, report of child maltreatment (Fryer & Miyoshi, 1994), pregnancy (Yamaguchi & Kandel, 1987), and childbirth (Fergusson et al., 1984).

What should you do if you have no compelling argument for initiating the clock at a particular point in time? Researchers in this situation typically use an arbitrary start time, as long as that time is unrelated to event occurrence. In an experiment, for example, you might use the date of randomization (Peto et al., 1976) or the date of intervention (Berk & Sherman, 1988; Greenhouse, Stangl, & Bromberg, 1989). Or in a study of ongoing social interactions, you might start the clock at any convenient moment, for there is little hope of identifying a substantively meaningful start time in a long-term continuing process. Using this argument, Gardner and Griffin (1988) used a single arbitrarily selected

20-minute segment of interaction in their study of continuities and breaks in gazes between a husband and wife during conversations about their marriage.

9.2.3 Specifying a Metric for Time

Once you identify the beginning of time, you must select the units in which you will record its passage. Sometimes you can record time using thin precise units. In the alcohol relapse study, Cooney and colleagues (1991) counted the number of *days* between a patient's release from the hospital and his or her first episode of heavy drinking. Such precision allows distinctions among individuals who relapsed, for example, in 5, 15, and 25 days. Often, however, you will only be able to record time using a series of finite intervals. In the teacher turnover study, for example, time was recorded in *years*. In other settings, you might know the *semester* an adolescent dropped out of school (Graham, 1997) or the *month* when a patient was readmitted to the hospital (Mojtabai, Nicholson, & Neesmith, 1997). We distinguish between data recorded in thin precise units and those recorded in thicker intervals by calling the former *continuous time* and the latter *discrete time*.

Time should be recorded in the smallest possible units relevant to the process under study. No single metric is universally appropriate, and even different studies of the identical event might use different scales. Consider three studies of job turnover: one focusing on members of the U.S. Congress, another focusing on teachers, and a third focusing on clerical workers. Political scientists studying congressional turnover generally use two-year time intervals because, with the few exceptions that result from death and mid-term resignation, all representatives complete their two years of elected service. Educational researchers studying teacher turnover typically use one-year intervals because most teachers leave their jobs at the end of the school year. Organizational researchers studying clerical workers, in contrast, record time in weeks or months because many employees in these jobs stay for less than one or two years, rendering the wider intervals, useful in the other studies, too coarse to reflect meaningful variation in the timing of transitions.

Despite the obvious advantages of measuring time as precisely as possible, many researchers find themselves with discrete time data. Three related factors underlie this phenomenon. First, some events can *occur* only at discrete points in time. High school students, for example, can graduate at only a small number of preset times during the year. Once graduated, they can enter college only at another small number of registration periods. Second, although some events can *theoretically* occur across a wide range of times, many individuals do not experience them this way. Although employees with term contracts might be able to quit their jobs at any point in time, most (especially professionals) leave when their contracts expire. Professors leave colleges at the end of the academic year; physicians leave health clinics when their obligations end (see, e.g., Singer, Davidson, Graham & Davidson, 1998). Third, data collection constraints—especially in retrospective studies—often force researchers to use intervals to record the passage of time. Due to memory failure, respondents can usually supply dates and times only in ranges or round numbers, even if encouraged by interviewers to be more precise. In the suicide ideation study, for example, only 15 people recalled their age at first suicidal thought with more precision than a full year, even though all respondents were asked to report their age to the half-year. Similarly, in a retrospective study of the age (in months) when young children were first placed in day care, Singer, Fuller, Keiley, and Wolf (1998) found that only 8% of the parents reported placement at any time other than the child's first 6 months of life, his or her birthday (12, 24, 36, or 48 months precisely) or his or her half birthday (6, 18, 30, 42, or 54 months precisely). Even if this pattern reflects overt decision making to place children in care when they reach their birthday (or half birthday) and does not reflect rounding, the effect on the data is the same: time is measured in discrete intervals.

Distinguishing between continuous- and discrete-time data is more than a methodological detail. Almost every feature of survival analysis—parameter definition, model construction, estimation, and testing—depends on the metric for time. Although the earliest descriptive methods for event occurrence (e.g., life-table methods) were developed for discrete-time data, modern methods of analysis (e.g., Cox regression—also known as proportional hazards modeling) assume that time is recorded on a continuous scale. This emphasis arises from the fact that researchers in medicine and engineering—the areas in which survival methods were originally developed—can usually record event occurrence precisely. Epidemiologists can use death certificates to measure human lifetimes in days; industrial product engineers can measure a machine's lifetime in minutes (and sometimes seconds).

Unfortunately, continuous-time methods break down when event times are highly discretized due to a problem known as "ties" (Cox & Oakes, 1984). With continuous-time data, the probability that two or more individuals share an identical event time (are "tied") is infinitesimally small. Because the probability of a tie is small, actual ties are few, and those that do occur can be treated as little more than a methodological nuisance. In section 14.2.2, we offer various approaches for handling such ties, each predicated on the assumption that observations only "appear" tied because the metric for time, while continuous, is too coarse. But even these solutions fail as the number of ties increases (Hertz-Picciotto & Rockhill, 1997; Hsieh, 1995) and they are obviously inappropriate if events can occur only at specific moments (Allison, 1984; Singer & Willett, 1993; Raftery, Lewis, Aghajanian, & Kahn, 1996).

In discrete-time data, ties are pervasive. In the teacher turnover study, for example, there are only 12 times when teachers can leave. Because the sample includes nearly 4000 teachers, the data set has thousands of ties. In the suicide ideation study, in which there are 16 possible event times (age 6 through age 21), there is only one age (21) at which there are no ties. Why analyze your data using an adaptation of a method that assumes that ties do not exist when you know, a priori, that they will be commonplace? Even more to the point, why use a method that you know is likely to break down with the data that you actually have in your hand?

These concerns lead us to divide our presentation of survival methods into two broad sections—discrete-time methods (chapters 10 through 12) and continuous-time methods (chapters 13 through 15). Although beginning in discrete time prevents us from initially exploiting the survival analysis programs widely available in statistical packages, it has a pedagogic advantage that we believe outweighs this liability: it turns out that parameter definition, model structure, statistical analysis, and interpretation of results—in short, everything about survival analysis—is simpler and more comprehensible in discrete time. This advantage is considerable for it permits us to present most of the new concepts in a more easily understood framework and then exploit this knowledge when we move to the admittedly more commonplace continuous world. It has been our experience that researchers who learn survival analysis using this two-step approach develop a more intuitive understanding of the model, its assumptions, and its implementation.

9.3 Censoring: How Complete Are the Data on Event Occurrence?

Having outlined the methodological features necessary for phrasing research questions about event occurrence, it may appear straightforward to move directly on to a presentation of statistical models for survival analysis. After all, it would seem that anyone who has studied introductory statistics will know how to describe the distribution of a quantitative outcome—here, time to event—by computing statistics like the sample mean, the standard deviation, the range, and so on. Whether the distribution is symmetric or skewed, unimodal or bimodal, with many or few unusual values, it appears easy to tabulate the data, examine its behavior, and compute familiar single-number summaries.

But researchers studying event occurrence face an analytic difficulty not encountered by those studying other kinds of outcomes. It is a difficulty encapsulated in the question: What should you do with the data from those individuals who do not experience the target event during the period of data collection? In each of studies described so far, approximately one-third of the sample did not experience the target event while the researcher watched: 32 of the former alcoholics remained abstinent, 2704 of the newly hired teachers were still teaching, 116 of the undergraduates remained free of suicidal thoughts. How can you estimate a mean length of time to event, or any other statistic, when you don't even know *whether* some members of the sample experienced the event, let alone *when* they did so?

How should you analyze such data? Should you set aside the cases with unknown event times? Yet these people are an important subgroup of respondents—they are not a random subset of the sample, they are the ones *least likely* to experience the event. To include them in the computation of summary statistics, it appears that you would need to assign them a value for the outcome. But what value is appropriate? How can you impute an outcome value that is, by its very nature, unknown? Does assigning event times to those who do not experience the event during the period of data collection even make sense, given that some of them may *never* experience the target event at all? Although all teachers will eventually leave teaching, some of the ex-alcoholics will never relapse and some of the undergraduates will never contemplate suicide.

Thorny questions, indeed. No matter when data collection begins, and no matter how long it lasts, some sample members are likely to have unknown event times. Statisticians call this problem *censoring* and they label the people with the unknown event times *censored observations*. Because censoring is inevitable—and a fundamental conundrum in the study of event occurrence—we now explore it in detail, first describing how it arises, next describing its different forms, and then describing how researchers have (incorrectly) attempted to handle it analytically.

9.3.1 How and Why Does Censoring Arise?

Censoring occurs whenever a researcher does not know an individual's event time. There are two major reasons for censoring: (1) some individuals will *never* experience the target event; and (2) others will experience the event, but not during the study's data collection. Some of these latter individuals will experience the event shortly after data collection ends while others will do so at a much later time. As a practical matter, though, these distinctions matter little because you cannot distinguish among them. That, unfortunately, is the nature of censoring: it prevents you from knowing the very quantity of interest—*whether* and, if so, *when* the target event occurs for a subset of the sample.

Depending upon a study's design, censoring can occur at a single point in time or at multiple points in time. In a prospective follow-up of a single cohort, censoring will occur at only one point in time—at the end of data collection. In the alcohol relapse study, all 28 individuals who were abstinent two years after hospital release were censored at 720 days. In a prospective follow-up of multiple cohorts that ends in a single chronological year, censoring will occur at multiple points in time. In the teacher turnover study, which tracked all special educators newly hired between 1972 and 1978 through 1985, those in the first entry cohort (1972) were censored at 12 years while those in the last entry cohort (1978) were censored at 7. A similar pattern of censoring arises in retrospective studies of age-heterogeneous samples—like the suicide ideation study—but for a different reason. Because individuals are censored not by the *end* of data collection but by the *occurrence* of data collection, those who have not yet experienced the target event are censored at their current age. This means that the 23 18-year-olds who had never contemplated suicide before filling out the questionnaire were censored at age 18 while the 26 21-year-olds who were also free of suicidal thoughts were censored at age 21.

The amount of censoring in a study is related to two factors: (1) the rate at which events occur; and (2) the length of data collection. If the event is common and data collection sufficiently long, most people will experience the event during data collection and your sample will contain few censored cases. If the event is rare or data collection is curtailed by resource or practical constraints, censoring will be widespread. After following a sample of young boys for five years—from age 6 through age 10—to see whether and when they reported getting drunk for the first time, Masse and Tremblay (1997) found that only 52% of their sample was censored. But after following a sample of men for twice as long (11 years, from age 17 to age 28) to determine whether and when they used cocaine for the first time, Burton, Johnson, Ritter, and Clayton (1996) found that 76% of their sample was censored.

Censoring can be minimized by design, but it can rarely be eradicated. Researchers conducting prospective studies can minimize censoring by following respondents for a longer period of time, but this comes at increased cost. Researchers conducting retrospective studies can minimize censoring by selecting older respondents and asking them to reflect back on a longer period of time, but this yields noisier recall data. And in both types of studies, people who *never* experience the target event will remain censored regardless of design. The bottom line in research on event occurrence: Censoring is inevitable.

9.3.2 Different Types of Censoring

Although the term "censored" refers to any individual with an unknown event time, there are actually several different types of censoring. Methodologists make two major types of distinctions: first, between *noninformative* and *informative* censoring mechanisms, and second, between *right-* and *left-*censoring. We discuss each of these below.

Noninformative versus Informative Censoring

A noninformative censoring mechanism operates independent of event occurrence and the risk of event occurrence. If censoring is under an investigator's control, determined in advance by design—as it usually is—then it is noninformative. In the teacher turnover study, censoring is noninformative because it occurs in a single calendar year (1985) for all 3941 special educators. So, too, in the suicide ideation study—censoring is noninformative because it occurs at the same chronological time (the day of data collection) for all 391 respondents. In both these examples, censoring occurs because data collection ends, not because of actions taken by study participants. We can therefore assume that all individuals who remain in the study after the censoring date are representative of everyone who *would have remained in the study* had censoring not occurred.

If censoring occurs because individuals have experienced the event or are likely to do so in the future, the censoring mechanism is informative. Consider the alcohol relapse study, which included two censoring mechanisms. The first was the end of data collection, when 28 individuals were still abstinent; this censoring mechanism is noninformative. The second was the attrition of four individuals who were lost to follow up after being abstinent for one year. If they were lost because they moved out of town (or due to some other unforeseen random occurrence), then this censoring mechanism is noninformative. But if they were lost because they started drinking again and stopped notifying investigators of their whereabouts, then this censoring mechanism is *informative*— people with censored data are likely to have experienced the event. Under these circumstances, we can no longer assume that those people who remain in the study after this time are representative of all individuals who would have remained in the study had censoring not occurred. The noncensored individuals differ systematically from the censored individuals.

No statistical method can produce unbiased analyses of event occurrence data if the censoring mechanism is informative. The problem, of course, is that when censoring is not determined by design, you have no way of knowing whether it is due to a random occurrence (e.g., the individual just stopped participating in the study) or to impending event occurrence (e.g., the treatment was no longer effective, so the individual stopped participating). The validity of a survival analysis rests on the assumption that censoring is noninformative, either because it occurs at random or because it occurs at a time dictated by design. If the process you are studying is prone to informative censoring mechanisms—as is common when studying treatment efficacy because individuals not responding tend to drop out—you must work hard to eliminate these problems before they arise. Attrition is the bane of longitudinal research, especially when it is systematic. Throughout what follows, we assume that the censoring mechanisms are noninformative.

Right- versus Left-Censoring

Right-censoring arises when an event time is unknown because event occurrence is not observed. Left-censoring arises when an event time is unknown because *the beginning of time* is not observed. All the censoring mechanisms described so far have led to right-censoring. Because this type of censoring is the one typically encountered in practice, and because it is the type for which survival methods were developed, references to censoring, unencumbered by a directional modifier, usually refer to right-censoring.

How do left-censored observations arise? Often they arise because researchers have not paid sufficient attention to identifying the beginning of time during the design phase. If the beginning of time is defined well—as that moment when all individuals in the population are eligible to experience the event but none have yet done so—left-censoring can be eliminated. It is for this reason that we emphasized the need for a careful specification of the beginning of time in the previous section. Unlike right-censoring, which usually exists regardless of design, left-censoring can generally be eradicated by thoughtfulness.

Some researchers nevertheless do find themselves with left-censored data. Those studying the occurrence of potentially repeatable events are especially susceptible. This was the dilemma faced by Fichman (1989) in his prospective study of absenteeism among coal miners. During a single calendar year, Fichman studied the length of each coal miner's many "attendance spells," which began when a miner returned to work after an absence and stopped when the miner was absent another day. On average, each of the 465 miners generated 11 attendance spells of known length. In addition, each generated one right-censored spell (the time between his last return to work and the end of data collection) and one left-censored spell (the time between his last return during the *previous* calendar year—the date of which is unknown—and his first absence during the year of data collection).

Left-censoring presents challenges not easily addressed even with the most sophisticated of survival methods (Hu & Lawless, 1996). Little progress has been made in this area since Turnbull (1974, 1976) offered some basic descriptive approaches and Flinn and Heckman (1982) and Cox and Oakes (1984) offered some directions for fitting models under a restrictive set of assumptions. The most common advice, followed by Fichman, is to set the left-censored spells aside from analysis. After all, he argued, there is nothing sacred about identifying January 1 as the "beginning of time." He therefore eliminated all left-censored spells by redefining the beginning of time as the date when each miner returned to work after the first absence of the calendar year. Redefining the beginning of time to coincide with a precipitating event—here a return from an absence—is often the best way of resolving the otherwise intractable problems that left-censored data pose. Whenever possible, we suggest that researchers consider such a redefinition or otherwise eliminate left-censored data through design.

In what follows, we typically assume that all censoring occurs on the right. In section 15.6, however, we describe what you can do when your data set includes what are known as *late entrants into the risk set*. You are most likely to encounter late entrants if you study *stock samples*, age-heterogeneous groups of people who *already* occupy the initial state when data collection begins—for example, a random sample of adults who have yet to experience a depressive episode. Your plan is to follow everyone for a fixed period of time—say ten years—and record whether and when sample participants experience their first episode. Because each sample member was a different age when data collection began, however, the ten years you cover do not cover the same ten years of peoples' lives: you follow the 20-year-olds until they are 30, the 21-year-olds until they are 31, the 22-year-olds until they are 32, and so on. For an outcome like depression onset, it makes little sense to clock "time" using chronological years (2009, 2010, etc). Instead, you would like to clock "time" in terms of *age*, but you do not observe everyone during the identical set of

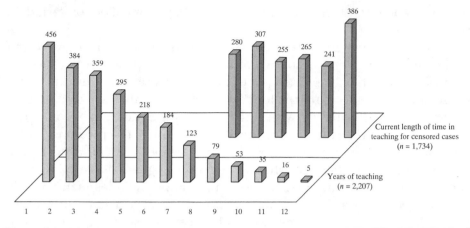

Figure 9.1. Distribution of the number of years in teaching by censoring status, for the 3941 special educators.

ages. The flexible strategies described in section 15.6 are useful for addressing this complication, and ultimately allow you to consider a range of alternative metrics for clocking time.

9.3.3 How Does Censoring Affect Statistical Analysis?

Censoring's toll can be seen in figure 9.1, which presents data from the teacher turnover study. The front row summarizes the sample distribution of the number of years in teaching for the 2207 special educators with known event times. Each teacher left at some point prior to the end of data collection: 456 left by the end of their first year, 384 left by the end of their second, and so on. The back row presents the distribution of *current length of time in teaching* for the 1734 special educators with censored event times. Although you probably should not label these latter data "event times" (because these teachers have yet to leave) statisticians refer to them as "censored event times"—the times when each individual had *yet* to experience the target event. Some of these teachers will leave the year after they were censored, while others will leave at a much later point in time.

Notice how different the two sample distributions are. The distribution for the 2207 teachers with known events times is skewed, peaking in year 1 and declining thereafter. The distribution for the 1734 teachers with censored event times is almost uniform, with no values less than 7 and approximately equal numbers of teachers at 8, 9, 10, and 11 years. The peak in the censored data distribution is at 12 years—the maximum observable length of stay in teaching given that data collection lasted 13 years.

This discrepancy in distributions—between those with known event times and those with censored event times—is typical. In prospective studies like this one, the discrepancy arises because the individuals with censored event times have the longest "lives"; here, they are those teachers who were still teaching when data collection ended. In retrospective studies, such as the suicide ideation study, a similar discrepancy results, but for a different reason. Those event histories are censored at the moment of data collection when, in age-heterogeneous samples, individuals are of different ages. Unless your sample includes many people who are as young as other sample members were at the earliest event times, the censored event times generally exceed most of the known event times.

Given these two disparate distributions, how can we adequately summarize how long teachers teach? Researchers faced with problems like these have adopted a variety of ad hoc approaches, none completely satisfactory. Some have simply set aside all individuals with censored event times (e.g., Abedi & Benkin, 1987; Siegfried & Stock, 2001). Researchers who exclusively study only those people who experience the target event are implementing this strategy, whether they recognize it or not. It is easy to fall into this trap—at first glance, it seems reasonable to use a sample of former teachers to ask how long the average teacher teaches. Were we to construct such a sample with these data, we would estimate the mean career duration to be 3.7 years.

But analyzing only those individuals who have actually experienced the target event excludes a crucial subgroup: here, the 1734 teachers still teaching. In fact, the length of time that the average teacher teaches must be greater than 3.7 years because that estimate relies exclusively on data from those who left early in their careers. To learn about "true" average career length and understand when teachers are especially likely to leave, data from both the censored and the uncensored cases must be incorporated simultaneously in the analysis. Censored career histories provide important information, especially about the probability that teachers will stay in the profession for long periods of time. Censored cases must not be excluded, even though we do not know when (or in other studies, even whether) they will ultimately experience the target event.

Cognizant of these concerns, some investigators incorporate the censored cases into their analyses by imputing event times. The most popular approach is to assign the censored cases the event time they possess at the end of data collection (e.g., Frank & Keith, 1984). Applying this strategy to our teacher career data (e.g., assigning a career length of 7 years to the 280 teachers censored in year 7, etc.) yields an estimated mean career duration of 7.5 years. While suitably longer than the former underestimate of 3.7 years, this estimate, too, cannot be correct. Imputing event times for censored cases simply changes all "nonevents" into "events" and further assumes that all these new "events" occur at the earliest time possible—that is, at the moment of censoring. Surely these decisions are most likely wrong.

Given these problems, some researchers proceed more cautiously. Rather than imputing event times, they abandon the "When?" question entirely, setting aside most of the information they possess about duration. They then dichotomize all individual event histories at some particular (and often arbitrary) point in time and ask *whether* the event has occurred *by* that time (e.g., Sargeant, Bruce, Florio, & Weissman, 1990) or by each of several successive points in time (e.g., Myers, McCauley, Calderon, & Treder, 1991). If we apply this tradition to our teacher career data, we might distinguish between teachers who leave before and after five years of service.

Although this approach is better than data elimination or incorrect imputation, dichotomization of event times does not resolve the censoring dilemma; it simply obscures it from view. Despite its simplicity, there are at least five major problems with dichotomization. First, dichotomization eliminates meaningful variation in event times by collapsing together individuals who left at many different points in their careers. Teachers who leave teaching after one year in the classroom undoubtedly differ from those who leave after four, yet they are all indistinguishable in an analysis dichotomized at year 5. Second, any particular dividing line, even one ostensibly relevant to the process under study, is arbitrary. Why dichotomize at year 5, rather than at years 3 or 7? Third, seemingly contradictory conclusions can result from nothing more than changes in the time chosen for dichotomization. Dichotomizing at year 2 in our teacher career data, for example, we conclude that men and women are almost equally likely to leave teaching (quit rates of 19% and 22%, respectively) but dichotomizing at year 8, we find an enormous sex differential (quit rates of 57% and 70%). Fourth, dichotomization discards information about later event occurrence. If we dichotomize at year 5, for example, we discard the known event times of the 495 teachers who left teaching between years 6 and 12. And fifth, once you have dichotomized, you can no longer address the fundamental question: "How long does the average teacher teach?"

Clearly, an alternative approach is needed, one that deals evenhandedly with both known and censored event times. After all, censored observations *do* tell you something about event occurrence, or more precisely about event *nonoccurrence*. If an observation is censored, you know that the individual did *not* experience the target event by the time of censoring. It is this insight—that, in telling us about event *nonoccurrence*, the censored cases do provide some information about event occurrence—that leads to a comprehensive strategy for incorporating censored cases into analyses, as we discuss in the next chapter.

10

Describing Discrete-Time Event Occurrence Data

Time does not change us, it just unfolds us.

—Max Frisch

Most empirical researchers are so comfortable conducting descriptive analyses that they rarely imagine that familiar statistical workhorses—such as means and standard deviations—may not always be suitable. As explained in chapter 9, censoring makes standard statistical tools inappropriate even for simple analyses of event occurrence data. A censored event time provides only partial information: it tells you only that the individual did not experience the target event by the time of censoring. In essence, it tells you more about event *nonoccurrence* than about event occurrence (the latter, of course, being your primary interest). Traditional statistical methods provide no ready way of simultaneously analyzing observed and censored event times. Survival methods do.

In this chapter, we present a framework for describing discrete-time event occurrence data. In addition to our primary agenda—which involves demonstrating how to implement the new methods and interpret their results—we have a secondary agenda: to lay the foundation for model building, the focus of chapters 11 and 12. As we will show, the conceptual linchpin for all subsequent survival methods is to approach the analysis on a period-by-period basis. This allows you to examine event occurrence sequentially among those individuals eligible to experience the event at each discrete point in time.

We begin in section 10.1 by introducing the *life table*, the primary tool for describing event occurrence data. Then, in section 10.2, we introduce three essential statistical summaries of the life table—the hazard function, the survivor function, and the median lifetime—and demonstrate that these ingenious statistics, which deal evenhandedly with censored and noncensored cases, are intuitively appealing as well. In section

10.3, we apply the new techniques to four empirical studies, to help you develop intuition about the behavior, interpretation, and interrelationships of life table methods. Sampling variation is the topic of section 10.4, where we show how to estimate standard errors. We conclude, in section 10.5, by showing you how to compute all these summaries using standard cross-tabulation programs available in every major statistical package.

10.1 The Life Table

The fundamental tool for summarizing the sample distribution of event occurrence is the *life table*. As befits its name, a life table tracks the event histories (the "lives") of a sample of individuals from the beginning of time (when no one has yet experienced the target event) through the end of data collection. Table 10.1 presents a life table for the special educator data set introduced in section 9.1.2. Recall that this study tracked the careers of 3941 teachers newly hired in the Michigan public schools between 1972 and 1978. Everyone was followed until 1985, when data collection ended. Defining the "beginning of time" as the teacher's date of hire, research interest centers on whether and, if so, when these special educators stopped teaching.

Divided into a series of rows indexing time intervals (identified in columns 1 and 2), a life table includes information on the number of individuals who:

- Entered the interval (column 3)
- Experienced the target event during the interval (column 4)
- Were censored at the end of the interval (column 5)

Here, these columns tally, for each year of the career, the number of teachers employed at the beginning of the year, the number who stopped teaching during the year, and the number who were censored at the end of the year (who were still teaching when data collection ended). We discuss the remaining elements of the life table in section 10.2.

Taken together, these columns provide a narrative history of event occurrence over time. At the "beginning of time," when everyone was hired, all 3941 teachers were employed. During the first year, 456 teachers quit, leaving 3485 (3941 − 456) to enter the next interval, year 2. During the second year, 384 teachers quit, leaving 3101 (3485 − 384) to enter the next interval, year 3. During the 7th year, censoring begins to affect the narrative. Of the 2045 special educators who taught continuously for 7 years, 123 quit by the end of that year and 280 were

Table 10.1: Life table describing the number of years in teaching for a sample of 3941 special educators

Year	Time interval	Number			Proportion of	
		Employed at the beginning of the year	Who left during the year	Censored at the end of the year	Teachers at the beginning of the year who left during the year	All teachers still employed at the end of the year
0	[0, 1)	3941	—	—	—	1.0000
1	[1, 2)	3941	456	0	0.1157	0.8843
2	[2, 3)	3485	384	0	0.1102	0.7869
3	[3, 4)	3101	359	0	0.1158	0.6958
4	[4, 5)	2742	295	0	0.1076	0.6209
5	[5, 6)	2447	218	0	0.0891	0.5656
6	[6, 7)	2229	184	0	0.0825	0.5189
7	[7, 8)	2045	123	280	0.0601	0.4877
8	[8, 9)	1642	79	307	0.0481	0.4642
9	[9, 10)	1256	53	255	0.0422	0.4446
10	[10, 11)	948	35	265	0.0369	0.4282
11	[11, 12)	648	16	241	0.0247	0.4177
12	[12, 13)	391	5	386	0.0128	0.4123
		⇧ Risk set			⇧ Hazard function	⇧ Survivor function

⇦ Median lifetime

censored. This left only 1642 teachers (2045 − 123 − 280) to enter their 8th year and of these, 79 quit by the end of that year and 307 were censored. In the later rows of the life table, censoring exacts a heavy toll on our knowledge about event occurrence. Among the 648 special educators still teaching at the beginning of year 11, for example, only 16 quit by the end of the year but 241 (15 times as many) were censored. All told, this life table describes the event histories for 22,668 "person-years:" 3941 year 1's, 3485 year 2's, up through 391 year 12s.

Like all life tables, table 10.1 divides continuous time into a series of contiguous intervals. In column 1, we label these intervals using ordinal numbers; in column 2, we define precisely which event times appear in each. The intervals in table 10.1 reflect a standard partition of time, in which each interval *includes* the initial time and *excludes* the concluding time. Adopting common mathematical notation, [brackets] denote inclusions and (parentheses) denote exclusions. Thus, we bracket each interval's initial time and place a parenthesis around its concluding time, writing these 13 intervals as $[0, 1)$, $[1, 2)$, $[2, 3)$, . . . , $[12, 13)$. Teachers are hired at time 0, so the 0th interval refers to that period of time between contract signing and the first day of school, a period when no event can occur. Each subsequent interval—labeled 1 through 12—refers to a specific year of the career. We define a year as that period of time between the first day of school in the fall and the end of the associated summer. The first day of school for the following academic year falls into the next interval. Under this partition, any event occurring between the first day of year 1 up to (but excluding) the first day of year 2 is classified as occurring during year 1. Any event occurring during the second year—be it on the first day of school or the last day of summer—is classified as occurring during year 2.

When devising your own life tables, you should select the temporal partition most relevant for your chosen time metric and for the way in which events unfold. More generally, we represent any arbitrary division of time by using the letter t to denote time and the subscript j to index time periods. We then write a series of general time intervals as $[t_0, t_1)$, $[t_1, t_2)$, . . . , $[t_{j-1}, t_j)$, $[t_j, t_{j+1})$ . . . , and so on. Any event occurring at t_1 or later but before t_2, is classified as happening during the first time interval $[t_1, t_2)$. The jth time interval, written as $[t_j, t_{j+1})$, begins immediately at time t_j and ends just before time t_{j+1}. No events can occur during the 0th interval, which begins at time 0 and ends just before t_1, the first observable event time. Conceptually, this interval represents the "beginning of time."

The next column of the life table displays information on the number of individuals who enter each successive time period. Statisticians use the term *risk set* to refer to this pool: those *eligible* to experience the event during that interval. For intervals in which no one is censored—here, the early years of the career—identification of the risk set is straightforward. Each year's risk set is just the prior year's risk set minus those individuals who experienced the event during the prior year. The year 4 risk set (2742), for example, is just the year 3 risk set (3101) diminished by the 359 teachers who quit during their third year. In those intervals when censoring occurs—in our example, during the later years of the career— the risk set declines because of both event occurrence and censoring. The year 9 risk set (1256), for example, is just the year 8 risk set (1642) diminished by the 79 teachers who quit during year 8 and the 307 teachers who were censored at the end of year 8.

An essential feature of the risk set's definition is that it is inherently irreversible: once an individual experiences the event (or is censored) in one time period, he or she drops out of the risk set in all future time periods. Irreversibility is crucial, for it ensures that everyone remains in the risk set only up to, and including, the last moment of eligibility. The risk set for year 9, for example, comprises only those individuals (1256 of the original sample of 3941) who taught continuously for at least 9 years. Individuals who left or were censored in a previous year are not "at risk" of leaving in year 9 and are therefore excluded from the risk set in this period and all subsequent periods.

Why is the concept of a risk set important? If censoring is non-informative (as described in section 9.3), we can assume that each interval's risk set is representative of all individuals who *would* have been at risk of event occurrence in that interval had everyone been followed for as long as necessary to eliminate all censoring (that could be eliminated). To understand the implications of this assumption, consider the risk sets in years 7 (2045) and 8 (1642). If censoring is noninformative, the 1642 teachers in the year 8 risk set are representative of that subset of the 2045 teachers who *would* have entered their eighth year if we could have observed them through that point in time (that is, were there no censoring). For this to be true, the 280 teachers censored at the end of year 7, who could not be observed in year 8, must be no different from the 1642 teachers who *were* observed in year 8. Under this assumption, we can generalize the behavior of the 1642 people in the year 8 risk set back to the entire population of teachers who would have entered their eighth year. This allows us to analyze event occurrence among the members of each year's risk set yet generalize results back to the entire population.

10.2 A Framework for Characterizing the Distribution of Discrete-Time Event Occurrence Data

Having described how the life table tallies data about event occurrence over time, we now introduce three invaluable statistical summaries of this information: the *hazard function*, the *survivor function*, and the *median lifetime*.

10.2.1 Hazard Function

The fundamental quantity used to assess the risk of event occurrence in each discrete time period is known as *hazard*. Denoted by $h(t_{ij})$, discrete-time hazard is the *conditional probability that individual i will experience the event in time period j, given that he or she did not experience it in any earlier time period.*[1] Because hazard represents the risk of event occurrence in each discrete time period among those people eligible to experience the event (those in the risk set) hazard tells us precisely what we want to know: whether and when events occurs.

We can formalize this definition by adopting some notation. Let T represent a discrete random variable whose values T_i indicate the time period j when individual i experiences the target event. For a teacher who leaves in year 1, $T_i = 1$; for a teacher who leaves in year 8, $T_i = 8$. Methodologists typically characterize the distribution of a random variable like T by describing its probability density function, the probability that individual i will experience the event *in* time period j, $\Pr[T_i = j]$, or its cumulative density function, the probability that individual i will experience the event *before* time period j, $\Pr[T_i < j]$. But because event occurrence is inherently conditional—an event can occur only if it has not already occurred —we characterize T by its *conditional probability density function:* the distribution of the probability that individual i will experience the event in time period j *given that he or she did not experience it at any time prior to j.* This is algebraically equivalent to the probability that the event will occur in the current time period, given that it must occur now, or sometime in the future, as follows:[2]

$$h(t_{ij}) = \Pr[T_i = j | T_i \geq j]. \tag{10.1}$$

The set of discrete-time hazard probabilities expressed as a function of time—labeled $h(t_{ij})$—is known as the population *discrete-time hazard function.*

We cannot overemphasize the importance of the conditionality inherent in the definition of *hazard.* Individual i can experience the event in time period j if, and only if, he or she did *not* already experience it any prior period. Conditionality ensures that hazard represents the probability of event occurrence among those individuals *eligible* to experience the event in that period—those in the risk set. As people experience events, they drop out of the risk set and are ineligible to experience the event in a later period. Because of this conditionality, the hazard probability for individual *i* in time period *j* assesses his or her *unique* risk of event occurrence *in* that period.

Notice that each individual in the population has his or her own discrete-time hazard function. This is similar to the way we specified individual growth models, by allowing each person to have his or her own true growth trajectory. Here, we specify that each individual, whom we ultimately distinguish from other members of the population on the basis of predictors (e.g., gender and subject specialty), has a hazard function that describes his or her true risk of event occurrence over time. In chapter 11, when we develop statistical models for predicting discrete-time hazard, we specify the relationship between parameters characterizing each person's hazard function and predictors. For now, because we are simply describing the distribution of event occurrence for a random sample of individuals from a homogeneous population among whom we are not (yet) distinguishing, we drop the subscript *i* (that indexes individuals) and write the discrete-time hazard function for a random individual in this population as $h(t_j)$.

Although this definition of hazard may appear far removed from sample data, examination of column 6 of the life table reveals that it is a commonsense summary of event occurrence. Column 6 presents the proportion of teachers teaching at the beginning of each year who left by the end of the year. Phrased more generally, it presents the proportion of each interval's risk set that experiences the event during that interval. Among these 3941 special educators, .1157 ($n = 456$) left by the end of their first year. Of the 3485 who stayed more than one year, .1102 ($n = 384$) left by the end of their second. Notice that these proportions, just like the definition of hazard, are conditional. Each represents the fraction of that year's risk set that leaves *that year*. This allows the proportions to be computed easily in every year, regardless of censoring. Among the 2045 teachers who taught continuously for six years, for example, .0601 ($n = 123$) left by the end of their seventh; of the 948 who taught continuously for 9 years, .0369 ($n = 35$) left at the end of their tenth.

What is the relationship between the population definition of discrete-time hazard in equation 10.1 and these sample proportions? Quite simply, these proportions are maximum likelihood estimates of the discrete-time hazard function (Singer & Willett, 1993). They are also the discrete limit of the well-known Kaplan-Meier estimates of hazard for continuous-time data (Efron, 1988). More formally, if we let *n events$_j$* represent the number of individuals who experience the target event in time period *j* and *n at risk$_j$* represent the number of individuals at risk during time period *j*, we estimate the value of discrete-time hazard in time period *j* as:

$$\hat{h}(t_j) = \frac{n\ events_j}{n\ at\ risk_j}. \tag{10.2}$$

Thus, we estimate $\hat{h}(t_1)$ to be 0.1157, $\hat{h}(t_2)$ to be .1102, and so on. Because no one is eligible to experience the target event during the initial time interval, here [0, 1), $h(t_0)$ is undefined.

The magnitude of hazard in each time interval indicates the risk of event occurrence in that interval. When examining estimated values of discrete-time hazard, remember that:

- *As a probability, discrete-time hazard always lies between 0 and 1.*
- *Within these limits, hazard can vary widely.* The greater the hazard, the greater the risk; the lower the hazard, the lower the risk.

Examining the estimated hazard function for the special educator data displayed in table 10.1, we see that in the first four years of teaching, hazard is consistently high, exceeding .10. This indicates that over 10% of the teachers still teaching at the beginning of each of these years leaves by the end of the year. After these initial "hazardous" years, the risk of leaving declines steadily over time. By year 8, hazard never exceeds 5%, and by year 10, it is just barely above 0.

A valuable way of examining the estimated discrete-time hazard function is to graph its values over time. The top panel of figure 10.1 presents the kind of plot we consider most useful, and we often display such a plot in lieu of tabling estimated hazard probabilities. Although some methodologists present discrete-time hazard functions as a series of lines joined together as a step function, we follow the suggestions of Miller (1981) and Lee (1992) and plot the discrete-time hazard probabilities as a series of points joined together by line segments. Plots like these can help you to:

- *Identify especially risky time periods*—when the event is particularly likely to occur
- *Characterize the shape of the hazard function*—determining whether risk increases, decreases, or remains constant over time

In this study, like many other studies of employee turnover, the estimated hazard function peaks in the first few years and declines thereafter.

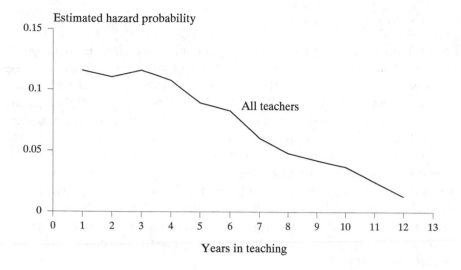

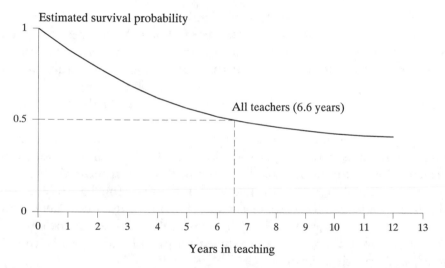

Figure 10.1. Estimated sample hazard and survivor functions for the 3941 special educators (with estimated median lifetime in parentheses on the plot of the sample survivor function).

Novice special educators, or those with only a few years of experience, are at greatest risk of leaving teaching. Once they gain experience (or perhaps, tenure), the risk of leaving declines. We present additional examples of estimated hazard functions in section 10.3, after we finish introducing the remaining elements of the life table.

10.2.2 Survivor Function

The *survivor function* provides another way of describing the distribution of event occurrence over time. Unlike the hazard function, which assesses the *unique* risk associated with each time period, the survivor function *cumulates* these period-by-period risks of event occurrence (or more properly, *nonoccurrence*) together to assess the probability that a randomly selected individual will "survive"—will *not experience the event*. Formally denoted by $S(t_{ij})$, the survival probability is defined as the *probability that individual i will survive past time period j*. For this to happen, individual i must *not* experience the target event in the jth time period *or in any earlier period*. In terms of T, the random variable for time, this implies that teacher i will still be teaching at the end of year j; in other words, T_i exceeds j. We therefore write the survival probability for individual i in time period j as:

$$S(t_{ij}) = \Pr[T_i > j] \tag{10.3}$$

and we refer to the set of survival probabilities expressed as a function of time—$S(t_{ij})$—as that individual's *survivor function*. As before, when we do not distinguish people on the basis of predictors, we write the survivor function for a random member of the population without the subscript i as $S(t_j)$.

How does a survivor function behave over time? At the beginning of time, when no one has yet experienced the event, everyone is surviving, and so by definition, its value is 1. Over time, as events occur, the survivor function declines toward 0 (its lower bound). In those time periods when hazard is high, the survivor function drops rapidly. In those time periods when hazard is low, it declines slowly. But unlike the hazard function, which can increase, decrease, or remain the same between adjacent intervals, the survivor function *will never increase*. When passing through time periods when no events occur, the survivor function simply remains steady at its previous level.

There are two ways of using sample data to compute maximum likelihood estimates of the population survivor function. The direct method, presented in the last column of table 10.1, can be used only in those intervals that precede the first instance of censoring. Although this limitation renders this method impractical for everyday use, we begin with it for its pedagogic value. To understand this method, think about what it means to "survive" through the end of an interval. For this to happen, a teacher must still be teaching by the end of that year. Column 6 captures this idea by presenting the proportion of *all* teachers (that is, of all 3941 teachers) still teaching by the end of each year. We see that .8843 (3485/3941) of the entire sample teach (survive) more than one year, .7869 (3101/3941) teach more than two years, and .5189 (2045/3941) teach more than six years. More generally we write:

$$\hat{S}(t_j) = \frac{n \text{ who have not experienced the event by the end of time period } j}{n \text{ in the data set}}.$$

In year 7 and beyond, we can no longer compute these proportions because we do not know the event times of the censored teachers. We therefore have no way of knowing how many people did not experience the target event by the end of each of these later time periods.

The alternative method, which can be used regardless of censoring, proceeds indirectly, capitalizing on the information about event occurrence contained in the estimated hazard function. The idea is that, for each interval, the estimated hazard probability tells us not only about the probability of event occurrence but also about the probability of *nonoccurrence*, which in turn tells us about survival. Let us first review the logic in the early years, before censoring takes its toll. The year 1 estimated hazard probability of .1157 tells us that $(1 - .1157)$ or .8843 of the original sample survives through the end of the first year. Similarly, the year 2 estimated hazard probability of .1102 tells us that $(1 - .1102)$ or .8898 of those special educators who enter year 2 survive through the end of that year. But because only .8843 of the original sample actually *enters* their second year, only .8898 of the .8843, or .7869 of the original sample, survives through the end of year 2. We therefore estimate the survival probability for year 2 to be $(.8898)(.8843) = .7869$, a value identical to that obtained from the direct method.

We can use the indirect method to estimate values of the survivor function in any year, even in the presence of censoring. The estimated survival probability for year j is simply the estimated survival probability for the previous year multiplied by one minus the estimated hazard probability for that year:

$$\hat{S}(t_j) = \hat{S}(t_{j-1})[1 - \hat{h}(t_j)]. \tag{10.4}$$

For example, we estimate that .5189 of all teachers survive through the sixth year of teaching. Because the estimated hazard probability for year 7 is .0601, we estimate that .9399 of those in the seventh-year risk set will *not* leave teaching that year. An estimate of the survival probability at the end of year 7 is thus $(.5189)(.9399) = .4877$. We have used this formula to estimate the survival probabilities for years 7 through 12 in table 10.1 (shown in italics). Plotted values appear in the lower panel of figure 10.1.

The estimated survivor function provides maximum likelihood estimates of the probability that an individual randomly selected from the population will "survive"—*not experience the event*—through each successive time period. In figure 10.1, notice that unlike the hazard function, which is presented only in the first time interval and beyond, the survivor function takes on the value 1.0 for interval 0—the origin of the time axis. As events occur, the estimated survivor function drops, here to .6958 by year 3, to .5656 by year 5, to .4877 by year 7, and to .4446 by year 9. Because many teachers stay for more than 12 years, the estimated survivor function does not reach zero, ending here at .4123. An estimated 41% of all special educators teach for more than 12 years; by subtraction, an estimated 59% leave in 12 years or less.

Notice that our estimate of the percentage of teachers still teaching after 12 years (41%) differs from the percentage still teaching at the end of data collection, 44% (1734/3941). Although small in this data set, this differential can be large, and it speaks volumes about what happens during analysis. Until the first censored event time, we can compute the percentage of the sample who survive *directly* so that these two percentages are identical. Once censoring occurs, we can no longer estimate the survivor function directly; we must estimate indirectly based upon those individuals who remain in the risk set. The beauty of survival analysis is that, under the assumption of independent censoring, we can use the risk set to estimate what would have happened to the entire remaining population were there no censoring. For example, although we know about year 12 event occurrence for only those 391 special educators in the first entry cohort who taught for 12 years, we can use these data to estimate what *would* have happened to teachers in the later cohorts were they to teach for 12 consecutive years. It is through this extrapolation that we generalize our sample results (including data on the censored individuals) back to the entire population.

Before leaving this discussion of the survivor function, let us resolve one small detail about its estimation. Use of equation 10.4 in any time interval requires an estimate of the function in the previous interval. Is there any way to eliminate this dependence, allowing the survivor function to be estimated solely on the basis of the hazard function? To see that the answer is yes, use equation 10.4 to write an expression for the sample survivor function in year $(j-1)$:

$$\hat{S}(t_{j-1}) = \hat{S}(t_{j-2})\left[1 - \hat{h}(t_{j-1})\right].$$

By repeatedly substituting this type of formula into equation 10.4 until time 0, when $S(t_0) = 1.0$, we find:

$$\hat{S}(t_j) = [1 - \hat{h}(t_j)][1 - \hat{h}(t_{j-1})][1 - \hat{h}(t_{j-2})] \dots [1 - \hat{h}(t_1)]. \qquad (10.5)$$

In other words, each year's estimated survival probability is the successive product of the complement of the estimated hazard probabilities across *this* and *all previous* years. For example, an estimate of the year 7 survival probability is $(1 - .0601)(1 - .0825)(1 - .0891)(1 - .1076)(1 - .1158)$ $(1 - .1102)(1 - .1157) = .4877$. Equation 10.5 allows us to estimate the survivor function directly from the estimated hazard function. Unfortunately, censoring prevents us from working in the opposite direction, estimating the hazard function directly from the estimated survivor function.

10.2.3 Median Lifetime

Having characterized the *distribution* of event times using the hazard and survivor functions, we often want to identify the distribution's center. Were there no censoring, all event times would be known, and we could compute a sample mean. But because of censoring, another estimate of central tendency is preferred: the *median lifetime.*

The estimated median lifetime identifies that value of T for which the value of the estimated survivor function is .5. It is the point in time by which we estimate that half of the sample has experienced the target event, half has not. Examining the estimated survivor function presented in column 6 of table 10.1, we know that the estimated median lifetime falls somewhere between year 6 (when an estimated .5189 of the teachers are still working at the end of the year) and year 7 (when this proportion drops below .5 to .4877). Because our metric for time is discretized at the year level, one way to report this conclusion would be to write that the average special educator leaves after completing six, but not a full seven, years of teaching.

Another way to estimate the median lifetime is to use interpolation. Interpolation is most useful when comparing subsamples, especially if medians fall in the same time interval. Even if this happens, the subsamples rarely have identical estimated survivor functions, suggesting that a median computed without interpolation is too coarse to characterize the differing survival experiences of the groups.

Following Miller (1981), we linearly interpolate between the two values of $S(t_j)$ that bracket .5. Let *m* represent the time interval when the sample survivor function is just above .5 (here, year 6), let $\hat{S}(t_m)$ represent the value of the sample survivor function in that interval, and let $\hat{S}(t_{m+1})$ represent its value for the following interval (when it must be just below .5), we estimate the median lifetime as:

$$\text{Estimated median lifetime} = m + \left[\frac{\hat{S}(t_m) - .5}{\hat{S}(t_m) - \hat{S}(t_{m+1})}\right]((m+1) - m). \qquad (10.6)$$

For the special educators, we compute the estimated median length of stay to be:

$$6 + \frac{(.5189 - .5)}{(.5189 - .4877)} = 6.6$$

In the lower panel of figure 10.1, we graphically illustrate this interpolation by drawing a line parallel to the time axis when the estimated survivor function equals .50 and by then dropping a perpendicular from the estimated survivor function to the time axis to identify the corresponding value of T.

Unlike the biased estimates of mean duration presented in section 9.3.3, the estimated median lifetime of 6.6 years correctly answers the question "How long does the average teacher teach?" We now see that the answer is not the mean 3.7 years we calculated by setting aside all censored observations nor the mean 7.5 years we calculated by treating all censored event times as known event times. Notice, too, that although we derive this estimate through a circuitous route—by first estimating a hazard function, then a survivor function, and finally a median lifetime—our answer is expressed in a comprehensible metric. Perhaps because physicians now routinely provide estimated median lifetimes to patients following diagnosis of an illness or initiation of treatment, you can use these summaries in other fields to communicate results. Although it is wise to remind your audience that this estimate is just a *median*—half the teachers stay for less than 6.6 years, the other half stay longer (or in some studies, may never experience the target event)—the statistic has much intuitive appeal.

What should you do if the estimated survivor function does not reach .5? This tells you that less than half of the population is predicted to experience the target event by the last time in the life table. This dilemma arises in studies of short duration or of rare (or less common) events, such as the onset of mental illness or illicit drug use. Although we can estimate a different percentile of the survivor function (say the 75th percentile), more often researchers present *cumulative survival rates*, values of the estimated survivor function after pre-specified lengths of time. In medical research, one-year, three-year, and five-year survival rates are common. In your own study, choose benchmarks suitable for the metric for time and the rate at which events occur. When studying teachers' careers, for example, estimated five- and ten-year survival rates (here 57% and 43% respectively) are common.

10.3 Developing Intuition About Hazard Functions, Survivor Functions, and Median Lifetimes

Developing intuition about these sample statistics requires exposure to estimates computed from a wide range of studies. To jump-start this process, we review results from four studies that differ across three salient dimensions—the type of event investigated, the metric used to record discrete time, and most important, the underlying profile of risk—and discuss how we would examine, and describe, the estimated hazard functions, survivor functions, and median lifetimes.

The four panels of figure 10.2 provide the basis for our work.

- Panel A presents data from Hall, Havassy, and Wasserman's (1990) study of relapse to cocaine use among 104 former addicts released from an in-patient treatment program. After 12 weekly follow ups, 62 people had relapsed; 42 (40.4%) were censored (drug-free).
- Panel B presents data from Capaldi, Crosby, and Stoolmiller's (1996) study of the grade when a sample of at-risk adolescents males had heterosexual intercourse for the first time. Among 180 boys tracked from seventh grade, 54 (30.0%) were still virgins (were censored) when data collection ended in 12th grade.
- Panel C describes the age at first suicide ideation for the 391 undergraduates in Bolger and colleagues' (1989) study introduced in section 9.1.3. Recall that 275 undergraduates reported having previously thought about suicide; 116 (29.7%) were censored (had not yet had a suicidal thought).
- Panel D describes how long female members of the U.S. House of Representatives remain in office. This data set tracks the careers of all 168 women who were elected between 1919 and 1996, for up to eight terms (or until 1998); the careers of 63 (37.5%) were censored.

10.3.1 Identifying Periods of High and Low Risk Using Hazard Functions

Hazard functions are the most sensitive tool for describing patterns of event occurrence. Unlike survivor functions, which *cumulate* information

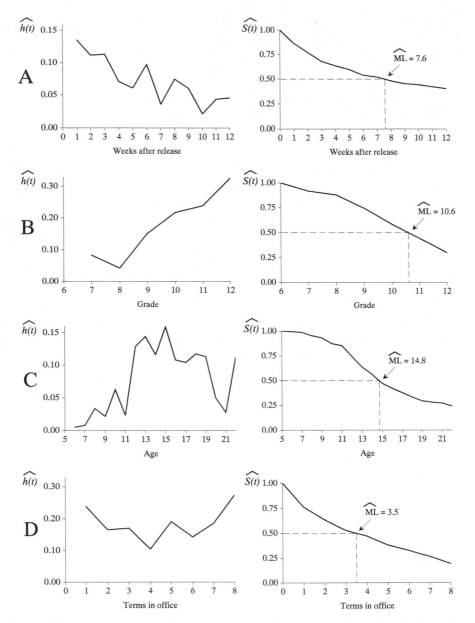

Figure 10.2. Estimated hazard functions, survivor functions, and median lifetimes from four studies: Panel A: Time to cocaine relapse after release from an in-patient treatment program. Panel B: Grade at first heterosexual intercourse for males. Panel C: Age at first suicide ideation. Panel D: Duration of uninterrupted congressional careers for female representatives.

across time, hazard functions display the *unique* risk associated with each time period. By examining variation over time in the magnitude of the hazard function, we identify when events are particularly likely, or unlikely, to occur.

We describe variation in the magnitude of the hazard function by locating its distinctive peaks and troughs. Peaks pinpoint periods of elevated risk; troughs pinpoint periods of low risk. When identifying peaks and troughs, be sure to look beyond minor period-to-period differences that may reflect nothing more than sampling variation (discussed in section 10.4). The goal is to learn about "the big picture"—the general profile of risk over time. Imagine stepping back from the estimated hazard functions in figure 10.2, glossing over the small inevitable zigzags between adjacent time periods, focusing instead on the function's overall *shape*. From this vantage point, the pronounced peaks and troughs appear at different locations along the time axis. The top two hazard functions have a single distinctive peak and a single distinctive trough—they are *monotonic*. The bottom two hazard functions have multiple distinctive peaks or troughs—they are *nonmonotonic*. Although the length of data collection affects our ability to identify peaks and troughs (in ways we will soon describe), let us examine each of these functions in greater detail, describing broadly how risk rises and falls over time.

Aside from sampling variation, the hazard function in Panel A peaks immediately after the "beginning of time" and declines rather steadily thereafter. Recently treated former cocaine addicts are most likely to relapse shortly after leaving treatment. Over time, as they acclimate to life outside the hospital, the risk of relapse declines. Monotonically decreasing hazard functions like these—also shown in figure 10.1 for the teacher turnover data—appear throughout the social and behavioral sciences. To some extent, this preponderance reflects social scientists' fascination with recurrence and relapse. Whether the target event is substance abuse (e.g., Hasin et al., 1996), mental illness (e.g., Mojtabia et al., 1997), child abuse (e.g., Fryer & Miyoshi, 1994), or incarceration (e.g., Harris & Koepsell, 1996; Brown, 1996), risk of recurrence is highest immediately after treatment, identification, or release. Monotonically decreasing hazard functions arise when studying other events as well. Two examples are Hurlburt, Wood and Hough's (1996) study of whether and when homeless individuals find housing and Diekmann, Jungbauer-Gans, Krassnig, and Lorenz's (1996) study of whether and when drivers respond aggressively to being blocked by a double-parked car.

The hazard function in Panel B is also monotonic but in the opposite direction: it begins low and increases over time. Few boys initiated intercourse in either seventh or eighth grade. Beginning in ninth grade, the risk of first intercourse increases annually among those who remain virgins. In ninth grade, for example, an estimated 15.0% of the boys who had not yet had sex do so for the first time; by 12th grade, 31.7% of the remaining virgins (admittedly only 45.1% of the original sample) do likewise. Monotonically increasing hazard functions are common when studying events that are ultimately inevitable (or near universal). At the "beginning of time," few people experience the event, but as time progresses, the decreasing pool of individuals who remain at risk succumb. Keifer (1988) found a similar pattern when characterizing the time it takes to settle a labor dispute, as did Campbell, Mutran, and Parker (1987), who studied how long it takes workers to retire.

Some hazard functions display multiple peaks or troughs. Panel C suggests that the risk of suicide ideation is low during childhood, peaks during adolescence, and then declines to near (but not quite) early childhood levels in late adolescence. (Do not pay much attention to the apparent increase in the last time period, for it is little more than sampling variation.) Diekmann and Mitter (1983) found a similar type of hazard function when they asked a sample of young adults to retrospectively report whether and if so when they had ever shoplifted. They found that the age at first shoplift varied widely, from age 4 to 16, with a peak during early adolescence—ages 12 to 14. In a different context, Gamse and Conger (1997) found a similar shape hazard function when following the academic careers of recipients of a postdoctoral research fellowship. The hazard function describing time to tenure was low in the early years of the career, peaked in years 6 through 8, and declined thereafter.

The U-shaped pattern in Panel D is nicknamed the "bathtub" hazard function. Risk is high at two different moments: immediately after the "beginning of time" and again, at the end of time. In studies of human lifetimes, especially in developing countries, the high initial risk reflects the effects of infant mortality while the later high risk reflects the effects of old age. Here, we find a similar pattern. Congresswomen are at greatest risk of leaving office at two points in their careers: immediately after their first election and then after having served for a long period of time (seven or eight terms). In the middle period—between the second and sixth terms—the effects of incumbency reign, with relatively few continuing representatives stepping down or losing an election.

Nonmonotonic hazard functions, like those in Panels C and D, generally arise in studies of long duration. This design dependency arises for a simple reason: in brief studies, the particular time associated with the middle peak (or trough) appears on the far right of the time axis—erroneously suggesting a monotonically increasing (or decreasing) hazard.

The problem is that when data collection is brief (and brief can be years!), we have no way of knowing what happens in time periods that occur *after* the end of data collection. To find a reversal, indicated by the multiple peaks (or troughs), data collection must be of a sufficient length.

This illustrates the need for a caveat whenever describing hazard functions: *Be sure that the time indicated at the end of the time axis has substantive meaning as an "end of time."* If not, use extreme caution when identifying (and describing) the varying pattern of risk. In Panel B, for example, had Capaldi and colleagues (1996) followed the 56 young men who had not had sex by the end of 12th grade, they might have found that the annual risk of initiation peaks even later, after high school. Or had we followed the congresswomen in Panel D for only four terms, we would have concluded that risk monotonically decreases over time. Although this conclusion is accurate for the first four terms, we would not want to erroneously generalize this short-term finding. Statements about periods of elevated (or diminished) risk must always be tied to statements about the range of time actually studied. Failure to do so is tantamount to extrapolating (through silence) beyond the range of the data.

What happens if the hazard function displays no peaks or troughs? When hazard is flat, risk is unrelated to time. Under these circumstances, event occurrence is independent of duration in the initial state implying that events occur (seemingly) at random. Because of age, period, and cohort effects—all of which suggest duration dependence—flat hazard functions are rare in the social and behavioral sciences. Two interesting examples, however, are whether and when couples divorce following the birth of a child (Fergusson, Horwood, & Shannon, 1984) and whether and when elementary school children shift their attention away from their teacher (Felmlee & Eder, 1984).

10.3.2 Survivor Functions as a Context for Evaluating the Magnitude of Hazard

As is apparent in figure 10.2, all survivor functions share a common shape, a monotonically nonincreasing function of time. At the beginning of time, each takes on the value 1.0. Over time, as events occur, each drops toward 0. Because of censoring, and because some individuals may never experience certain events no matter how long data collection lasts, few estimated survivor functions fall to zero. The value of the survivor function at the "end of time" estimates the proportion of the population that will survive past this last observed period.

Studying the concurrent features of the estimated hazard and survivor functions in figure 10.2 reveals a great deal about their interrelationship. Examining the four panels we see that:

- *When hazard is high, the survivor function drops rapidly*—as in the early time periods in Panels A and D.
- *When hazard is low, the survivor function drops slowly*—as in the early time periods in Panels B and C and the later time periods of A and C.
- *When hazard is zero, the survivor function remains unchanged.* Although not shown here, if $h(t_j) = 0$, $S(t_j)$ will be identical to $S(t_{j+1})$.

In general, large values of hazard produce great changes in survivorship; small values produce little change.

If the survivor function is simply a cumulative reflection of the magnitude of the peaks and troughs in the hazard function, of what *additional* value is it? One advantage is its intuitive appeal, which renders it useful when communicating findings. More important, though, the survivor function provides a context for evaluating the period-by-period risks reflected in the hazard function. Because the survivor function cumulates these risks to estimate the fraction of the population remaining in each successive time period, its value indicates the proportion of people exposed to each period's hazard. If the estimated survival probability is high when hazard is high, many people are affected; if it is low, even if hazard is high, there are few people left to experience the elevated risk.

The practice of using the survivor function to provide a context for evaluating the magnitude of hazard is similar to epidemiologists' practice of studying both prevalence and incidence. Incidence measures the number of new events occurring during a time period (expressed as a proportion of the number of individuals at risk), whereas prevalence cumulates these risks to identify the total number of events that have occurred by a given point in time (also as a proportion; see, e.g., Kleinbaum, Kupper, & Morgenstern, 1982; Lilienfeld & Stolley 1994). Stated this way, we can see that incidence and prevalence correspond directly to hazard and survival: hazard represents incidence, survival represents cumulative prevalence. Epidemiologists rely on incidence when identifying the risk factors associated with disease occurrence because prevalence confounds incidence with duration—conditions of longer duration may be more prevalent even if they have equal or lower incidence rates. But epidemiologists also recognize that prevalence, like survival, has an advantage: it assesses the extent of a problem at a particular point in time. Estimates of prevalence thereby provide a context for evaluating the magnitude of incidence. In survival analysis, estimates of the survivor function provide a similar context for evaluating the magnitude of hazard.

The consequence of this argument is that the survivor function indicates whether the elevated risks in periods of high hazard are likely to affect large numbers, or small numbers, of people. At the extreme, if risk is high among a very small group, the times of greatest risk may not be the times when most events occur. If hazard is increasing while the risk set is decreasing, a high hazard may have little effect. In both the age at first intercourse study and the congressional turnover study, the last periods are those with the highest hazards. But the risk sets in these periods are smaller than the risk sets in the earlier periods, so the elevated hazard may indicate that fewer total events take place in these periods than did in the earlier periods with lower hazards. For example, three times as many congresswomen ($n = 19$) leave office in their second term, when hazard is .17, than leave in their eighth term ($n = 6$), when hazard is .27.

This irregular correspondence between hazard and the number of events does not indicate a flaw in the concept of hazard; rather, it underscores the need for examining the survivor function. Hazard is inherently conditional: it only describes the risk of event occurrence among those at risk. Be sure to reassert this conditionality periodically so that you do not mistakenly conclude that more events occur in time periods when fewer events actually occur.

10.3.3 Strengths and Limitations of Estimated Median Lifetimes

Unlike hazard and survivor functions, which describe the *distribution* of event times, the median lifetime identifies the distribution's location or "center." Examining the estimated median lifetimes displayed in figure 10.2, for example, we see that the average former addict relapses 7.6 weeks after treatment (Panel A), the average at-risk adolescent male initially has heterosexual intercourse midway through the second semester of tenth grade (Panel B), the average teenager has had a suicidal thought by age 14.8 years old (Panel C), and the average U.S. congresswoman remains in office for 3.5 terms (Panel D).

When examining a median lifetime, we find it helpful to remember three important limitations on its interpretation. First, it identifies only an "average" event time; it tells us little about the distribution of event times and is relatively insensitive to extreme values. Second, the median lifetime is not necessarily a moment when the target event is *especially* likely to occur. For example, although the average congresswoman remains in office for just under four terms, hazard is actually low during the fourth term. Third, the median lifetime reveals little about the distribution of risk over time; identical median lifetimes can result from dramatically different survivor and hazard functions.

We illustrate these insights in figure 10.3, which presents estimated hazard functions, survivor functions, and median lifetimes for four hypothetical data sets. We constructed these data sets purposefully, with the goal of highlighting the difficulties inherent in the interpretation of median lifetimes. In each data set, comprising ten time intervals, the estimated survival probability in period 5 is exactly .50 so that the estimated median lifetime is precisely 5.0.

Notice the dramatic differences in the accompanying hazard and survivor functions. Although all four data sets have the same estimated median lifetime, few researchers examining these panels would conclude that the studies had anything in common. In Panel A, hazard begins low, rises steadily until its peak in period 5, and then declines steadily until period 10. This type of situation, in which the estimated median lifetime of 5.0 coincides with the period of greatest risk, is what most people initially believe the estimated median lifetime suggests.

But before concluding that an estimated median lifetime tells us anything about the *shape* of the hazard or survivor function, examine Panel B. The first half of the hazard function in this panel is identical to that of Panel A—it begins low and climbs steadily to its peak in period 5. After that point in time, however, hazard remains high, at the same value as in time period 5. Yet the median lifetime remains unchanged because its computation depends solely on the early values of the estimated survivor function (before it reaches .50). Its later values have no effect whatsoever on the calculation.

The remaining two data sets present even more extreme relationships between profiles of risk and median lifetimes. In Panel C, hazard begins high and declines steadily over time. Here, the estimated median lifetime of 5.0 corresponds to a low risk period and the distribution of risk over the entire time axis looks entirely different from that presented in Panels A and B (although the estimated medians of the survivor functions are identical). A similar conclusion comes from examining Panel D. Here, hazard is constant across time and yet the estimated median lifetime falls in the same exactly place on the time axis.

What conclusions should you draw from this exercise?

• *Never draw inferences about the hazard or survivor functions on the basis of an estimated median lifetime.* All this statistic does is identify one

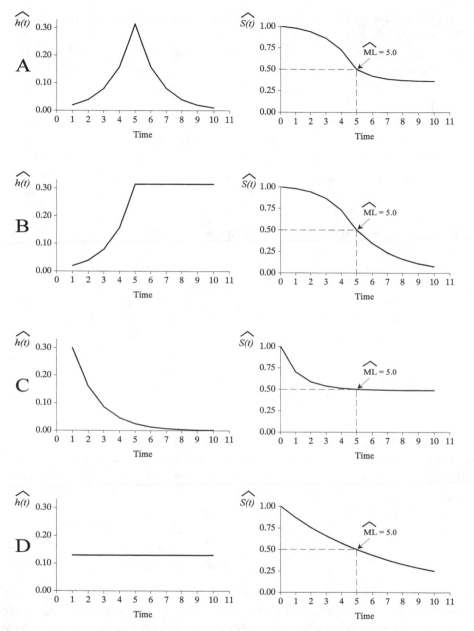

Figure 10.3. Learning to interpret median lifetimes. Results from four hypothetical data sets constructed so that each has the same estimated median lifetime (5.0), but dramatically different estimated hazard and survivor functions.

particular point—albeit a meaningful one—along the estimated survivor function's path.

- *Never assume that the time corresponding to the estimated median lifetime is one of particularly high risk.* The estimated median lifetime tells us nothing about hazard in that, or any other, time period.
- *Remember that a median is just a median—nothing more than one estimate of the location of a distribution.* If you want to know more about the *distribution* of event occurrence, the hazard function and, to a lesser extent, the survivor function are more useful.

10.4 Quantifying the Effects of Sampling Variation

When examining estimated hazard functions, we suggested that small period-to-period fluctuations were likely due to sampling variation. We now quantify the degree of sampling variation by computing the standard errors of the estimated hazard probabilities (section 10.4.1) and survival probabilities (section 10.4.2).

10.4.1 The Standard Error of the Estimated Hazard Probabilities

Consider the population value of hazard in the jth time period, $h(t_j)$. Using equation 10.2, we can estimate this parameter as that fraction of time period j's risk set (n_j) who experience the target event in that period. Because this estimate is simply a sample proportion, its standard error can be estimated using the usual formula for estimating the standard error of a proportion:

$$se(\hat{h}(t_j)) = \sqrt{\frac{\hat{h}(t_j)(1 - \hat{h}(t_j))}{n \ at \ risk_j}}. \qquad (10.7)$$

The left side of table 10.2 presents estimated hazard probabilities and their accompanying standard errors for the special educator data. We present these estimates to seven decimal places so that you can confirm our calculations with precision. Notice that all the standard errors are very small, never exceeding .007, indicating that each hazard probability is estimated very precisely. Precision is a direct consequence of the large number of teachers being tracked and the relatively low annual exit rates. The hazard probability for year 1 is estimated using a risk set of 3941 and even the much-diminished risk set for year 11 has 391 members. If your

Table 10.2: Calculating standard errors for estimated hazard and survival probabilities

Year	n_j	Hazard function		Survivor function		
		Estimated hazard probability	Standard error	Estimated survivor probability	Term under the square root sign	Standard error
1	3941	0.1157067	0.0050953	0.8842933	0.0000332	0.0050953
2	3485	0.1101865	0.0053041	0.7868561	0.0000687	0.0065235
3	3101	0.1157691	0.0057455	0.6957625	0.0001109	0.0073288
4	2742	0.1075857	0.0059173	0.6209084	0.0001549	0.0077282
5	2447	0.0890886	0.0057588	0.5655925	0.0001948	0.0078958
6	2229	0.0825482	0.0058289	0.5189038	0.0002352	0.0079589
7	2045	0.0601467	0.0052576	0.4876935	0.0002665	0.0079622
8	1642	0.0481120	0.0052812	0.4642295	0.0002973	0.0080048
9	1256	0.0421974	0.0056726	0.4446402	0.0003324	0.0081067
10	948	0.0369198	0.0061243	0.4282242	0.0003728	0.0082686
11	648	0.0246913	0.0060961	0.4176508	0.0004119	0.0084764
12	391	0.0127877	0.0056821	0.4123100	0.0004450	0.0086981

initial sample size is smaller and the rate of decline in the risk set steeper, the standard errors for each successive time interval will be larger.

How can you develop your intuition about the magnitude of these standard errors? Although a precise answer involves simultaneous examination of the numerator and denominator of equation 10.7, we can examine each component separately to develop two general ideas:

- *The closer hazard is to .50, the less precise the estimate; the closer hazard is to 0 (or 1), the more precise.* The numerator in equation 10.7 is at its maximum when the estimated value of hazard is .50 and it declines as it goes toward either 0 or 1. Because the estimated value of hazard is usually below .5 (less than half of a risk set experiences the event), we can simplify this statement to say that larger values of hazard are usually measured less precisely and smaller values measured more precisely (for the same size risk set).

- *The larger the risk set, the more precise the estimate of hazard; the smaller the risk set, the less precise.* Because the size of the risk set appears in the denominator of equation 10.7, the estimated standard error will be larger in those time periods when fewer people are at risk. As the risk set declines over time, later estimates of hazard will tend to be less precise than earlier estimates.

Why, then, do we not observe a more dramatic increase in the standard error for hazard in the special educator data in table 10.2? This stability results from two phenomena. First, the estimated value of hazard declines over time, so the general increase in standard error that accompanies a decrease in the size of the risk set is counterbalanced by the decrease in the standard error that accompanies hazard's decline. Second, although the size of the risk set declines over time, even the last time period in this data set contains 391 individuals. Were the later risk sets smaller, we would observe a more noticeable increase in the standard error of hazard.

10.4.2 Standard Error of the Estimated Survival Probabilities

Estimating the standard error of a survival probability is a more difficult task than estimating the standard error of its associated hazard probability. This is because unlike hazard, which is estimated as the fraction of the risk set that experience the target event in any given period, the survival probability is estimated as a product of (1-hazard) for this *and all previous* time periods. Estimating the standard error of an estimate that is itself the product of several estimates is a difficult statistical task. Indeed, it is so difficult that statisticians rarely recommend that you estimate the standard error of the survival probabilities directly; rather, you can do almost as well by relying on what is known as Greenwood's approximation.

In an early classic paper on life tables, Greenwood (1926) demonstrated that the standard error of the survival probability in time period j can be approximated as:

$$se\left(\hat{S}(t_j)\right) = \hat{S}(t_j)\sqrt{\frac{\hat{h}(t_1)}{n_1\left(1-\hat{h}(t_1)\right)} + \frac{\hat{h}(t_2)}{n_2\left(1-\hat{h}(t_2)\right)} + \cdots + \frac{\hat{h}(t_j)}{n_j\left(1-\hat{h}(t_j)\right)}} \qquad (10.8)$$

The summation under the square root involves all time periods up to *and including* the time period of interest. The standard error of the estimated survivor function in time period 1 involves only the first term; the standard error in time period 2 involves only the first two terms. As the estimated survivor function in time period j depends upon the estimated hazard function in that time period as well as estimates from all preceding time periods (as shown in equation 10.5), it should come as no surprise that its standard error also involves the estimated values of hazard in all preceding time periods.

The standard errors of the estimated survival probabilities for the special educator data are shown in table 10.2. Each is very small, never even reaching 0.01. This suggests that our individual estimates of the survival probabilities, just like our individual estimates of the hazard probabilities, are quite precise. But as an approximation, Greenwood's formula is accurate only asymptotically. Harris and Albert (1991) suggest that these standard errors should not be trusted for any time period in which the size of the risk set drops below 20.

10.5 A Simple and Useful Strategy for Constructing the Life Table

Having demonstrated the value of the life table, we now address the practical question: How can you construct a life table for *your* data set? For preliminary analyses, it is easy to use the prepackaged routines available in the major statistical packages. If you choose this approach, be sure to check whether your package allows you to: (1) select the partition of time; and (2) ignore any "actuarial" corrections invoked due to continuous-time assumptions (that do not hold in discrete time). When event times have been measured using a discrete-time scale, actuarial corrections (discussed in chapter 13) are inappropriate. Although most packages clearly document the algorithm being used, we suggest that you double-check by comparing results with one or two estimates computed by hand.

Despite the simplicity of preprogrammed algorithms, we prefer an alternative approach for life table construction. This approach requires construction of a *person-period data set*, much like the person-period data set used for growth modeling. Once you create the person-period data set, you can compute descriptive statistics using any standard cross-tabulation routine. The primary advantage of this approach rests not with its use for the descriptive analyses outlined in this chapter but in its use for model building. As will become apparent in chapters 11 and 12, the person-period data set is an integral tool for the systematic fitting of discrete-time hazard models. And as will become apparent in chapters 14 and 15, it also forms the conceptual foundation for fitting certain continuous time hazard models as well. An associated website presents code for creating person-period data sets in several major statistical packages.

10.5.1 The Person-Period Data Set

Like the person-period data set used for growth modeling, the person-period data set used for discrete-time survival analysis has multiple lines of data for each person under study. An important difference, however, is that the person-period data set used for growth modeling has a separate record for each time period when an individual is *observed*, whereas the person-period data set for discrete-time survival analysis has a separate record for each time period when an individual is *at risk*.

Researchers often store event history data in a "person-oriented" file, in which each individual's data appears on a single record. Each record contains all the data ever collected for that person. As you collect additional longitudinal data, you add variables to the file. If you think of this file as a spreadsheet, with individuals indexed in rows and variables in columns, over time the file grows in width but never in length. A person-period data set, in contrast, spreads the data for each individual across multiple records, each record describing a specific time period. With each additional wave of data collection, the person-period data set grows in length. A person-period data set grows in width only if new variables, not assessed on a previous occasion, are added to the protocol.

Figure 10.4 illustrates the conversion from a person-oriented data set (in the left panel) to a person-period data set (in the right panel) using three individuals from the special educator study. The first two teachers have known event times (they stayed 3 and 12 years, respectively); the third was censored at 12 years. The person-oriented data set describes these teachers' event histories using two variables:

- *Event time* (here, T). For the first two teachers with known event times, T_i is set to that time (3 and 12, respectively). For the third teacher, who was still teaching when data collection ended, T_i is also set to 12 (the last time period when the event *could* have occurred).
- *Censoring indicator* (here, *CENSOR*). For teachers with known event times (subjects 20 and 126), $CENSOR = 0$; for teachers with censored event times (subject 129), $CENSOR = 1$.

As there are 3941 teachers in this sample, the data set has 3941 records.

In the person-period data set, each individual has a separate record for each discrete-time period when he or she was at risk of event occurrence. Because individuals first become at risk during year 1, $j = 1$ is the first time period recorded in the person-period data set for each teacher. (In other data sets, it may be meaningful to count time from another origin; if so, simply set the values of this variable accordingly.) Teacher 20, who taught for three years, has three records, one for the first, second, and third years of teaching. Teacher 126, who taught for 12 years has 12 records, one per year, as does teacher 129 (who was still teaching when

"Person-Level" data set   "Person-Period" data set

ID	T	CENSOR
20	3	0
126	12	0
129	12	1

ID	PERIOD	EVENT
20	1	0
20	2	0
20	3	1
126	1	0
126	2	0
126	3	0
126	4	0
126	5	0
126	6	0
126	7	0
126	8	0
126	9	0
126	10	0
126	11	0
126	12	1
129	1	0
129	2	0
129	3	0
129	4	0
129	5	0
129	6	0
129	7	0
129	8	0
129	9	0
129	10	0
129	11	0
129	12	0

Figure 10.4. Conversion of a person-level data set into a person-period data set for three special educators from the teacher turnover study.

data collection ended, thereby remaining at risk of event occurrence in all 12 years). The values of the variables in the person-period data set reflect the status of person i on that variable in the jth period. Referring to the right panel of figure 10.4, the simplest person-period data set includes:

- A *period variable*, here *PERIOD*, which specifies the time-period j that the record describes. For teacher 1, this variable takes on the values 1, 2, and 3 to indicate that this teacher's three records describe her status in these three years. For the other two cases, *PERIOD* takes on the values 1 through 12, to indicate that those are the years represented in the twelve records.
- An *event indicator*, here *EVENT*, which indicates whether the event occurred in that time period (0 = no event, 1 = event). For each

person, the event indicator *must* be 0 in every record except the last. Noncensored individuals experience the event in their last period, so *EVENT* takes on the value 1 in that period (as in the third record for teacher 20 and in the 12th record for teacher 126). Censored individuals (such as teacher 129) never experience the event, so *EVENT* remains at 0 throughout.

If you want to include data on substantive predictors that might be associated with event occurrence (such as a teacher's gender, salary, or classroom assignment), these variables could be easily added to the person-period data set. We discuss this extension at length in section 11.3.

Person-period data sets have many more records than their corresponding person-oriented data sets because they have one record for each time period that an individual is at risk of event occurrence. Since 3941 teachers were at risk in year 1, 3485 were at risk in year 2, 3101 were in risk in year 3, up through the 391 who were at risk in year 12, this person-period data set will have $3941 + 3485 + \cdots + 391 = 24{,}875$ records. We can also compute the number of records for which $EVENT = 1$ by subtracting the number of censored cases (those who will *never* receive the value 1) from the size of the original risk set (n). In this example, because 1734 of the original 3941 teachers have censored event times, we know that EVENT will take on the value 1 in only $(3941 - 1734) = 2207$ of the records and the value 0 in the remaining $(24{,}875 - 2207) = 22{,}668$.

10.5.2 Using the Person-Period Data Set to Construct the Life Table

All the life table's essential elements can be computed through cross-tabulation of *PERIOD* and *EVENT* in the person-period data set. Any statistical package can produce the output, as displayed in table 10.3. To numerically verify the accuracy of this approach, compare these entries to the life table in table 10.1. Below, we explain why this approach works.

The cross-tabulation of *PERIOD* by *EVENT* in the person-period data set produces a *J* by 2 table. Each row *j* describes the event histories of those people in the risk set during the *j*th time period. The number of cases in the row (the TOTAL column in table 10.3) identifies the size of that period's risk set. This is because an individual contributes a record to a period if, and only if, he or she is at risk of event occurrence in that period. The column labeled $EVENT = 1$ indicates the number of people experiencing the event in the *j*th period. This is because the variable *EVENT* takes on the value 1 only in the particular time period when the individual experiences the event. In all other time periods, *EVENT* must

Table 10.3: Cross-tabulation of event indicator (*EVENT*) and time-period indicator (*PERIOD*) in the person-period data set to yield components of the life table

PERIOD	EVENT = 0	EVENT = 1	Total	Proportion EVENT = 1
1	3,485	456	3,941	0.1157
2	3,101	384	3,485	0.1102
3	2,742	359	3,101	0.1158
4	2,447	295	2,742	0.1076
5	2,229	218	2,447	0.0891
6	2,045	184	2,229	0.0825
7	1,922	123	2,045	0.0601
8	1,563	79	1,642	0.0481
9	1,203	53	1,256	0.0422
10	913	35	948	0.0369
11	632	16	648	0.0247
12	386	5	391	0.0128
Total	22,668	2,207	24,875	

be 0 (as shown in the adjoining column). Given that the table provides period by period information about the size of the risk set and the number of people who experienced the event, it should come as no surprise to find that the row percentage (shown in the last column) estimates hazard, as shown in equation 10.2. Taken together, then, the cross-tabulation provides all information necessary for constructing the life table.

Notice how this cross-tabulation reflects the effects of censoring. In each of the first seven time periods, when no data are censored, the number of individuals for whom *EVENT* = 0 (the number *not* experiencing the event) is identical to the number of individuals at risk of event occurrence in the next time period. For example, because 2742 teachers did not leave teaching in year 3, these same 2742 teachers were eligible to leave teaching in year 4. After year 7, however, this equivalence property no longer holds: the number of individuals at risk of event occurrence in each subsequent year is *smaller* than the number who did not experience the event in the previous year. Why? The answer reflects the effects of censoring. The discrepancy between the number of events in the jth period and the number at risk in the $(j+1)$st period indicates the number of individuals censored in the jth time period. For example, because 1563 teachers did not leave teaching in year 8 but only 1256 were at risk of leaving in year 9, we know that $(1563 - 1256) = 307$ were censored at the end of year 8 (as shown in table 10.1).

The ability to construct a life table using the person-period data set provides a simple strategy for conducting the descriptive analyses outlined in this chapter. This strategy yields appropriate statistics regardless of the amount, or pattern, of censoring. Perhaps even more important, the person-period data set is the fundamental tool for fitting discrete-time hazard models to data, using methods that we describe in the next chapter.

11

Fitting Basic Discrete-Time
Hazard Models

To exist is to change, to change is to mature.

—Henri Bergson

As you summarize your event history data by constructing life tables, plotting sample hazard and survival probabilities, and estimating median lifetimes, you may find yourself asking *Why?* Why do some teachers leave teaching after only one year, while others stay in the career for ten? Why do some teens contemplate suicide while others never have a suicidal thought?

We address questions like these—questions about *why* events occur at different times for different people—by fitting statistical models of hazard to data. In specifying a particular hazard model, you make hypotheses about how you think the risk of event occurrence is systematically related to predictors. In estimating the model's parameters and evaluating its goodness-of-fit, you gather evidence for the tenability of your hypotheses and you quantify predictor effect size. But just as the definition of hazard depends on whether time has been measured continuously or discretely, so, too, does the form of the statistical model used to represent the relationship between hazard and predictors. In this chapter and the next, we present statistical models of hazard for data collected in discrete time. The relative simplicity of these models makes them an ideal entrée into the world of survival analysis. In subsequent chapters, we extend these basic ideas to situations in which event occurrence is recorded in continuous time.

Good data analysis involves more than using a computer package to fit a statistical model to data. To conduct a credible discrete-time survival analysis, you must: (1) specify a suitable model for hazard and understand its assumptions; (2) use sample data to estimate the model parameters; (3) interpret results in terms of your research questions; (4) evaluate model fit and test hypotheses about (and/or construct confidence intervals for) model parameters; and (5) communicate your findings. In this chapter, we illustrate this entire process using the "age at first intercourse" study introduced in section 10.3. Our goal is to explicate the essential concepts and illustrate the logical steps involved in fitting basic discrete-time hazard models to data. This sets the stage for our subsequent discussion of how to evaluate the assumptions underpinning the model and how to extend it flexibly across many circumstances in chapter 12.

11.1 Toward a Statistical Model for Discrete-Time Hazard

When we developed the earlier multilevel model for change, we explored our data by examining empirical growth trajectories coded separately by values of a predictor (for example, the plots for boys and girls in figure 2.7). This helped us see that a two-level model, expressing parameters from each individual's growth trajectory as a function of predictors, could represent a hypothesized relationship between growth over time and a predictor. Using a similar exploratory approach, we now use plots of sample hazard and survivor functions to help motivate the creation of a statistical model for studying hypothesized relationships between event occurrence and predictors in discrete-time.

Throughout this chapter, we use Capaldi et al.'s (1996) data on the grade of first heterosexual intercourse for a sample of 180 at-risk boys (introduced in section 10.3). In figure 10.2 (Panel B), we saw that the hazard probability associated with first heterosexual intercourse increased fairly steadily over time so that by the end of data collection, in twelfth grade, 70.0% of the boys had already engaged in sex. Now we explore the hypothesis that grade at first intercourse is systematically related to the boys' early childhood experiences—specifically whether they lived with both of their biological parents throughout their formative years. To ensure that this predictor is exogenous, predating event occurrence, we focus on parental transitions that occurred *prior* to seventh grade, the first time period in which a boy could have reported having had sex. This predictor, which we call *PT*, takes on two values: 0 for boys who lived with both biological parents ($n = 72$, 40.0% of the sample) and 1 for boys who experienced one or more parenting transitions ($n = 108$, 60.0%).

11.1.1 Plots of Within-Group Hazard Functions and Survivor Functions

Plots of sample hazard functions and survivor functions estimated separately for groups distinguished by their predictor values are invaluable

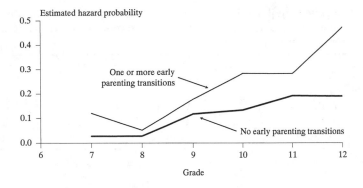

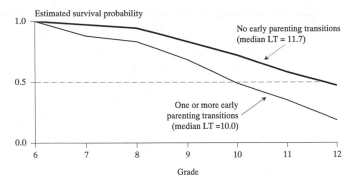

Figure 11.1. Estimated hazard functions and survivor functions for grade at first heterosexual intercourse, by presence or absence of parenting transitions prior to 7th grade.

exploratory tools. If a predictor is categorical, like *PT*, construction of these displays is straightforward. If a predictor is continuous, you should just temporarily categorize its values for plotting purposes. Figure 11.1 presents estimated hazard functions (top panel) and survivor functions (bottom panel) by levels of *PT*. The bold functions are for boys with no parenting transitions ($PT = 0$); the fainter functions are for boys with one or more transitions ($PT = 1$). Because we refer to the estimated values of these functions throughout this chapter, table 11.1 presents the life tables on which the displays are based (top and middle panels), as well as estimated hazard and survival probabilities for the pooled sample (bottom panel).

Let's begin by examining the within-group hazard functions. Even though these plots present summary statistics, not individual data points, we examine them for the same reasons we examine scatterplots: to help us formulate statistical models suitable for representing the process

Table 11.1: Life table describing the grade at first heterosexual intercourse, by presence or absence of parenting transitions prior to seventh grade (top and middle panels) and overall (bottom panel)

| | Number who . . . | | | | |
Grade	Were at risk (virgins) at the beginning of the grade	Had sex during the grade	Were censored at the end of the grade	Hazard probability	Survival probability
No Parenting Transitions ($PT = 0$)					
7	72	2	0	0.0278	0.9722
8	70	2	0	0.0286	0.9444
9	68	8	0	0.1176	0.8333
10	60	8	0	0.1333	0.7222
11	52	10	0	0.1923	0.5833
12	42	8	34	0.1905	0.4722
One or More Parenting Transitions ($PT = 1$)					
7	108	13	0	0.1204	0.8796
8	95	5	0	0.0526	0.8333
9	90	16	0	0.1778	0.6852
10	74	21	0	0.2838	0.4907
11	53	15	0	0.2830	0.3519
12	38	18	20	0.4737	0.1852
Overall, Regardless of Parenting Transitions					
7	180	15	0	0.0833	0.9167
8	165	7	0	0.0424	0.8778
9	158	24	0	0.1519	0.7444
10	134	29	0	0.2164	0.5833
11	105	25	0	0.2381	0.4444
12	80	26	54	0.3250	0.3000

under investigation and to discover unusual features in the data. When examining plots like these, you should ask two major questions:

- *What is the shape of the hazard function for each group?* When are the peaks and troughs? Do they occur at similar, or different, times for each group?
- *Does the relative level of the hazard function differ across groups?* Is the hazard function for one group consistently higher than for the other(s)? Is the relative magnitude of the differential in hazard between groups consistent over time, or does it vary over time?

Answers to these questions help inform hazard model specification and parameter estimation. As we will see, the basic discrete-time hazard model assumes that the *shape* of the hazard function is similar across groups, but that its *relative level* differs. If exploratory analysis suggests a violation of either of these assumptions, we need a modified model like those presented in chapter 12.

What do we see in these subsample hazard functions in figure 11.1? Examining their *shape*, we find a strong similarity: both begin low (in 7th and 8th grade) and climb steadily thereafter. Between 8th and 11th grade, for example, the sample hazard probability for boys with no parenting transitions increases sixfold, from .0286 to .1923; for boys with a transition, it increases fivefold, from .0526 to .2830. Although the *precise* locations of the peaks and troughs differ slightly across groups, their relative temporal positions are similar—a trough at the "beginning of time" and a peak at the "end of time." It is this gestalt impression, not the individualities of each group's particular estimated hazard function, that we look for when exploring graphs of sample summary data. To focus on the precise details of each function would be unwise, given the effects of sampling variation.

Next examine each sample hazard function's *relative level.* Notice the substantial differential between the groups—in each time period, the hazard function for boys who experienced a parenting transition is *higher* than that for boys who did not. This means that, in each grade between 7th and 12th, the conditional probability of having initial intercourse is greater for boys not raised with both biological parents. Is this differential in level consistent over time? Once again, we seek an overall impression, not precise quantification. Although the relative magnitude of the differential varies somewhat across periods—in 9th grade it is 51% higher among boys with a parenting transition (.1778 vs. .1176) whereas in 12th grade it is 149% higher (.4737 vs. .1905)—experience suggests that this amount of period-to-period variability is not extreme.

What do we look for when exploring the corresponding within-group sample survivor functions? The survivor function does not tell us about the risk of event occurrence in any given period because it cumulates hazard (or more precisely the complement of hazard) across this and all previous periods. When evaluating the effect of predictors this disadvantage has an advantage: Because of cumulation, within-group survivor functions are ideal summaries of the predictors' *compounded* effects. But when moving from examining hazard to examining survivor functions, be aware of the inevitable *reversal* in their relative rankings: the group with the *higher* hazard function (the group with the greater probability of event occurrence) will have the *lower* survivor function (the lower probability of survival). Here, the survivor function for boys who experienced a parenting transition is *lower* than that for boys raised with their two biological parents (because the hazard function is *higher*).

What do these within-group sample survivor functions tell us about the cumulative effects of the predictor *PT*? The overwhelming message is that its effect is large. By as early as 9th grade, only 68.5% of the boys who experienced a parenting transition remained virgins in comparison to 83.3% for the boys raised by both biological parents. By 12th grade, only 18.5% had not yet had intercourse compared to 47.2% of boys in the two-parent group. These sizable differentials translate into dramatic differences in median lifetimes. For boys with a parental transition, the estimated median grade at first intercourse is 10.0; for boys raised with their two biological parents, the median is just under 12 (11.7). Capaldi and colleagues (1996) assert (and most parents would agree) that this nearly two-year delay—from the beginning of 10th grade until the end of 11th grade—is substantial, with potentially serious psychological, social, and health consequences.

11.1.2 What Kind of Statistical Model Do These Graphs Suggest?

To postulate a statistical model to represent the relationship between the population discrete-time hazard function and predictors, we must deal with two complications apparent in these displays. One is that any hypothesized model must describe the shape of the *entire discrete-time hazard function* over time, not just its value in any one period, in much the same way that a multilevel model for change characterizes the shape of entire individual growth trajectory over time. A second complication is that, as a conditional probability, the value of discrete-time hazard must lie between 0 and 1. Any reasonable statistical model for hazard must recognize this constraint, precluding the occurrence of theoretically impossible values.

Before presenting the formal model specification, which we do in section 11.2, we now presage its salient features using sample data. By proceeding this way, we hope to motivate the representation we adopt and indicate how its parameters address substantive questions about the effects of predictors.

The Bounded Nature of Hazard

Let us first consider how we can deal with hazard's upper and lower bounds of 0 and 1. The easiest solution, recommended for reasons both practical (see, e.g., Mosteller & Tukey, 1977) and theoretical (see, e.g., Box & Cox, 1964), is *transformation*—expressing a variable on a different scale. Transformation can improve distributional behavior (for example,

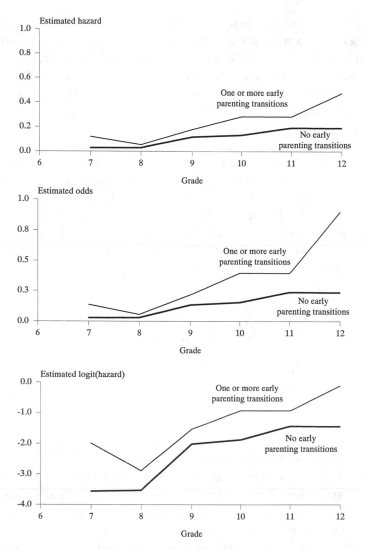

Figure 11.2. Re-expressing estimated hazard functions on different scales: Panel A: hazard. Panel B: odds. Panel C: logit (hazard).

variables with skewed distributions can be transformed to symmetry), and in the case of discrete-time survival analysis, can also: (1) prevent specification of inadmissable values; and (2) render disparate values of hazard more easily comparable.

For (conditional) probabilities like the values of discrete-time hazard, Cox (1972) recommends two transformations: the *odds* and *log odds* (commonly referred to as *logits*) transformations.[1] The effect of both these transformations is shown in figure 11.2, which plots the within-group hazard functions for the first intercourse data in three ways. The top panel presents the raw functions, displayed on a scale of 0 to 1 (expanded from figure 11.1). The middle panel transforms the hazard probabilities to odds, which here also happen to range between 0 and 1. The bottom panel presents the same functions on a logit scale. We now examine these displays, exploring how each transformation affects the shapes and levels of the within-group functions.

In general, odds compare the relative magnitude of two complementary probabilities: the probability that an event *will* occur and the probability that it *will not* occur. If the probability that an event will occur is .50, then the probability that it will *not* occur is also .50, and we say that the associated odds of occurrence are 50:50 or 1 to 1. If the probability of event occurrence is .80, the probability that the event will not occur is .20, and the associated odds are 80:20 or 4 to 1. Generically, to express odds as a single number (not as a ratio), we compute the following quotient:

$$odds = \frac{probability}{1 - probability}.$$ (11.1)

In the case of discrete-time survival analysis, where hazard is a conditional probability—the probability that an event will occur in any time period *given* that it has not occurred in earlier periods—if hazard is .80 in a particular time period, then the (conditional) odds of event occurrence in that period are 4.0 (the phrase "to 1" is implicit). If hazard is .20, odds are 0.25, and so on. Notice that the value 1.0 is the "center" of the odds scale. In any time period, if a hazard probability is greater than .50, the conditional odds of event occurrence are greater than 1.0; if hazard probability is below .50, as is common when studying event occurrence, the conditional odds of event occurrence are less than 1.0.

The effect of the odds transformation on the distance between hazard functions depends on the magnitude of hazard. Comparing the first two panels of figure 11.2, we see that when hazard is small, less than approximately .15 (as in the first three time periods), taking odds has little effect on the separation of the sample hazard functions. This is because when hazard is small, the denominator in equation 11.1 is close to 1 and the value of odds is approximately equal to the value of hazard. When hazard is large, the effect of the transformation is greater—notice the dramatic difference in the 12th grade values in the two panels. The net effect of taking odds here is to widen the gap between the two within-group hazard functions, accentuating the differences in level found on the raw scale.

The odds metric is not without problems, however. First, although odds can take on any nonnegative value, the odds statistic is still bounded below by the value 0. This allows for the possibility that a linear model that represents odds as a function of predictors could lead to logically impossible (i.e., negative) predicted values. Second, the odds scale is asymmetric. A raw difference of .50 in odds translates into a different magnitude of effect depending upon the value of odds. Taking the natural logarithm of odds—log_e(odds), commonly referred to as a *logit transformation*—as we do in the bottom panel of figure 11.2, ameliorates these difficulties.

The effect of the logit transformation also depends on the magnitude of the hazard itself. If two values of hazard are small (close to 0) the logit transformation increases the distance between them; if two values of hazard are large (far from 0) the logit transformation decreases the distance between them. Comparing the middle and bottom panels of figure 11.2, we find that on the logit scale, the distance between the two functions in the early time periods (when hazard and odds are both small) is larger and the distance between the two functions in the later time periods (when hazard and odds are larger, closer to .50) is smaller. When working with log odds, don't be dismayed by negative values; logit hazard is unbounded and negative whenever hazard is less than .50. The relative ordering of risk remains the same, however: the higher the value of logit hazard (the closer it is to 0), the greater the risk of event occurrence.

The changing size of the gap between the within-group functions illustrates the secondary benefit of the logit transformation: in comparison to the raw scale, the logit scale renders the distance between functions more comparable over time. This is not a coincidence, but a well-known feature of the transformation that makes it so popular. In the top and middle panels, the size of the gap between the functions *increases* over time. In the logit scale in the bottom panel, the size of the gap is fairly stable over time. The small differences in raw hazard (and odds) in the 7th and 8th grades translate into larger differences in logit hazard while the large difference in raw hazard (and odds) in 12th grade diminishes substantially, so that it is nearly comparable to the size of the gap in all other time periods. Taking logits stabilizes the gap between the two functions over time, facilitating comparison (and ultimately modeling).

What Statistical Model Could Have Generated These Sample Data?

Having expressed hazard on a logit scale, now examine the bottom plot in figure 11.2 and ask: What statistical model should we use to represent

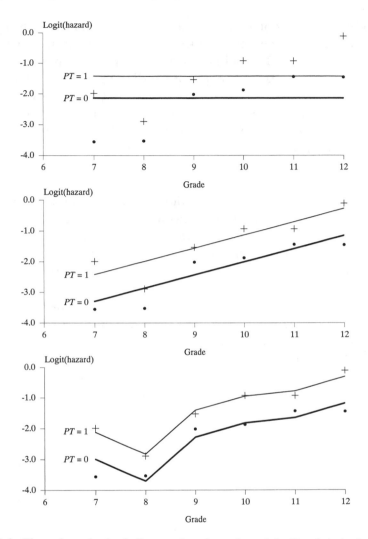

Figure 11.3. Three hypothesized discrete-time hazard models. Panel A: logit hazard is horizontal with time. Panel B: logit hazard is linear with time. Panel C: logit hazard is completely general with time.

the population relationship between logit hazard and its predictors (here, just *PT*). What functional form should the model have? Will a linear model suffice or is a more complex function necessary? How should the predictor's effect be represented in the proposed model?

Figure 11.3 depicts three alternative population models for discrete time logit hazard in this example. We display the models graphically, not algebraically, to highlight their representation, not parameterization. To facilitate comparison of the model and the within-group hazard functions, the panels include symbols denoting the latter's values (+ for $PT =$ 1; • for $PT = 0$). This juxtaposition allows you to think of these plotted symbols as "data points" or "observed values" that might have been generated by the underlying hypothesized model. Although we do not adopt all of them, we present these three models to emphasize that there is no single and uniquely suitable discrete-time hazard model. The term refers to a broad class of models; these examples are only three.

Similarities across these models highlight three assumptions inherent in the broad class of discrete-time hazard models presented in this chapter:

- *For each value of the predictor, there is a postulated logit hazard function.* If the predictor is dichotomous, we postulate that the population comprises *two* functions; if the predictor is continuous, we postulate that the population comprises as many hazard functions as there are predictor values. (For models with more than one predictor, we postulate the existence of as many functions as there are *combinations* of predictor levels.)
- *Each of these logit hazard functions has an identical shape,* although there is great flexibility in the specification of that shape. We constrain the shape of each function to be the same for all predictor values. Within this constraint, we have great flexibility in the choice of shapes. The logit hazard functions can be flat (the top panel), linear (the middle panel), completely general (the bottom panel), or any other shape imaginable.
- *The distance between each of these logit hazard functions is identical in every time period.* Regardless of the common *shape* of the postulated logit hazard functions, the differences in *level* for the different values of the predictors remain the same. The gap cannot be larger in some periods and smaller in others. This means that effect of the predictor on the log odds of event occurrence is hypothesized to be constant over time.

These three features of the discrete-time hazard model are *assumptions* about the population that may, or may not, hold in practice. Assumptions allow us to postulate models, estimate parameters, and evaluate goodness of fit. As with any assumption, we retain some healthy skepticism about its tenability. For now, though, we ask you that you hold this skepticism in abeyance. In chapter 12, we discuss how to evaluate these assumptions and how to relax them when necessary.

As you examine figure 11.3, focus on the correspondence between the plotted and hypothesized values. Although the degree of correspondence varies across panels, all three models seem plausible to a greater or lesser degree. The less constrained the model, of course, the better the fit. In the top panel, where the hypothesized model constrains population logit hazard to be flat over time, we find the largest differences between observed and hypothesized values. In the middle panel, where we postulate a linear relationship between logit hazard and time, this discrepancy diminishes substantially. In the third model, where we place no constraints on the hypothesized functions' shape, allowing them to be completely general (although constant in shape across groups), the difference between observed and hypothesized values virtually disappears.

A major task in specifying a suitable model for discrete-time hazard is to select a suitable representation for the common shape of the logit hazard function. This is similar to the task in growth modeling of selecting a parametric form for the common shape of the individual growth trajectory. Although theory may provide a guide, prespecification of a particular common shape can be difficult. For this reason, in survival analysis, we typically begin with a general specification that places no constraints on the shape of the logit hazard function (as in the bottom panel of figure 11.3). Letting the "data speak for themselves" allows the common shape to reflect the overall log odds of event occurrence in each time period. In chapter 12, we introduce methods for deciding whether a more constrained representation might suffice.

What do these models say about the effect of the predictor PT? Notice that, in each panel, regardless of the shape of the common hypothesized logit hazard function, the gap between them is identical across time periods. This means that each model postulates that the effect of the predictor PT essentially acts to "shift" one entire logit hazard function vertically relative to the other. To formalize the idea, we:

- *Label one of the logit hazard functions as the "baseline," the logit hazard function obtained when the value of the substantive predictor is 0.* (If there is more than one substantive predictor, it is the logit hazard function obtained when *all* are 0.) In this example, the baseline function is the logit hazard function for boys without an early parenting transition (for whom $PT = 0$).

- *Quantify the size of the shift between the baseline function and the other function obtained with a one-unit increment in the value of the predictor.* The effect of the predictor is to shift the entire baseline function vertically. The size of the gap between the two functions obtained for a one-unit difference in the value of the predictor measures the size of the predictor's effect. If the gap is narrow, the baseline function is hardly shifted and the predictor's effect is small. If the gap is wide, the baseline function is shifted sub-stantially, and the predictor's effect is large.

Conceptually, then, our model ascribes any vertical displacement in logit hazard to predictors in much the same way as we ascribe differences in mean levels of a continuous outcome to predictors in a linear regression model. All we need now is a strategy for specifying this model alge-braically in a way that: (1) is flexible enough to allow the shape of the common logit hazard function to take on a different value in each time period; and (2) includes shift parameters that indicate the size of the vertical displacement in logit hazard resulting from one unit differences in the value of the predictors.

11.2 A Formal Representation of the Population Discrete-Time Hazard Model

Let us now step back from the specifics of this data set and consider the general problem. In section 10.1, when we defined $h(t_{ij})$, the population discrete-time hazard function for individual i in time period j, we assumed that the population was *homogeneous*: individuals were not distinguished from each other on the basis of predictors. Now, we introduce the possibility that the population is *heterogeneous*: that different individuals, distinguished on the basis of their values of selected predictors, may have different hazard functions. This is known as *observed heterogeneity*, the hypothesis that individuals will have *different* hazard functions if they have different values for observed predictors. (Individuals who have identical values on observed predictors will still have identical hazard functions.)

We introduce observed heterogeneity into the definition of hazard by first identifying predictors hypothesized to be associated with event occurrence. Some will be *time-invariant*, like *PT*; others will be *time varying*, like the boy's drug use (or subsequent occurrence of a parenting transi-tion). As in the multilevel model for change, time-varying predictors can take on different values in each time period (although we assume that their values remain constant within periods). To write the model gener-ally, so that it can include predictors of both types, let $X_{1ij}, X_{2ij}, \ldots, X_{Pij}$ represent the P predictors and x_{pij} denote individual i's values for the pth predictor in time period j. If the predictor is time-invariant, it simply takes on the same value in each period; if it is time varying, it can take on dif-ferent values. We introduce observed heterogeneity into the definition of the hazard function by writing:

$$h(t_{ij}) = \Pr[T_i = j | T_i \geq j \ and \ X_{1ij} = x_{1ij}, X_{2ij} = x_{2ij}, \ldots, X_{Pij} = x_{Pij}]. \quad (11.2)$$

The population value of discrete-time hazard for person i in time period j is the probability that he or she will experience the target event in that time period, *conditional* on no prior event occurrence *and* his or her particular values for the P predictors in that time period.

Equation 11.2 indicates that the population value of hazard for individual i in time period j depends upon his or her values for the P predictors; it does not specify the functional form of that dependence. This is where the ideas of section 11.1.2 come in. There, we saw that it makes sense to describe the relationship between a predictor and the entire population logit discrete-time hazard function by focusing on one logit hazard function (a "baseline") and vertically shifting this baseline function by a constant amount in each time period per unit difference in the predictor. To specify the model, we need a way of expressing the baseline function so that it displays the flexible shape we desire. Were we constraining logit hazard to be linear with time (as in the middle panel of figure 11.3), we might write the baseline function using terms similar to the level-1 growth model (by including a component like $\beta_{0i} + \beta_{1i}TIME$). But to invoke the most flexible representation, we do not use a linear specification (or even a quadratic or cubic). Instead, we use a set of "time indicators," dichotomies whose values index the many discrete time periods.

The easiest way to understand what the time indicators represent is to return to the person-period data set (introduced in section 10.4) and focus on the variable *PERIOD*, which identifies the particular time period a given record represents. Because the event can occur in one of J time periods, the standard dummy variable representation of *PERIOD* yields J "time indicators," $D_{1ij}, D_{2ij}, \ldots, D_{Jij}$. For simplicity, we drop the subscripts i and j and write the time indicators as $D_1, D_2, \ldots D_J$. Each time indicator is set to 1 in the time period it represents and 0 elsewhere. For example, $D_1 = 1$ in the first time period and 0 thereafter, $D_2 = 1$ in the second time period and 0 in all other periods, as shown below.

PERIOD	D_1	D_2	$\ldots$	D_{J-1}	D_J
1	1	0	0	0	0
2	0	1	0	0	0
$\ldots$	0	0	$\ldots$	0	0
$J-1$	0	0	0	1	0
J	0	0	0	0	1

Because *PERIOD* identifies the *J* time periods when an event can occur, we often refer to the collective set of time indicators using the conceptual label "*TIME*." This nomenclature highlights an apparent paradox of the discrete-time hazard model: time, the conceptual *outcome*, is actually the fundamental *predictor*. This seeming anomaly reflects our reformulation of the research question from "What is the relationship between event times and predictors?" to "What is the relationship between the risk of event occurrence in each time period and predictors?" This reformulation is vital, for it is by answering the second question that we answer the first.

Using the time indicators as well as the substantive predictors (the *X*'s), we can write the population discrete-time hazard model as:

$$\text{logit } h(t_{ij}) = [\alpha_1 D_{1ij} + \alpha_2 D_{2ij} + \cdots + \alpha_J D_{Jij}]$$
$$+ [\beta_1 X_{1ij} + \beta_2 X_{2ij} + \cdots + \beta_P X_{Pij}]. \tag{11.3}$$

As expected, the model's left side presents a transformed version of hazard. When we write a statistical model in this manner, the transformation used is called the *link function*—the function that "links" predictors to outcomes (McCullagh & Nelder, 1989). The discrete-time hazard model in equation 11.3 uses a logit link. The model's right side is composed of two sets of terms separated by brackets. The first set of terms, the α's multiplied by their respective time indicators, act as multiple intercepts, one per period. As a group, these parameters represent the baseline logit hazard function, the value of logit hazard when all *P* substantive predictors are 0. The second set of terms, the β's multiplied by their respective substantive predictors, represent the shift in the baseline logit hazard function corresponding to unit differences in the associated predictors.

The representation in equation 11.3 provides the desired flexible representation for the baseline logit hazard function (through the terms in the first set of brackets) while constraining the effect of each predictor to be constant across time periods (through the terms in the second set of brackets). It also facilitates parameter interpretation in that:

- *Each intercept parameter, α_1, α_2 . . . , α_J, represents the value of logit hazard (the log odds of event occurrence) in that particular time period* for individuals in the "baseline" group.
- *Each slope parameter, β_1, β_2, . . . , β_P, assesses the effect of a one unit difference in that predictor on event occurrence,* statistically controlling for the effects of all other predictors in the model.

When writing a discrete-time hazard model, especially one with no time-varying predictors, we often make two minor changes to the subscripts in the hopes of increasing clarity with little loss of generality:

$$\text{logit } h(t_j) = [\alpha_1 D_1 + \alpha_2 D_2 + \cdots + \alpha_J D_J]$$
$$+ [\beta_1 X_1 + \beta_2 X_2 + \cdots + \beta_P X_P]. \tag{11.4}$$

This representation eliminates: (1) the subscript i, indexing individuals, because its presence is implicit; and (2) the subscript j, indexing time periods, from the right side of the equation because it is either redundant (for the time indicators) or implicit (for the substantive predictors). When writing a model for a specific data set, we usually renumber the subscripts for the α's and the time indicators as well. For the grade at first intercourse data, for example, a model for the main effect of PT is:

$$\text{logit } h(t_j) = [\alpha_7 D_7 + \alpha_8 D_8 + \cdots + \alpha_{12} D_{12}] + \beta_1 PT, \tag{11.5}$$

so that the subscripts on the α's and the time indicators now identify the time periods when events can occur (grades 7 through 12) instead of the general periods 1 through J.

11.2.1 What Do the Parameters Represent?

The discrete-time hazard model includes two types of parameters: the α's, which represent the baseline logit hazard function, and the β's, which assess the effects of substantive predictors. We can clarify these parameters' roles and interpretations by examining the link between the population discrete-time hazard model in equation 11.5 and the graphical display in the bottom panel of figure 11.3.

Substituting the two possible values of PT into equation 11.5 yields two submodels:

$$\text{when } PT = 0: \text{logit } h(t_j) = [\alpha_7 D_7 + \alpha_8 D_8 + \cdots + \alpha_{12} D_{12}]$$
$$\text{when } PT = 1: \text{logit } h(t_j) = [\alpha_7 D_7 + \alpha_8 D_8 + \cdots + \alpha_{12} D_{12}] + \beta_1 \tag{11.6}$$

The first model in equation 11.6 indicates that when $PT = 0$, the entire logit hazard function can be written using just the α parameters and their time indicators. Because PT is the only substantive predictor, and it takes on the value 0 for this group, this submodel represents the *baseline logit hazard function* (the value of logit hazard when *all* predictors in the model are 0). The second submodel indicates that when $PT = 1$, the population logit hazard function can be written by shifting this entire baseline function vertically by the constant amount, β_1.

How does the first model in equation 11.6 actually represent the baseline logit hazard function? Notice that unlike a familiar regression model, the model contains no single stand-alone intercept, a parameter not multiplied by any predictor.[2] Instead, the parameters α_7 through α_{12} act like multiple intercepts, one per time period, indicating the value of the outcome in each particular period. We can interpret these parameters as intercepts because of the way we have defined the time indicators. In seventh grade, for example, only $D_7 = 1$, so that all other terms disappear leaving the population value of logit hazard in 7th grade to be α_7. Similarly, in eighth grade, only $D_8 = 1$ so that all terms except that involving D_8 disappear, leaving the population value of logit hazard in eighth grade to be α_8. More generally, we write:

$$\text{For the baseline group: logit } h(t_j) = \alpha_j. \qquad (11.7)$$

Because each α parameter represents the conditional log odds of event occurrence in that time period for individuals in the baseline group, taken together, the α's represent the baseline logit hazard function.

The population parameter β_1 represents the hypothesized influence of the predictor PT on the logit hazard function. To clarify its interpretation, compare the two submodels in equation 11.6. When $PT = 0$, the model contains no β_1, and the value of logit hazard in each time period is given in equation 11.7. When $PT = 1$, this "slope" parameter increases the value of logit hazard in every time period by β_1:

$$\text{When the substantive predictor} = 1: \text{logit } h(t_j) = \alpha_j + \beta_1. \qquad (11.8)$$

The parameter β_1 therefore quantifies the difference in the population value of logit hazard per unit difference in the predictor. Because PT is a dichotomy, β_1 quantifies the increment in the log odds of first intercourse in every time period for boys who experienced a parenting transition.

This algebraic decomposition maps directly onto the graphical representation in the bottom panel of figure 11.3. The top panel of figure 11.4 clarifies this correspondence by displaying the values of the time indicators $(D_7 - D_{12})$ in each time period (on the axis for *GRADE*) and the hypothesized values of the population parameters $\alpha_7 - \alpha_{12}$ and β_1 (on the plot). Because the α's indicate the value of logit hazard when the predictor $PT = 0$, α_7 represents the value of logit hazard in seventh grade for this group, α_8 represents the value in eighth grade, and so on, as shown using the square plotting symbol (■). As a group, then, the α's represent the baseline logit hazard function. The population parameter β_1 (shown using an arrow ↑) assesses the effect of one unit difference in the value of the predictor. As this predictor takes on only two values (0 and 1), β_1 represents the size of the vertical shift in the entire logit hazard function for the second group of boys (those with a parenting transition). Notice that we display β_1 only once on the plot because the size of the gap is *identical* across time periods.

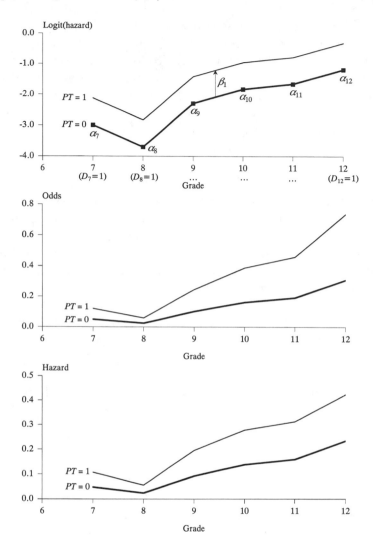

Figure 11.4. Understanding the parameters included in the population discrete-time hazard model by superimposing the hypothesized model on the estimated logit hazard functions for the age at first intercourse data. Panel A is for Model 11.3 with the effect of a single dichotomous predictor *PT*. Panels B and C re-express the hypothesized discrete-time hazard model on the scale of odds and hazard.

What happens if we have more than one substantive predictor, some of which are continuous, not categorical? Notice that when writing the general discrete-time hazard model in equation 11.3, we made no assumptions about the number, or type, of predictors. We write the model generally because, as in regression, within limits imposed by identification and power, both categorical and continuous predictors (and inter-actions) can be included. Later in this chapter, we consider the effects of a second predictor, *PAS*, which assesses the parents' level of antisocial behavior during the child's formative years. Like *PT*, this variable is time-invariant, measured before the child entered seventh grade, but unlike *PT* it is continuous, not dichotomous. Across the full sample, it ranges from −1.7 to +2.8 with a mean of 0. We write a discrete-time hazard model including the main effects of both these predictors as:

$$\text{logit } h(t_j) = [\alpha_7 D_7 + \alpha_8 D_8 + \cdots + \alpha_{12} D_{12}] + \beta_1 PT + \beta_2 PAS \qquad (11.9)$$

The set of α's remains known as the baseline logit hazard function, but the addition of a second predictor changes the identity of the baseline group. The baseline in this model is the *subset* of boys who lived with both biological parents ($PT = 0$) and who scored precisely 0 on the antisocial behavior measure ($PAS = 0$). Because the change in the identity of the baseline group changes the interpretation of the α's, we expect that when we fit this (or any other) model to these data, the estimated values of the α's will differ as well. As in any statistical model, consistent use of a symbol to represent a parameter does not imply that the parameter consistently represents the same population quantity.

The change in the identity of the baseline group raises a question concerning the wisdom of using the term "baseline" to describe a feature of a hazard model. The term, used in both discrete-time and continuous-time survival analysis, is just a label; it need not refer to the hazard function for an actual (or even plausible) group. If 0 is a valid value for all substantive predictors, the baseline group is plausible. If it is not valid for just one substantive predictor, the baseline group does not actually exist in practice. Lack of plausibility is of little concern, however. As in regression, where the single intercept will not represent a plausible value if a model includes a predictor that cannot be 0, the *multiple intercepts* in the discrete-time hazard model need not be directly interpretable either. Regardless, they function as place-holders; hence the label "baseline." (If 0 is not a valid value for a substantive predictor, we compute other predicted values of hazard by substituting alternative values of the predictor that are valid. We discuss this further in section 11.5 when describing how to plot the results of a *fitted* discrete-time hazard model.)

Interpretation of the slope parameters (here, β_1 and β_2) is altered by the inclusion of multiple predictors in ways you would expect. Each represents the effect of a one-unit difference in its associated predictor while statistically controlling for the other predictors in the model. β_1 assesses the effect of parenting transitions controlling for parental

Table 11.2: Formulas for re-expressing logits and odds into odds and probabilities

Original scale	Desired scale	Use the transformation
Logit	Odds	$Odds = e^{logit}$
Odds	Probability	$Probability = \dfrac{odds}{1+odds} = \dfrac{e^{logit}}{1+e^{logit}}$
Logit	Probability	$Probability = \dfrac{1}{1+e^{-logit}}$

antisocial behavior and β_2 assesses the effect of the parental antisocial behavior, controlling for parental transitions.

11.2.2 An Alternative Representation of the Model

The discrete-time hazard model in equations 11.3 and 11.4 expresses logit hazard as a linear function of J time indicators and P substantive predictors. We begin with this representation because we believe it is easier to understand a model that links a *transformed* outcome to a linear function of predictors than an equivalent model that links a *raw* outcome to a *nonlinear* function of predictors. But the discrete-time hazard model can be expressed in alternative ways by applying an *inverse transformation* to both sides of the equation. As we have gone from probabilities (in the form of hazard), through odds, and finally to logits, we need three inverse transformations: from logits to odds, from odds to probabilities, and from logits directly to probabilities. Table 11.2 presents formulas for these three transformations.

The formulas in table 11.2 can be applied to any value of odds or logits. Using the first transformation, a value of -2.0 on a logit scale becomes a value of $e^{-2.0} = 0.1353$ on an odds scale. Using the second transformation, this value of odds can be transformed into a probability: $0.1353/(1 + 0.1353) = 0.1192$. The third transformation permits direct conversion from a logit scale to a probability (without passing through the odds scale), so that we find, for example, that $1/(1 + e^{-(-2.0)})$ is 0.1192.

Inverse transformations can also be applied to entire equations like the discrete-time hazard model in equation 11.3. Applying the third transformation in table 11.2 we have:

$$h(t_{ij}) = \frac{1}{1+e^{-\{[\alpha_1 D_{1ij}+\alpha_2 D_{2ij}+\cdots+\alpha_J D_{Jij}]+[\beta_1 X_{1ij}+\beta_2 X_{2ij}+\cdots+\beta_P X_{Pij}]\}}}. \tag{11.10}$$

Notice that, expressed in this way, the hypothesized discrete-time hazard model is just a variant of a standard logistic regression model. The difference is that instead of a single intercept, this model has multiple intercepts, one per time period. This representation highlights an important feature of the postulated model: By specifying a linear relationship between predictors and logit hazard we imply a *nonlinear* relationship between predictors and *raw hazard*.

How does this nonlinear relationship manifest itself? We address this question by plotting the population discrete-time hazard model on different scales. The top panel of figure 11.4 uses a logit scale. The remaining panels apply the transformations of table 11.2 to yield displays on an odds scale (the middle panel) and a hazard scale (the bottom panel). Although perhaps not apparent, all that changes from panel to panel is the *scale* of the *Y* axis. *All three graphs present the identical model.*

To understand the consequences of the nonlinearity inherent in the discrete-time hazard model, focus initially on the top two panels. Although the models are *identical*, the impression they create about the distance between the functions in each panel differs:

- *When hazard is expressed on a logit scale (the top panel), the distance between functions is identical in every time period.* This is an assumption of our model that seems reasonable in this data set and in many others.
- *When hazard is expressed on an odds scale (the middle panel), the distance between functions is not identical; instead, one function is a constant magnification (or diminution) of the other.* Although the *actual distance* between functions differs across periods, the *relative distance* does not. The *ratio* of the value of one function in a given period is a constant multiple of the value of the other in that same time period.

Closely examining the panels, we see that in those time periods when logit hazard is far from 0 (7th and 8th grades), the distance between functions has shrunk whereas in those time periods when logit hazard is closer to 0 (11th and 12th grades), the distance has grown. This change is guaranteed by the nature of the inverse transformation for converting logits into odds. Antilogging logit-hazard profiles with a constant vertical separation will yield odds profiles that are constant multiples of each other. Because these profiles are *proportional*, the assumption of constant vertical separation in a logarithmic scale (here, the log odds scale) is known as the *proportionality assumption*, and a model that invokes this assumption about odds is known as a *proportional odds model*. We examine the proportionality assumption in more detail in chapter 12; for now, simply note that a discrete-time hazard model expressed using a logit transformation assumes that the population odds profiles, computed at all possible values of the predictor (or combinations of predictors), are proportional to each other.

Is the pair of population hazard functions in the bottom panel also proportional? Although they may appear so, any perceived proportionality is only approximate. The ratio of hazards in 12th grade is actually *smaller*, for example, than the comparable ratio in 7th. You perceive approximate proportionality because the discrepancy between odds and hazard is small when odds are small (because the denominator of the transformation, (1 + odds), is close to 1 when odds are small). A discrete-time hazard model based on a logit transformation does not assume proportional *hazards* but rather proportional *odds*. As we show in section 12.2, we can also develop a discrete-time hazard model using a different transformation—the *complementary log-log transformation*—that yields a *proportional hazards model*.

Ironically, then, a discrete-time hazard model, which can be expressed as a *linear* relationship between *logit* hazard and predictors or as a *proportional* relationship between *odds* and predictors, has no simple corresponding interpretation for hazard in its raw, untransformed state. This realization has two practical consequences when exploring the relationship between discrete-time hazard and predictors:

- *Use caution when examining within-group hazard functions as recommended at the beginning of this chapter* (as in the top panel of figure 11.1). Focus on gestalt impressions of differences in level. Because the model specifies that the *odds* of an event occurring will be proportional across levels of a predictor, the differential in level will be greater when hazard is high and smaller when hazard is low.
- *Plot within-group hazard functions not just on a raw scale, but also on odds and logit scales.* Plots like those in figure 11.2 are informative for model specification. Look for constant differentials in logit hazard, proportional differentials in odds, and approximately proportional differentials in raw hazard.

Although the hazard scale remains the easiest scale for communicating results, the two alternative scales (odds and log odds) can be especially helpful when conducting your analysis.

11.3 Fitting a Discrete-Time Hazard Model to Data

With data collected on a random sample of individuals from a target population, you can easily fit a discrete-time hazard model, estimate its parameters using maximum likelihood methods, and evaluate goodness-of-fit. Readers seeking the mathematical details—construction and maximization of the likelihood function and the statistical properties of the resulting estimates—should consult one of the many technical papers that present these results (e.g., Allison, 1982; Laird & Oliver, 1981; Efron, 1988; Singer & Willett, 1993). In this section, we explore the *concepts* underlying this process. Our goal is to show that although the model may appear complex and unfamiliar, it can be fit using software that *is* familiar by applying standard logistic regression analysis in the *person-period data set*. Use of familiar software facilitates analysis and allows you to capitalize on everything you already know about fitting statistical models to data.

11.3.1 Adding Predictors to the Person-Period Data Set

To fit the discrete-time hazard model to data, you must add some variables to the person-period data set. When introduced in section 10.5, the person-period data set contained just two variables: *PERIOD* (indexing time periods) and *EVENT* (signifying event occurrence). Now it must include the set of time indicators and the substantive predictors you wish to examine.

Figure 11.5 illustrates these modifications using three boys from the first intercourse study. The first two boys (subjects 193 and 126) have known event times (9th and 12th grades); the third (subject 407) was censored at 12th grade. In addition to the two event occurrence variables (*T* and *CENSOR*), the person-oriented data set (in the top panel) contains time-invariant predictors, whose values do not change over time (here, *PT* and *PAS*) and time-varying predictors, whose values *may* change over time. Each time-varying predictor must be represented by *J* variables. Here, *DRUG7* through *DRUG12* assess substance use in each grade on a scale from 0 (total abstinence) to 4 (heavy use). *Values* of time-varying predictors can be missing for periods subsequent to event occurrence. For this reason, boy 193, who had intercourse in ninth grade, has data for *DRUG7* through *DRUG9*, but missing values for *DRUG10* through *DRUG12* (denoted using a dot). As there are 180 boys in this sample, the data set has 180 records.

In the person-period data set, on the other hand, individual i has J_i records, one for each discrete-time period when he or she was at risk of event occurrence. For boy 193, who had intercourse in ninth grade, $J_i = 3$. For boys 126 and 407, $J_i = 6$, but for different reasons: one had intercourse in the sixth time period (12th grade); the other completed the

"Person-Level" data set

ID	T	CENSOR	PT	PAS	DRUG7	DRUG8	DRUG9	DRUG10	DRUG11	DRUG12
193	9	0	1	1.16	0.67	0	1.33	.	.	.
126	12	0	1	0.12	1	0	0	0	0	0
407	12	1	0	-0.96	0	0	0.33	1.33	1.33	1.67

"Person-Period" data set

ID	PERIOD	D7	D8	D9	D10	D11	D12	EVENT	PT	PAS	DRUG
193	7	1	0	0	0	0	0	0	1	1.16	0.67
193	8	0	1	0	0	0	0	0	1	1.16	0
193	9	0	0	1	0	0	0	1	1	1.16	1.33
126	7	1	0	0	0	0	0	0	1	0.12	1
126	8	0	1	0	0	0	0	0	1	0.12	0
126	9	0	0	1	0	0	0	0	1	0.12	0
126	10	0	0	0	1	0	0	0	1	0.12	0
126	11	0	0	0	0	1	0	0	1	0.12	0
126	12	0	0	0	0	0	1	1	1	0.12	0
407	7	1	0	0	0	0	0	0	0	-0.96	0
407	8	0	1	0	0	0	0	0	0	-0.96	0
407	9	0	0	1	0	0	0	0	0	-0.96	0.33
407	10	0	0	0	1	0	0	0	0	-0.96	1.33
407	11	0	0	0	0	1	0	0	0	-0.96	1.33
407	12	0	0	0	0	0	1	1	0	-0.96	1.67

Figure 11.5. Re-expressing the original person-level data set as a person-period data set: results for three boys included in the age at first intercourse study.

study a virgin, thereby remaining at risk for all six periods. Because the "beginning of time" in this study is seventh grade, $j = 7$ is the first period recorded for each person in the person-period data set. The value of each variable in this data set reflects the status of person i on that variable in period j. Referring to the bottom panel of figure 11.5, the person-period data set includes:

- A *period indicator*, here *PERIOD*, which specifies the time-period j that the record describes.
- A *set of time indicators*, here D_7 through D_{12}, that also identify the particular time-period described in the record.
- An *event indicator*, here *EVENT*, which indicates whether the target event occurred in that time period (0 = no event, 1 = event).
- *Time-invariant predictors*, here *PT* and *PAS*, whose values remain constant across an individual's multiple records. *PT* remains at 1 for the three records of boy 193 and the six records of boy 126, and it remains at 0 for the six records of boy 407. *PAS* is similarly stable across records.
- *Time-varying predictors*, here *DRUG*, which record the values of the variable in the time period relevant for the record. The original six variables *DRUG7–DRUG12* become a single variable in the person-period data set, whose values reflect the level of

drug use in period j. For example, the values of *DRUG7–DRUG9* for boy 193 (0.67, 0.00, and 1.33) become the values of *DRUG* for the three records of this case. This case's missing values for *DRUG10–DRUG12* are unproblematic because the person-period data set does not include records for these unobserved periods.

Because this study tracked only 180 individuals for up to only six years, this person-period data set is relatively small, with only 822 records.

Adding substantive predictors to the person-period data set is an essential precursor to model fitting. But constructing this data set has another advantage as well: it can be used to estimate the within-group hazard and survivor functions needed for exploratory analysis. Following the identical strategies presented in section 10.4.2 for a full sample, you can obtain all the information necessary for estimating within-group hazard and survivor functions using three-way cross-tabulations. To construct the life tables presented in table 11.1, for example, all we did was cross-tabulate the substantive predictor *PT* by the structural predictors *PERIOD* and *EVENT* in the person-period data set.

11.3.2 Maximum Likelihood Estimates for the Discrete-Time Hazard Model

As explained in section 3.4 in the context of the multilevel model for change, the method of maximum likelihood seeks those estimates of the population parameters—here, the α's and the β's—that maximize the likelihood of observing the sample data. Building on that general presentation, below we describe conceptually how ML estimates are computed for the discrete-time hazard model.

The likelihood function for the discrete-time hazard model expresses the probability that we would observe the specific pattern of event occurrence actually observed—the exact pattern of 0's and the 1's for the variable *EVENT* in the person-period data set. Because each individual has J_i observations, one per period of risk, each contributes J_i terms to the likelihood function. As a result, the likelihood function is composed of as many terms as there are *records* in the person-period data set.

The particular *value* that person i contributes for time period j depends on his or her true value of hazard in that time period, $h(t_{ij})$, and whether he or she experienced the event in that time period ($EVENT_{ij}$). We determine the value of each contribution using the following logic.

- *The probability that individual* i *experiences the event in time period* j, *given no previous event occurrence, is* $h(t_{ij})$. In the time period when individual i experiences the event (when $EVENT = 1$), he or she contributes $h(t_{ij})$ to the likelihood function.

- *The probability that individual* i *does* not *experience the event in time period* j, *given no previous event occurrence, is* $(1 - h(t_{ij}))$. In all those time periods when individual i does *not* experience the event (when $EVENT = 0$), he or she contributes $(1 - h(t_{ij}))$ to the likelihood function.

Individuals who are censored only contribute terms of the second type. Individuals who experience the target event contribute one term of the first type and $J_i - 1$ terms of the second type.

To write the likelihood function in such a way that each contribution is appropriately included, we use a mathematical "trick" that capitalizes on the properties of exponents. The variable $EVENT$ takes on only two values, 0 and 1. Recall that if we raise any quantity x to the first power (x^1), we simply get x, and if we raise x to the 0th power (x^0), we simply get 1. We therefore write the likelihood function for the discrete-time hazard model as:

$$\text{Likelihood} = \prod_{i=1}^{n} \prod_{j=1}^{J_i} h(t_{ij})^{EVENT_{ij}} (1 - h(t_{ij}))^{(1 - EVENT_{ij})}. \quad (11.11)$$

Although perhaps daunting, all equation 11.11 does is algebraically codify the logic just outlined. The two product signs (the Π's) ensure that the likelihood function multiplies the contributions of each record in the person-period data set across all individuals (through the first product sign) and all time periods for each individual (through the second product sign). Because the event indicator must be either 0 *or* 1, but never both, only *one* of the two terms contributes to the likelihood function in each person's record. In a time period when the event *does* occur for individual i, only the first term remains (because the second term becomes 1: $(1 - h(t_{ij}))^0 = 1$). In a time period when the event does *not* occur for individual i, only the second term remains (because the first term becomes 1: $h(t_{ij})^0 = 1$).

The likelihood function in equation 11.11 expresses the probability of observing the sample data on event occurrence that we actually observed (the values of the variable $EVENT$) as a function of unknown population parameters. As written, each unknown parameter is that person's population value of hazard in that time period $(h(t_{ij}))$. *Where are the unknown parameters we care about—the α's and the β's—that tell us about the baseline logit hazard function and the effects of substantive predictors?* Reflection on the model in section 11.2 reveals that these unknown parameters are here in equation 11.11, they are simply lurking in the guise of these population values of hazard! To make their appearance explicit, we rewrite the likelihood function in equation 11.11 by substituting in the values of $h(t_{ij})$ in the discrete-time hazard model in equation 11.3:

$$\text{Likelihood} = \prod_{i=1}^{n}\prod_{j=1}^{J_i}\left[\frac{1}{1+e^{-\{[\alpha_1 D_{1ij}+\alpha_2 D_{2ij}+\cdots+\alpha_J D_{Jij}]+[\beta_1 X_{1ij}+\beta_2 X_{2ij}+\cdots+\beta_P X_{Pij}]\}}}\right]^{EVENT_{ij}}$$

$$\times\left[1-\frac{1}{1+e^{-\{[\alpha_1 D_{1ij}+\alpha_2 D_{2ij}+\cdots+\alpha_J D_{Jij}]+[\beta_1 X_{1ij}+\beta_2 X_{2ij}+\cdots+\beta_P X_{Pij}]\}}}\right]^{(1-EVENT_{ij})}$$

$$(11.12)$$

Notice that both the α's and the β's, *and their respective predictors*, the D's and the X's, have appeared, showing that the likelihood function includes both observed variables (the D's, the X's, and *EVENT*) and unknown parameters (the α's and the β's).

As explained in section 3.4, statisticians routinely take the logarithm of the likelihood function so as to make the mathematics of estimation more tractable. To see how this works, examine the simpler representation of the likelihood function in equation 11.11, in which the unknown parameters appear through the population values of hazard, instead of the more complex representation in equation 11.12 in which the α's and the β's appear. We work with this representation only to simplify the equations—the same results accrue if we work with the specification in equation 11.12. Taking logarithms of both sides of equation 11.11, we have the log-likelihood function:

$$LL = \sum_{i=1}^{n}\sum_{j=1}^{J_i} EVENT_{ij}\log h(t_{ij})+(1-EVENT_{ij})\log(1-h(t_{ij})). \qquad (11.13)$$

Notice that products have turned into sums and exponents have turned into multipliers. Now, we "only" need to find those values of the α's and the β's (which allow us to compute the values of $h(t_{ij})$) that maximize the log-likelihood function.

Although simpler than the likelihood function itself, the maximum value of the log-likelihood function in equation 11.13 has no closed form solution either. Fortunately, though, there is an easy, practical solution to maximizing this log-likelihood. It turns out that the standard logistic regression routines widely available in all major statistical packages, *when applied appropriately in the person-period data set*, actually provide estimates of the parameters of the discrete-time hazard model that maximize the log-likelihood in equation 11.13. We provide the mathematical proof of this statement in a technical paper published elsewhere (Singer & Willett,

1993). From a data-analytic perspective, however, you can easily fit the discrete-time hazard model to data using the method of maximum-likelihood simply by regressing the event indicator, *EVENT*, on the time indicators and on the selected substantive predictors in the person-period data set using logistic regression analysis.

Before returning to the practical matters of fitting the model to data, which we do in the next section, let us conclude this section by addressing two concerns you may have about this approach: How can an analysis of the multiple records in a person-period data set yield appropriate parameter estimates, standard errors, and goodness-of-fit statistics when: (1) the sample size appears to have been inflated (the person-period data set is much larger than the number of individuals studied); and (2) the J_i records for each person in the person-period data set do not appear to be obtained independently of each other.

Precise resolution of these conundrums requires delving into technical topics beyond the scope of this book. Yet the intuition behind their resolution can be explained relatively simply: once you move from cross-sectional to longitudinal analysis, you must think of your data as existing within the *person-period* framework. The notion that each individual contributes only one record to an analysis is a holdover from the cross-sectional world. A person-period data set allows each person to contribute data whenever he or she is at risk. From this perspective, the analytic sample *should* be much larger than the number of individuals under study because we are not modeling a single random variable time (T) but rather the conditional probability of event occurrence for each person in each of his or her J_i periods of risk. It is not that the person-period data set is *too large*; rather, the person-oriented data set is *too small!*

As for the nonindependence of the multiple records within a person, that conundrum is resolved by remembering that the hazard function describes the *conditional* probability of event occurrence, where the conditioning (as shown in equation 11.2) depends upon the individual surviving until each particular time period (up through J_i) and his or her values for the substantive predictors in each time period. We therefore assume that all records in the person-period data set are conditionally independent.

11.3.3 Fitting the Discrete-Time Hazard Model to Data

To obtain maximum likelihood estimates of the population parameters in the discrete-time hazard model in equation 11.3, you simply use a logistic regression routine to regress the event indicator (*EVENT*) on the time indicators (D_1 through D_J) and the selected substantive predictors (X_1 through X_P) in the person-period data set. Any logistic regression routine will do, as all produce equivalent output, differing only in format and labeling. For the results presented in this chapter, we used the logistic regression routine in SAS (PROC LOGISTIC), but you should feel free to use the statistical software package with which you are most familiar (e.g., SPSS, SYSTAT, Stata).

To illustrate the process of fitting, interpreting, and testing statistical hypotheses for the discrete-time hazard model, in the following sections we compare the results of four models fitted to the 822 records in the person-period data set for the grade at first intercourse study:

$$\text{Model } A: \text{logit } h(t_j) = [\alpha_7 D_7 + \alpha_8 D_8 + \cdots + \alpha_{12} D_{12}]$$
$$\text{Model } B: \text{logit } h(t_j) = [\alpha_7 D_7 + \alpha_8 D_8 + \cdots + \alpha_{12} D_{12}] + \beta_1 PT$$
$$\text{Model } C: \text{logit } h(t_j) = [\alpha_7 D_7 + \alpha_8 D_8 + \cdots + \alpha_{12} D_{12}] \qquad + \beta_2 PAS$$
$$\text{Model } D: \text{logit } h(t_j) = [\alpha_7 D_7 + \alpha_8 D_8 + \cdots + \alpha_{12} D_{12}] + \beta_1 PT + \beta_2 PAS$$

$$(11.14)$$

Model A includes only the main effect of *TIME*; Model B includes the main effects of *TIME* and *PT*; Model C includes the main effects of *TIME* and *PAS*; and Model D includes the main effects of *TIME*, *PT*, and *PAS*. Results of model fitting are presented in table 11.3.

We recommend that you use four complementary strategies to facilitate interpretation of results: (1) compare the goodness-of-fit of alternative models so that you can decide which predictors should be included and which can be set aside; (2) examine parameter estimates and present numerical summaries of effect size; (3) compute, and then graphically display, fitted hazard and survivor functions at selected values of predictors; and (4) construct confidence intervals for interesting parameters. In the sections that follow, we address these four topics, beginning with parameter interpretation (section 11.4) and display construction (section 11.5) and continuing with a discussion of goodness of fit and hypothesis testing (section 11.6) and confidence interval construction (section 11.7). We once again adopt this unusual sequence, placing parameter interpretation and graphical display before inference, so as to continue our focus on model conceptualization. When analyzing your own data, you will undoubtedly test hypotheses first, turning to interpretation and display only after choosing a smaller subset of models.

Table 11.3: Results of fitting four discrete-time hazard models to the grade at first intercourse data ($n = 180$, n *events* = 126)

	Model A	Model B	Model C	Model D
Parameter Estimates and Asymptotic Standard Errors				
D_7	−2.3979***	−2.9943***	−2.4646***	−2.8932***
	(0.2697)	(0.3175)	(0.2741)	(0.3206)
D_8	−3.1167***	−3.7001***	−3.1591***	−3.5847***
	(0.3862)	(0.4206)	(0.3890)	(0.4231)
D_9	−1.7198***	−2.2811***	−1.7297***	−2.1502***
	(0.2217)	(0.2724)	(0.2245)	(0.2775)
D_{10}	−1.2867***	−1.8226***	−1.2851***	−1.6932***
	(0.2098)	(0.2585)	(0.2127)	(0.2647)
D_{11}	−1.1632***	−1.6542***	−1.1360***	−1.5177***
	(0.2291)	(0.2691)	(0.2324)	(0.2757)
D_{12}	−0.7309**	−1.1791***	−0.6421**	−1.0099***
	(0.2387)	(0.2716)	(0.2428)	(0.2811)
PT		0.8736***		0.6605**
		(0.2174)		(0.2367)
PAS			0.4428***	0.2964*
			(0.1140)	(0.1254)
Goodness-of-fit				
LL	−325.98	−317.33	−318.59	−314.57
Deviance	651.96	634.66	637.17	629.15
n parameters	6	7	7	8
AIC	663.96	648.66	651.17	645.15
BIC	681.00	668.54	671.05	667.87
Deviance-based Hypothesis Tests				
$H_0: \beta_{PT} = 0$		17.30*** (1)		8.02** (1)
$H_0: \beta_{PAS} = 0$			14.79*** (1)	5.51* (1)
Wald Hypothesis Tests				
$H_0: \beta_{PT} = 0$		16.15*** (1)		7.79** (1)
$H_0: \beta_{PAS} = 0$			15.10*** (1)	5.59* (1)

$\sim p < .10$; $* p < .05$; $** p < .01$; $*** p < .001$.

11.4 Interpreting Parameter Estimates

Fitting a discrete-time hazard model to data yields two types of parameter estimates: the $\hat{\alpha}$'s, which together provide estimated values of the baseline logit hazard function; and the $\hat{\beta}$'s, which assess the effects of substantive predictors. Table 11.3 identifies these estimates by their associated predictors—D_7 through D_{12}, *PT*, and *PAS*. Each of these is discussed in the sections below.

11.4.1 The Time Indicators

As a group, the $\hat{\alpha}$'s are maximum likelihood estimates of the baseline logit hazard function. The amount and direction of variation in their values describe the shape of this function and tell us whether risk increases, decreases, or remains steady over time. If the $\hat{\alpha}$'s are approximately equal, the risk of event occurrence is unrelated to time and the hazard function is flat. If the $\hat{\alpha}$'s decline over time—if they are larger in early periods compared to later periods—the baseline hazard function decreases over time. If the $\hat{\alpha}$'s increase over time—if they are smaller in early periods compared to later periods—the hazard function increases over time. The fairly steady increase over time in the magnitude of the $\hat{\alpha}$'s in each model in table 11.3 (from values far below 0 to ones near 0) shows that, in this sample of boys, the risk of first intercourse increases over time.

Precise interpretation of the $\hat{\alpha}$'s requires identification of the baseline group—those individuals for whom every substantive predictor in the model takes on the value 0. Each of the models in table 11.3 has a different baseline group. In Model B, it is the boys for whom $PT = 0$; in Model C, it is the boys for whom $PAS = 0$; and in Model D, it is the boys who meet both criteria. Model A is a special case. Because it includes no substantive predictors (such as PT or PAS), its baseline is the *entire sample*. The $\hat{\alpha}$'s for this model provide the estimated value of the logit hazard function for the entire group of boys.

Experience increases the ease with which you can directly examine numerical values of the $\hat{\alpha}$'s (which are expressed on a logit scale) and quantify the relative magnitude of risk in a given time period. But even with experience, few researchers find the logit metric appealing. We therefore routinely re-express the $\hat{\alpha}$'s using the inverse transformations of table 11.2. Although you can convert estimates into odds, it is more common to move directly into the metric of hazard by taking the antilogit of $\hat{\alpha}$:

$$\text{For the baseline group: } \hat{h}(t_j) = \frac{1}{1+e^{-\hat{\alpha}_j}}. \tag{11.15}$$

Table 11.4 presents the $\hat{\alpha}$'s for Model A expressed on both odds and hazard scales. A similar table could be constructed for any set of $\hat{\alpha}$'s. All that would change is the identity of the baseline group. For each time period j, the table identifies the predictor (D_j) associated with the population parameter α_j, the parameter estimate (from table 11.3), and the fitted values of odds and hazard. The table shows that in the full sample, in which individuals are not distinguished by predictors, the risk

Table 11.4: Interpreting the results of fitting an initial discrete-time hazard model including the main effect of *TIME*; re-expressing parameter estimates as fitted odds and fitted hazard probabilities

Time period	Predictor	Parameter estimate $(\hat{\alpha}_j)$	Fitted odds $e^{(\hat{\alpha}_j)}$	Fitted hazard $\dfrac{1}{1+e^{(-\hat{\alpha}_j)}}$
7	D_7	−2.3979	0.0909	0.0833
8	D_8	−3.1167	0.0443	0.0424
9	D_9	−1.7198	0.1791	0.1519
10	D_{10}	−1.2867	0.2762	0.2164
11	D_{11}	−1.1632	0.3125	0.2381
12	D_{12}	−0.7309	0.4815	0.3250

of first intercourse increases fairly steadily over time. In 7th grade, the fitted hazard probability is .0833, by 9th grade it has nearly doubled to 0.1519, and by 12th grade it has doubled again to 0.3250. Notice that these fitted values of hazard are identical to the sample estimates presented in the overall life table (Panel C of table 11.1). This identity holds because we are computing fitted values for a model with no substantive predictors.

11.4.2 Dichotomous Substantive Predictors

Model B adds the main effect of the time-invariant predictor *PT* to the main effect of *TIME*. As shown in table 11.3, the estimated coefficient for *PT* is 0.8736. Because *PT* is a dichotomy, which takes on the value 0 for boys who spent their early childhood years with both biological parents and 1 for those who experienced a parenting transition, the coefficient's positive sign indicates that, in every grade from 7th to 12th, boys in the latter group are at greater risk of initially having intercourse. The coefficient's magnitude tells us about the size of the risk differential measured on a logit scale. We estimate the vertical separation between the two logit hazard functions for these groups to be 0.8736.

As in logistic regression, we rarely interpret raw parameter estimates. More commonly, we antilog them, yielding an *odds ratio*, the ratio of the odds of event occurrence in two groups—one for which the predictor takes on the value 0 and another for which the predictor takes on the value 1 (see, e.g., Hosmer & Lemeshow, 2000; Long, 1997; Powers & Xie, 1999). To understand how this works, let us compute the odds of event occurrence in time period j for each group of boys. When $PT = 0$, the estimated odds that a boy will have intercourse in grade j are $e^{\hat{\alpha}_j D_j}$. When

$PT = 1$, the estimated odds are $e^{\hat{\alpha}_j D_j + \hat{\beta}_1}$. To compare these odds, we take the *ratio* of these estimates:

$$\text{Estimated odds ratio} = \frac{e^{\hat{\alpha}_j D_j + \hat{\beta}_1}}{e^{\hat{\alpha}_j D_j}} = \frac{e^{\hat{\alpha}_j D_j} e^{\hat{\beta}_1}}{e^{\hat{\alpha}_j D_j}} = e^{\hat{\beta}_1} \qquad (11.16)$$

Notice that the terms involving the subscript j (the time indicators and their parameter estimates) cancel out. Equation 11.16, therefore, shows that antilogging the single coefficient, $\hat{\beta}_1$, allows us to compare the odds of event occurrence for the two groups in *every time period*. Antilogging the coefficient for *PT* in Model B we find $e^{0.8736} = 2.40$. This tells us that, in every grade, the estimated odds of first intercourse are nearly two and one half times as high for boys who experienced a parenting transition in comparison to boys raised with both biological parents. In substantive terms, an odds ratio of this magnitude represents a substantial, and potentially important, effect.

As noted in section 11.1.2, odds ratios are symmetric about 1. If a dichotomous predictor is associated with an odds ratio of 1.0, the odds of event occurrence in the two groups are equal. If the odds ratio is greater than 1.0, the event is more likely to occur in the second group; if they are less than 1.0, it is less likely. Because of this symmetry, you can re-express an odds ratio using the *other group*—the one for whom the variable takes on the value 1—as the reference group. All you need do is compute the reciprocal of the odds ratio, which in this example is $1/2.40$ = 0.42. This tells us that the estimated odds of first intercourse for boys who did not experience a parenting transition are approximately 40% of the odds for boys who did. These complementary ways of reporting effect sizes are equivalent, although many people mistakenly perceive that an effect expressed in the metric of odds *appears* larger when expressed as a number greater than 1 than when expressed as a fraction.

11.4.3 Continuous Substantive Predictors

Model C adds the main effect of *PAS* to a model that already includes the main effect of *TIME*. As shown in table 11.3, the resulting parameter estimate is 0.4428. Its positive sign indicates that, in each grade, boys whose parents manifest higher levels of antisocial behavior are more likely to have intercourse. Its magnitude estimates the size of the vertical differential in logit hazard corresponding to a 1-unit difference in *PAS*.

To express an effect size for a continuous predictor using the metric of odds ratios, you must first decide whether a one-unit difference is appropriate. If so, as in this example where *PAS* ranges from -1.7 to $+2.8$, we can antilog the estimate (as for a dichotomous predictor) and interpret the resulting odds ratio. As $e^{(0.4428)} = 1.56$, we conclude that, in each grade, the estimated odds of first intercourse are just over 50% higher for boys whose parents score one unit higher on this antisocial behavior index.

If a one-unit difference is not meaningful, you can use a different comparative increment. Following the logic introduced in section 11.4.2, we compare the odds of event occurrence for two groups who differ by the general increment c on the variable *PAS* by computing:

$$\text{Estimated odds ratio} = \frac{e^{\hat{\alpha}_j D_j + \hat{\beta}_2 c}}{e^{\hat{\alpha}_j D_j}} = \frac{e^{\hat{\alpha}_j D_j} e^{\hat{\beta}_2 c}}{e^{\hat{\alpha}_j D_j}} = e^{\hat{\beta}_2 c}.$$

Antilogging yields the estimated odds ratio. For a two-unit difference in *PAS*, for example, we have $e^{(0.4428*2)} = 2.42$. This tells us that for each two-unit difference in *PAS*, the estimated odds of first intercourse in each grade are nearly two and half times as high, a value nearly identical to the estimated effect size for the dichotomous predictor *PT*.

11.4.4 Polytomous Substantive Predictors

Similar interpretive strategies can be extended to polytomies: variables that take on one of several discrete values. Whether nominal or ordinal, with categories representing a single value or a set of values, the strategy is the same: If a substantive predictor has Q categories, define a set of Q indicator variables, Z_1, Z_2, through Z_Q, one per level, and enter any $(q-1)$ of them into the model. There are many equivalent ways of coding the Q indicators. The most common strategy, used when constructing the time indicators in section 11.2, is to set $Z_q = 1$ for those cases taking on the qth value and 0 otherwise:

Category	Z_1	Z_2	. . .	Z_Q
1	1	0	0	0
2	0	1	0	0
. . .	0	0	. . .	0
Q	0	0	0	1

Under this coding scheme, parameter interpretation is similar to that for any a 0/1 predictor: the group representing the omitted level serves as the reference category for all comparative statements.

To illustrate, let us further explore the child's parental transition history, previously examined using the dichotomy *PT*. The researchers actually collected additional data that allow us to subdivide the 108 boys who experienced a parental transition into *three* groups: 26 who were raised by a single parent who remained single, 37 who were raised by a repartnered parent, and 45 who experienced two or more separations or repartnerings. This allows us to create three new variables—*PT1*, *PT2*, and *PT3*—to represent these subgroups. The group of boys for whom *PT* = 0 remains the reference group.

When we fit a discrete-time hazard model that includes the main effect of *TIME* and these three dichotomies, we obtain the following results for the substantive predictors:

Predictor	Parameter Estimate	Estimated Odds Ratio
PT1	0.6570	1.93
PT2	0.5688	1.77
PT3	1.3213	3.75

The comparative odds of first intercourse are nearly twice as high for boys who were raised by a single parent (*PT1*) or a repartnered parent (*PT2*) and nearly four times as high for boys who experienced two or more separations or repartnerings (*PT3*). The global variable *PT* collapses these three groups together to yield a parameter estimate of 0.8736, which is a weighted average of these three parameter estimates, and an estimated odds ratio of 2.40, which is a weighted average of these three estimated odds ratios. In section 11.6, we explore the question of whether this polytomous representation is superior to the dichotomous representation used so far.

11.5 Displaying Fitted Hazard and Survivor Functions

Graphic displays can be powerful tools for identifying and summarizing trends over time. We compute fitted hazard and survivor functions for particular values of substantive predictors by substituting the parameter estimates back into the discrete-time hazard models and obtaining predicted values of logit hazard. These straightforward calculations can be executed either within your statistical package (by outputting the parameter estimates from the logistic regression procedure) or in any spreadsheet program.

Table 11.5: Computing fitted hazard probabilities and survival probabilities from Model B for the two groups of boys for whom $PT = 0$ and $PT = 1$

| Time period | $\hat{\alpha}_j$ | $\hat{\beta}_1$ | Fitted value of | | | | | |
| | | | Logit hazard | | Hazard | | Survival | |
			$PT = 0$	$PT = 1$	$PT = 0$	$PT = 1$	$PT = 0$	$PT = 1$
7	−2.9943	0.8736	−2.9943	−2.1207	0.0477	0.1071	0.9523	0.8929
8	−3.7001	0.8736	−3.7001	−2.8265	0.0241	0.0559	0.9293	0.8430
9	−2.2811	0.8736	−2.2811	−1.4075	0.0927	0.1966	0.8432	0.6772
10	−1.8226	0.8736	−1.8226	−0.9490	0.1391	0.2791	0.7259	0.4882
11	−1.6542	0.8736	−1.6542	−0.7806	0.1605	0.3142	0.6094	0.3348
12	−1.1791	0.8736	−1.1791	−0.3055	0.2352	0.4242	0.4660	0.1928

11.5.1 A Strategy for a Single Categorical Substantive Predictor

Let us first consider the construction of graphic displays using Model B, which includes the single substantive predictor *PT*. As shown in table 11.5, all you need do is table the parameter estimates for the time indicators and manipulate these values to:

- *Derive fitted values of logit hazard at the chosen levels of the predictor,* by summing appropriate multiples of the parameter estimates.
- *Transform the fitted values of logit hazard into fitted values of hazard,* using the inverse transformations shown in the third row of table 11.2.
- *Compute the fitted values of the survivor function,* by substituting the fitted values of hazard into equation 10.5.
- *Compute fitted median lifetimes* using equation 10.6, which here yields values of 11.8 and 9.9.

The three panels of figure 11.6 display these results—the top panel presents the fitted logit hazard functions, the middle panel presents the fitted hazard functions, and the bottom panel presents the fitted survivor functions (with identification of the fitted median lifetimes).

Begin by examining the fitted hazard functions plotted on a logit scale. The discrete-time hazard model guarantees that these functions will be separated by a constant amount in each time period (here, 0.8736, the value of the estimated coefficient for *PT*). But a constant vertical separation in the metric of logit hazard does not produce a constant vertical separation in the metric of raw hazard (probability). Instead, in the

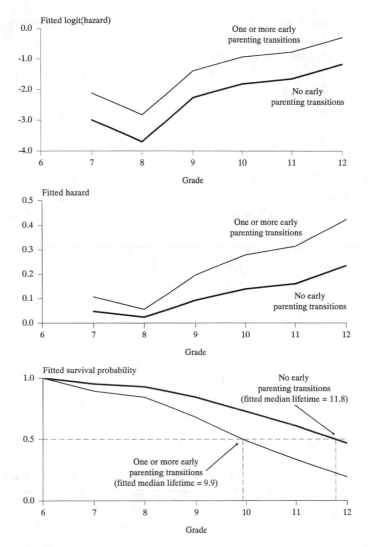

Figure 11.6. Fitted logit hazard, hazard, and survivor functions for the grade at first inter-course data, from Model B, which includes the main effects of *TIME* and *PT*.

middle panel of figure 11.6, we see that the top hazard function, for boys with a parenting transition, is an approximate magnification of the bottom hazard function for boys without a transition. Recall that we add the qualifier "approximate" because, as shown in section 11.2.2, the pro-portionality assumption for this model involves odds, not hazards. This assumption yields only an approximate degree of proportionality when expressed in the hazard scale.

The middle and bottom panels are perhaps even more effective devices for communicating results. Unlike numerical summaries, these graphs: (1) explicitly depict just how much higher the risk of first intercourse is in each grade among boys who have experienced a parenting transition; and (2) demonstrate how period-by-period differentials cumulate into a substantial difference in survivorship. The fitted median grade at first intercourse for boys with one or more transitions is 9.9 versus 11.8 for boys in the other group. By the end of ninth grade, we estimate that 68% of the boys who experienced a parenting transition remain virgins, in comparison to 84% of the boys without a transition. Taking the complement of these percentages, we estimate that 32% of the former group have had sex, in comparison to 16% of the latter group. By the end of 12th grade, these differentials have escalated: an estimated 81% of the boys who experienced a parenting transition had sex, in comparison to only 53% of the boys who did not. We find that summary statistics like these, although based on the results of complex analyses, can be communicated widely, especially if they are accompanied by graphic displays such as these.

11.5.2 Extending this Strategy to Multiple Predictors (Some of Which Are Continuous)

It is easy to display fitted hazard and survivor functions for models involving multiple predictors by extending these ideas in a straightforward manner. Instead of plotting one fitted function for *each* predictor value, select several *prototypical* predictor values (using strategies presented in section 4.5.3 and plot fitted functions for combinations of these values. However tempting it is to select many prototypical values for each predictor, we suggest that you limit the number of values lest the displays become crowded, precluding the very interpretation they were intended to facilitate.

Figure 11.7 presents prototypical hazard and survivor functions for Model D in table 11.3. (We do not use a logit scale because, as noted in section 11.5.1, we rarely use this scale for presentation.) To construct these graphs, we needed prototypical values for the continuous predictor *PAS*. We selected the values −1, 0, and 1 because: (1) they are round numbers; (2) 0 is the sample mean; and (3) −1 and +1 correspond roughly to the 10th and 90th percentiles. (In the person-level data set, the 10th percentile is −0.93 and the 90th percentile is +1.20.) To facilitate communication, we refer to these values as low, medium, and high *PAS*. Having chosen these values, computation proceeds as in table 11.5. The fitted functions for these six prototypical boys display all possible

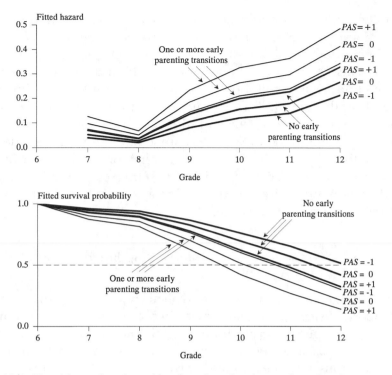

Figure 11.7. Fitted hazard and survivor functions for the grade at first intercourse data, from Model D, which includes the main effects of *TIME*, *PT*, and *PAS*.

combinations of the selected values of *PT* and *PAS*. As in figure 11.6, we use bold to represent the fitted functions when *PT* = 0 and fainter lines to represent those for *PT* = 1.

To reduce clutter, the fitted median lifetimes are not presented on the plot, although we facilitate their identification by including a dashed line at .50. We can augment this display with a table presenting the estimated median lifetimes for the chosen prototypical individuals:

Parent antisocial score	No parenting transition	One or more parenting transitions
Low (−1)	>12.0	10.7
Medium (0)	11.5	10.1
High (+1)	10.9	9.6

Although a table like this is unnecessary when displaying the results of a simple model with just two predictors, its utility increases as the number of predictors grows.

Coupled with the numerical results in table 11.3 and the tabular presentation of the fitted median lifetimes above, figure 11.7 provides compelling evidence about these predictors' effects. An average boy whose parents are high on the antisocial scale and who experienced a parenting transition in childhood has intercourse for the first time in the middle of ninth grade (9.6). In contrast, an average boy whose parents were low on the antisocial scale and who remained together throughout the boy's early childhood delays intercourse until after 12th grade. Controlling statistically for the effect of the other predictor, each of these variables has a large effect on the risk of first intercourse, suggesting that the early childhood environment in which a boy was raised can have profound effects during adolescence.

11.5.3 Two Cautions when Interpreting Fitted Hazard and Survivor Functions

A serious danger in displaying prototypical trajectories is that someone might incorrectly conclude that the predictors' effects vary over time. We have seen many people, even those well-schooled in the art and science of survival analysis, incorrectly conclude that a varying gap between fitted hazard functions indicates a time varying effect. Their reasoning is so straightforward that it appears unimpeachable: the only way that the gap between fitted hazard functions could vary over time is if the effect of the predictor varies over time.

Nothing could be further from the truth. We believe that the source of the confusion stems from the inappropriate, but understandable, extension of lessons learned when examining fitted regression lines (and fitted growth trajectories). When fitting linear models, increasing (or decreasing) gaps between fitted lines *do* indicate that the effect of the predictor displayed on the plot (here *PT* or *PAS*) differs according to levels of the predictor displayed on the *X* axis (here *TIME*). In regression, a varying gap between fitted lines is so intertwined with the concept of a statistical interaction that the hypothesis test for an interaction is sometimes labeled a test of parallelism.

In survival analysis, in contrast, a varying gap between fitted hazard functions does not imply a statistical interaction. Whenever there is a constant gap between fitted logit hazard functions (as specified by the model), there will be a varying gap between fitted hazard functions. Because the model expresses the linear effect of the predictor on logit hazard, you cannot draw a conclusion about the stability of an effect using graphs plotted on a raw hazard scale. In fact, the logic works in the *opposite* direction. If the size of the gap between fitted hazard functions is constant over time, effect of the predictor must *vary* over time! If you suspect that the effect of a predictor varies over time (as often happens in practice), it is straightforward to specify this behavior in ways that we introduce in the next chapter.

A second caution about interpreting fitted functions concerns the possibility that someone will incorrectly conclude that they depict sample estimates for subgroups of individuals, not fitted values from a statistical model. Fitted hazard and survivor functions (as in figure 11.6 and 11.7) differ from sample hazard and survivor functions (as in figure 11.1). Fitted values come from models that place constraints on predictors' effects. In the basic discrete-time hazard model, we constrain the vertical distance between pairs of fitted logit hazard functions to be constant over time. Although this assumption may be plausible it will rarely (if ever) be exactly true. As a result, fitted values will differ from sample values. Just as a fitted regression line will not pass through the sample means at every given value of a predictor, so, too, the fitted hazard functions will not pass through the sample hazard values in every time period. If the model fits the data well, the sample and fitted functions will be similar, but not coincident. The sample plots in figure 11.1 and the fitted plots in figure 11.6 are strikingly similar. We observe this resemblance precisely because this model fits the data particularly well. In other data sets, the fit may be of lower quality and the correspondence will break down. As this issue brings us to the broader issue of how well the discrete-time hazard model fits the data, we now address this latter issue directly.

11.6 Comparing Models Using Deviance Statistics and Information Criteria

We now introduce two important questions that we usually address before interpreting parameters and displaying results: Which of the alternative models fits better? Might a predictor's observed effect be the result of nothing more than sampling variation? Although we purposefully postponed these questions so as to emphasize model conceptualization and interpretation, we now address these topics. Because we have already introduced the basic ideas in the context of the multilevel model for change (sections 4.6 and 4.7), our discussion capitalizes on that earlier presentation.

11.6.1 The Deviance Statistic

We begin with the log-likelihood statistic, *LL*, a summary statistic routinely output (in some form) by any program that provides ML estimates.

As discussed in section 4.6, its relative magnitude across a series of models fit to the same set of data can be informative (although its *absolute* magnitude is not). The larger the LL statistic, the better the fit. In table 11.3, we see that LL is largest for Model D, which includes both substantive predictors and smallest for Model A, which includes none.

As in the multilevel model for change, we use the LL statistic not as an object of study in its own right but rather as the basis for computing a deviance statistic. For a given set of data, deviance quantifies how much worse the current model is in comparison to the best possible model you could fit, also known as the saturated model. For the discrete-time hazard model, the saturated model will reproduce every observed value of *EVENT* in the person-period data set. The only model that can do this must have as many parameters as there are data records. A model short by even just one or two parameters may not reproduce the sample data. A saturated model therefore has a one-to-one correspondence between ML estimates and observed values: just one set of data can produce the estimates and the estimates can come from just one set of data. Given this correspondence, the value of the likelihood function obtained by substituting these ML estimates for the unknown parameters must be 1. As the logarithm of 1 is 0, the LL statistic for the saturated model must be 0.

This allows us to define the deviance statistic for the discrete-time hazard model as:

$$\text{Deviance} = -2 \log \text{likelihood}_{\text{current model}} \qquad (11.18)$$

just as we did in equation 4.16 for the multilevel model for change. A deviance statistic for a discrete-time hazard model will always be greater than 0 (unless you compute it for the full model, in which case it will be exactly 0). The better the fit of the current model, the smaller its deviance. The second row of the second panel in table 11.3 presents deviance statistics for each of the four models fit. Model D, which has the most parameters, has the smallest deviance statistic; Model A, which has the fewest, has the largest.

But perhaps parsimonious Model A doesn't fit "so" badly? Perhaps Model D is unnecessarily complex? When models are nested, you can address questions like these using *likelihood ratio tests*, which compare deviance statistics for two competing models:

- A *current model*, which includes predictors whose effects you would like to evaluate.
- A *reduced model*, which does not include these predictors.

We formulate the likelihood ratio test generally by allowing each model to also include the time indicators (D_1 through D_J) and other substantive predictors (X_1 through X_P). Using equation 11.4 as a basis, we write the competing models as:

$$\text{Current model: } \operatorname{logit} h(t_j) = [\alpha_1 D_1 + \cdots + \alpha_J D_J] + [\beta_1 X_1 + \cdots + \beta_P X_P]$$
$$+ [\beta_{W_1} W_1 + \cdots + \beta_{W_k} W_k]$$
$$\text{Reduced model: } \operatorname{logit} h(t_j) = [\alpha_1 D_1 + \cdots + \alpha_J D_J] + [\beta_1 X_1 + \cdots + \beta_P X_P].$$

$$(11.19)$$

To evaluate whether the improvement in fit resulting from the added k predictors, W_1 through W_k, is nonnegligible, we test the compound null hypothesis:

$$H_0 : \beta_{W_1} = \beta_{W_2} = \cdots = \beta_{W_k} = 0, \qquad (11.20)$$

by comparing the deviance statistics of the two models. As in the case of the multilevel model for change, we can do so only if: (1) both models are fit to the identical data; and (2) the reduced model is *nested* within the current one. The constancy of data criterion requires elimination of all cases with any missing values for either model. The nesting criterion requires that you can write the reduced model by placing *constraints* on the values of the parameters in the current model. As framed in equation 11.19, the reduced model is nested within the current model because we can write the former by setting the β's associated with the W's in the latter to 0. Under the null hypothesis in equation 11.20, the difference in deviance statistics will (asymptotically) have a χ^2 distribution on k degrees of freedom: When the difference in deviance statistics is large, we reject the null hypothesis, concluding that we prefer the current model. When the difference in deviances is small, we fail to reject, concluding that the reduced model is not appreciably worse than, and is therefore preferable to, the less parsimonious full model.

11.6.2 Deviance-Based Hypothesis Tests for Individual Predictors

Comparing deviance statistics for pairs of nested models that differ only by a single substantive predictor permits evaluation of the "statistical significance" of that predictor. The tests presented in the third panel of table 11.3 focus on the two substantive predictors, *PT* and *PAS*.

Begin with the uncontrolled test for *PT* in the second column for Model B. Model B differs from Model A by 1 parameter, β_{PT}, which is associated with the single predictor *PT*. Subtracting deviance statistics (651.96 − 634.66) yields a difference of 17.30 (as shown). As the .1% critical value for a χ^2 distribution on one degree of freedom is 10.83, we reject the null hypothesis at the $p < .001$ level that in a model with no substantive predictors, $\beta_{PT} = 0$. To examine the uncontrolled effect of *PAS*, compare Models A and C. After establishing that A is nested within C (because we can write A by setting β_{PAS} to 0), we find a difference in deviance statistics of 14.79 (651.96 − 637.17). As this also exceeds the .1% critical value for a χ^2 distribution on one degree of freedom, we reject the null hypothesis at the $p < .001$ level that in a model with no substantive predictors $\beta_{PAS} = 0$.

In addition to testing the uncontrolled contribution of each substantive predictor, we can examine its *controlled* contribution holding constant the other's effect. To evaluate the effect of *PT* controlling for *PAS*, we compare the deviances of (the nested) models C and D. This yields a difference of 637.17 − 629.15 = 8.02 on one degree of freedom, which exceeds the 1% critical value of a χ^2 distribution on one degree of freedom (6.63). To evaluate the effect of *PAS* controlling for *PT*, we compare the deviances of (the nested) Models B and D. This difference in deviances (5.51) exceeds the 5% critical value for the χ^2 distribution on one degree of freedom. We therefore reject the controlled null hypotheses as well.

11.6.3 Deviance-Based Hypothesis Tests for Groups of Predictors

By using k predictors to formulate the likelihood ratio test in equations 11.19 and 11.20, we can use this same approach to compare models that differ by *groups of predictors*. As before, we require only that: (1) both models be estimated using the identical set of data; and (2) the reduced model be *nested* within the current model. In table 11.3, for example, we may test the compound null hypothesis $H_0: \beta_{PT} = \beta_{PAS} = 0$ by comparing Models A and D. As the difference in their deviance statistics—22.81—exceeds the .1% critical value of a χ^2 distribution on two degrees of freedom—13.82—we reject this compound null at the $p < .001$ level.

Joint tests are especially useful for evaluating effects of several *related* predictors. Sometimes the related predictors will collectively represent the same underlying construct (as might several measures of socioeconomic status). Other times, the related predictors will be a series of dichotomies representing a single polytomy, as in section 11.4.4. There, we fit a discrete-time hazard model that included three dichotomies—*PT1*, *PT2*, and *PT3*—which together indicate whether the boy was raised by a single parent, a repartnered parent, or a parent who experienced yet another transition. The parameter estimates for these three variables were 0.6570, 0.5688, and 1.3213, respectively.

Two models in table 11.3 are nested within this discrete-time hazard model. Model A is nested within it because it includes only the time indicators; setting the parameters for all three dichotomies to 0 yields this simpler specification. Model B is nested within it because setting the parameters for all three dichotomies to be *identical*—that is, setting β_{PT1} = β_{PT2} = β_{PT3}—yields this simpler specification. This latter comparison exemplifies an important type of nested model. The definition of nesting requires only that it is possible to obtain the reduced model by *placing constraints* on the parameters in the current model. Although the most popular type of constraint is to set the parameters to 0, a model that constrains these parameters to be identical is also nested.

Because both Models A and B are nested within this new model, we can compare the deviance statistics for the new model to each of these nested models. In doing so, we test two different null hypotheses. By comparing the new model to A, we test the null that all three parameters are equal to 0 (H_0: β_{PT1} = β_{PT2} = β_{PT3} = 0). By comparing the new model to B, we test the null that all three parameters are equal to some common parameter β_{PT} (H_0: β_{PT1} = β_{PT2} = β_{PT3} = β_{PT}).

What do we find? The deviance statistic for the new model, which has nine parameters (the six time indicators and the three dichotomies representing the different types of parental transitions), is 626.72. Comparing this value to that of Model A, we have a difference in deviances of $651.96 - 626.72 = 25.24$, which exceeds the 1% critical value of a χ^2 distribution on three degrees of freedom. This suggests that the three dichotomies should remain in our discrete-time hazard model. Comparing this value to that of Model B, we have a difference in deviances of $634.66 - 626.72 = 7.94$, which exceed the 5% point of the χ^2 distribution on two degrees of freedom. We reject the null hypothesis that all three parameters are identical (that H_0: β_{PT1} = β_{PT2} = β_{PT3} = β_{PT}) and conclude that we should consider adopting this expanded representation of the parental transition variable instead of the simpler representation that aggregates together the supplemental information on parenting transitions (using the single variable *PT*).

11.6.4 Comparing Nonnested Models Using AIC and BIC

As discussed in section 4.6.4, you may occasionally want to compare the goodness-of-fit of alternative models that are not nested. Your goal may not be to test a statistical hypothesis about a (set of) population parameter(s) but rather to choose between models that include *different* predictors. As long as the focal models are fit to the same set of data, you can make useful comparisons based on the information criteria AIC and BIC.

Having discussed their computation in section 4.6.4, we do not reiterate those details here. Suffice it to say that both penalize the log-likelihood statistic for the number of parameters present in the model, but that the BIC statistic also takes into account total sample size. If a model has p parameters, AIC = Deviance + $2p$. We use a similar calculation to compute BIC, but as it "corrects" for sample size, we must decide *which* sample size to use. Following the recommendations of Xie (1994) and Raftery (1995), we use the *number of events*, not the number of individuals or the number of records in the person-period data set. As 126 boys had intercourse during this study, we compute the BIC statistic for these models as Deviance + $(\ln(126))p)$ or Deviance + $4.84p$.

The last row of the second panel of table 11.3 presents AIC and BIC statistics for the four models. For each statistic, the smaller its value, the better the fit. Comparing these statistics we conclude that Model B, which uses the single predictor *PT* fits better than Model C, which uses the single predictor *PAS*, because the former has an AIC of 648.66 and a BIC of 668.54 whereas the latter has an AIC of 651.17 and a BIC of 671.05. We also conclude that Model D, which uses *two* predictors, fits better than any of the other models, despite its larger number of parameters, for it has an AIC of 645.15 and a BIC of 667.87.

11.7 Statistical Inference Using Asymptotic Standard Errors

An alternative strategy for testing hypotheses about predictors' effects is to compare a parameter estimate to its asymptotic standard error (*ase*). Like any standard error, the *ase* measures the precision of an estimate; the smaller the *ase*, the more precise the estimate. (The qualifier asymptotic indicates that these standard errors are only approximate, with the quality of the approximation improving as the sample size becomes infinitely large.) Because the sampling distribution of a maximum likelihood parameter estimate is asymptotically normal, we can use *ase*'s to test hypotheses about population values of model parameters and odds ratios, and construct confidence intervals around their associated estimates.

Given the appeal of using *ase*'s, you might wonder why we began our discussion of inference using a deviance-based approach that provides no facility for confidence interval construction? Our rationale is as follows. As the sample size tends toward infinity, hypothesis tests based on deviance statistics and *ase*'s become equivalent. But, in samples that are not infinitely large (in other words, in *all* actual analyses), the two tests are not equivalent, and in small samples, the discrepancy can be large. Although some statisticians (e.g., Rothenberg, 1984) suggest that neither approach is uniformly better, others (e.g., Hauck & Donner, 1977) suggest that the deviance-based approach is superior especially in "small" samples (where "small" refers to sizes that most empirical researchers would call large!). We therefore recommend that you conduct hypothesis tests using the deviance-based approach when possible, and for that reason, we present this strategy first.

Yet there are two powerful arguments for using the *ase* approach, especially when fitting hazard models to data. First, it is easy to implement. Unlike the deviance-based approach, which requires fitting multiple nested models and calculating differences in deviances (usually by hand), the *ase* approach is routinely offered in all statistical packages. Hypothesis testing from this perspective involves little more than determination of those parameters with sufficiently small p-values. Second, the *ase* approach allows you to construct confidence intervals around estimates of population parameters. The deviance-based approach does not provide for this option, requiring you to evaluate predictors' effects solely from a hypothesis testing perspective. For these reasons, we now describe how to use *ase*'s to test statistical hypotheses about interesting population parameters, and construct (asymptotic) confidence intervals around their associated estimates.

11.7.1 The Wald Chi-Square Statistic

As discussed in section 4.7, a Wald chi-square statistic compares a maximum likelihood parameter estimate to its asymptotic standard error in much the same way as a t-statistic in regression analysis compares a least-squares parameter estimate to its standard error. But unlike the t-statistic, which, under the null hypothesis that a parameter is 0, is simply the ratio of these two quantities, the Wald chi-square statistic squares this ratio yielding a test statistic that has a χ^2 distribution on one degree of freedom. More formally, in the following discrete-time hazard model:

$$\text{logit } h(t_j) = [\alpha_1 D_1 + \cdots + \alpha_J D_J] + [\beta_1 X_1 + \cdots + \beta_P X_P] + \beta_W W. \qquad (11.21)$$

We test the null hypothesis that $\beta_W = 0$ by squaring the ratio of the parameter estimate to its asymptotic standard error and comparing this result to a χ^2 distribution on one degree of freedom:

$$\text{Wald } \chi^2 = \left[\frac{\hat{\beta}_W}{ase(\hat{\beta}_W)} \right]^2 \sim \chi^2 \text{ on } 1 \, d.f. \qquad (11.22)$$

When the Wald chi-square statistic is large relative to critical values of the χ^2 distribution, we reject the null hypothesis; when it is small, we fail to reject.

The logistic regression analysis routines in all major statistical packages routinely output asymptotic standard errors, Wald chi-square statistics, and associated p-values. (Equivalently, some packages take the square root of the Wald chi-square statistic and present it as a z-statistic, comparatively evaluated against a standard normal distribution.) Table 11.3 presents ase's alongside their respective parameter estimates; Wald chi-square statistics appear in the bottom panel. For the two substantive predictors in Model D, we obtain Wald chi-square statistics of 7.79 ($p = .0053$) for PT and 5.59 ($p = .0181$) for PAS. As maximum likelihood theory leads us to expect, these values are close to those obtained when testing the identical null hypotheses using a difference in deviances approach. The other approach (shown in the third panel of table 11.3) yields test statistics of 8.02 and 5.51, respectively, that are also distributed as χ^2 on one degree of freedom. Because of the similarity in numeric values (both differences are less than .25), the two approaches lead to identical conclusions about rejection of the null hypothesis. With a larger sample size, the discrepancies in values would be smaller, and if the sample size were "infinite," the discrepancies would disappear.

As discussed in section 4.7, in the context of the multilevel model for change, Wald chi-square statistics can also be used to test simultaneous hypotheses about multiple predictors. Because these tests are not routinely supplied by statistical packages, and also because we prefer tests based on the deviance statistic (which are easy to conduct for multiple predictors), we do not present this approach here. Interested readers should consult Long (1997) or Hosmer and Lemeshow (2000).

11.7.2 Asymptotic Confidence Intervals for Parameters and Odds Ratios

Asymptotic standard errors can be used to construct asymptotic confidence intervals (ACIs) around estimates of population parameters. Focusing on the parameter β_W associated with the additional predictor W in the model in equation 11.21, we construct a 95% ACI for β_W as:

$$\hat{\beta}_W \pm 1.96\left[ase\left(\hat{\beta}_W\right)\right]. \tag{11.23}$$

A 95% ACI for β_{PT} in Model D of table 11.3 is $0.6605 \pm 1.96(0.2367)$, which yields $(0.1966, 1.1244)$. A 95% ACI for β_{PAS} in this same model is $0.2964 \pm 1.96(0.1254)$, which yields $(0.0506, 0.5422)$. As with any confidence interval, the true value of the population parameter may not fall within the ACI. Because these intervals are *asymptotic*, their accuracy improves with increasing sample size.

Because the ACI in equation 11.23 focuses on a parameter that we rarely interpret directly, we usually construct a different ACI to examine a predictor's effect. As in section 11.4, where we exponentiated parameter estimates and interpreted effects as odds ratios, we usually construct an ACI for this population parameter instead. To construct an ACI for the population odds ratio associated with the predictor W in the model in equation 11.21, you exponentiate the limits of equation 11.23:

$$e^{\left[\hat{\beta}_W \pm 1.96[ase(\hat{\beta}_W)]\right]}. \tag{11.24}$$

Equation 11.24 yields a 95% ACI for the population odds ratio associated with a one-unit difference in the predictor W. For the two substantive predictors in Model D, for example, we find that the 95% ACI for the odds ratio for PT is ($e^{0.1966}$, $e^{1.1244}$) or (1.22, 3.08), and the 95% ACI for the odds ratio associated with the predictor PAS is ($e^{0.0506}$, $e^{0.5422}$) or (1.05, 1.72).

Notice that neither of these ACIs covers 1. When an odds ratio is 1, the odds of event occurrence for two groups are equal. Just as you can test the null hypothesis that $\beta_W = 0$ by determining whether the 95% ACI for that parameter covers 0, so, too, may you test the null hypothesis that the population odds ratio associated with a one-unit increment in W is 1 by determining whether its ACI covers 1. Examining the 95% ACIs for these two parameters, we see that neither is likely to have an odds ratio as low as 1.0, but the odds ratio for a one-unit difference in PAS may be as high as nearly 2.0, while the odds ratio comparing boys with and without a parenting transition may be as high as 3.0.

Unlike many other confidence intervals with which you may be familiar (including those in equation 11.23), the ACI for an odds ratio *is not symmetric* about its maximum likelihood estimate (known as the "point estimate" of the population parameter). The midpoint of the ACI for the odds ratio for PT is 2.15, in comparison to the point estimate of 1.94 (= $e^{0.6605}$), and the midpoint of the ACI for the odds ratio for PAS is 1.39, in comparison to its point estimate of 1.35 (= $e^{0.2964}$). In other words, the maximum likelihood estimate will not fall precisely in the middle of this ACI. Asymmetry is guaranteed by the formula for computing the ACI for the odds ratio in equation 11.24. Exponentiating the *symmetric* limits in equation 11.23 must produce limits that are *asymmetric*. Because of the way the exponential transformation works, the point estimate will be nearer to the limit closer to 1 than it will be to the limit further from 1. When the point estimate is greater than 1, as here, it will be closer to the lower ACI limit; when the point estimate is less than 1, as when the associated parameter estimate is negative, it will be closer to the upper ACI limit.

The asymmetry of the ACI for the population odds ratio may be disturbing initially, but it has no effect on the interval's interpretation. In addition, the interval has an advantage not shared by symmetric intervals: the ACI in equation 11.24 will never yield a theoretically impossible value. Because the ACI for the odds ratio exponentiates the limits for the parameter estimate, and exponentiation will always yield a positive number, it is impossible to derive an ACI limit for the odds ratio that is not positive. By working with a statistical model designed explicitly for discrete-time data, we ensure that the confidence limits for all parameters fall within the range of plausibility.

12

Extending the Discrete-Time Hazard Model

Some departure from the norm will occur as time grows more open about it.

—John Ashbery

Like all statistical models, the basic discrete-time hazard model invokes assumptions about the population that may, or may not, hold in practice. Because no model should be adopted without scrutiny, we devote this chapter to examining its assumptions, demonstrating how to evaluate their tenability and relax their constraints when appropriate. In doing so, we illustrate practical principles of data analysis and offer theoretical insights into the model's behavior and interpretation.

In section 12.1, we revisit our original specification for the main effect of *TIME* in the discrete-time hazard model—which, in the previous chapter, was specified using a system of time indicators—and we compare it with other specifications that constrain the shape of the baseline hazard function in different ways. In doing so, we show how the discrete-time hazard model can become more parsimonious, often with little sacrifice in goodness-of-fit. In section 12.2, we re-examine the logit link that we used to relate hazard to predictors in the previous chapter and compare it to an alternative—the *complementary log-log* link—which yields an important correspondence with the continuous time hazard models that we will describe subsequently. We deal with time-varying predictors in section 12.3, showing how to include them in the discrete-time model and discussing inferential difficulties that their inclusion raises. In sections 12.4 through 12.6, we examine three important assumptions embedded in the discrete-time hazard model—the linear additivity assumption ("all predictors operate only as main effects"); the proportionality assumption ("the effects of each predictor are constant over time"); and the no unobserved heterogeneity assumption ("population hazard depends only on predictor values"). We conclude, in section 12.7, by describing analytic strategies for "residual" analysis to accompany model fitting. Throughout, our goal is to offer sensible ways of evaluating whether standard assumptions hold and to show how they can be relaxed when they do not. Together, the model and its extensions provide a flexible way of exploring relationships between discrete-time event occurrence and predictors that is applicable in all but the most unusual circumstances.

12.1 Alternative Specifications for the "Main Effect of *TIME*"

The predictors selected to represent the "main effect of *TIME*" in the discrete-time hazard model affect the model's ability to capture the shape of the baseline (logit) hazard function. Aware of this, in chapter 11, we used a completely general specification for *TIME* by including J dummy variables, $D_1, D_2, \ldots, D_J$, to identify the discrete-time periods in the hazard model as predictors. This specification not only puts no constraints on the shape of the baseline hazard, it is also: (1) easily *interpretable*—each parameter, α_j, represents the population value of logit hazard in time period j for the baseline group; (2) *informative*—patterns in the values of the α_j's describe the temporal shape of hazard; and (3) *consistent with life table estimates*—fitted values from a model with no substantive predictors reproduce estimates obtained using the life table approach.

Use of a completely general specification for *TIME* is an analytic decision, not an integral feature of the model. Nothing about the model or its estimation *requires* adoption of this, or any other, particular specification for *TIME*. For example, even before writing the model in section 11.2 formally, we presented figure 11.3, which suggested that, in the population from which the sample was drawn, logit hazard might be modeled as constant over time (top panel) or linear over time (middle panel). The completely general specification that we eventually adopted, while reasonable, is often not essential.

Why might you consider alternative specifications? As Efron (1988) and Fahrmeir and Wagenpfeil (1996) argue, the completely general specification of *TIME*: (1) lacks parsimony—if J is large, it requires the inclusion of many unknown parameters in the model; and (2) yields fitted hazard functions that can fluctuate erratically across consecutive time periods, owing to nothing more than sampling variation. Goodness-of-fit being equal, it is sensible to prefer parsimonious "well-behaved" models over those that are excessively overparameterized and erratic.

Ideally, as in the choice of an individual growth model, your specification for *TIME* in the discrete-time hazard model should be motivated by a combination of theory, previous research, and exploratory analysis. Less persuasively, post hoc inspection of parameter estimates from the completely general specification may also suggest a structure. The paramount criterion for decision making is goodness-of-fit. The goal is to identify a temporal specification that fits well and is parsimonious, too. A secondary criterion is the degree to which the parameters can be interpreted.

In this section, we present some alternative specifications for the main effect of *TIME* and discuss how to select among them. Although we recommend that you routinely explore such alternatives in any analysis, serious consideration is essential under three circumstances:

- *When your study involves* many *discrete time periods*, because the data collection period is long or because time is less coarsely discretized. When this happens, the general specification can call for exceedingly many dummy predictors. This decreases the overall statistical power of your analyses and, as we show in section 12.3.1, decreases the power to discern whether the effect of substantive predictors varies over time.
- *When hazard is expected to be near zero in some time periods*, because the risk of event occurrence is low. Under the completely general specification for time, when few or no events occur, in even just one period, maximum likelihood model-fitting algorithms may fail to converge, coefficient stability decrease, and parameter estimates take on implausible or impossible values.
- *When some time periods have small risk sets*, because either the initial sample is small or the effects of hazard and censoring dramatically diminish the size of the risk set over time. When the risk set is small, the number of events occurring in any time period is also likely to be small, making it likely that the hazard will be near zero, leading to the problems in fitting we have described immediately above.

We begin, in section 12.1.1, with a set of polynomial options. In section 12.1.2, we offer criteria for selecting among these alternatives, and in section 12.1.3, we describe how to interpret the resultant parameter estimates.

12.1.1 An Ordered Series of Polynomial Specifications for *TIME*

Interestingly, even though event occurrence has been recorded in discrete time, it is still possible to treat the predictor "*TIME*" as though it has a continuous specification in the discrete-time hazard model (provided, of course, that you restrain your interpretation of fitted hazard to the range of *TIME*'s observed values). Most person-period data sets—including those in figures 10.4 and 11.5—already include such a continuous specification, embodied in the variable *PERIOD*, whose values identify the time period that the record describes. The system of dummy predictors used in chapter 11 to represent the main effect of *TIME* could easily be replaced en masse by the simple predictor *PERIOD*, if we believed that logit hazard was indeed related linearly to time.

Once you accept the view that *TIME* can be specified as a continuous predictor in a discrete-time hazard model, then choosing the completely general specification of its effect—using a system of dummy predictors—may not be the logical starting point for model fitting. Instead, you might first hypothesize that logit hazard is perhaps constant over time, or that it varies linearly over time, or that it peaks (or bottoms out) in one or more periods. Alternatively, you might expect logit hazard to vary smoothly during one epoch and then shift abruptly to a different smooth functional form in a later epoch. How can such features be represented in the hazard model?

The practical dilemma is that for any set of data, there exist an infinite number of potential specifications. As we described earlier when specifying individual growth models, these include: transformations in the ladder of powers (section 6.2); polynomial forms (section 6.3); and truly nonlinear trajectories (section 6.4). So, too, a smooth specification can include one or more discontinuities (as in section 6.1). For the empirical researcher, the array of options can be dizzying (or paralyzing).

Rather than catalog the (infinitely) many possibilities here (!), we focus on a subset of model specifications that we find particularly useful: the ordered set of polynomials shown in table 12.1. We focus on these not because one of them always emerges as "best," but because, taken together, they encompass a wide array of alternatives. Sometimes, you might also investigate the utility of the *logarithm* of *TIME* as a predictor of logit hazard, but this specification is often indistinguishable from a linear specification if the number of time periods is relatively small (as it usually is). If you have reason to believe that log(*TIME*)—or any other specification, for that matter—is superior, we encourage you to investigate its performance using the strategies we present below.

Let's begin with the last row of the table, which presents the completely general specification of *TIME*. Because its fitted hazard function will be identical to the life table estimates, this model will always have the lowest (best) deviance statistic in comparison to any other model (without substantive predictors). No model will *ever* have a lower deviance (although

Table 12.1: Selected smooth polynomial representations for the main effect of *TIME* in a discrete-time hazard model

Order of polynomial	Behavior of logit hazard	n parameters	Model
0	Constant	1	$\text{logit } h(t_j) = \alpha_0 ONE$
1	Linear	2	$\text{logit } h(t_j) = \alpha_0 ONE + \alpha_1(TIME_j - c)$
2	Quadratic	3	$\text{logit } h(t_j) = \alpha_0 ONE + \alpha_1(TIME_j - c)$ $+ \alpha_2(TIME_j - c)^2$
3	Cubic	4	$\text{logit } h(t_j) = \alpha_0 ONE + \alpha_1(TIME_j - c)$ $+ \alpha_2(TIME_j - c)^2 + \alpha_3(TIME_j - c)^3$
4	Three stationary points	5	$\text{logit } h(t_j) = \alpha_0 ONE + \alpha_1(TIME_j - c)$ $+ \alpha_2(TIME_j - c)^2 + \alpha_3(TIME_j - c)^3$ $+ \alpha_4(TIME_j - c)^4$
5	Four stationary points	6	$\text{logit } h(t_j) = \alpha_0 ONE + \alpha_1(TIME_j - c)$ $+ \alpha_2(TIME_j - c)^2 + \alpha_3(TIME_j - c)^3$ $+ \alpha_4(TIME_j - c)^4 + \alpha_5(TIME_j - c)^5$
Completely general		J	$\text{logit } h(t_j) = \alpha_1 D_1 + \cdots + \alpha_J D_J$

the "cost" of the large number of parameters may make this model less attractive). As the best fitting model, the completely general specification of *TIME* provides an invaluable anchor on the continuum of goodness-of-fit. The goodness-of-fit of *any* other model that includes only the main effect of *TIME* can never be better.

Now consider the first model in the table, a zero-order polynomial that eliminates the effect of *TIME* altogether. By eliminating all predictors, this *constant* model constrains logit hazard to be identical across all J periods. In fact, the constant value of logit hazard, α_0, will be a weighted average of the values of all the α's in the completely general specification. However, because the constant model contains only one parameter, regardless of the number of periods, it has the largest (worst) deviance statistic of any model. So this model occupies the other extreme on the goodness-of-fit continuum.

Notice that in specifying the constant model, we associate α_0 with an explicit predictor labeled *ONE*. As befits its name, *ONE* takes on the value 1 for *every* record in the person-period data set. Although use of an explicit constant like this may be unfamiliar, all computer programs that include an intercept in any statistical model by default use such a predictor. Because the completely general specification has no single intercept, the use of *ONE* facilitates programming. By including *ONE* whenever you want to include an explicit intercept, you can consistently exclude the "default" intercept from *all* discrete-time hazard models you fit.

The remaining models in table 12.1 contain polynomial specifications for the main effect of *TIME*. As discussed in section 6.3.1, the order of the polynomial determines the number of stationary points (and therefore bends) in the function. A pth order polynomial has $(p - 1)$ stationary points. A 1st-order polynomial—a line—has no stationary points. A 2nd-order polynomial—a quadratic—has one; a cubic has two; a 5th-order polynomial has four. We explore 4th- and 5th-order polynomials here because their multiple stationary points allow us to test whether, and how many times, the logit hazard function hits peaks and troughs. We do not usually adopt these complex forms in practice. If your analyses suggest the need for a polynomial of this order, perhaps the completely general specification—or another alternative—should be used.

As in the multilevel model for change, we can also facilitate interpretation by subtracting a centering constant c from *TIME*. If $c = 0$, α_0 represents the value of logit hazard in time period 0, a value that rarely exists. If $c = 1$, α_0 represents the value of logit hazard in period 1, a far more common value. More generally, if we set c to any other value, α_0 represents the value of logit hazard in that period.

How will the goodness-of-fit of these polynomial models compare to the extremes provided by the constant and completely general specifications? Because each—from the linear through the fifth order—is nested within each subsequent model, the deviance statistic for a later model will be no larger than (no worse than) than that for all earlier models. If the added term is unnecessary, the deviance statistic will remain unchanged. If the additional term improves the fit, the deviance statistic will decrease (improve) as well. Each model will have a deviance statistic no larger than (no worse than) the constant specification. No polynomial model will have a deviance statistic as small as (as good as) the completely general specification (unless it is a Jth order polynomial).

The net result is that when we compute the deviance statistic for every model in table 12.1, *for any particular set of data*, it will never increase across the rows. Moreover, unless each extra term adds nothing to the prediction, the deviance statistic will consistently decrease from row to row. We illustrate this behavior in table 12.2, which presents the results of fitting the seven models in table 12.1 to data collected by Gamse and Conger (1997). As part of a study of the National Academy of Education–Spencer Foundation Post-Doctoral Fellowship Program, they followed the careers of 260 semifinalists and fellowship recipients who took an academic job after earning a doctorate. Each was tracked for up to nine years to see whether and, if so, when they received tenure. Of the

Table 12.2: Comparison of alternative smooth polynomial representations for the main effect of *TIME* in a baseline discrete-time hazard model for the academic tenure data ($n = 260$, $n\ events = 166$)

Representation for *TIME*	n parameters	Deviance	Difference in deviance in comparison to . . .		AIC	BIC
			Previous model	General model		
Constant	1	1037.57	—	**206.37** (8)	1039.57	1042.68
Linear	2	867.46	**170.11** (1)	**36.26** (7)	871.46	877.68
Quadratic	3	836.30	**31.16** (1)	5.10 (6)	842.30	851.63
Cubic	4	833.17	3.13 (1)	1.97 (5)	841.17	853.61
Fourth order	5	832.74	0.43 (1)	1.54 (4)	842.74	858.29
Fifth order	6	832.73	0.01 (1)	1.53 (3)	844.73	863.39
General	9	831.20	—	—	849.20	877.19

260, 166 (63.8%) received tenure during data collection; 94 (36.2%) were censored.

Focus on the first three columns of the table. Each row indicates the particular specification for *TIME* being used, the number of parameters in the model, and the deviance statistic. As expected, the deviance statistic for the constant model is the largest (1037.57). With each successive polynomial, the deviance statistic drops from a high of 867.46 (for the linear specification) to a low of 832.73 (for the fifth-order polynomial). This latter value closely approximates that for the general specification, 831.20, which is guaranteed to be the smallest. Before considering whether these differences in deviance are sufficient to warrant use of an alternative specification for *TIME*, let us examine the corresponding fitted logit hazard functions. Doing so not only highlights the behavior of logit hazard, it also offers a graphical means of comparing the fit of competing specifications. The top panel of figure 12.1 presents five fitted functions—for the general model (faint lines), the constant, linear, and cubic models (dashed lines) and the quadratic model (bold lines). We do not present fitted functions for the fourth- and fifth-order polynomials because, as suggested by their deviance statistics, we cannot distinguish them from the cubic.

When comparing fitted hazard functions derived from alternative specifications for *TIME*, try to discern whether any of them reproduces the completely general specification well enough to confirm its suitability. For these data, the constant hazard function is clearly inappropriate—it fits the data poorly. The linear model is also unsuitable, yielding fitted

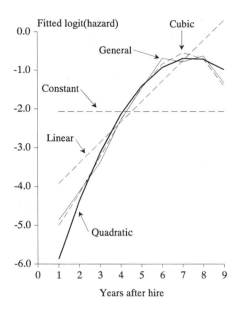

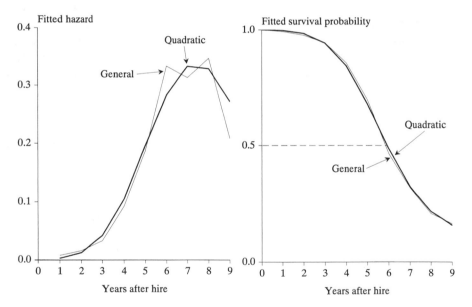

Figure 12.1. Understanding what it means to smooth the "main effect of time." Fitted logit hazard functions, hazard functions, and survivor functions for time to first tenure, using alternative specifications for the main effect of time.

values that are too high in the early periods, too low in the middle periods, and too high again in the later periods. The quadratic model resembles the general specification (although it is a little low in the beginning). The cubic model fits better, but its superior fit (registering the additional early trough) may not be "worth" the additional parameter. As these considerations lead us once again to the comparison of deviance statistics in light of the number of parameters in the model, we now describe how to make these comparisons formally.

12.1.2 Criteria for Comparing Alternative Specifications

The decline in the deviance statistic across models indicates that fit improves with increasing complexity of the temporal specification. To evaluate the magnitude of this decline, we must also account for the increased number of parameters in the model. You should not adopt a more complex specification if it fits no better than a simpler one. But if an alternative specification is (nearly) as good as the most general one, it may be "good enough." At the same time, we would not want an alternative that performs measurably worse than we know we can do.

As shown in the fourth and fifth columns of table 12.2, we address these issues by comparing each model's deviance statistic to that of: (1) its predecessor; and (2) the completely general specification. From section 11.6, we know that if a reduced model is nested within a current model, we can test hypotheses about the latter's extra parameters by comparing the difference in their deviance statistics to a critical value drawn from χ^2 distribution with degrees of freedom equal to the number of parameters added between the reduced and current model.[1] We therefore recommend that you:

- *Compare deviance statistics across consecutive models to test the impact of each new polynomial term.* As each model adds just one parameter, this difference can be compared to a critical value from a χ^2 distribution on one degree of freedom. By this process, you can identify the lowest order polynomial for which no higher order one fits better—that is, identify the model that provides the last test that rejects.
- *Compare the deviance statistic for each polynomial specification to the completely general specification.* For a model with k parameters, evaluate the difference in deviance between it and the completely general model against a critical value from a χ^2 distribution on $J - k$ degrees of freedom. If you fail to reject, the alternative is

viable, for it is more parsimonious than the completely general model, and fits nearly as well. If not, the alternative is not "good enough," no matter how much better it is than a lower order polynomial.

As you are testing multiple hypotheses with the same set of data, be careful not to base decisions rigidly on p-values. In table 12.2, we use bold type to denote a hypothesis test that rejects at the .05 level.

AIC and BIC statistics, which penalize the deviance statistic for the presence of additional parameters, provide other benchmarks for comparison. Although these information criteria yield no formal hypothesis test, we have found, across a large number of data sets, that the specification with lowest, or nearly lowest, AIC or BIC is often the most attractive.

What do we find when we apply these criteria to the tenure data? As shown in table 12.2 (and as suggested by the plots in figure 12.1), the linear specification is superior to the constant ($\chi^2 = 170.10$, 1 $d.f.$) and the quadratic is better still ($\chi^2 = 31.158$, 1 $d.f.$). No higher order polynomial appears necessary, as all yield differences in deviance of less than 3.84 (the .05 critical value of the χ^2 distribution on 1 $d.f.$) when compared the model preceding them in the list. We also ask whether any polynomial (especially the quadratic) is nearly as good as the completely general model. Although neither the constant nor the linear models are adequate (note the large differences in deviance), the quadratic performs well with only three parameters (instead of nine). The difference in deviance statistics for these two models is 5.10 (far less than the .05 critical value of a χ^2 distribution on 6 $d.f.$).

All signs therefore point to the superiority of the quadratic specification, which fits nearly as well as the general model, but with fewer parameters. To further evaluate its performance, the bottom panels of figure 12.1 present fitted hazard and survivor functions for the quadratic (using bold lines) and the completely general (using faint lines) specifications. The fitted hazard function for the quadratic specification works well in *all* time periods. Its seemingly poor performance in the early periods (in the top plot) is minimized in the move to a hazard scale. Because taking antilogits bounds the fitted values of hazard between 0 and 1, it tightens the distance between the problematic values when logit hazard is small. Also notice that the fitted hazard function for the quadratic is *smoother* than that for the general specification. The latter has two peaks, in years 6 and 8, with an apparent drop in year 7, while the former has a single peak in year 7 (with years 8 and 6 providing the next highest values). The quadratic seems more satisfactory in the sense that small period-to-period fluctuations are smoothed, revealing a single peak in year 7—precisely what we would expect to find when modeling time to tenure. Finally, notice that the fitted survivor functions for the quadratic and general models are nearly coincident. Because survivor functions cumulate hazards over time, subtle differences between fitted hazard functions yield only modest differences in fitted survivor functions. Estimated median lifetimes for these two models (shown using the dotted line) are virtually identical—5.8 for the general specification and 5.9 years for the quadratic.

Our clear preference is for the quadratic specification. But before taking this resolution as unambiguous, examine the AIC and BIC statistics. Although the BIC is at its minimum for the quadratic, the AIC statistic is at its minimum for the *cubic* (841.17). Although this difference is trivial, the cubic has some appeal. As shown in the top panel of figure 12.1, it fits somewhat better, particularly in the early years. Moving to an odds scale ameliorates this superiority, but some researchers might be attracted to this more complex model. To support this decision, they would cite the 3.13 difference in deviance between the two, which, although not "statistically significant," is noticeable (especially in contrast to the subsequent trivial differences in deviance in table 12.2). We mention this alternative not because we recommend it for these data, but rather because it highlights the ambiguity of data analysis.

Although decision rules cannot substitute for judgment, intuition, and common sense, we nevertheless conclude by offering two guidelines for selecting among alternative specifications:

- *If a smooth specification works nearly as well as the completely general one, appreciably better than all simpler ones, and no worse than all more complex ones, consider adopting it.*
- *If no smooth specification meets these criteria, retain the completely general specification.*

If this decision process leads you to a polynomial specification, then you can interpret the model's parameters easily, as we now discuss.

12.1.3 Interpreting Parameters from Linear, Quadratic, and Cubic Specifications

One advantage of a polynomial specification is that you can often interpret its parameters directly. As discussed in section 6.3.1 in the context of the multilevel model for change, with practice you can examine the coefficients in a polynomial function and describe a fitted function's shape without even constructing a plot. In this section, we reiterate general principles, but some readers might want to review our earlier presentation before moving on. (In what follows, notice that we work directly with the raw parameters expressed on a logit hazard scale. Although you can transform and interpret them on a hazard scale, we find that this becomes cumbersome with higher order polynomials.)

Begin with the simplest possibility, the linear specification. As in any linear specification, the intercept, α_0, represents the value of the outcome (here, logit hazard) when the predictor is 0. As the predictor in this model is $TIME - c$, α_0 represents the value of logit hazard in time period c. The slope parameter, α_1, is unaffected by the subtraction of c from $TIME$. It represents the increase (or decrease) in logit hazard per unit difference in $TIME$. As these interpretations are straightforward, we omit illustration.

In the quadratic model, the intercept, α_0, still measures the value of logit hazard in time period c. So, too, the "slope parameter," α_1, still measures the increase (or decrease) in logit hazard per unit of $TIME$, but only at one particular instant, time c. Remember that a quadratic has no *single* rate of change—the differential in logit hazard per unit time varies over time. The curvature parameter, α_2, identifies whether the logit hazard function is convex, with a trough ($\cup$), or concave, with a peak ($\cap$). If α_2 is positive, the hazard function is convex to the time axis; if negative, it is concave. This parameter can also be used to determine the time at which logit hazard reaches its peak (or trough). By calculus (differentiating with respect to $TIME$ and setting the derivative to 0), we can show that the time that the function reaches its peak (or trough) is $[c - \frac{1}{2}(\alpha_1/\alpha_2)]$. As with any polynomial, the timing of the peak (or trough) may not fall within the range of the data.

We illustrate this interpretation using the tenure data. Setting c to 5 and fitting a quadratic discrete-time hazard model, we obtain: logit $\hat{h}(t_j)$ $= -1.4107ONE + 0.6089(TIME_j - 5) - 0.1256(TIME_j - 5)^2$. The fitted value for logit hazard in year 5 is -1.4107 and the instantaneous rate of change in logit hazard in year 5 is 0.6089. Because the parameter estimate for the quadratic term is negative, we know that the hazard function is concave to the time axis, reaching a peak. Computing the time value for the flipover point, $[5 - \frac{1}{2}(0.6089/(-0.1256))] = 7.4240$, we find that the peak occurs midway through year 7. As the peak in raw hazard occurs at the same moment as that of the peak in logit hazard, we estimate that risk of receiving tenure is highest in the middle of year 7.

Cubic functions involve not just one bend, but two. As before, the intercept, α_0, represents the value of logit hazard in time period c. The sign of the slope parameter, α_1, indicates the direction of change in logit hazard over time in time period c—if α_1 is positive, hazard is increasing;

if α_1 is negative, hazard is decreasing. Every cubic curve has one trough and one peak, although the bends need not occur within the range of the data. The sign of α_3 indicates whether the function first hits a peak or a trough. If α_3 is positive, the cubic has an early peak and a late trough; if it is negative, it has an early trough and a late peak. To identify when the function reaches these bends, we again use calculus (differentiating the model with respect to *TIME*, setting the derivative to 0, and solving the resulting quadratic equation), to find:

$$\text{Times of peaks and troughs} = c + \frac{-\alpha_2 \pm \sqrt{\alpha_2^2 - 3\alpha_1\alpha_3}}{3\alpha_3}. \tag{12.1}$$

We illustrate these interpretations by fitting a discrete-time hazard model with a cubic temporal specification to the tenure data, to find:

$$\text{logit } \hat{h}(t_j) = -1.4555ONE + 0.7471(TIME_j - 5) - 0.1094(TIME_j - 5)^2$$
$$- 0.0187(TIME_j - 5)^3.$$

The fitted value for logit hazard in year 5 is -1.4555, a value close to the -1.4107 of the quadratic. As α_1 is positive, we know that the function is increasing over time in year 5. Because the coefficient for the cubic term is negative, we know that the hazard function has an early trough and a late peak. We identify these times by substituting into equation 12.1 to find -1.0877 and 7.1875. The early trough falls outside the range of the data. The late peak of 7.1875 corresponds closely to the peak identified in the quadratic model (7.4240). This close correspondence should come as no surprise given that the quadratic and cubic models had very comparable goodness-of-fit statistics.

12.2 Using the Complementary Log-Log Link to Specify a Discrete-Time Hazard Model

When we introduced the discrete-time hazard model, we argued that we needed to transform the outcome—discrete-time hazard—to preclude derivation of theoretically impossible fitted hazard values. The logit transformation represented a natural choice because it allowed us to: (1) specify the model using familiar terminology; (2) use widely available software for estimation; and (3) exploit interpretive strategies with which many empirical researchers are comfortable.

Just like the choice of a completely general specification for the main effect of *TIME*, use of a logit link is an analytic *decision*. Nothing about the way in which the model is postulated or fit requires the adoption of this, or any other, particular link function. Before logistic regression analysis became so popular, for example, statisticians routinely analyzed binary outcome data using a *probit* link, which maps probabilities onto the inverse of the cumulative standard normal distribution. In fact, the inverse of *any* cumulative distribution function can serve as a link function; logits owe their popularity to their computational and interpretive convenience (Collett, 1991; Hosmer and Lemeshow, 2000).

For discrete-time hazard, one other link function deserves consideration: the *complementary log-log* transformation, abbreviated *clog-log*. For a given probability, the complementary log-log is:

$$\text{clog-log} = \log(-\log(1 - \text{probability})) \tag{12.2}$$

Like the logit transformation, the clog-log transformation maps probabilities onto a new scale with no upper or lower bound. But while the logit transformation yields the logarithm of the *odds of event occurrence*, the clog-log transformation yields *the logarithm of the negated logarithm of the probability of event nonoccurrence*. Below, we explain when and why the clog-log transformation is useful (section 12.2.1), we develop a discrete-time hazard model based upon it (section 12.2.2), and we provide advice for researchers trying to choose between these alternative specifications (section 12.2.3).

12.2.1 The Clog-Log Transformation: When and Why It Is Useful

Figure 12.2 depicts the effect of the complementary log-log transformation by displaying its values (on the vertical axis) for a range of underlying hazard probabilities (on the horizontal axis). For comparison, the figure also displays the same values of hazard transformed using the logit transformation. When hazard is small, say below .20, both transformations yield similar values. At higher values of hazard, the transformations diverge. Once hazard is greater than .50 (an admittedly high value), the two sets of transformed values are quite disparate.

To understand the impact of the clog-log transformation and how it differs from that of the logit transformation, examine the *relative* distances between pairs of transformed values. Particularly, notice that the slope of the clog-log versus probability curve is less steep at higher values of hazard than the slope of the logit vs. probability curve. This means that the "distance" between pairs of values on the clog-log scale per "unit" difference in hazard probability gets consistently smaller at higher values of hazard, compared to their companions on the logit scale. Also, unlike the logit transformation, which is symmetric on either side of the "focal"

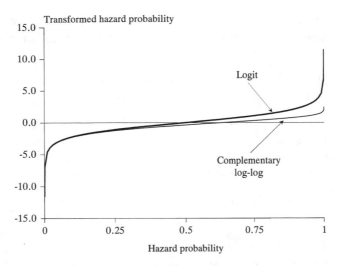

Figure 12.2. Comparing effects of the complementary log-log and logit transformations. Transformed values are plotted for raw probabilities ranging from 0.00001 to 0.99999.

hazard probability of .5, the clog-log transformation is *asymmetric*. You can demonstrate this numerically by examining two particular values of hazard, .20 and .30, and their complements, .80 and .70. Comparing the clog-log values corresponding to the first two hazard probabilities (−1.50 and −1.04) to those of the second two (.48 and .19), we find that neither their difference, nor their ratio, is constant. Finally, note, too, that there is no decent "round" number associated with the "central" hazard probability of .5—while .5 transforms to 1 on an odds scale and 0 on a logit scale, it becomes −0.3665 on the clog-log scale (hardly a memorable value).

All this may lead you to question the new transformation's utility. But as we show in section 12.2.3, it does have one fundamental advantage: it provides a discrete-time statistical model for hazard that has a built-in *proportional hazards* assumption, and not a *proportional odds* assumption (as in the case of the logit link). This would be completely unremarkable except for one thing: it provides a conceptual parallelism between the clog-log discrete-time hazard model and the models that we will ultimately describe for continuous-time survival analysis. As we show in chapters 14 and 15, the most popular continuous-time hazard model—the Cox regression model—also invokes a proportionality assumption in *hazards* not *odds*. Using a clog-log link here provides a discrete-time analog for the future-preferred continuous-time hazard model. Researchers who seek consistency of assumption in their hazard modeling regardless of the way in which time is recorded, may find this parallelism compelling.

12.2.2 A Discrete-Time Hazard Model Using the Complementary Log-Log Link

Any discrete-time hazard model postulated using a logit link can be rewritten using a clog-log link, simply by substituting transformations of the outcome. For example, we can write a general discrete-time hazard model for J time periods and P substantive predictors as:

$$\text{clog-log } h(t_j) = [\alpha_1 D_1 + \alpha_2 D_2 + \cdots + \alpha_J D_J] \\ + [\beta_1 X_1 + \beta_2 X_2 + \cdots + \beta_P X_P]. \tag{12.3}$$

In positing this model, we invoke assumptions similar to those articulated in section 11.1.2 for a model with a logit link (cf. equation 11.4). The difference here is that the assumptions describe the behavior of clog-log hazard, not logit hazard. Specifically, we assume that: (1) for each combination of predictor values, there is a postulated clog-log hazard function; (2) each of these clog-log hazard functions has an identical shape; and (3) the distance between each of these clog-log hazard functions is identical in every time period. As you would expect, all extensions of a discrete-time logit hazard model—for example, use of a smooth polynomial for *TIME*—can be applied directly to the clog-log version, too.

Usually, and particularly when risks are low, models fit using either link function yield virtually identical results. We can presage this close correspondence by examining transformed sample hazard functions for the grade of first intercourse data of chapter 11. Based on figure 11.2, figure 12.3 presents transformed hazard functions by values of the dichotomous predictor *PT,* parental transition status. For each group of boys, the graph presents the sample functions plotted on a logit scale (the dashed lines) and clog-log scale (the solid lines). As we would expect from figure 12.2, the pairs of transformed functions are more similar in the early grades (when hazard is small) and diverge over time (as hazard increases). Note, too, that the equidistance assumption could easily be posited between either pair of functions—it is not as if this assumption appears more (or less) tenable for functions expressed on a clog-log scale in comparison to the logit scale.

Most major statistical packages allow you to fit a discrete-time hazard model with a clog-log link. All you need do is specify this link when invoking the logistic regression routine within the person-period data set. Implementation is easy, but be sure to specify the model in terms of event *occurrence,* not *nonoccurrence,* in other words, the *EVENT* indicator must

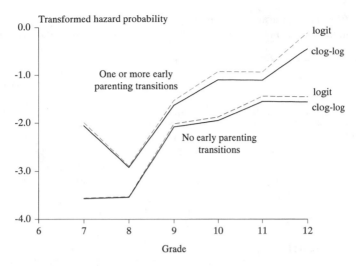

Figure 12.3. Sample hazard functions for the grade of first intercourse data displayed on different scales. Sample functions are expressed on a logit scale (the dashed lines) and complementary log-log scale (the solid lines).

be properly coded for the package's procedure. The need for extra caution is due to the asymmetry of the clog-log transformation. Unlike the logit link, which allows you to simply reverse the signs of all parameter estimates if you inadvertently model event *nonoccurrence*, there is no equivalent conversion for the clog-log link. The entire model differs if the event indicator is reverse coded (behavior you can, and should, confirm with your statistical package). Apart from this caution, model fitting proceeds as when a logit link is used; so, too, does goodness-of-fit and hypothesis testing.

Table 12.3 presents the results of fitting a pair of discrete-time hazard models to the grade at first intercourse data: the first with a clog-log link, the second with a logit link. (The logit link reproduces Model B of table 11.3.) Notice the numerical similarity of the parameter estimates—especially for the time indicators—and the near coincident deviance statistics (which differ by only 0.151). Numerical similarity is common when fitting identical models with alternate link functions (and net risks of event occurrence are low), and suggests that choice of a link function should depend on other considerations.

As with any link, we can derive a fitted baseline hazard function by applying the appropriate inverse transformation to the parameter estimates and the time indicators (see section 11.4). The inverse transformation for the clog-log link can be obtained by re-expressing equation 12.3 in terms of the original probability:

Table 12.3: Comparing parameter estimates and fitted baseline hazard functions for two discrete-time hazard models fit to the grade of first intercourse data ($n = 180$, n events $= 126$)

| | Parameter estimates | | Fitted baseline hazard | |
| | | | Clog-log $1 - e^{-e^{(\hat{\alpha})}}$ | Logit $1/(1 + e^{(-\hat{\alpha})})$ |
	Clog-log	Logit		
D_7	−2.9733	−2.9943	0.0498	0.0477
D_8	−3.6592	−3.7001	0.0254	0.0241
D_9	−2.3156	−2.2811	0.0940	0.0927
D_{10}	−1.9001	−1.8226	0.1389	0.1391
D_{11}	−1.7621	−1.6542	0.1578	0.1605
D_{12}	−1.3426	−1.1791	0.2298	0.2352
PT	0.7854	0.8736	—	—
Deviance	634.501	634.662		

One model uses a complementary log-log link; the other a logit link.

$$\text{probability} = 1 - e^{\left(-e^{(\text{clog-log})}\right)} \tag{12.4}$$

The fourth column of table 12.3 presents the fitted baseline hazard function obtained by applying this inverse transformation to the model's parameter estimates. For comparison, the table also presents the fitted baseline hazard function obtained from the model fit using a logit link.

Notice the similarity in fitted baseline hazard functions. The only time period in which we perceive an appreciable difference is the last, when hazard exceeds .20, and even then, the difference is trivial. In general, fitted hazard functions from models estimated with both link functions will be indistinguishable unless hazard is high, once again suggesting that the quality of the estimates does not provide a rationale for selecting one of these link functions over the other.

When interpreting the effects of substantive predictors, however, the interpretation *does* depend on the link. In both cases, we antilog parameter estimates, but whereas an antilogged coefficient from a model with a logit link is an odds ratio, an antilogged coefficient from a model with a clog-log link is a *hazard ratio*. Antilogging the parameter estimate of 0.7854 for the predictor PT in the clog-log model yields 2.19. In every grade from 7th through 12th, we estimate the hazard of first intercourse for boys who experienced a parenting transition to be 2.2 times the hazard for their peers raised with both biological parents. This interpretation contrasts with that from the model with a logit link, which suggests that the *odds* of first intercourse are 2.4 times the odds for boys who experienced a parenting transition.

Which estimate is "right"? Is hazard multiplied by 2.2 or are odds multiplied by 2.4? An epidemiologist would be quick to remind you that odds-ratios and hazard ratios (also known as relative risks) are distinct, but interrelated, measures of effect size. As befits their names, one compares odds; the other hazards. Numerically, we have:

$$\text{hazard ratio} = \frac{h(t_j: \text{group 1})}{h(t_j: \text{group 2})},$$

$$\text{odds ratio} = \frac{h(t_j: \text{group 1})/[1 - h(t_j: \text{group 1})]}{h(t_j: \text{group 2})/[1 - h(t_j: \text{group 2})]}.$$

When hazard is very small, the bottom terms in the odds ratio's numerator and denominator approach 1, making the two ratios nearly identical. When hazard is larger, these bottom terms cannot be ignored, and the ratios will differ. For these data, where the baseline hazard varies from 0.05 to 0.25, the denominators cannot be ignored and hence the estimates of effect size differ. Given the similarity in goodness-of-fit, it is difficult to argue that one estimate is "right" and the other is "wrong." The real issue is really the choice of statistical model, a topic that we now address.

12.2.3 Choosing between Logit and Clog-Log Links for Discrete-Time Hazard Models

As the foregoing discussion illustrates, empirical evidence rarely indicates the superiority of one link function over another. The goodness-of-fit of models with competing link functions will differ only if hazard is relatively high during one or more time periods. Unless events are especially common, or discretization especially coarse, hazard will be low and fitted models similar.

How, then, should you choose between alternative link functions? In his landmark paper on survival analysis, Cox (1972) used a logit link to posit the discrete-time hazard model, and this approach enjoys widespread popularity today. Among its many advantages are its familiarity, ease of implementation, and ease of interpretation. These advantages were especially true when no software existed for fitting models specified with a clog-log link, but even now—with software abundant—the overwhelming popularity of logistic regression analysis for binary outcomes cannot be ignored.

If implementation issues are moot, you must base your decision on other issues. The primary advantage of the clog-log link is that in invoking a proportional hazards assumption it yields a direct analog to the continuous time hazard model. This was noted as early as 1978 by Prentice and Gloeckler, who demonstrated the internal inconsistency in Cox's (1972) original paper. If you believe that the underlying metric for time is truly continuous and that the only reason you observe discretized values is due to measurement difficulties, a model specified with a clog-log link has much to recommend it. This suggests that the clog-log link is most attractive when analyzing *interval-censored* data—data in which events unfold in continuous time, but your information about event occurrence is restricted to discrete-time intervals (see, e.g., Allison, 1995; Hosmer & Lemeshow, 2000).

If data are collected in truly discrete time, the clog-log specification has no particular advantage. As Beck (1999); Beck, Katz, and Tucker (1998); and Sueyoshi (1995) argue, the proportional hazards assumption is no more sacred than the proportional odds assumption, and while consistency across models is noble, so, too, is simplicity (which decreases the chances of mistake). After all, the data "don't know" whether they were generated by a proportional hazards or proportional odds model. And why do we need to invoke consistent assumptions in discrete and continuous time? When fitting linear and logistic regression models to noncensored outcomes, we routinely use different link functions: the normal link for the linear regression model and the logit link for the logistic regression model. Hence, for truly discrete-time data, the simpler logit link may suit eminently well.

12.3 Time-Varying Predictors

We introduced a variety of types of time-varying predictors in section 5.3 and discussed their advantages in the context of the multilevel model for change. As you would expect, similar advantages accrue when including time-varying predictors in the discrete-time hazard model. In their study of employee turnover, Harrison, Virick, and William (1996), for example, examined the effect of job performance using two internal time-varying predictors. The researchers found effects for both concurrent sales volume and *monthly rate of change* in sales volume, concluding that more productive employees and those whose sales were on an upward trajectory were less likely to leave. When studying precursors of seasonal affective disorder (SAD), Young et al. (1997) used an ancillary time-varying predictor—hours of sunshine—to examine the effect of exposure to natural light on disease onset. By exploring variation in time to SAD relapse among individuals living in different locations, the researchers found evidence to support the protective effects of sunlight.

Discrete-time survival analysis adapts naturally to the inclusion of time-varying predictors. Because models are fit using a person-period data set, a time-varying predictor simply takes on its appropriate value for each person in each period. This intuitive behavior allowed us to include a time-varying predictor in section 11.3.1 when we explained how to construct the person-period data set. Even though we were focusing then on time-invariant predictors, we included the time-varying predictor *DRUG*, whose values recorded the boy's reported level of drug use in each specific grade (see figure 11.5).

In this section, we demonstrate how to include time-varying predictors in the discrete-time hazard model. We begin, in section 12.3.1, by examining the assumptions implicit in a model that includes time-varying predictors. In section 12.3.2, we show how to fit the model to data and interpret results. We conclude, in section 12.3.3, with some words of caution about interpreting the effects of time-varying predictors. In addition to the issues of reciprocal causation raised in section 5.3.4, this model raises other complications that can compromise your ability to draw clear convincing conclusions.

12.3.1 Assumptions Underlying a Model with Time-Varying Predictors

Conceptually, you need no special strategies to include time-varying predictors in a discrete-time hazard model. The basic model (e.g., equation 11.3) allows for their inclusion because each variable has two subscripts—i indexing individuals and j indexing time periods. To simplify, we suppressed these subscripts in equation 11.4 and beyond. We now return to the original specification, retaining subscripts i and j when appropriate. For example, we can write a model with logit link including two predictors, one time-invariant (X_{1i}) and the other time-varying (X_{2ij}), as:

$$\text{logit } h(t_{ij}) = [\alpha_1 D_{1ij} + \alpha_2 D_{2ij} + \cdots + \alpha_J D_{Jij}] + [\beta_1 X_{1i} + \beta_2 X_{2ij}]. \tag{12.5}$$

The model stipulates that individual i's value of logit hazard in time period j depends upon his or her value of X_1, which is constant across all occasions, and his or her value of X_2 in time period j.

Given how straightforward this appears, you might ask why we devote a special section of our book to time-varying predictors in discrete-time hazard models. You might also recognize that we have already included some special time-varying predictors in all previous models: the predictors representing the "main effect of *TIME*." The values of these variables—be they dummies or polynomials—vary *within* people. The smooth specifications in table 12.1 even include the subscript j ($TIME_j$) so as to clarify this predictor's time-varying nature.

But most time-varying predictors are special, differing from those that represent the main effect of *TIME* and those that are time-invariant. Across rows in the person-period data set, these latter predictors are orderly, with values specifiable entirely in advance. Most time-varying predictors, in contrast, are not so orderly; they can take on *any* value in *any* period. This possibility—that a predictor can take on different values in different time periods—affects the way we conceptualize the model's assumptions.

We illustrate these ideas using data from Wheaton, Rozell, and Hall (1997), who examined the link between stressful life experiences and the risk of psychiatric disorder. Using a random sample of adults, ages 17 to 57, in metropolitan Toronto, the researchers conducted a structured interview that allowed them to determine whether, and if so at what age (in years), each individual first experienced a depressive episode. Among the 1393 respondents, 387 (27.8%) experienced a first onset between ages 4 and 39. Using the same interview, the researchers also ascertained whether, and if so at what age, each respondent first experienced 19 traumatic events, including major hospitalization, physical abuse, and parental divorce. Here, we focus on one of these stressors, first parental divorce, experienced by one-tenth of the sample ($n = 145$) at risk of an initial depressive episode. The time-varying predictor PD_{ij} indicates whether the parents of individual i divorced during, or before, time period j. In the time periods *before* the divorce, $PD_{ij} = 0$; in time periods coincident with, or subsequent to, the divorce, $PD_{ij} = 1$. Coding PD_{ij} in this way allows its effect to capture both the immediate and long-term impacts of parental divorce.

As in any discrete-time survival analysis, our first step is to select a specification for the main effect of *TIME*. Because this person-period data set is huge (36,997 records) and sparse (only 387 events occurred), a completely general specification—using 36 time dummies—performs poorly, yielding a baseline hazard function that fluctuates erratically in the later periods and is inestimable in the one early period (age 5) when no one reported initial onset. Exploration of the possible polynomial specifications of section 12.1 reveals that a cubic function of *TIME* fits nearly as well as the completely general specification ($\chi^2 = 34.51$, 32 *d.f.*, $p > .25$) and is measurably better than a quadratic ($\chi^2 = 5.83$, 1 *d.f.*, $p < .05$). We therefore postulate that:

$$\text{logit } h(t_{ij}) = \alpha_0 ONE + \alpha_1 (AGE_{ij} - 18) + \alpha_2 (AGE_{ij} - 18)^2$$
$$+ \alpha_3 (AGE_{ij} - 18)^3 + \beta_1 PD_{ij}. \tag{12.6}$$

Individual i's value of logit hazard in time period j is a function of his or her age and parental divorce status in that particular time period.

In specifying this model, we implicitly make three assumptions about the relationship between the risk of event occurrence and the time-varying predictor. These assumptions, given below, are restatements of the three assumptions presented in chapter 11 for models with a single time-invariant predictor: (1) for each value of the predictor, there is a postulated logit hazard function; (2) each of these functions has an identical shape; and (3) the distance between each of these functions is identical in every time period. When a model includes a time-varying predictor, we cannot say "for each value of the predictor" because the predictor may take on *different* values in *different* time periods. We therefore assume:

1. *For each value of the predictor in time period* j, *there is a postulated value of logit hazard.* Notice that we refer to the value of logit hazard for each value of the predictor in time period j. Interpretation of a time-varying predictor's effect must be tied to its values at each point in time.

2. *Joining consecutive postulated values of logit hazard for* constant *values of the time-varying predictor yields logit hazard functions with identical shapes.* The shape can be jagged or smooth, but the only hazard functions that have identical shapes are those that join postulated values of logit hazard corresponding to *constant* values of the time-varying predictor. Functions joining values of logit hazard corresponding to *varying* values of the time-varying predictor will not have identical shapes at all!

3. *The distance between each of these logit hazard functions is identical in every time period.* Although the values of the time-varying predictor *may* vary over time, its *effect* on logit hazard in each time period is constant. For time-invariant and time-varying predictors alike, we assume time-*invariant* effects on logit hazard.

These purposefully general statements apply equally to both time-varying and time-invariant predictors. For predictors of the latter type, these assumptions reduce to those in chapter 11. For models with several predictors of either type, simply modify these statements so that they describe the behavior of the postulated values of logit hazard at all possible *combinations* of values of the predictors.

Let's examine each assumption, beginning with the first, which says that to interpret a predictor's effect, you must tie all statements to the predictor's value in each time period. To explore this assumption, consider a second discrete-time hazard model that adds a time-invariant gender indicator, *FEMALE*:

$$\text{logit } h(t_{ij}) = \alpha_0 ONE + \alpha_1(AGE_{ij} - 18) + \alpha_2(AGE_{ij} - 18)^2$$
$$+ \alpha_3(AGE_{ij} - 18)^3 + \beta_1 PD_{ij} + \beta_2 FEMALE_i. \qquad (12.7)$$

Adding a time-invariant predictor allows us to contrast the comparisons that β_1 and β_2 assess. First consider β_2, which assesses the differential risk of initial depression onset (measured on a logit hazard scale) for women in comparison to men (controlling, of course, for parental divorce status). Because *FEMALE* is time-invariant, we need not tie our interpretation to the predictor's value in a time period—its value in *any* period is its value in *all* periods. β_1, in contrast, assesses the differential in risk among people who recently, or previously, experienced a parental divorce in comparison to those whose parents are still married (controlling for the effects of gender). Because *PD* is time-varying, its effect does not contrast static groups of people—men and women—but rather people who differ by unit values on the predictor at each point in time (here, respondents with divorced and married parents). Although the comparison's *name* remains the same—it contrasts individuals who have, and have not, experienced a parental divorce—the people who constitute the comparison in each time period differ.

By including a time-varying predictor we thereby compare *different* groups of people at different times. As individuals experience a parental divorce, they *switch* group membership (although for the version of *PD* we use here, each person can switch only once, and in only one direction). Not so with time-invariant predictors. In evaluating the effects of *FEMALE*, β_2 consistently compares the risk differential for the *same* groups people in every time period. Even though the number of men and women being compared diminishes over time as the risk set decreases (through event occurrence and censoring), no member of the risk set ever *switches* group membership: once a woman, always a woman.

Does it make sense to compare *different* groups of people at different times? This flexibility is what allows the model to capture the essence of a time-varying predictor's effect—as the predictors' values change, the groups compared change as well. To understand why this makes sense, consider the *sample* proportions of the risk set who experienced the target event (here, depression onset) in each time period (here, ages 4 through 39) as a function of each person's time-dependent status on the time-varying predictor (here, *PD*). Proportions like these are easily obtained by cross-tabulating three variables in the person-period data set—the period indicator (here, *AGE*), the time-varying predictor (here, *PD*), and the event indicator, here *DEPRESS*). Although you might think it wise to examine sample hazard functions for individuals who share *temporal patterns* of time-varying predictors, we rarely do so, for two reasons. First, the values of most time-varying predictors manifest so many possible temporal patterns that each unique group of individuals with the same profile is too small, yielding erratic estimates. Even a constrained dichotomy like *PD* manifests 37 possible patterns of value over time—one for each of the 36 ages when a parental divorce could occur, as well as a 37th pattern (consistent 0's) for those whose parents remained married. For a dichotomy that can swing freely between its two values, 2^J temporal patterns are possible, and for continuous predictors, the number of patterns usually exceeds the number of individuals under study! Second, the model does not postulate a relationship between predictor *patterns* and logit hazard. The model expresses individual's *i*'s value of logit hazard in time period *j* as a function of the values of the predictors—both time-varying and time-invariant—in *that* time period. Sample hazard functions for individuals who share particular temporal patterns may be interesting, but they are not what you model.

Figure 12.4 presents, by age, the proportions of the risk set that experienced an initial depressive episode as a function of their parental divorce status.[2] We plot the proportions on a raw scale in the top panel and a logit scale in the bottom panel, using •'s for those whose parents are married in the period (*PD* = 0) and +'s for individuals who experienced a parental divorce in this, or any previous, period (*PD* = 1). Over and above the effect of parental divorce, notice that the proportion experiencing an initial depressive episode is low during childhood, accelerates during the teens, peaks in the early twenties, and declines somewhat thereafter. Beyond this temporal pattern, the proportions are higher among those whose parents divorced (the +'s) than among those whose parents are still married (the •'s).

How does the hypothesized model reflect these patterns? To clarify the link between the model, its assumptions, and the sample data, we superimpose a hypothetical specification of equation 12.6 on each display. The curves join consecutive population values of (logit) hazard for constant values of the time-varying predictor. As postulated in assumption 2, this yields logit hazard functions with identical shapes (which, of course, differ in shape when examined on a hazard scale). The lower function indicates the hypothesized risk of initial depression onset for individuals with a constant value of *PD* = 0 over time, those who *never* experienced a

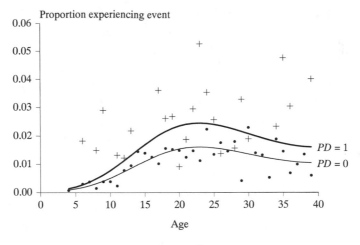

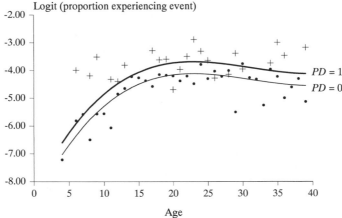

Figure 12.4. Understanding the link between sample data and the postulated model when examining the effects of time-varying predictors. Proportions of the risk set at each age who experienced an initial depressive episode at that age, as function of their parental divorce status at that age, expressed on a raw scale (top panel) and logit scale (bottom panel). •'s are for those whose parents remain married in the period; +'s are for individuals who experienced a parental divorce in this, or any previous, period.

parental divorce (until age 39, at least). The upper function indicates the hypothesized risk of initial depression onset for individuals with a constant value of $PD = 1$, those whose parents divorced when they were age 4 or younger. As postulated in assumption 3, the distance between these logit hazard functions is identical in every time period. In equation 12.6, the α's describe each function's identical shape and β_1 assesses the magnitude of the constant vertical distance between them.

Notice that we do not present explicit population (logit) hazard functions for individuals with *varying* values of *PD* over time. We do not do so because assumptions 2 and 3 refer to groups of individuals with *constant* values of the time-varying predictor. Yet the model in equation 12.6 allows us to hypothesize the existence of other population hazard functions resulting from all possible temporal patterns of value for the time-varying predictor, one for each *pattern*. Here, we hypothesize the existence of 35 additional population functions, one for each of the 35 ages (from 5 to 39) when parents could divorce.

What do these additional functions look like? Not surprisingly, they coincide with *portions* of the two functions shown in figure 12.4. As postulated in assumption 1, logit hazard takes on a distinct value for each value of the time-varying predictor in time-period *j*. Because *PD* is dichotomous, only two population values of (logit) hazard are possible in each period. In periods before a parental divorce, the values are those on the *bottom* function. In the year of a parental divorce, the values are those on the *upper* function. And because of the way we have coded *PD*—it indicates whether an individual's parents divorced at *this* or any *previous* age—people remain on the upper function from the time of parental divorce forward. For people whose parents divorced when they were 15, for example, they initially coincide with the lower curve, jump to the upper curve at age 15, and remain there in perpetuity. For people whose parent divorced when they were 32, they follow the lower curve for longer, until age 32 when they, too, jump to the upper curve (and stay there in perpetuity). And given the way we have postulated the model, we also assume that after age 32, the (logit) hazard functions for these two groups are indistinguishable from each other *and* from those for anyone who experienced a parental divorce at any earlier age.

This discussion demonstrates that population (logit) hazard functions for individuals with varying values of time-varying predictors will *not* share a common shape. Even in this example, in which every population (logit) hazard function begins low, jumps up, and remains at this second level in perpetuity, the timing of the jump, and hence the shape of the function, differs across individuals in the population. Time-varying predictors whose behavior is less constrained than *PD* can yield population (logit) hazard profiles with even more disparate shapes. To illustrate, consider that:

- If a time-varying predictor is continuous, and some members of the risk set have *increasing* values while others have *decreasing* values, the population (logit) hazard profiles may *cross*! The key to understanding this paradox is to remember that the assumptions of identical shapes and equidistance hold only for individuals who take on time-invariant values of the time-varying predictors. When the values of these predictors vary, these assumptions need not hold.

- If a time-varying predictor is dichotomous, but it can take on one of its values on only one occasion, one of the population (logit) hazard profiles, while easily drawn, is theoretically impossible! If we coded parental divorce as a short-term event, occurring only in the time period when the divorce occurs, a population (logit) hazard profile for individuals with a *constant* value of 1 cannot exist. Hypothetically, though, a function for this constant value can still be drawn.

The display of selected functions that are identical in shape and equidistant does not imply that *all* possible logit hazard functions share these properties. So, too, the ability to plot a particular function does not imply that the function describes the risk of event occurrence for a realistic group of people. To accurately interpret a time-varying predictor's effect, you must understand how the predictor actually varies with time in practice.

Finally, consider assumption 3: that the effect of the time-varying predictor is identical over time. We discuss this assumption briefly because it is identical to an assumption invoked in all discrete-time hazard models: In each time period, regardless of predictor type, unit differences in a predictor's values correspond to constant differences in logit hazard. In the model in equation 12.6, which compares people whose parents have divorced with those who have not, this differential is always β_1. The magnitude of this effect does not depend upon either: (1) the respondent's age; or (2) the respondent's age at parental divorce. In adopting this specification, we assume that the effect of parental divorce on the risk of initial depression onset is the same for a child, a teenager, or an adult. So, too, we assume that the effect of parental divorce is the same whether the divorce occurred when the respondent was a child, teenager, or adult. This assumption of time-invariant effects—for both time-varying and time-invariant predictors—is, of course, just an assumption. For now, we continue to cleave to its validity. In section 12.4, we examine its tenability and relax it when necessary.

12.3.2 Interpreting and Displaying Time-Varying Predictors' Effects

You can fit a discrete-time hazard model with time-varying predictors using exactly the same strategies presented in chapter 11. In the person-period data set, use a logistic regression routine to regress the event indicator on variables representing the main effect of *TIME* and the desired predictors. Hypotheses can be tested by comparing deviance statistics for nested models that include, and exclude, particular predictors. Fitting the model in equation 12.7 to the depression data, we find:

$$\operatorname{logit} \hat{h}(t_{ij}) = -4.5866ONE + 0.0596(AGE_{ij} - 18) - 0.0074(AGE_{ij} - 18)^2$$
$$+ 0.0002(AGE_{ij} - 18)^3 + 0.4151PD_{ij} + 0.5455FEMALE_i.$$

$$(12.8)$$

Comparing deviance statistics for the two models that individually exclude *PD* and *FEMALE* shows that each predictor is statistically significant controlling for the effect of the other (differences of 5.93 and 26.19, respectively, both exceeding the .05 critical value of 3.84 in a χ^2 distribution on 1 *d.f.*).

How do we interpret the magnitude of these effects? Antilogging the coefficient for *FEMALE* ($e^{0.5455} = 1.73$) we estimate that, controlling for the effect of parental divorce, the odds of initial depression onset are 73% higher for women. Antilogging the coefficient for *PD* ($e^{0.4151} = 1.51$), we estimate that, controlling for the effect of gender, at every age from 4 to 39, the odds of initial onset are about 50% higher for individuals who experienced a concurrent, or previous, parental divorce.

Notice how we modify our interpretation for time-varying predictors. For a time-invariant predictor like *FEMALE*, the increased risk of depression onset among women takes it toll in every time period. Interpretations need not be tied to any single time period or set of time periods. For a time-varying predictor like *PD*, the increased risk that results from parental divorce "kicks in" only in those time periods concurrent with, or subsequent to, the breakup. In every period from the divorce on, the odds of experiencing an initial depressive episode are one-half higher. In earlier periods, before the divorce occurs, these individuals—who are later at greater risk of depression—are not at any greater risk than peers whose parents remain married for years to come. Time-bounding the comparative statement reinforces the notion that a time-varying predictor's effect must be tied to its value in each time period. Equivalent time-bounding modifications are made for continuous time-varying predictors.[3]

Plots of fitted hazard and survivor functions for prototypical individuals provide another vehicle for conveying a time-varying predictor's effect. When offering strategies for selecting prototypical values for time-invariant predictors (see section 4.5.3), we suggested you select substantively interesting values that cover a suitable range of variation. Selection of prototypical values for time-varying predictors is far more complex because you must specify the predictor's values in not just one, but in all, time periods. We therefore suggest that you begin by selecting either:

- *Time-invariant values.* For the time-varying predictor *PD*, create two prototypical individuals: one who *never* experienced a parental divorce ($PD_j = 0$, for all *j* periods) and one who experienced a parental divorce during, or before, the first time period ($PD_j = 1$ for all *j* periods).
- *Time-varying values that correspond to substantively interesting patterns of temporal variation.* Among the 153 respondents who experienced a parental divorce before age 40—regardless of whether they experienced a depressive episode—the median (respondent) age at (parental) divorce was 10, with lower and upper quartiles of 4 and 16. This leads to the creation of four prototypes: one who never experienced a parental divorce (representing 90% of the sample) and three others who experienced a divorce at an early (4), middle (10), and later age (16).

When a time-varying predictor is dichotomous, like *PD*, these strategies are relatively straightforward. When a predictor is continuous, these strategies can become complex, as you must consider the possibility that the predictor will take on *many* different values in each time period. For now, we focus on dichotomies; at the end of this section, we adapt these strategies for the continuous case.

How can you decide between these approaches? When logically plausible, we recommend time-invariant values because the resultant fitted hazard functions reflect the assumptions of the model (in that, on a logit scale, they have an identical shape and are equidistant). For time-varying predictors whose values *cannot* remain constant over time, however, the resultant displays may be theoretically impossible and potentially misleading. In these situations, we recommend selecting values that reflect substantively interesting patterns of temporal variation, for at least the fitted functions will be plausible and meaningful. Unfortunately, these fitted hazard functions, even when plotted on a logit scale, will not be equidistant sharing an identical shape. We have seen this lead to misinterpretation and the misimpression that assumptions are violated. In the extreme, if values for one prototypical individual move in one direction (e.g., 0, 0, 0, 1, 1, 1) while those for a second move in the other (e.g., 1, 1, 1, 0, 0, 0), the fitted (logit) hazard functions may cross, creating the erroneous impression that the effect of the predictor *reverses* itself over time, when in fact, it remains constant. Regardless of your approach, remember that the decision to present fitted functions for a small

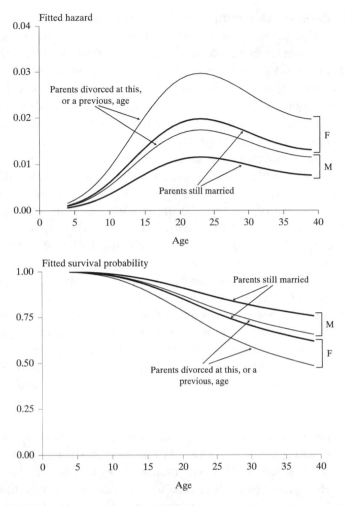

Figure 12.5. Displaying the results of fitting a discrete-time hazard model with a time-varying predictor. Fitted hazard and survivor functions for men and women who vary with respect to their parental divorce status. The darker functions are for individuals who did not experience a parental divorce before age 40; the fainter functions are for individuals who experienced an initial parental divorce at age 4 or earlier.

number of prototypical patterns does not imply that these are the only ones possible. As when presenting fitted hazard and survivor functions for continuous time-invariant predictors, the presentation of a few functions does not negate the existence of many more.

In figure 12.5, we present four fitted hazard functions, two for women and two for men. If we treat men who never experienced a parental divorce as a baseline group—they have the lowest hazard function here—

the display depicts the elevated risk of initial depression onset for men whose parents divorced before age 4 (those on the upper male hazard function) and women, regardless of whether and when they experienced a parental divorce. When describing this display, we would also emphasize the existence of many *implicit* fitted hazard functions—for men and women who experienced a parental divorce at any time between ages 4 and 39. Each of these groups would initially follow the *bottom* function for their gender (while their parents are still married) and would then jump to the *upper* function for their gender in the year of the parental divorce. For this type of predictor, then—a dichotomy whose values follow a unidirectional path—each pair of fitted hazard functions in figure 12.5 provides an *envelope* of all those possible—the bottom two provide an envelope for men; the top two provide an envelope for women. Focusing attention on just these two sets reduces clutter and highlights the most extreme contrasts possible.

Because survivor functions cumulate complementary values of hazard across successive time periods, plots of fitted survivor functions provide a different window on the effects of time-varying predictors. In the bottom panel of figure 12.5, we present fitted survivor functions—for men and women—that describe the two extremes for each group. Within gender, the top function is for a person whose parents remained married until age 39; the bottom is for a person who experienced a parental divorce by age 4. What about the fitted survivor functions for individuals who experienced a parental divorce between age 5 and 39? For each of the 35 additional fitted hazard functions not shown in the top panel there is a corresponding fitted survivor function not shown in the bottom panel. The two pairs of fitted survivor functions in figure 12.5 also provide an envelope for them, but *this* envelope behaves differently than that for fitted hazard functions. Because survivor functions *cumulate* the effects of hazard, the 37 fitted survivor functions display *distinct* values in all time periods subsequent to the parental divorce. As individuals experience a parental divorce, they do not jump from one fitted survivor function to the other (as they do with the fitted hazard functions). Fitted survivor functions remain coincident only until the (respondent) age of parental divorce. When parents divorce, the fitted survivor function drops, taking a distinct path shared only by those who also experienced a parental divorce at that age. Their fitted survivor function never joins the others (for this predictor) at any time in the future. In contrast, the fitted hazard functions for men who experienced a parental divorce at any earlier age all eventually converge.

What should you do if the time-varying predictor is continuous? Above all else, do not base your selection on the *observed* distribution of a predictor in each time period. To understand why, we describe a study we conducted of the career durations of 5100 newly hired teachers in North Carolina (Murnane, Singer, & Willett, 1989). Teachers' time-varying salary was a statistically significant predictor of their stay-or-leave decisions: those paid lower salaries (adjusted for inflation) were more likely to leave. When selecting prototypical values for display, we did not use the person-period data set to compute the average salary paid to the teachers *remaining* each year. Although easy to obtain, this information is not useful because the average salary paid to teachers in the data set in later years is *higher* simply because of the effect of salary on job duration: many of poorly paid teachers already left! If we used the "average" salary among remaining teachers to present results, the fitted values of hazard in the later years would be artificially low.

We therefore used an alternative approach, relying on information *external* to the data set. Using state-level data, we constructed an "average salary stream for North Carolina teachers." Selecting the first year of data collection as our base year, we computed the average salary paid to a first-year teacher in that year, the average salary for a second-year teacher in the following year, and so on. All salaries were adjusted for inflation (as the time-varying predictor had been). We then presented fitted hazard functions for teachers in the average salary stream and for two other prototypes: one consistently paid $2000 *less* and another consistently paid $2000 more. We chose these $2000 increments because they were (1) within the range of the salaries offered; (2) substantively interesting; and (3) easily communicated.

If you lack an external source for creating a prototypical stream of values, you have two options. One is to use the distribution of the predictor in the first time period before anyone experiences any events. You can then construct a prototypical stream by postulating systematic period-to-period changes from the initial values. A second option, available when event occurrence does not remove an individual from the overall *data* set (although it obviously removes them from the person-period data set for survival analysis), is to use the predictor's distribution in the full data set. This is precisely what we did when we described the distribution of respondent age at parental divorce earlier in this section. Although 145 of the 1393 respondents experienced a parental divorce while at risk of an initial depressive episode, another 8 experienced a subsequent parental divorce. The descriptive statistics about respondent age at parental divorce given earlier were based on the full set of $145 + 8 = 153$ individuals.

Some readers may perceive an irony in this discussion. We caution against selecting prototypical values based on the observed distribution of a predictor over time, yet we use these very data for model fitting. If the data are not "good enough" for the modest task of prototypical value selection, how can they be "good enough" for the important task of model fitting? The resolution of this conundrum lies in another assumption of the discrete-time hazard model: no unobserved heterogeneity. The model assumes that all heterogeneity in hazard across individuals is "observed," embodied in its predictor effects. While this assumption cannot be true—no model is perfect and all omit some important predictors—as a practical matter, we proceed as if it is. (We discuss this assumption further in section 12.6.) As the risk set decreases owing to event occurrence, its composition changes, changing the distribution of both time-varying *and time-invariant* predictors. The distribution of *any* predictor in later time period will differ from its distribution at the beginning of time. When selecting prototypical values of time-invariant predictors, we did not confront this issue because we examined these predictors' distribution only at the beginning of time. Were we to examine their distribution in later periods, they, too, would differ. Under the assumption of no unobserved heterogeneity, we postulate that, in any time period, those individuals in the risk set who share particular values of the predictors (whether time-varying or time-invariant) are representative of everyone in the population who would share these values at this point in time. This allows the changing risk set to form a valid basis for model fitting despite its unsuitability for prototypical value selection.

12.3.3 Two Caveats: The Problems of State and Rate Dependence

When we introduced the use of time-varying predictors in the multilevel model for change, we devoted an entire section (section 5.3.4) to the problem of reciprocal causation. When including time-varying predictors in a hazard model, not only do all of *those* problems apply, two new problems crop up: *state dependence* and *rate dependence.*

A time-varying predictor is *state-dependent* if its values at time t_j are affected by an individual's *state* (event occurrence status) at time t_j: $EVENT_{ij}$. A time-varying predictor is *rate-dependent* if its values at time t_j are affected by the individual's value of *hazard* (the "rate") at time t_j: $h(t_{ij})$. To illustrate this distinction, imagine studying the relationship between marital dissolution and two time-varying predictors: employment status and spousal satisfaction. Both predictors are likely negatively associated with the risk of divorce. But what is the connection between predictors and outcome? Yes it makes sense that employment is good for marital

stability (or at least that unemployment is bad) and certainly spousal satisfaction must also be a plus. Yet a model that links contemporaneous
values of these time-varying predictors with marital status cannot confirm
the direction of the link. Isn't it possible that employment is affected by
marital status, since married individuals are more likely to be working? If
so, your inferences are clouded by state dependence. And isn't it possible that spousal satisfaction is affected by someone's risk of divorce, since
marital instability must decrease satisfaction? If so, your inferences are
clouded by rate dependence.

Although defined and ancillary time-varying predictors generally
escape problems of rate and state dependence, contextual and internal
predictors do not. The core problem is that these types of predictors can
be potentially influenced either by an individual's event status at time t_j
or his hazard at time t_j. Given that contextual and internal time-varying
predictors are so important, what should you do?

As suggested in section 5.3.4, use theory as a guide, play your own
harshest critic, and determine whether your inferences are clouded by
these problems. Second, if your data allow, recode the time-varying predictors so that their coding in time period j reflect their values in *previous* time periods. As with the multilevel model for change, a hazard model
does not require contemporaneous coding. Most researchers use this
approach without thinking. Yet in some contexts, it might be more logical
to use *lagged predictors* to link *prior* predictor status with current outcome
status. Your hypothesized models may be more compelling and less prone
to inferential problems.

Lagged predictors are simple to construct and easy to interpret. In the
depression onset study, suppose we had wanted to study the link between
parental and child depression. We would create a time-varying predictor
indicating whether the parent had experienced a depressive episode at
each respondent age. Next, we would create a series of lagged predictors
whose values indicated whether the respondent's parent experienced a
depression in the previous year (a lag of one period), the year before that
(a lag of two periods) and perhaps even the year before that (a lag of
three periods). The first lagged predictor would assess the impact of
parental depression in the year after the episode, the second lagged predictor would assess the impact in the year after that, and so on. Lagged
predictors make it difficult to argue rate or state dependence because
an individual's depression status or risk of depression onset at any age
is unlikely to have affected his or her parental depression status one to
three years earlier.

The popularity of lagged predictors is increasing as researchers confront the inferential dilemmas of state and rate dependence. In their

study of the link between adolescent family structure and the risk of a premarital birth, Wu and Martinson (1993) lagged all time-varying predictors by one year. They argued that although the occurrence, or risk, of a premarital birth might increase the risk of parental transitions in a contemporaneous, or subsequent, year, it is difficult to imagine this could happen a full year earlier, before either the woman (or her parents) could know about the birth. Similarly, in their 12-week study of relapse among former cocaine, nicotine, and alcohol users, Hall, Havassy, and Wasserman (1990) lagged all time-varying predictors by one week. These researchers argued that it is the individual's status in the prior week that is the substantively interesting predictor of relapse in the next.

Lagging, of course, is no panacea. First, it may require imputation for the *first* time period because few researchers collect data that describe the values of predictors occurring before the first possible event occurrence. Although you could discard the first time period entirely and begin your analyses one period later, this wasteful approach can discard potentially important information. Second, lagged predictors can be less compelling theoretically. It would be a shame not to explore the contemporaneous link between parental and child depression onset, for example, even if the investigation is clouded by issues of state and rate dependence. Resolution of a *potential* interpretive dilemma may be too high a price to pay if the resulting analyses divert attention from more substantively interesting issues.

To illustrate, consider an extreme example of this last point. In recent years, researchers have begun to explore *anticipatory effects*, in which a predictor in a *later* time period may be associated with event occurrence in an *earlier* period. Although this may seem backwards, the idea is more plausible than it first appears. In the depression onset study, for example, Wheaton and colleagues (1997) actually hypothesized that the risk of depression onset begins to rise in the year or two *prior* to a parental divorce. Similarly, Yamaguchi and Kandel (1985) hypothesized that the risk of cocaine onset begins to drop in the year immediately preceding marriage. Both sets of researchers found empirical evidence to support these anticipatory effects by coding a time-varying predictor's values to reflect *future* status. To respond to the obvious criticisms of reciprocal causation, each argued that the value of the predictor (parental divorce or marriage), while reflecting status at a later point in time actually describes a longer *process* that began prior to the official transition. Neither contemporaneous nor lagged predictors would have uncovered this effect. Our parting words are simple: in variable construction, as in all things analytic, methodological advice must always be supplemented with common sense and substantive theory.

12.4 The Linear Additivity Assumption: Uncovering Violations and Simple Solutions

All the hazard models postulated so far have invoked a familiar, but restrictive, assumption: that unit differences in the value of a predictor (whether time-invariant or time-varying) correspond to fixed differences in logit hazard. This is known as the "linear additivity assumption" because we postulate that a predictor's effect does not depend upon: (1) the values of other predictors in the model (i.e., the effect is *additive*); or (2) the position of the unit difference along its scale (i.e., the effect is *linear*). When studying the relationship between salary and job duration, for example, the first assumption stipulates that a $1 difference has the same effect for everyone distinguished by other predictors in the model—for men and women, staff and management. The second assumption stipulates that this $1 difference has the same effect across the entire salary spectrum. This constrains the differential in risk for employees making $25,000 versus $25,001 to be identical to that for employees making $125,000 versus $125,001.

We expect that all readers are comfortable evaluating the tenability of these assumptions in other statistical models and making corrections when appropriate. Indeed, we did not even raise these issues when discussing the multilevel model for change as we expect you can figure out how to address these issues in that context. When fitting discrete-time hazard models, however, you do not have access to many of the individual level tools you typically use to evaluate these assumptions—plots of individual data, plots of residuals *vs.* predictors, and so on. Because the focus on hazard causes you to analyze group level summaries, model violations can be more difficult to discern. We therefore devote this section to introducing practical strategies for diagnosing and correcting violations of the linear additivity assumption.

12.4.1 Interactions Between Substantive Predictors

The gulf between substantive theory and empirical research is great when it comes to the study of statistical interactions. Almost every discussion of behavior—from psychology to psychiatry to political science to economics—suggests that predictors' effects may differ depending upon an individual's background, culture, and environment. Yet most of the models presented in substantive journals emphasize main effects. Perhaps this "main effects" bias is understandable. Most researchers collect data on *so* many predictors that it is difficult enough to examine all *main* effects, let alone the many pair-wise and higher order interactions. Add to this the fact that statistical power available for detecting interaction effects in any sample is lower than that available for detecting main effects, and exploration of interactions seems an unrewarding task.

We do not advocate fishing expeditions. Open searches for interactions can be counterproductive, leading to the discovery of many "effects" that result from little more than sampling variation. But there *are* at least two circumstances when a guided search for interactions is crucial:

- *When theory (or common sense!) suggests that two (or more) predictors will interact in the prediction of the outcome.* If you hypothesize the existence of interactions a priori, your search will be targeted and efficient.
- *When examining the effects of "question" predictor(s), variables whose effects you intend to emphasize in your report.* You need to be certain that these predictors' effects do not differ according to levels of other important predictors, lest you misrepresent your major findings.

With this in mind, we now demonstrate how to (1) explore your data for the possibility of statistical interactions; and (2) include the additional appropriate terms when necessary.

We illustrate these ideas using data from Keiley and Martin (2002), who examined the effect of child abuse (*ABUSED*) on the risk of first juvenile arrest. In a sample of 1553 adolescents—887 of whom were abused—342 were arrested between ages 8 and 18. Because *ABUSED* was their focal predictor, Keiley and Martin wanted to ensure that they did not erroneously conclude that its effect held equally for all individuals when, in fact, it might differ across groups. The researchers therefore investigated the interaction between *ABUSED* and all other predictors, finding that its effect differed by the youth's race.

How might you uncover a hint of a statistical interaction during exploratory analysis? Begin by recalling that, in some ways, interaction effects are like main effects—the only difference is that they indicate that the effect of one predictor is *conditioned* or *moderated* by levels of another. So, just as we explore main effects by examining within-group sample hazard functions, we explore interaction effects by examining within-group functions defined by *combinations* of predictors.

The top panel of figure 12.6 presents sample logit hazard functions computed separately for the two levels of the predictor *ABUSED*. The left side presents results for White youths (two-thirds of the sample); the right side presents results for Black youths. In examining plots like these, we initially ask the same two questions that we ask when examining any set of sample hazard functions: (1) What is the shape of the function for

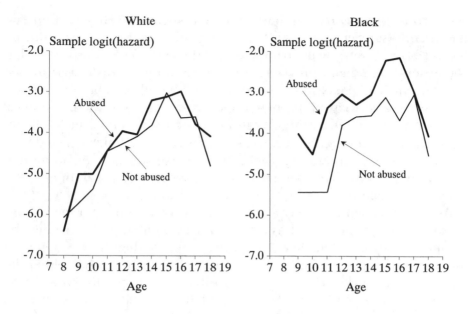

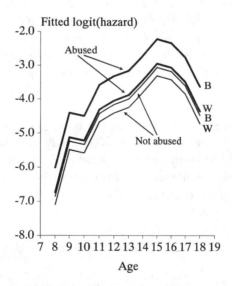

Figure 12.6. Uncovering statistical interactions between substantive predictors. The top panel presents *sample* (logit) hazard functions for age at first juvenile arrest for youths who have (dark lines) and have not (dashed lines) been abused, by race. The bottom panel presents *fitted* (logit) hazard functions for White and Black youths as a function of their abuse status.

each group? and (2) Does its level differ across groups? Here we find that the risk of first juvenile arrest is low during childhood, accelerates during the teen years, and peaks between ages 14 and 17. We also find that youngsters who have been abused are consistently at greater risk of first arrest than their peers who have not been abused. But because we are looking for symptoms of an interaction, we supplement these two questions with a third: Does the magnitude of the differential in level vary across groups? Among White youths, the risk differential associated with abuse is notable, but among Black youths, it is profound. A differential magnitude in risk like this suggests a possible statistical interaction between predictors.

As you would expect, we can test for an interaction by creating a cross-product term and adding it as a predictor to a model that includes the component main effects. By comparing the goodness-of-fit of nested models that include, and exclude, the cross-product, you test whether the interaction is statistically significant. Applying this approach to the juvenile arrest data, we fit two models: one that includes 11 time indicators and the main effects of *ABUSED* and *BLACK*; and a second that adds the cross-product *ABUSED* by *BLACK*. As the difference in deviance statistics between these models is 4.05 ($p < .05$, 1 *d.f.*), we conclude that there is a statistically significant interaction between *ABUSED* and *BLACK* in the prediction of the occurrence of juvenile arrest.

As in linear (or logistic) regression, we interpret interaction effects by simultaneously considering *all* the constituent parameters, for the cross-product term *and* its main-effect components. The parameter estimates for *ABUSED*, *BLACK*, and *ABUSED* × *BLACK* are 0.3600, 0.2455, and 0.4787, respectively. As both predictors are dichotomies, this leads to consideration of four prototypical individuals, who represent all possible combinations of *ABUSED* and *BLACK*:

Prototype	*ABUSED*	*BLACK*	Combined Parameter Estimates	Estimated Odds Ratio
White/not abused	0	0	$0 \times 0.3600 + 0 \times 0.2455$ $+ 0 \times 0.4787 = 0.0000$	1.00
White/ abused	1	0	$1 \times 0.3600 + 0 \times 0.2455$ $+ 0 \times 0.4787 = 0.3600$	1.43
Black/not abused	0	1	$0 \times 0.3600 + 1 \times 0.2455$ $+ 0 \times 0.4787 = 0.2455$	1.28
Black/ abused	1	1	$1 \times 0.3600 + 1 \times 0.2455$ $+ 1 \times 0.4787 = 1.0842$	2.96

The first prototype—a White youth who was not abused—is the "baseline group" for whom all substantive predictors are 0. The other prototypes represent different combinations of predictors. In the column labeled "Combined Parameter Estimates," we multiply each coefficient—for *ABUSED*, *BLACK*, and *ABUSED* × *BLACK*—by the prototypical values and total the result. Because both predictors here are 0/1 dichotomies, this calculation is simple. With continuous predictors, the prototypes will reflect other specific values but the procedure will be the same.

This table illustrates the statistical interaction between abuse and race. Among Whites, the odds of first juvenile arrest are 43% higher if the youth was abused. Among youths who have not been abused, the odds of first juvenile arrest are 28% higher for Blacks in comparison to Whites. But it is Black youths who have been abused who are at especially great risk of first juvenile arrest. In comparison to their white peers who were not abused, their odds of first arrest are nearly three times as high. This super-elevated risk is much greater than the sum of each of the separate risks associated with abuse or race. In other words, it is the *co-occurrence* of these circumstances—being a Black youth who has been abused—that augers particularly ill for life chances.

We can also examine fitted hazard (and survivor) functions for these prototypes. The bottom panel of figure 12.6 presents the fitted hazard functions (on a logit scale for consistency). This display reinforces the low risk of first arrest among both Black and White youths who have not been abused (as represented by the two dashed lines). Whites who have been abused face a somewhat elevated risk of arrest in comparison to both nonabused groups. But Black youths who have been abused are at much greatest risk. Their fitted (logit) hazard function is further away from the other functions than it would have been if each predictor had only been represented by a main effect.

12.4.2 Nonlinear Effects

Like their "main effects" bias, empirical researchers also display a "linear" bias, in that most fitted models reported in substantive journals incorporate predictors in a raw untransformed state. But why should predictors have only linear effects? Shouldn't a $1000 raise have a bigger effect for an employee at the low end of a salary scale than for an employee at the high end? Like the additivity assumption, the linearity assumption may indeed hold, yet it is always wise to review its tenability as well.

There are two general strategies for exploring the linearity assumption. The simplest approach—although somewhat blind—is to fit additional models, replacing the raw predictor with a re-expressed version. Although the additional models also invoke a linearity constraint, use of re-expressed predictors guarantees that the effects represent nonlinear relationships for the *raw* predictors. The ladder of power (section 6.2.1) provides a dizzying array of options. The second approach is to categorize each continuous variable into a small number of groups, create a series of dummy variables representing group membership, and visually examine the *pattern of parameter estimates for consecutive dummies* to deduce the appropriate functional form. If the pattern is linear, retain the predictor in its raw state; if not, explore an alternative specification.

As the first approach is straightforward, we illustrate the second, using the depression onset data presented in section 12.3. In addition to the effects of gender (*FEMALE*) and time-varying parental divorce (*PD*), we now examine the effect of a third predictor, number of siblings (*NSIBS*). Stress researchers hypothesize that children from larger families are less prone to depression, and if this is true, *NSIBS* should be negatively associated with risk.[4] To investigate this hypothesis, table 12.4 presents the results of fitting three discrete-time hazard models to these data, each using a different strategy for representing the sib-size effect.

Begin with Model A, which finds the expected negative effect of *NSIBS*. Inverting the estimate's antilog ($e^{-0.0814}$) for ease of interpretation yields 1.08. For every extra sibling, the estimated odds of *not* experiencing an initial depressive episode are 8% higher. This model constrains the effect of sib-size to be linear. If we use the model to compare two pairs of prototypical individuals—two from small families (with one or two siblings) and two from large families (with six or seven siblings)—the predicted differential in risk of initial depression onset associated with one additional sibling will be identical.

Might the protective benefit of siblings be unequal across the family size spectrum? With Model A, we cannot know because we constrain the sib-size effect to be linear. To investigate the possibility that its effect is not linear, Model B replaces *NSIBS* with a system of dummy variables. Creation of the system is straightforward, *once* you identify the categories to use. In identifying categories, we suggest you using a small number of groups (between four and eight) that are:

- *Equally spaced*, which helps you use the resultant parameter estimates to evaluate linearity.
- *Equally sized*, which prevents creation of very small groups with potentially unstable estimates.

Because *NSIBS* is highly skewed (ranging from 0 to 26(!), with a median of 2), we cannot optimize on both criteria. Were the sample small and

Table 12.4: Strategies for testing for nonlinear effects

	Model A	Model B	Model C
Parameter Estimates and Asymptotic Standard Errors			
ONE	−4.3587***	−4.5001*	−4.4828***
	(0.1216)	(0.2067)	(0.1087)
(AGE-18)	0.0611***	0.0615***	0.0614***
	(0.0117)	(0.0117)	(0.0117)
(AGE-18)^2	−0.0073***	−0.0073***	−0.0073***
	(0.0012)	(0.0012)	(0.0012)
(AGE-18)^3	0.0002*	0.0002*	0.0002*
	(0.0001)	(0.0001)	(0.0001)
PD$_j$	0.3726*	0.3727*	0.3710*
	(0.1624)	(0.1625)	(0.1623)
FEMALE	0.5587***	0.5596***	0.5581***
	(0.1095)	(0.1095)	(0.1095)
NSIBS	−0.0814***		
	(0.0223)		
1 OR 2 SIBS		0.0209	
		(0.1976)	
3 OR 4 SIBS		0.0108	
		(0.2100)	
5 OR 6 SIBS		−0.4942~	
		(0.2545)	
7 OR 8 SIBS		−0.7754*	
		(0.3437)	
9 OR MORE SIBS		−0.6685~	
		(0.3441)	
BIGFAMILY			−0.6108***
			(0.1446)
Goodness-of-fit			
Deviance	4124.29	4117.98	4118.78
n parameters	7	11	7
AIC	4138.29	4139.98	4132.78

~$p < .10$; * $p < .05$; ** $p < .01$; *** $p < .001$.
Results of fitting three discrete-time hazard models to the age at depression onset data ($n = 1393$, n *events* = 387). Each model includes the effect of time-varying parental divorce (PD), FEMALE, and family size, but the predictors used to represent family size differ. Model A uses the continuous predictor NSIBS; Model B categorizes NSIBS into five groups; and Model C splits NSIBS into just two groups using the dichotomy BIGFAMILY.

stability an issue, we might emphasize the latter criteria, but since the sample is large, we can emphasize the former. Thus, we create five predictors: *1 OR 2 SIBS* ($n = 672$), *3 OR 4 SIBS* ($n = 330$), *5 OR 6 SIBS* ($n = 159$), *7 OR 8 SIBS* ($n = 72$), and *9 OR MORE SIBS* ($n = 62$). The omitted "reference" category represents "only" children ($n = 98$).

We evaluate the linearity assumption by examining the pattern of parameter estimates (and accompanying standard errors) for the system of dummies. With equally spaced predictor categories, a linear effect will lead to successive estimates being "equidistant." The five coefficients in Model B (in light of their asymptotic standard errors) are *not* equidistant, suggesting nonlinearity. Instead, they fall into two broad groups: one for small- to mid-size families—*1 TO 2 SIBS* and *3 TO 4 SIBS*—and another for large families—*5 OR 6 SIBS, 7 OR 8 SIBS,* and *9 OR MORE SIBS.* Because only children are the omitted group, the minuscule and statistically nonsignificant coefficients for the first two predictors (0.0209 and 0.0108) indicate that we cannot distinguish between respondents with no siblings, one or two siblings, or three or four siblings. But *large* families *are* different. Each of the other three coefficients (−0.4942, −0.7754, and −0.6685) has a much lower risk. Comparing these coefficients to each other, however, we find few differences in magnitude, suggesting that they, too, may be combined into a second group.

Notice that we do not emphasize the goodness-of-fit of these models, nor do we suggest adopting Model B over A. As often happens when we replace a continuous predictor with a system of dummies, the difference in deviance (here, 6.31) is too small to justify the additional parameters (here four, $p > .20$). Yet the coefficients in Model B suggest that Model A may be misleading. The effect of family size is not smooth and continuous but rather jagged and disjunctive. This suggests the need for a new *dichotomous* predictor, one that contrasts small to mid-size families with large ones.

Model C uses this new sib-size predictor, *BIGFAMILY*, which contrasts respondents with five or more siblings (21% of the sample) and everyone else. Notice its large and statistically significant effect, which enables Model C to have a superior AIC statistic. Antilogging its coefficient yields 0.54, which indicates that the estimated odds of initial depression onset among individuals raised in large families are nearly half as large as those for peers (of the same gender and parental divorce status) raised in small to mid-size families. Large families confer a protective benefit, but the effect is *nonlinear*. Additional siblings in a small to mid-size family (with five or fewer children) confer no benefit, nor do additional siblings in large families (with six or more children). But there *is* a difference in risk for these two groups—those from large families face nearly half the risk of an initial depression onset.

The ability to investigate, and relax, the linearity assumption frees investigators from placing inappropriate constraints on predictors' effects. Sometimes you will use continuous specifications while other times you will use dichotomies (or polytomies). Given that we just moved from a continuous to a categorical variable, we conclude with a word of caution. We do not advocate routinely categorizing continuous variables using procedures like "median splits." Whenever possible, keep continuous predictors continuous. If your exploration of nonlinearity suggests few problems, retain the continuous specification. If nonlinearity can be ameliorated using a smooth polynomial, that, too, is preferable. (In this example, neither a quadratic nor cubic sib-size specification fit nearly as well as *BIGFAMILY*.) But there are times, like these, when an effect is simply *not* smooth, and in these circumstances, judicious categorization, even dichotomization, is better than using a model whose assumptions are not met.

12.5 The Proportionality Assumption: Uncovering Violations and Simple Solutions

All the discrete-time hazard models postulated so far invoke another common, but restrictive, assumption: that each predictor has an identical effect in every time period under study. This constraint, known as the *proportionality assumption*, stipulates that a predictor's effect does not depend on the respondent's duration in the initial state. In the depression onset study, for example, we assume that the effect of sib-size is identical in childhood, adolescence, and adulthood. In the teacher career study, we assume that the effects of salary are identical for novice and veteran educators alike.

Yet is it not possible, even likely, that the effects of some predictors will *vary* over time? The effect of family size might be more pronounced during childhood, when individuals live at home, than during adulthood, when day-to-day interaction with siblings is rare. Or the effect of salary might be more pronounced during teachers' initial years in the classroom, when the job is most difficult, rather than during their later years, when the comforts of tenure and allure of a pension may be reward enough. In these situations, and in many others, a time-varying effect seems not just plausible, but inevitable.

In this section, we show how to: (1) examine the tenability of the proportionality assumption; and (2) relax it, when appropriate, by including *interactions with time* (also known as *time-dependent* or *duration-dependent* effects). Our primary goal is to convince you that violations are an exciting analytic opportunity, not a methodological nuisance. After all, why *should* a predictor's effect be constant over time? Shouldn't the effects of some predictors depend on a respondent's duration in the initial state? Too many researchers remain tethered to their computer program's default options, which typically constrain predictors' effects to be time-invariant and hazard models to be proportional. Although this assumption may hold, blind faith is little guarantee. We encourage readers to embrace the methods presented in this section because detection of a violation is simple and correction is simpler still, usually leading to richer and more subtle conclusions.

12.5.1 Discrete-Time Hazard Models That Do Not Invoke a Proportionality Assumption

There are dozens of ways of violating the proportionality assumption. To provide a sense of the range of possibilities, let us begin with a simple discrete-time hazard model model with one time-invariant predictor:

$$\text{logit } h(t_{ij}) = [\alpha_1 D_{1ij} + \alpha_2 D_{2ij} + \cdots + \alpha_J D_{Jij}] + \beta_1 X_{1i}. \tag{12.9}$$

As written, one-unit differences in X_1 correspond to identical differences in logit hazard—β_1—in all J time periods. This implies that all population logit hazard profiles are equidistant. In Panel A of figure 12.7, we present a pair of population logit hazard functions that meets this assumption. (Recall that the term "proportionality assumption" stems from the fact that were we to plot these functions on an *odds* scale, the upper would be a *magnification* of the lower, making them proportional.)

Now consider the possibility that the effect of X_1 is not this consistent over time. In the remaining panels of figure 12.7, we present three of the many alternatives ways in which this might happen:

- *A predictor's effect may* increase *over time.* The two population logit hazard functions in Panel B are relatively indistinguishable initially but grow increasingly distinct over time. This means that the predictor is unimportant initially but becomes more important with age.
- *A predictor's effect may* decrease *over time.* The two population logit hazard profiles in Panel C are disparate initially but become increasingly similar as time goes on. This means that the predictor is important at the beginning of time but becomes less critical with age.
- *A predictor's effect may be* particularly pronounced *in some time periods.* The population logit hazard profiles in Panel D are distinct in some periods and nearly coincident in others. This means that the predictor is associated with event occurrence in some, but not all, time periods.

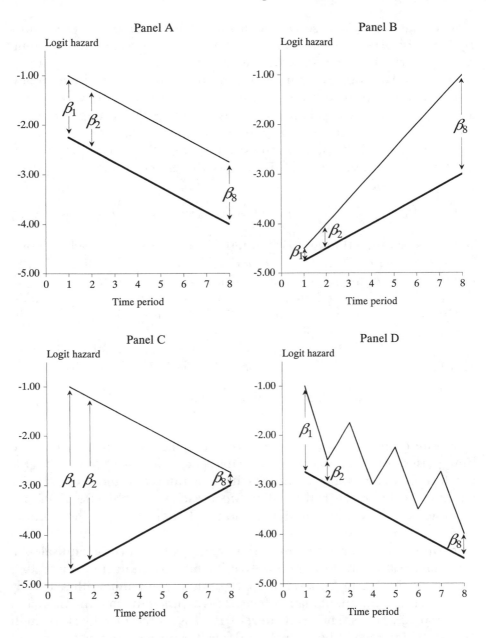

Figure 12.7. Population logit hazard functions that do, and do not, adhere to a proportionality assumption. A: a population that adheres to the proportionality assumption. B: a population in which the effect of the predictor *increases* over time. C: a population in which the effect of the predictor *decreases* over time. D: a population in which the effect of the predictor *differs erratically* over time.

Notice that each of these characterizations focuses on the predictor's *effect*, not its *values*. This is because both time-invariant and time-varying predictors can have time-varying effects.

What kind of population model is required to represent a time-varying effect adequately? To suggest a suitable model, we label the vertical separation between each pair of functions in figure 12.7 using a series of parameters, β, one per period: β_1 indicates the size of the vertical separation in period 1, β_2 indicates the size in period 2, etc. One population model that captures this behavior allows the effect of X_1 to differ from period to period:

$$\text{logit } h(t_{ij}) = [\alpha_1 D_{1ij} + \cdots + \alpha_J D_{Jij}] + \beta_1 X_{1i} D_{1ij} + \cdots + \beta_J X_{1i} D_{Jij}. \qquad (12.10)$$

Because each time dummy (D_1 through D_J) is 1 only once, each β represents the effect of X_1 in that time period. By allowing the effect of X_1 to differ over time, the population logit hazard profiles need not be equidistant. Were we to plot these functions on an odds scale, the upper would *not* be a magnification of the lower. We therefore say that the model in equation 12.10 is *nonproportional.*

One way of understanding the nonproportional model in equation 12.10 is to decompose it into its J period-by-period components:

$$\begin{aligned} &\text{In time period 1: logit } h(t_{i1}) = \alpha_1 + \beta_1 X_{1i} \\ &\text{In time period 2: logit } h(t_{i2}) = \alpha_2 + \beta_2 X_{1i} \text{ and so on.} \dots \end{aligned} \qquad (12.11)$$

The time dummies disappear because each takes on the value 1 in the time period indicated, 0 otherwise. This specification—which highlights the fact that each time period has its own intercept and its own shift parameter—helps clarify the link between each β and the effect of X_1: β_1 represents the effect of X_1 in time period 1, β_2 represents its effect in time period 2, and so on.

Another way of understanding this model is to notice its equivalence with statistical interaction models. We include an interaction by adding a cross-product term (or a set of cross-products) to a model that already includes the component main effects. This is precisely what the model in equation 12.10 does. The main effect of *TIME* is represented by D_1 through D_J and the interaction between X_1 and *TIME* is represented by terms $X_1 D_1$ through $X_J D_J$. Where is the main effect of X_1? To see that its presence is implicit, recall that we could parameterize the main effect of *TIME* by either: (1) eliminating one time indicator and including a stand-alone intercept; or (2) retaining all time indicators and excluding the intercept. In writing equation 12.10, we faced a similar choice: (1) eliminate one cross-product term and include the main effect of X_1 (which would create its explicit presence); or (2) retain all cross-product terms and exclude the main effect (which conceals its implied presence). For consistency, we adopted the second approach, which places the β's and the α's on equal interpretive footing (as can be seen in equation 12.11). Each β_j indicates the effect of X_1 in time period j, just as each α_j indicates the value of logit hazard when all predictors are 0.

The realization that a nonproportional hazards model can be conceptualized as a model that contains an interaction between a predictor and *TIME* has two important consequences. First, it yields a useful nomenclature: A *main effects hazards model* adheres to the proportionality assumption; an *interaction with time hazards model* does not. Second, it provides a simple strategy for testing the proportionality assumption: compare the difference in deviance statistics for these two models to a χ^2 distribution on $J - 1$ degrees of freedom. If the interaction with time model is sufficiently superior to the main effects model, we can justify the additional parameters, and we adopt the nonproportional model in lieu of the original model with the untenable assumption.

What happens if the improvement in fit is insufficient? Although failure to reject may indicate the assumption's viability, it may also indicate the use of an overly general interaction with time model. The specification of equation 12.10 is fine if the predictor's effect is erratic, but if it differs smoothly over time (as in Panels B and C of figure 12.7), it is wasteful, requiring too many parameters. We therefore routinely follow up this test in two ways:

- *By examining the* pattern *of estimates for the interaction with time terms.* If they systematically *increase* or *decrease*, this suggests a *smooth* interaction with time specification.
- *By explicitly testing the goodness-of-fit of more parsimonious specifications for the interaction with time.* Instead of postulating that the predictor's effect *differs* over time, postulate a specific functional form for the changing effect.

In both cases, we focus on linear and stage-wise patterns, unless theory points toward another alternative.

How can we specify a more parsimonious interaction with time model? One common option is to postulate that the effect of X_1 changes linearly over time:

$$\text{logit } h(t_{ij}) = [\alpha_1 D_{1ij} + \cdots + \alpha_J D_{Jij}] + \beta_1 X_{1i}$$
$$+ \beta_2 X_{1i}(TIME_j - c). \tag{12.12}$$

In this model, β_1 assesses the effect of X_1 in time period c and β_2 describes how this *effect* linearly increases (if β_2 is positive) or decreases (if β_2 is negative) across consecutive time periods. By comparing deviance statistics for this model to the main effects model in equation 12.9, we test the null hypothesis that the effect of X_1 does not differ linearly over time.

Another parsimonious specification postulates that the effect of X_1 differs across two (or more) broadly defined epochs. For example, we can create a dichotomous variable *LATE*, whose values indicate whether a time period occurs during the later (rather than earlier) epoch and write:

$$\text{logit } h(t_{ij}) = [\alpha_1 D_{1ij} + \cdots + \alpha_J D_{Jij}] + \beta_1 X_{1i} + \beta_2 X_{1i} LATE_{ij}. \quad (12.13)$$

This type of model is most useful when the J discrete time periods can be grouped into a small number of meaningful eras: pre-adolescence vs. adolescence, pre-tenure vs. post-tenure, high school vs. college.

Parsimonious nonproportional models have strengths and weaknesses. On the plus side, the use of just one additional predictor yields a more powerful test of proportionality. If the time-varying effect is smooth (as in equation 12.12) or phased (as in equation 12.13), the model in equation 12.10 is unnecessarily general, yielding a test whose power decreases as the number of time periods increases. On the down side, the increased power accrues only if the functional form hypothesized is correct. As this is unknowable, these models may be too constrained. It is for this reason that we always begin with the general model in equation 12.10. Doing so decreases the possibility of adhering to the proportionality assumption simply because the alternatives examined were too constrained.

12.5.2 Investigating the Proportionality Assumption in Practice

Having outlined the ways in which the proportionality assumption can be violated and corrected, we now show how to implement these strategies in practice. We do so using five waves of data provided by Graham (1997) who tracked the mathematics course-taking history of 3790 high school students from tenth grade through the third semester of college. Among the 1875 boys and 1915 girls who were enrolled in a mathematics class in 10th grade, only 93 men and 39 women took a mathematics class in each of the next five terms: 11th grade, 12th grade, and the first three semesters of college. Suspecting that the groups terminated their mathematics careers at different points in time, Graham assessed not only whether women were at greater risk of "leaving" mathematics but also whether the magnitude of the gender differential differed over time. A constant differential would imply a stable male/female discrepancy; a *varying* gender differential would imply that there are periods when the gender differential is especially large (or small).

The top left panel of figure 12.8 presents the within-gender sample (logit) hazard functions. For both males and females, the risk of terminating one's mathematics career zig-zags over time, peaking twice, in 12th grade and at the end of the second semester of college. Over and above this pattern, females are consistently at greater risk of termination in every term under study. Also notice that the magnitude of the gender differential *varies* over time: it is smallest in 11th grade and greatest in the third semester of college, taking on intermediate values in the terms in between. This last observation—unequal distances between sample logit hazard functions—suggests that the proportionality assumption may not hold.

Table 12.5 presents the results of fitting three models to these data: a main effects model (A), the general interaction-with-time model (B), and a linear interaction-with-time model (C). Fitted hazard functions for each—plotted on a logit scale—are presented in the remaining panels of figure 12.8. Model A, the main effects model, is the most parsimonious and poorest fitting. As shown in figure 12.8, this model constrains the main effect of *FEMALE* to be identical in each time period, yielding fitted hazard functions that are equidistant on a logit scale (and that would be magnifications and diminutions of each other when plotted on an odds scale). Exponentiating the coefficient for *FEMALE* ($e^{0.3786} = 1.46$), we estimate that the odds that a woman will end her mathematics career in any term are nearly 50% higher than are the odds for a man. By postulating that *FEMALE* has only a main effect, we do not allow the odds of termination to differ across terms. Instead, we constrain the odds to be identical in 11th grade, 12th grade, and each of the first three semesters of college, *whether they are or are not.*

Model B, the general interaction with time model, allows the gender differential to differ in each term. This model is so unconstrained that the differential could even reverse itself and be positive in one term and negative in the next. Although B fits better than A, a comparison of deviance statistics suggests that the improvement is insufficient to justify four extra parameters ($\chi^2 = 8.04$, 4 *d.f.*, $p > .05$).

Were we to end our investigation here, we would conclude that there was insufficient evidence to suggest a violation. But examination of the parameter estimates for the interaction terms in B suggests that this decision might be premature. Notice how the estimates rise over time, beginning at 0.1568 in 11th grade, rising to 0.4187 in 12th grade, and rising again in each of the first three semesters of college, peaking in the third semester at 0.6008. (When *J* is large, a plot of the parameter estimates

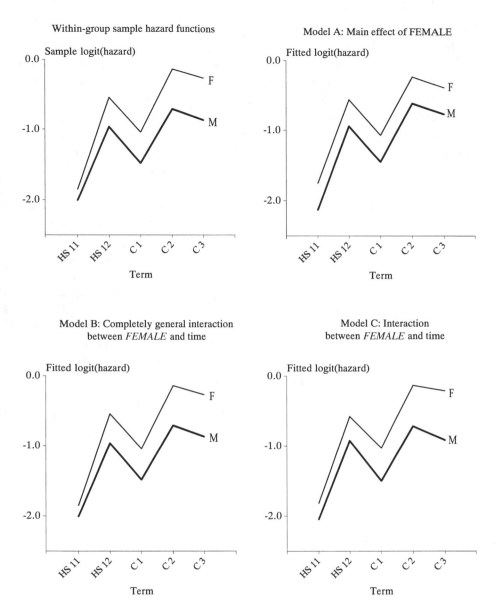

Figure 12.8. Sample hazard functions and alternative fitted discrete-time hazard models that invoke, and relax, the proportionality assumption. For the mathematics dropout data, the upper left panel presents sample hazard functions by gender, plotted on a logit scale. The remaining three panels present fitted models: A: the main effect of *FEMALE*. B: a completely general interaction between *FEMALE* and *TIME*. C: an interaction between *FEMALE* and linear *TIME*.

Table 12.5: Strategies for testing for interactions with time

	Model A	Model B	Model C
Parameter Estimates and Asymptotic Standard Errors			
HS11	−2.1308***	−2.0077***	−2.0459***
	(0.0567)	(0.0715)	(0.0646)
HS12	−0.9425***	−0.9643***	−0.9255***
	(0.0479)	(0.0585)	(0.0482)
COLL1	−1.4495***	−1.4824***	−1.4966***
	(0.0634)	(0.0847)	(0.0665)
COLL2	−0.6176***	−0.7100***	−0.7178***
	(0.0757)	(0.1007)	(0.0861)
COLL3	−0.7716***	−0.8690***	−0.9166***
	(0.1428)	(0.1908)	(0.1557)
FEMALE	0.3786***		0.2275**
	(0.0501)		(0.0774)
FEMALE × HS11		0.1568	
		(0.0978)	
FEMALE × HS12		0.4187***	
		(0.0792)	
FEMALE × COLL1		0.4407***	
		(0.1158)	
FEMALE × COLL2		0.5707***	
		(0.1445)	
FEMALE × COLL3		0.6008*	
		(0.2857)	
FEMALE × (TIME − 1)			0.1198*
			(0.0470)
Goodness-of-fit			
Deviance	9804.31	9796.27	9797.81
n parameters	6	10	7
AIC	9816.31	9816.27	9811.81

$\sim p < .10$; $* p < .05$; $** p < .01$; $*** p < .001$.

Results of fitting three discrete-time hazard models to the mathematics career data ($n = 3790$, n events $= 3684$). Model A includes the main effect of FEMALE; Model B includes the completely general interaction between FEMALE and TIME; Model C allows the effect of FEMALE to vary linearly over TIME.

over time can be helpful in detecting patterns.) The coefficient for FEMALE in Model A (0.3786) is a weighted average of these estimates. Their systematic increase in Model B points toward a different nonproportional model in which the effect of FEMALE is allowed to interact with linear TIME.

This is precisely what is done in Model C, which includes the main effects of TIME and FEMALE, as well as their interaction. Comparing its deviance statistic to Model A, we find a difference of 6.50 on 1 *d.f.* ($p =$ 0.0108). We therefore *reject* the proportionality assumption and conclude that the interaction with linear time model is preferable to A. Is Model C just as good as B? The deviance statistics for these two models differ by only 1.54 (3 *d.f.*, $p > .25$). We therefore conclude that a parsimonious specification fits no worse than a general one.

How do we interpret the gender differential implied by Model C? Because we have centered *TIME* at 1, the coefficient for *FEMALE* (0.2275) estimates the differential in time period 1, which here is 11th grade. Antilogging yields 1.26, which leads us to estimate that in 11th grade, the odds of ending one's mathematics course-taking career are 26% higher for females. Dividing the coefficient by its asymptotic standard error (0.0774) yields a test of the null hypothesis of no gender differential in 11th grade, which we reject ($p = 0.0033$). The coefficient for the *FEMALE* by (*TIME* − 1) interaction assesses the period-by-period increment in risk as students progress through school. In 12th grade, the estimated odds ratio is $e^{(0.2275+0.1198)} = 1.42$; in the first semester of college, it rises to $e^{(0.2275+2(0.1198))} = 1.60$ and by the third semester of college, it rises further to $e^{(0.2275+4(0.1198))} = 2.03$. These results confirm the unsuitability of the main effects model (Model A), which constrains these odds ratios to be constant at 1.46. During high school, the gender differential is smaller than this; during college, it is greater. This growing gap is clearly illustrated in the fitted hazard functions for Model C shown in figure 12.8.

Before leaving this example, reexamine the fitted functions for Model B shown in figure 12.8. Because this model allows the effect of *FEMALE* to differ completely over time, the distance between functions differs entirely across periods. Now compare these *fitted functions* with the sample functions directly above. It is no accident that these two plots are identical. Just as fitted hazard functions from a model with no substantive predictors reproduce sample hazard functions, fitted hazard functions from a general interaction with time model reproduce within-group sample hazard functions (for categorical predictors). This correspondence provides another way of understanding what we do when we examine the tenability of the proportionality assumption. The general specification does not "smooth" the sample functions in any way. The main effects model "smoothes" the functions until they are equidistant (in logit hazard space). The parsimonious specifications take a middle stance, smoothing the functions to an intermediate degree. From this perspective, in examining the proportionality assumption, we assess whether, and how, the sample hazard functions for each predictor's effect should be smoothed over time.

12.6 The No Unobserved Heterogeneity Assumption: No Simple Solution

All the hazard models discussed in this book—both the discrete-time models we are discussing now and the continuous-time models we will soon introduce—impose an additional assumption to which we have alluded: the assumption of no unobserved heterogeneity. Every model assumes that the population hazard function for individual i depends only on his or her predictor values. Any pair of individuals who share identical predictor profiles will have identical population functions. (We note that this is not the first time you have invoked an assumption like this. You routinely invoke a similar assumption when you fit any logistic regression model, even though it is no more likely to hold!)

Many data sets will not conform to this assumption. As in the multi-level model for change (and regular regression for that matter), pairs of individuals who share predictor profiles are very likely to have different outcomes. Even the most prescient among us will be unable to identify *every* predictor associated with each person's population hazard function. If our model omits one or more important predictors, we have *unobserved heterogeneity*.

Unobserved heterogeneity can have serious consequences. In their classic 1985 paper, Vaupel and Yaskin elegantly demonstrate what they call "heterogeneity's ruses"—the ability of unobserved heterogeneity to create the misimpression that a hazard function follows a particular form, when in fact it may not. Although the types of ruses can be elaborate and complex, we can illustrate the basic problem using the special educator career data introduced in chapters 9 and 10. As shown in figure 10.1, the sample hazard function declines steadily over time. In describing this function in section 10.2.1, we wrote: "Novice special educators, or those with only a few years of experience, are at greatest risk of leaving teaching. Once they gain experience (or perhaps, tenure) the risk of leaving declines."

Because of the possible existence of unobserved heterogeneity, this simple compelling conclusion may be wrong! To illustrate how this could happen, consider the following thought experiment. Suppose that our initial sample includes equal proportions of three different types of special educators, those at:

- *High risk*—they may have not wanted to be teachers in the first place, but entered the profession because there were job opportunities.
- *Medium risk*—they wanted to work, but their commitment to

teaching is lower. As a result, they are consistently susceptible to exploring other options (e.g., taking another job, returning to school, or quitting to raise a family).

- *Low risk*—they would like to teach until retirement.

Further suppose that for each group, the risk of leaving teaching is *constant over time*. This means that the population hazard function for each teacher is *flat*—its level reflects his or her consistent level of risk. The only difference between teachers is the level of their hazard function: high, medium, or low.

What would the *overall hazard function* for this heterogeneous group of teachers look like if we ignored the fact that they came from three disparate groups? (This is what we do if we omit the (admittedly unobserved) predictor that distinguishes the three groups.) Immediately after hire, the hazard function would fall in the middle of the three levels of risk. Over time, teachers in the high-risk group would be most likely to leave, those in the medium-risk group would be moderately likely to leave, while those in the low-risk group would be likely to stay. Because of these differential levels of risk, the composition of the risk set changes substantially over time—as high-risk teachers leave, the risk set becomes composed of greater fractions of medium and especially low-risk teachers. The net result is that the overall hazard function—evaluated across the entire set of teachers—will decline over time, owing to nothing more than the changing composition of the risk set.

Unobserved heterogeneity raises a serious interpretive problem. If it exists—and you have no way of knowing with certainty whether it does—the observed pattern of risk may not reflect the true pattern of risk. Fortunately, the effect of unobserved heterogeneity is itself consistent: it always leads to hazard functions that appear to decline over time. As a result, if you find sample hazard functions that *increase over time*, as did those for the age at first intercourse data in chapter 11 and the tenure and depression onset data in this chapter, you are probably safe. On the other hand, if you find sample hazard functions that *decrease over time*—a common pattern when studying employee turnover, recidivism, and other events in which individuals face a high initial risk—interpretation can be ambiguous. A declining hazard function may suggest exactly what we intimated in chapter 10 for the special educator data *or* it may suggest a data set plagued by unobserved heterogeneity (Follmann & Goldberg, 1988).

Is it possible to fit a hazard model that accounts for unobserved heterogeneity? As you might expect, doing so requires that we have either additional data (for example, data on repeated events within individuals) or that we invoke other—perhaps less tenable—assumptions about the distribution of event times or errors (Aalen, 1988; Heckman and Singer, 1984; Vaupel, Manton, & Stallard, 1979; Scheike & Jensen, 1997; Mare, 1994). As a result, most empirical researchers—and we—proceed ahead, if not ignoring the problem, at least not addressing it. In the remainder of this book, we assume that all heterogeneity is observed and attributable to the predictors included in our models.

How problematic is it to turn a blind eye? Xue and Brookmeyer (1997) demonstrate that, even in the face of unobserved heterogeneity, we can interpret parameters in a discrete-time hazard model as *population averaged* coefficients, which describe behavior in a cohort (instead of *subject-specific* coefficients, which describe what happens to an individual over time). Gail, Wieand, and Piantadosi (1984) also suggest that the tests for model parameters remain valid. Of course, if you discover a hazard function that declines steadily over time, you should at least consider whether your inferences about the underlying causes of that decline are credible or whether they are undermined by the possibility of unobserved heterogeneity. If you would like to go further than this and investigate the effects of unobserved heterogeneity, we suggest that you begin with the references provided in the previous paragraphs.

12.7 Residual Analysis

In this chapter, we have examined the tenability of many assumptions inherent in the discrete-time hazard model and provided strategies for extending the model when these assumptions are unmet. But before concluding that your model is sound, you should ascertain how well it performs for individual cases. As in regular regression, we address this question by examining residuals.

Residuals compare—usually through subtraction—an outcome's "observed" value to its model-based "expected" value. For a discrete-time hazard model, a simple difference will not suffice because each person has not a single outcome but a set of outcomes—one for each time period when he or she was at risk. This suggests the need for a residual defined at the *person-period* level. A further complication is that the observed outcome in every time period has a value of either 0 or 1 while its expected value—the predicted hazard probability—lies between these extremes. By definition, the model underpredicts in any person-period when the event occurs and overpredicts in all others. This makes it perhaps unwise to define a residual, even at the person-period level, by the subtraction of observed and expected values.

Statisticians have developed many kinds of residuals for assessing the prediction of dichotomous outcomes (Hosmer & Lemeshow, 2000; Collett, 1991), some of which can be adapted for use in discrete-time hazard modeling. Because a full discussion of this topic is beyond scope of this book, here we focus on one type of residual—the *deviance residual*—that we find especially useful.

For individual i in time period j, the deviance residual is defined as:

$$DEV_{ij} = sign(EVENT_{ij} - \hat{h}_{ij})$$
$$\times \sqrt{-2[EVENT_{ij} \log(\hat{h}_{ij}) + (1 - EVENT_{ij}) \log(1 - \hat{h}_{ij})]}, \quad (12.14)$$

where "*sign*( )" indicates that the residual takes the sign (negative or positive) of the argument in parentheses. Deviance residuals are so named because, when squared, they represent an individual's contribution to the deviance statistic for that time period. The sum of the squared deviance residuals across all the records in a person-period data set yields the deviance statistic for the specified model.[5] Most software programs that fit logistic regression models can output deviance residuals to the person-period data set.

The length of equation 12.14 might suggest that deviance residuals are a complex function of an individual's event occurrence status and fitted hazard probability. But because the event indicator must be either 0 or 1, only one term under the square root can be non-zero, which allows us to write:

$$DEV_{ij} = \begin{cases} -\sqrt{-2\log(1 - \hat{h}_{ij})}, & \text{if } EVENT_{ij} = 0 \\ \sqrt{-2\log(\hat{h}_{ij})}, & \text{if } EVENT_{ij} = 1. \end{cases} \quad (12.15)$$

The absolute value of a deviance residual indicates how well the model fits that person's data for that period. Large absolute values identify person-period records whose outcomes are poorly predicted. The sign of a deviance residual is determined solely by event occurrence: if individual i experiences the event in that period, the residual is positive; otherwise it is negative. As a result, most deviance residuals are negative, except for those in time periods when noncensored individuals experience the target event.

We illustrate how to examine deviance residuals using the grade of first sexual intercourse data analyzed in chapter 11. In Model D of table 11.3, we expressed the risk of sexual initiation as a function of *TIME* and two predictors: parental transitions (*PT*) and parent antisocial score (*PAS*). Table 12.6 presents deviance residuals from this fitted model for eight boys. Each has as many residuals as he has person-period records. Subject

Table 12.6: Deviance residuals from Model D of table 11.3 for 8 boys in the grade at first intercourse data set

| | | | | | | | Deviance residuals | | | | | SS |
ID	PT	PAS	GRADE	CENSOR	GR7	GR8	GR9	GR10	GR11	GR12	deviance
22	1	-.6496	12	0	-.4117	-.2944	-.5840	-.7176	-.7748	1.4145	3.7133
112	1	-.6609	12	1	-.4111	-.2940	-.5831	-.7166	-.7737	-.9563	2.6220
166	1	2.7814	11	0	-.6615	-.4807	-.9108	-1.0903	1.1914	—	4.1064
89	0	-.0752	11	0	-.3248	-.2314	-.4645	-.5752	1.8624	—	4.1740
102	1	.6049	8	0	-.4913	2.3695	—	—	—	—	5.8558
87	1	2.6779	7	0	1.8176	—	—	—	—	—	3.3038
67	1	2.2747	12	0	-.6180	-.4477	-.8559	-1.0294	-1.1007	1.0430	4.6741
212	0	-.9618	12	1	-.2857	-.2032	-.4098	-.5090	-.5524	-.6958	1.3393

465

22, who had intercourse for the first time in 12th grade, has six residuals, one for each of grades 7 through 12. Boys 112 and 212 also have six residuals because they, too, were followed through 12th grade (when they were censored). The other boys in the table have fewer residuals because each initiated intercourse in an earlier grade. As expected, deviance residuals are negative in all the time periods before event occurrence and remain negative in the final period for boys who have not yet had sex. The deviance residual is positive only if, and when, the event occurs.

Unlike regular residuals, with deviance residuals we do not examine their distribution nor plot them versus predicted outcome or observed predictor values. We eschew such displays because they will manifest uninformative deterministic patterns. Instead, we examine deviance residuals on a case-by-case basis, generally through the use of *index plots*—sequential plots by *ID*. The top panel of figure 12.9 presents an index plot of the deviance residuals for Model D. Most of the residuals are negative, except for those records in which the target event occurs. When examining index plots like these, look for *extreme* observations—person-period records with extraordinarily large residuals. Large residuals suggest person-period records with poor model fit. Although this plot includes five extreme residuals—identified using *ID* numbers—none is so far away that it seems especially problematic.

Index plots of deviance residuals can be unsatisfying because they do not *jointly* consider the quality of model fit across each person's multiple records. One way of addressing this concern is to *aggregate* each person's residuals into a single summary. Rather than summing their (absolute) values, we square and then sum. Although this strategy has no formal statistical justification, it has the salutary effect of yielding an individual-level summary of model fit measured in a meaningful metric: the individual's contribution to the deviance statistic.

We display the sum of squared deviance residuals for each individual in Model D in the final column of table 12.6 and the bottom panel of figure 12.9. If we sum these values across all 180 boys, they would total 629.15, the deviance statistic for the model. Notice that although both plots in figure 12.9 identify the same problematic cases, the cases are more prominent in the bottom panel. Collapsing across each person's multiple records is a powerful tool for identifying individuals whose outcomes are poorly predicted. In this data set, for example, examination of the records for the five identified cases reveals that these boys initiated intercourse very early (grades 7 or 8) despite their low risk (attributable to their parental transition status and/or antisocial score). Indeed, if we delve into the data for the 20 boys with the highest sum of squared deviance residuals, we find that 18 (90.0%) had initiated intercourse in

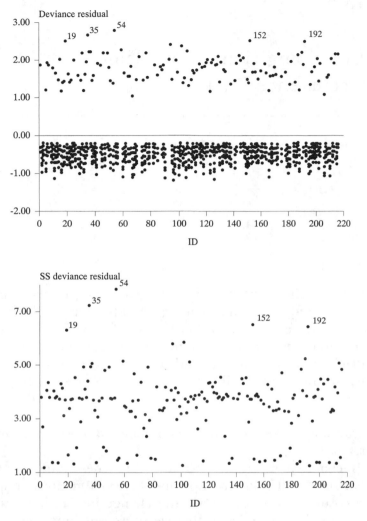

Figure 12.9. Index plots of deviance residuals for the grade at first intercourse data. The top panel presents residuals computed at the person-period level. The bottom panel presents residuals aggregated to the person level.

ninth grade or earlier. Among the remaining 160, in contrast, only 27 (16.9%) had done so. In this way, examination of the sum of the squared deviance residuals suggests that Model D may not be very effective at predicting early initiation of sexual intercourse. This suggests that it would be prudent to search for additional variables that might be better predictors of early initiation risk.

13

Describing Continuous-Time Event Occurrence Data

Time has no divisions to mark its passing.

—Thomas Mann

Researchers can often record event occurrence using a fine-grained time metric. When studying relapse among recently treated alcoholics, for example, Cooney and colleagues (1991) used life history calendars to determine—to the *nearest day*—whether and, if so, when former drinkers began drinking again. When studying the reactions of frustrated motorists, Diekmann and colleagues (1996) used on-site observation to determine—to the *nearest hundredth of a second*—whether and, if so, when drivers blocked at a green light started to honk their car horns. We say that measurements like these have been obtained in what we think of as *continuous time.*

In this chapter, we present strategies for describing continuous-time event data. Although survivor and hazard functions continue to form the cornerstone of our work, the change in the time scale from discrete to continuous demands that we revise our fundamental definitions and modify estimation strategies. In section 13.1, we identify salient properties of continuous-time data and redefine the survivor and hazard functions as required. In section 13.2, we estimate these functions using a pair of simple strategies—called the *discrete-time* and *actuarial* methods—that require the continuous event times to be grouped into intervals. In section 13.3, we introduce a superior approach—the Kaplan-Meier method—that does not require the artificial grouping of data but that yields estimates of only the survivor, not the hazard, function.

In the remainder of the chapter, we offer solutions to the core conundrum embedded in continuous-time event data: our inability to estimate the hazard function well. This conundrum is a concern as it leads some

468

researchers to conclude that they should not even *try* to ascertain the pattern of risk over time. In section 13.4, we offer one possible solution by introducing a close mathematical relative of the hazard function that *is* easy to estimate in continuous time: the *cumulative* hazard function. As its name implies, this function "cumulates" hazard over time, assessing the total amount of risk faced by participants between the beginning of time and each observed event time. In section 13.5, we offer a second solution that uses the cumulative hazard function to generate *approximate* values of the hazard function by a method known as *kernel smoothing*. We conclude, in section 13.6, by applying these new strategies to data from four empirical studies. This sets the stage for chapter 14, in which we specify and fit statistical models for testing the effects of predictors on event occurrence in continuous time.

13.1 A Framework for Characterizing the Distribution of Continuous-Time Event Data

Variables measured with greater precision contain more information than those measured with less precision. Ordinal scales are more inform-ative than nominal scales; interval scales are better still. These familiar principles apply equally in the study of event occurrence. When possi-ble—and appropriate, given the way in which events unfold over time—we recommend that you record event occurrence using the finest metric possible. Finer distinctions, as long as they can be made reliably, lead to more subtle interpretations and more powerful analyses.

Unfortunately, a switch from discrete- to continuous-time survival analysis is not as trivial as you might hope. In discrete time, the defini-tion of the hazard function is intuitive, its values are easily estimated, and simple graphic displays can illuminate its behavior. In continuous time, although the survivor function is easily defined and estimated, the hazard function is not. As explained below, we must revise its definition and develop new methods for its estimation and exploration.

13.1.1 Salient Features of Continuous-Time Event Occurrence Data

We begin by contemplating what it means to assess event occurrence using a truly continuous metric. For the moment, set aside the realities of data collection and imagine that, for every noncensored individual, you could identify the precise *instant* when the target event occurs: Joseph Wright was arrested at 9:20 A.M on the 32nd day after his release from prison. Jane Kendall took her first drink at 6:19 P.M on the 129th day after her discharge from an alcohol treatment program.

What would the distribution of a sample of such event times look like? We find it helpful to visualize a time line with explicit demarcations that denote every possible instant when the event could occur. Our initial demarcations might be weeks, days, hours, or even minutes, but if time is truly continuous, we need not stop there. Each minute can be divided into seconds and each second can be subdivided into 10ths, 100ths, and 1000ths of seconds. Each 1000th of a second can be divided into nanoseconds and each of these can be subdivided further still. Ultimately, we must conclude that there exist an *infinite* number of possible instants when the target event can occur. Every data collection period—finite though it may be—contains an *infinite* number of such instants.

Because continuous time is infinitely divisible, the distribution of event times displays two highly salient properties:

- *The probability of observing* any particular event time *is infinitesimally small.* In continuous time, the probability that an event will occur *at any specific instant* approaches 0. The probability may not reach 0, but as time's divisions become finer and finer, it becomes smaller and smaller.
- *The probability that two or more individuals will share the same event time is also infinitesimally small.* If the probability of event occurrence at each instant is infinitesimally small, the probability of co-occurrence (a "tie") must be smaller still.

These two properties stand in stark contrast with what we have come to expect in discrete time. In the latter case, if events can occur only in a small number of time periods, and some events *do* occur, we expect the probability of event occurrence in at least some periods to be non-zero and ties to be pervasive.

Distributional properties like these can be inferred only by invoking statistical theory. We cannot illustrate them with data because they describe expectations in the limit, not sample behavior. Real data require measurement and measurement lead to rounding—into weeks, days, or some other unit. Rounding alters these expectations because it increases the probability of observing each rounded event time as well as the probability of observing a tie. In the extreme, if we round continuous event times using a very coarse metric, we return to the world of discrete time! The important feature of continuous time is that we can invoke the assumption—that may, or may not, hold in practice—that these distributional properties hold because time is infinitely divisible.

Table 13.1: Known and censored (*) event times for 57 motorists blocked by another automobile (reaction times are recorded to the nearest hundredth of a second)

1.41	2.12	2.54	2.83	3.14	3.56	4.18	4.71*	6.03	12.29
1.41*	2.19	2.56	2.88	3.17	3.57	4.30*	4.96	6.21*	13.18
1.51	2.36*	2.62	2.89	3.21	3.58	4.44	5.12*	6.30	17.15*
1.67	2.48	2.68	2.92	3.22	3.78	4.51	5.39	6.60*	
1.68	2.50	2.76*	2.98	3.24	4.01*	4.52	5.73	7.20	
1.86	2.53	2.78*	3.05*	3.46*	4.10	4.63*	5.88*	9.59	

To concretize these inherently unobservable ideas, examine table 13.1, which presents the data from Diekmann and colleagues' (1996) horn-honking study. At a busy intersection in Munich, Germany, the researchers recorded the reaction times—to the nearest hundredth of a second—of 57 motorists purposefully blocked at a green light by a Volkswagen Jetta. Forty-three motorists honked their horns; 14 did not. An asterisk (*) identifies the censored event times.

Because time was recorded using such a fine metric, the distribution of event times in table 13.1 almost manifests both of the distributional properties outlined above. The probability of observing any *specific* event time must be infinitesimally small because we observe only 56 of the infinite number of distinct times when an event could occur. There is only one tie, at 1.41, the earliest recorded event time. Although this tie may have resulted from a measurement "floor"—the recording device may not have been able to assess a honk before 1.41 seconds—we need not invoke this argument to rationalize the tie's appearance. Real data sets will include ties even if theory states that events cannot co-occur. The explanation is that had a finer metric been used—say, 1000ths of seconds—we could distinguish between these values. Such is the nature of continuous time: there always exists a finer division of time that renders co-occurrence impossible.

As we introduce methods for describing continuous-time event data,we will invoke these distributional properties often, using them to explain why and how we modify discrete-time analytic strategies for the new continuous-time context. Although data sets rarely manifest both properties exactly, the possibility that they *could* provides impetus for revision. If it appears peculiar that such a seemingly minor change could create a need to overhaul an entire methodological framework, perhaps an analogy will help. Just as we use different models and estimation strategies when analyzing continuous and dichotomous noncensored outcomes (e.g., linear vs. logistic regression), so, too, do we need different methods when analyzing continuous and discrete censored event occurrence data. As you begin to understand continuous-time survival analysis, knowledge of discrete-time survival analysis will facilitate your learning transition, just as knowledge of linear regression undoubtedly provided you with a solid foundation when it came to learning more sophisticated analytic methods.

13.1.2 The Survivor Function

Let us begin with the survivor function, for its definition remains similar across the time metrics. The fundamental difference in the definition across the two worlds is the obvious respecification of the random variable used to represent time, which now must be continuous. Now we let T be a *continuous* random variable whose values T_i indicate the precise instant when individual i experiences the target event. Although it is common to represent the potential values of T using just the lower case letter t, here, to maintain parallelism with the discrete-time specification, we add the subscript j, to write t_j. In continuous time, then, t_j clocks the infinite number of possible *instants* when the target event could occur. For a motorist who honks 2.98 seconds after the light turns green, $T_i = 2.98$; for a motorist who honks after 7.20 seconds, $T_i = 7.20$.

Respecification of T allows us to retain the basic structure of the survivor function's definition. In continuous time, the survival probability for individual i at time t_j is the probability that his or her event time, T_i, will exceed t_j:

$$S(t_{ij}) = \Pr[T_i > t_j]. \tag{13.1}$$

Individual i's set of survival probabilities over time—$S(t_{ij})$—is still known as his or her survivor function. When we do not distinguish individuals on the basis of predictors, we remove the subscript i, letting $S(t_j)$ represent the survivor function for a randomly selected member of the population. As expected, the initial value for the continuous-time survivor function at time t_0 is 1.

Because the definition in equation 13.1 is essentially identical to its discrete-time counterpart in equation 10.3, we need no new strategies for interpretation. We therefore turn to the hazard function, which does *not* retain the same definition or *meaning* as the metric for time changes from discrete to continuous.

13.1.3 The Hazard Function

The hazard function assesses the *risk*—at a particular moment—that an individual who has not yet done so will experience the target event. In discrete time, the moments are time periods, which allows us to express hazard as a conditional probability. In continuous time, the moments are the infinite number of infinitesimally small instants of time that exist within any finite time period, a change that requires us to alter our definition.

Why must we change the definition of hazard? The problem is that the concept of probability falls apart for a continuous random variable like T. The failure has nothing to do with either survival analysis or censoring. Rather, it is a consequence of the fact that if there exist an infinite number of instants when an event *can* occur, the probability that an event *does* occur at any particular instant must approach 0 as the units of time get finer. At the limit, in truly continuous time (unbridled by the practicalities of measurement), the probability that T takes on any specific value t_j has to be 0. This means that we can no longer define hazard as a (conditional) probability because it would be 0 at all values of t_j. If we did not modify our definition of hazard, we would be left with the meaningless statement that all continuous-time hazard functions are constant at 0, regardless of the event under study or the risk of event occurrence over time.

To develop a sensible definition of hazard in continuous time, we must resolve a basic dilemma: hazard should quantify risk at particular *instants*, but mathematically, we can quantify risk only by cumulating together instants to form *intervals*. Statisticians resolve this dilemma by recognizing that instants and intervals can be thought of as one and the same *if the intervals are so small that we can think of them as instants*. Although this may seem like semantic hair-splitting, the argument rests on a simple realization: when we divide a finite period of time into smaller and smaller units, we inevitably create a corresponding series of *intervals*.

To see how intervals and instants are related, once again let us divide a finite time period—say, one year—into increasingly smaller units. Each year contains 365 one-day intervals (366 in a leap year), but if we divide each day into hours, we have $365 \times 24 = 8,760$ one-hour intervals. Dividing each of these into minutes, we have $365 \times 24 \times 60 = 525,600$ one-minute intervals, which ultimately lead to $365 \times 24 \times 60 \times 60 = 31,536,000$ one-second intervals. As we use finer and finer units (tenths of seconds, nanoseconds, and beyond), we eventually find that this finite time period includes an *infinite* number of *intervals*, each so narrow that it appears to be an *instant*. In essence, then, instants and intervals are one and the same if the width of the interval *approaches*, but never quite reaches, 0.

Mathematicians codify this argument by letting the symbol Δt ("delta t") represent the vanishing width of each of these infinitesimally small intervals. This allows us to write the jth time interval as $[t_j, t_j + \Delta t)$, where the opening bracket indicates that the instant t_j falls just *inside* the interval and the closing parenthesis indicates that the next instant, $t_j + \Delta t$, falls just *outside*. To assess individual i's hazard at time t_j, we are inclined at first to compute the probability that his or her event time, T_i, falls in the jth interval (conditional on survival until the start of the interval) *as the interval's width* Δt) *approaches*, but does not reach, 0:

$$\text{limit as } \Delta t \to 0 \text{ of } \{\Pr[T_i \text{ is in the interval } (t_j, t_j + \Delta t)|T_i \geq t_j]\},$$

where the opening phrase "limit as $\Delta t \to 0$" indicates that we evaluate the conditional probability in the brackets as the interval width moves closer and closer to 0. At its core, notice the parallelism between this definition and that for discrete-time hazard in equation 10.1.

But moving from instants to intervals creates a problem. The magnitude of this conditional probability depends upon the interval's (small and diminishing) width. The probability in a one-second interval will differ from the probability in a one-nanosecond interval. Hazard's definition must account for the interval's width, a goal achieved logically by division. Dividing this conditional probability by the interval width, Δt, we define individual i's continuous-time hazard at time t_j to be:

$$h(t_{ij}) = \text{limit as } \Delta t \to 0 \left\{ \frac{\Pr[T_i \text{ is in the interval } (t_j, t_j + \Delta t)|T_i \geq t_j]}{\Delta t} \right\}. \quad (13.2)$$

As before, the collection of individual i's values of hazard over time—$h(t_{ij})$—is his or her hazard function. And when we do not distinguish individuals on the basis of predictors, we write the hazard function for a randomly selected member of the population as $h(t_j)$.

Because the definitions of hazard differ in continuous and discrete time, their interpretations differ as well. Most important, *continuous-time hazard is not a probability*. Instead, it is a *rate*, assessing the conditional probability of event occurrence *per unit of time*. No matter how tempted you might be to use the nomenclature of probability to describe rates, in continuous time, please resist the urge. Rates and probabilities are not the same, and so the interpretive language is not interchangeable.

Although the concept of a rate may seem unfamiliar, rates are a common part of everyday life. When you say that it takes one hour to travel 60 miles, you are saying that your *rate of travel* is 60 miles *per hour*. When you say that your annual salary is $60,000, you are saying that your *rate of pay* is 60,000 dollars *per year*. A crucial feature of rates is that they must be attached to a denominator with an explicit unit of measurement. A distance of 60 or a salary of 60,000 is meaningless unless attached to the appropriate unit of time. How else would you know whether the 60 miles were traveled in an hour, a minute, or a day or whether the $60,000 were earned per year, per month or per week. Similarly, changing the unit of time allows identical rates to be stated in different ways. Sixty miles per hour is identical to one mile per minute. Sixty thousand dollars per year is identical to 5000 dollars per month. When describing any rate—including a hazard rate—we must always specify the units in which time is measured.

One way of developing an intuition about continuous-time hazard rates is to think about them as providing information about the expected number of events that occur in a finite period of time. If an event is repeatable—for example, getting a parking ticket—and hazard is constant, we can apply this logic to an individual's cumulative experience. If you have a monthly hazard rate of 0.10 for getting a parking ticket, for example, you can anticipate 0.10 tickets per month. But since you face the same monthly hazard of 0.10 January through December, we can multiply your monthly hazard rate by 12 to compute an annual hazard rate of 1.2. This tells you that you can expect to receive 1.2 parking tickets per year. For nonrepeatable events, we use the same logic but instead of associating the cumulative experience with an individual, we associate it with a hypothetical population of individuals who share a common hazard function. If the constant monthly hazard of menarche among seventh-grade girls is .04, for example, each month we expect 4% of the remaining premenarcheal girls to menstruate.

An important difference between continuous-time hazard rates and discrete-time hazard probabilities is that rates are not bounded from above. Although neither can be negative, rates can easily exceed 1.0. You may not have noticed, but we introduced a rate greater than 1 in the previous paragraph when we stated that a monthly hazard rate of .10 is identical to an annual hazard rate of 1.2. The possibility that continuous-time hazard rate can exceed 1 has serious consequences because it requires that we revise the statistical models that incorporate the effects of predictors. We cannot posit a model in terms of *logit* hazard (as in discrete time) because that transformation is defined only for values of hazard between 0 and 1. As a result, when we specify continuous-time hazard models in chapter 14, our specification will focus on the *logarithm* of hazard, a transformation that *is* defined for all values of hazard greater than 0.

13.2 Grouped Methods for Estimating Continuous-Time Survivor and Hazard Functions

In principle, in continuous time, we would like to estimate a value for the survivor and hazard functions at every possible instant when an event could occur. In practice, we can do so only if we are willing to adopt constraining parametric assumptions about the distribution of event times. To support this approach, statisticians have identified dozens of different distributions—Weibull, Gompertz, gamma, and log-logistic, to name a few—that event times might follow, and in some fields—industrial product testing, for example—parametric estimation is the dominant mode of analysis (see, e.g., Lawless, 1982).

In many other fields, including most of the social, behavioral, and medical sciences, nonparametric methods are more popular. The fundamental advantage of nonparametric methods is that we need not make constraining assumptions about the distribution of event times. This flexibility is important because: (1) few researchers have a sound basis for preferring one distribution over another; and (2) adopting an *incorrect* assumption can lead to erroneous conclusions. With a nonparametric approach, you essentially trade the *possibility* of a minor increase in efficiency if a particular assumption holds for the guarantee of doing nearly as well for most data sets, regardless of its tenability.

For decades, in a kind of mathematical irony, statisticians obtained nonparametric estimates of the continuous-time survivor and hazard functions by grouping event times into a small number of intervals, constructing a life table, and applying the discrete-time strategies of chapter 10 (with some minor revisions noted below). In this section, we describe two of the most popular of these grouped strategies: the *discrete-time* method (section 13.2.1) and the *actuarial* method (section 13.2.2). Before beginning, understand that we present these grouped methods first not because they are superior to the ungrouped Kaplan-Meier method of section 13.3—for they are not—but because they are easily understood, simply computed, and, as a result, widely used.

13.2.1 Constructing a Grouped Life Table

Grouped estimation strategies begin with a life table that partitions continuous time into a manageable number of contiguous intervals. When choosing a partition, you should seek one that is: (1) substantively meaningful; (2) coarse enough to yield stable estimates; and (3) fine enough to reveal discernible patterns. Although intervals of width 1 are intuitive and simple, other options may be preferable for certain time metrics. If time is measured in months, for example, quarterly (3-month), semi-annual (6-month), or yearly (12-month) intervals have appeal. If time is measured in weeks, monthly intervals might be better. All intervals need not be the same width, and it may be prudent to use wider intervals at later times to obtain risk sets of adequate size. If you use the life table

Table 13.2: Life table for the horn-honking data with discrete-time and actuarial estimates of the hazard and survivor functions ($n = 57$, n events $= 42$)

Time interval	n at risk	n events	n censored	$\hat{p}(t)$	Discrete-time estimates[a] $\hat{S}(t)$	Discrete-time estimates[a] $\hat{h}(t)$	Actuarial estimates[a] $\hat{S}(t)$	Actuarial estimates[a] $\hat{h}(t)$
					1.0000	—	1.0000	—
[1, 2)	57	5	1	0.0877	0.9123	0.0877	0.9115	0.0926
					(.0375)	(.0375)	(.0378)	(.0414)
[2, 3)	51	14	3	0.2745	0.6619	0.2745	0.6537	0.3294
					(.0632)	(.0625)	(.0643)	(.0868)
[3, 4)	34	9	2	0.2647	0.4867	0.2647	0.4754	0.3158
					(.0683)	(.0757)	(.0690)	(.1039)
[4, 5)	23	6	4	0.2609	0.3597	0.2609	0.3396	0.3333
					(.0673)	(.0916)	(.0680)	(.1342)
[5, 6)	13	2	2	0.1538	0.3044	0.1538	0.2830	0.1818
					(.0674)	(.1001)	(.0674)	(.1280)
[6, 7)	9	2	2	0.2222	0.2368	0.2222	0.2122	0.2857
					(.0673)	(.1386)	(.0666)	(.2000)
[7, 8)	5	1	0	0.2000	0.1894	0.2000	0.1698	0.2222
					(.0685)	(.1789)	(.0654)	(.2208)
[8, 18)	4	3	1	0.7500	0.0473	0.0750	0.0243	0.1500
					(.0444)	(.1317)	(.0331)	(.0573)

[a] Each cell presents parameter estimates and standard errors.

routine in a computer package, consider overriding the default partition as it may not be optimal for your particular situation.

Table 13.2 presents a grouped life table for the horn-honking data of table 13.1. The table divides the 18-second observation period into eight intervals: the first seven are 1-second long; the last is 10 seconds long. We collapse the data from seconds 8 through 18 because only four motorists were at risk by the beginning of the 8th second. The opening bracket and closing parenthesis indicates that each interval *includes* the initial time and *excludes* the terminal time. As in discrete-time, we refer to interval $[t_j, t_{j+1})$ as the jth interval.

13.2.2 The Discrete-Time Method

The discrete-time estimator of the continuous-time survivor function is obtained by applying the discrete-time principles of section 10.2.2 to the data in the grouped life table. Although the computational algorithm is identical, the nomenclature changes as a consequence of hazard's redefinition.

To explain how the discrete-time method works, we introduce a new quantity, $p(t_j)$, the conditional probability that a member of the risk set at the beginning of interval j will experience the target event during that interval. In discrete time, we labeled this quantity "hazard," but now we use the neutral term "*conditional probability*" to distinguish it from a continuous time *hazard rate*. Letting n at risk$_j$ represent the number of individuals at risk at the beginning of interval j and n events$_j$ indicate the number of individuals who experience the event during that interval, we estimate $p(t_j)$ to be:

$$\hat{p}(t_j) = \frac{n \; events_j}{n \; at \; risk_j}. \tag{13.3}$$

Sample estimates for the horn-honking data appear in the fifth column of table 13.2. Although we need these conditional probabilities to estimate the survivor and hazard functions, we rarely interpret them directly as their magnitude depends upon their associated interval's width.

As in chapter 10, we obtain the discrete-time estimator of the survivor function at time t_j (that is, the probability of surviving past interval t_j) by multiplying the successive probabilities of surviving through each interval, from the 1st to the jth. Each of these probabilities, in turn, is just the complement of the conditional probability of event occurrence during the interval. We therefore write the discrete-time estimator of the continuous-time survivor function for the jth interval as:

$$\hat{S}(t_j) = (1 - \hat{p}(t_1))(1 - \hat{p}(t_2)) \cdots (1 - \hat{p}(t_j)). \tag{13.4}$$

Owing to the definition of the survivor function in equation 13.1, we associate the survival probability for the jth interval with the instant at the *end* of the interval (that is, with the instant just before t_{j+1}).

Discrete-time estimates of the continuous-time survivor function for the horn-honking data appear in the sixth column of table 13.2; a plot appears in the upper left hand panel of figure 13.1. Because the first-second survival probability of 0.9123 is relatively high—given the lower values that follow—we conclude that motorists give the blocking car a "grace period," an initial moment to move. The immediate steep decline in survivorship that follows—to 0.6619, 0.4867, and 0.3597 by the end of the fourth second—suggests that this reprieve is short-lived. Using equation 10.6 to linearly interpolate between the event times at the ends of the relevant intervals, we estimate the median time to horn-honk to be 3.92 seconds!

We can also estimate the continuous-time hazard function using the grouped data, but because hazard is now a *rate*, not a probability, we need

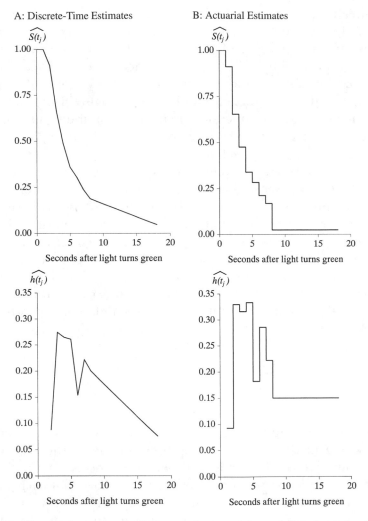

Figure 13.1. Grouped estimates of the survivor and hazard functions for the horn-honking data computed using two methods.

a modified approach. Equation 13.2 defines the hazard rate as the limit of the conditional probability of event occurrence in a (vanishingly small) interval divided by the interval's width. A logical estimator is thus the ratio of the conditional probability of event occurrence in an interval to the interval's width. Letting $width_j$ denote the width of interval j (i.e., $t_{j+1} - t_j$), the discrete-time estimator of the continuous-time hazard rate is:

$$\hat{h}(t_j) = \frac{\hat{p}(t_j)}{width_j}. \tag{13.5}$$

When intervals are of width 1 (like the first eight in table 13.2), hazard is identical to the conditional probability of event occurrence. Otherwise, hazard is the conditional probability divided by the interval's width, or the "average probability" of event occurrence *per unit of time*. Estimates for the horn-honking data appear in the seventh column of table 13.2; a plot appears in the lower left hand panel of figure 13.1. As presaged by the estimated survivor function, hazard is low in the first second, rises dramatically in the second second, remains high through the fourth second, when it begins to decline.

Because we calculate discrete-time estimates using the methods of chapter 10, it makes sense that we compute their standard errors using identical methods as well (equations 10.7 and 10.8). As a result, the standard errors—shown in table 13.2—manifest properties similar to those discussed in chapter 10. Most important, the larger the risk set, the smaller the standard error. Because the risk set is largest at t_0 and diminishes over time, standard errors generally increase. In the horn-honking data, where the initial sample is small and events occur in quick succession, almost all estimates have relatively large standard errors, which then grow over time.

13.2.3 The Actuarial Method

The actuarial method—often referred to as the life-table method—uses a strategy similar to the discrete-time method. The fundamental difference concerns assumptions about the distribution of event occurrence and censoring in each interval. The discrete-time method ignores this issue, assuming that all events and censoring occur at the interval's *endpoint*. The actuarial method, in contrast, assumes that each interval's events and censoring are distributed equally throughout the interval. Invoking this *actuarial assumption* is tantamount to assuming that events and censoring occur *at random* during the interval. This leads us toward an estimate based on information presumed to be available at the interval's *midpoint*.

A simple way of demonstrating how the actuarial assumption works is to redefine what it means to be "at risk." Redefinition changes the size of each interval's risk set, thereby changing the magnitude of the resulting estimates. Depending on the function to be estimated, a different redefinition is used.

For the survivor function, we ask: What does it mean to be "at risk of surviving" past the end of an interval? Because a censored individual is no longer "at risk of surviving" once censoring occurs, we redefine each interval's risk set to account for the censoring we assume to occur equally throughout. This implies that half the censored individuals would no longer be at risk half-way through, so we redefine the number of individuals "at risk of surviving past interval j" to be:

$$n' \ at \ risk_j = n \ at \ risk_j - \frac{n \ censored_j}{2}.$$

The actuarial estimate of the survivor function is obtained by substituting $n' \ at \ risk_j$ for $n \ at \ risk_j$ in the discrete-time formulas just presented in section 13.2.2 (equations 13.3 and 13.4).

Actuarial estimates of the survivor function for the horn-honking data appear in the eighth column of table 13.2; a plot appears in the upper right hand panel of figure 13.1. We plot the actuarial estimates as a step function associating each survival probability with its entire time interval. This practice is standard, even though estimates are based on data assumed to be available at each interval's *midpoint*. This makes the actuarial estimate appear choppier than the discrete-time estimate, although the actual values—tabulated in table 13.2—are similar. In fact, because $n' \ at \ risk_j$ cannot exceed $n \ at \ risk_j$, the actuarial estimate of the survivor function is "more conservative" as it will never exceed the discrete-time estimate.

The decision to associate each actuarial estimate of the survivor function with an entire interval has an unusual impact on the estimation of median lifetimes. If we linearly interpolate between the event times at the ends of the relevant intervals, we find a median of 3.86 seconds. Yet statisticians rarely suggest interpolation for the same reason as they plot the estimates as a step function. Even though the survival probabilities are estimated using data assumed to be available at each interval's *midpoint*, the resultant values are associated with every moment in the interval, including the *last* (as in figure 13.1). This suggests that the estimated median lifetime is simply the event time at the *end* of the interval when the sample survivor function first hits (or dips below) .50. For the horn-honking data, in which the survival probability is 0.6547 at the end of the third interval and 0.4754 at the end of the fourth, this yields an estimate of 4 seconds. Although admittedly coarser than the interpolated value of 3.86, many empirical researchers prefer the noninterpolated value because it is more conservative.

To estimate the hazard function using the actuarial approach, we again redefine what it means to be "at risk." Now, however, we ask about the "risk of event occurrence" *during* the interval, not the "risk of survival" *past* the interval. This change of definition suggests that each interval's risk set should be diminished not just by censoring but also by event occurrence, because either eliminates the possibility of subsequent event occurrence. Because categorization continues to prevent us from knowing precisely when people leave the risk set, we assume that exits are scattered at random throughout the interval. This implies that half these individuals are no longer at risk of event occurrence halfway through, so we redefine the number of individuals "at risk of event occurrence" in interval j to be:

$$n'' \text{ at risk}_j = n \text{ at risk}_j - \frac{n \text{ censored}_j}{2} - \frac{n \text{ events}_j}{2}.$$

The actuarial estimator of the continuous-time hazard function is then obtained by substituting n'' *at risk$_j$* for *n at risk$_j$* in discrete-time formulas of section 13.2.2 (equations 13.3 and 13.5).

Actuarial estimates of the hazard function for the horn-honking data appear in the last column of table 13.2; a plot appears in the lower right hand panel of figure 13.1. We plot these estimates as a step function, each step indicating the hazard rate for the entire interval. Once again, this gives the plot a choppier appearance than the discrete-time estimate, although the pattern of risk over time is, of course, very similar. In fact, because n'' *at risk$_j$* cannot exceed *n at risk$_j$*, the actuarial estimate of hazard will always exceed the discrete-time estimate, unless both are 0 and the issue is moot.

Because actuarial estimates use modified risk set definitions, we must compute their standard errors using modified versions of the formulas introduced in chapter 10. The appropriate formulas, developed by Gehan (1969), are implemented in all software packages that offer actuarial estimates. Results for the horn-honking data appear below their associated parameter estimates in table 13.2. Because these standard errors closely mirror those for the discrete-time estimates—except that they are typically larger—we omit discussion of their behavior.

We now raise the obvious question: Which grouped method is superior? Although neither is as good as the Kaplan-Meier approach we introduce next, each has advantages. Discrete-time estimates are easier to interpret, especially when all intervals are of width 1 (because then they are identical to the conditional probability of event occurrence). Actuarial estimates adjust (albeit simplistically) for the computational categorization of time and are "more conservative" in that they underestimate survivorship and overestimate risk. In practice, the absolute difference between the two is usually small, and the relative difference in their values over time—which reflects the *shape* of each function—is smaller still (aside from visual differences attributable to plotting conventions). Your choice is largely a matter of convenience, dictated usually by the options in your statistical package. For the kinds of broad-brush insights these methods provide, either will suffice.

13.3 The Kaplan-Meier Method of Estimating the Continuous-Time Survivor Function

A fundamental problem with grouped estimation methods is that they artificially categorize what is now, by definition, a continuous variable. Different categorizations yield different estimates, and for truly continuous data, categorization makes little sense. Shouldn't it be possible to use the *observed* data—the actual event times—to describe the distribution of event occurrence? This compelling idea underlies the Kaplan-Meier method, named for the statisticians who demonstrated (in 1958) that the intuitive approach—also known as the *product-limit method*—has maximum likelihood properties as well. Below, we explain how this approach works and why it is preferable.

The Kaplan-Meier method is a simple extension of the discrete-time method with a fundamental change: instead of rounding event times to construct the intervals, capitalize on the raw event times and construct intervals so that each contains just *one observed event time* (as shown in table 13.3). Each Kaplan-Meier interval begins at one observed event time and ends just before the next. In the horn-honking study, the first three event times of 1.41, 1.51, and 1.67 lead to two intervals: [1.41, 1.51) and [1.51, 1.67). By convention, we also construct an *initial interval*, here [0, 1.41), which begins at t_0 and ends immediately *before* the first event. The final interval begins at the last event time (here 13.18) and ends at either this time (if there are no larger censored values) or infinity if the largest event time is censored (as it is here). If an individual is censored at an observed event time—like the person censored at 1.41—we "break the tie" by assuming that the event preceded censoring. This places the tied censored case within the observed event time's interval. Tied observed times are similarly placed in their common interval.

The Kaplan-Meier estimate of the survivor function is obtained by applying the discrete-time estimator of section 13.2.2 to the data in these intervals. All statistical packages include a routine for computing and plotting the estimates. Numerically, the process is simple: first compute the conditional probability of event occurrence (column 7) and then successively multiply the complements of these probabilities together to obtain the Kaplan-Meier estimate of the survivor function (column 8). Because the Kaplan-Meier estimator of the survivor function is identical

Table 13.3: Kaplan-Meier (product limit) estimates for the horn-honking data ($n = 57$; n *events* = 42)

Interval	[Start	End	n at risk	n events	n censored	$\hat{p}(t)$	$\hat{S}(t)$	$se[\hat{S}(t)]$	Width	$\hat{h}_{KM}(t)$
0	0.00	1.41	57	0	0	—	1.0000		—	
1	1.41	1.51	57	1	1	0.0175	0.9825	0.0174	0.10	0.1750
2	1.51	1.67	55	1	0	0.0182	0.9646	0.0246	0.16	0.1138
3	1.67	1.68	54	1	0	0.0185	0.9467	0.0299	0.01	1.8500
4	1.68	1.86	53	1	0	0.0189	0.9289	0.0343	0.18	0.1050
5	1.86	2.12	52	1	0	0.0192	0.9110	0.0380	0.26	0.0738
6	2.12	2.19	51	1	0	0.0196	0.8931	0.0412	0.07	0.2800
7	2.19	2.48	50	1	1	0.0200	0.8753	0.0441	0.29	0.0690
8	2.48	2.50	48	1	0	0.0208	0.8570	0.0468	0.02	1.0400
9	2.50	2.53	47	1	0	0.0213	0.8388	0.0492	0.03	0.7100
10	2.53	2.54	46	1	0	0.0217	0.8206	0.0514	0.01	2.1700
11	2.54	2.56	45	1	0	0.0222	0.8023	0.0534	0.02	1.1100
12	2.56	2.62	44	1	0	0.0227	0.7841	0.0552	0.06	0.3783
13	2.62	2.68	43	1	0	0.0233	0.7659	0.0569	0.06	0.3883
14	2.68	2.83	42	1	2	0.0238	0.7476	0.0584	0.15	0.1587
15	2.83	2.88	39	1	0	0.0256	0.7285	0.0599	0.05	0.5120
16	2.88	2.89	38	1	0	0.0263	0.7093	0.0614	0.01	2.6300
17	2.89	2.92	37	1	0	0.0270	0.6901	0.0626	0.03	0.9000
18	2.92	2.98	36	1	0	0.0278	0.6710	0.0637	0.06	0.4633
19	2.98	3.14	35	1	1	0.0286	0.6518	0.0647	0.16	0.1788
				⋮	⋮					
27	3.57	3.58	26	1	0	0.0384	0.5121	0.0692	0.01	3.8400
28	3.58	3.78	25	1	0	0.0400	0.4916	0.0694	0.20	0.2000
				⋮	⋮					
35	4.96	5.39	14	1	1	0.0714	0.3349	0.0683	0.43	0.1660
36	5.39	5.73	12	1	1	0.0833	0.3070	0.0681	0.34	0.2450
37	5.73	6.03	11	1	1	0.0909	0.2791	0.0674	0.30	0.3030
38	6.03	6.30	9	1	1	0.1111	0.2481	0.0666	0.27	0.4115
39	6.30	7.20	7	1	1	0.1429	0.2126	0.0659	0.90	0.1588
40	7.20	9.59	5	1	0	0.2000	0.1701	0.0650	2.39	0.0837
41	9.59	12.29	4	1	0	0.2500	0.1276	0.0611	2.70	0.0926
42	12.29	13.18	3	1	0	0.3333	0.0851	0.0535	0.89	0.3745
43	13.18	∞	2	1	1	0.5000	0.0425	0.0403	—	—

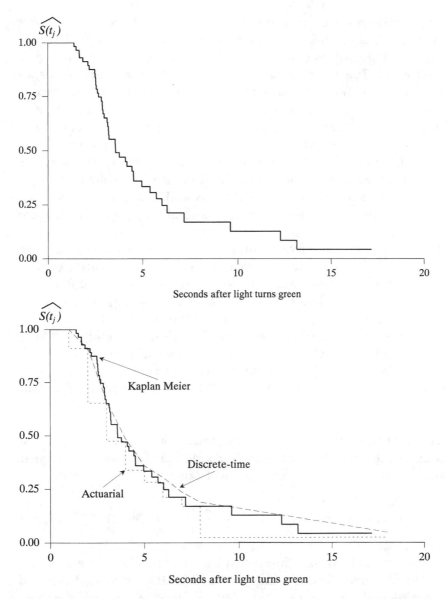

Figure 13.2. Kaplan-Meier estimates of the continuous time survivor function for the horn-honking data. The top panel presents the estimates themselves; the bottom panel compares these estimates to the discrete-time and actuarial estimates plotted in figure 13.1.

to the discrete-time estimator of chapter 10, its standard errors (column 9) are estimated using the same formula (equation 10.8).

The top panel of figure 13.2 plots the Kaplan-Meier estimate of the survivor function for the horn-honking data. As for actuarial estimates, we plot Kaplan-Meier estimates as a step function that associates the estimated survival probability with the entire interval. If the largest event time is censored, as it is here (17.15), we extend the step for the last estimate out to that largest censored value. When event occurrence is recorded using a truly continuous metric, ties will be rare, and the resulting plot will appear relatively smooth, at least when event occurrence is common. As the risk set diminishes, or when event occurrence is rare, the plotted survivor function will appear more jagged. Ironically, then, our plots of the estimated continuous-time survivor function appear somewhat discrete while those of the estimated discrete-time survivor function appear somewhat continuous. Although it is possible to use a step function for plotting in discrete time as well, doing so would make it more difficult to discern the function's general shape and to compare functions across groups.

Having estimated the survivor function, it is easy to estimate the median lifetime. The standard approach, implemented in software packages, is to identify the first observed event time when the value of the estimated survivor function either precisely hits, or dips below, 0.50. Examining table 13.3, we find that the estimated median time to horn-honk is 3.58 seconds, because the estimated survival probability is 0.5121 in the [3.57, 3.58) interval and .4916 in the [3.58, 3.78) interval.

An alternative, albeit less common, approach to median lifetime estimation is offered by Miller (1981), Lee (1992), and others. Arguing that the median lifetime *must* be earlier than the value conventionally identified whenever the estimated survival probability is not precisely 0.5 at an observed event time, they suggest linear interpolation between the two bracketing event times. For the horn-honking data, we apply the linear interpolation formula of equation 10.6 to find:

$$\text{Estimated median lifetime} = 3.57 + \left(\frac{.5121 - .5000}{.5121 - .4916}\right)(3.58 - 3.57) = 3.5759.$$

In this data set, of course, interpolation yields little apparent gain because the observed event times bracketing the estimated survival probability of 0.50 are so close anyway. In other data sets, however, interpolated estimates can be valuable—especially when comparing subsamples—if the distance between observed event times in that temporal vicinity is greater.

How does the Kaplan-Meier estimate of the survivor function compare to those obtained using discrete-time and actuarial methods? To facilitate comparison, the lower panel of figure 13.2 displays all three estimates for the horn-honking data. Because it updates most frequently—every time an event occurs—the Kaplan-Meier estimate is the most refined. Actuarial estimates are generally the smallest because they diminish the risk set in an attempt to account for the distribution of censoring. Discrete-time estimates periodically coincide with the Kaplan-Meier estimates—especially when an event occurs near the end of a pre-specified coarse interval—but they ignore potentially meaningful detail about the distribution of event occurrence within intervals. Perhaps the only drawback of the Kaplan-Meier method is that when plotted for subgroups—as a precursor to statistical modeling—the "drops" occur at different times reflecting the observed event times within each subgroup. Although varying drop locations can make visual comparison trickier, this is a trivial criticism of a maximum likelihood estimate that provides great resolution with few assumptions. The conclusion: Kaplan-Meier estimates are unquestionably the best of the three.

If Kaplan-Meier estimates are so superior, why did we bother with the discrete-time and actuarial methods? Unfortunately, the answer is simple: there is no Kaplan-Meier estimator of hazard. This increases the appeal of the grouped methods—coarse as they are—for they provide estimates of hazard, albeit categorized and crude. And because it is wise to examine "parallel" estimates computed using the *same* approach, you may wish to examine survivor functions estimated using discrete-time or actuarial methods knowing full well that they are inferior to the Kaplan-Meier approach.

Although there is no Kaplan-Meier estimate of hazard, consider what happens when we apply the discrete-time estimator of hazard in equation 13.5 to the data in the fine Kaplan-Meier intervals. Using the subscript KM to denote these estimates, we have:

$$\hat{h}_{KM}(t_j) = \frac{\hat{p}_{KM}(t_j)}{width_j}.$$

The resulting Kaplan-Meier "type" hazard estimates appear in the last column of table 13.3. Because the risk set diminishes steadily while the number of events in each interval remains constant at 1 (unless there are observed ties), the numerator of this equation—the $\hat{p}_{KM}(t_j)$—inevitably rises.[1] Dividing by the interval width (in the penultimate column) "averages" these increasing values across their respective intervals. But because the interval width varies widely (and is itself a function of the distribution of event times), the resulting estimates vary from one interval to the next. Their values are usually so erratic that pattern identification is near impossible. As a result, few software packages offer these estimates, although they do form the basis of other descriptive methods, as we now explain.

13.4 The Cumulative Hazard Function

The inability to estimate the continuous-time hazard function well represents a major analytic stumbling block for survival analyses of continuous-time data. From discrete time, you know how meaningful the hazard function is and why it is essential to examine its behavior graphically over time. Without knowledge of its shape, overall and by levels of predictors, model fitting can become a "black box," disconnected from data. Although grouped estimates of hazard can be helpful, they are too coarse to provide the necessary insights; Kaplan-Meier type estimates of hazard are simply too erratic to be meaningful.

This is where the *cumulative hazard function* comes in. Denoted by $H(t_{ij})$, the cumulative hazard function assesses, at each point in time, the *total amount of accumulated risk* that individual i has faced from the beginning of time until the present. More formally, at time t_j, individual i's value of cumulative hazard is defined as:

$$H(t_{ij}) = \underset{\text{between } t_0 \text{ and } t_j}{cumulation}[h(t_{ij})], \tag{13.6}$$

where the phrase "cumulation between t_0 and t_j" indicates that cumulative hazard totals the infinite number of specific values of $h(t_{ij})$ that exist between t_0 and t_j.[2] Unlike the hazard function, which is difficult to estimate well in continuous time, the cumulative hazard function can be estimated simply using nothing more than byproducts of the Kaplan-Meier method.

Because $H(t_{ij})$ literally *cumulates* hazard, examination of its changing level over time tells us about the shape of the underlying hazard function. Although cumulation prevents it from describing the *unique* risk at each particular instant—that, after all, is what the hazard function assesses—comparison of its changing levels allows us to deduce this information. Below, we show how to use the cumulative hazard function to learn about the shape of the underlying hazard function. If you are wondering why we introduce this important new function only now, so late in the book, the answer is simple: we first need it in the context of continuous-time event data, where we cannot estimate hazard well.

13.4.1 Understanding the Meaning of
Cumulative Hazard

Many people find the concept of cumulative hazard elusive because it is measured in a metric difficult to quantify. We can make general statements—if individual i has been relatively risk free, $H(t_{ij})$ will be "low";

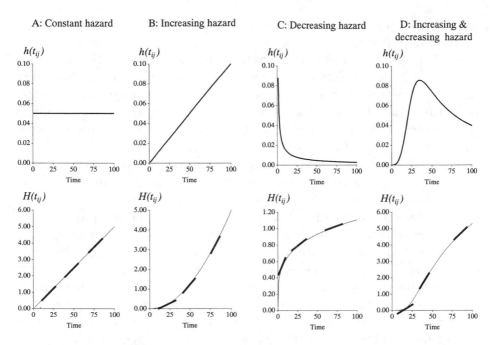

Figure 13.3. Population hazard functions and cumulative hazard functions reflecting four different profiles of risk over time. Panel A: constant hazard. Panel B: increasing hazard. Panel C: decreasing hazard. Panel D: nonmonotonic hazard.

if individual i has faced great and persistent risk, $H(t_{ij})$ will be "high"—but we cannot say exactly which values are "high" and which are "low." Cumulative hazard is not a probability (even in discrete time) nor is it a rate. Because a value of cumulative hazard has little meaning on its own, we rarely tabulate the function, preferring to explore its behavior graphically over time.

In subsequent sections, we examine estimated cumulative hazard functions from various studies in continuous time, distilling what they tell us about the hazard function and its behavior over time. In preparation for this work, in this section, we examine *hypothetical cumulative hazard functions* obtained by invoking different assumptions about the distribution of event occurrence over time (see the bottom row of figure 13.3). Although sample functions will never be as smooth, idealized representations like these allow us to accompany each cumulative hazard function with a related graph that we cannot obtain from real data: the associated hazard function (in the top row). Working in an idealized arena like this allows us to concretize the link between cumulative hazard and hazard so that when we move on to sample data, we will have a foundation for drawing conclusions about hazard on the basis of cumulative hazard alone.

By definition, the cumulative hazard function begins at 0 and *increases over time*—sometimes rapidly, sometimes slowly. Only during those rare periods—never even shown in figure 13.3—when an individual faces no further risk (i.e., when $h(t_{ij}) = 0$), will $H(t_{ij})$ remain constant (at its previous level). This seemingly inevitable rise in cumulative hazard often causes some people to conclude (incorrectly) that all cumulative hazard functions "look alike" and that it is difficult to draw conclusions about the underlying hazard function from this aggregated summary. As we show below, however, nothing could be further from the truth.

To deduce the shape of the underlying hazard function, study how the *rate of increase* in the cumulative hazard function *changes* over time. Each change in its rate of increase reflects a corresponding change in the level of the hazard function itself. To see this correspondence, examine Panel A, where the relationship between hazard and cumulative hazard is clearest. Because $h(t_{ij})$ is constant, $H(t_{ij})$ increases linearly, as the same fixed amount of risk—the constant value of hazard—is added to the prior cumulative level in each successive instant. Now imagine working backwards, from cumulative hazard to hazard, for this is what you must do during data analysis. To work in this direction, you must "guesstimate" the *rate of increase in* $H(t_{ij})$ at different points in time. We find it helpful to superimpose line segments on the plot (as we do in figure 13.3) that summarize the function's rate of increase during different periods. Because the cumulative hazard function in Panel A is linear, each line segment has an identical slope, which tells us that the rate of increase in $H(t_{ij})$ is identical regardless of t—it remains unchanged. If the rate of increase in cumulative hazard is constant over time, the hazard function must be constant as well, as it is in the upper graph.

Next examine Panel B, where the hazard function increases linearly over time. At each successive instant, the cumulative hazard function increases more rapidly—it *accelerates*—because a larger amount of risk is added to the prior cumulated level. Acceleration can be deduced from the increasing slopes of the three line segments drawn on the graph *tangent* to $H(t_{ij})$ at three points in time. The slope of each segment assesses the rate of increase in $H(t_{ij})$ at the moment when the segment *touches* the curve. Because each segment's slope is increasingly steep, the rate of increase in $H(t_{ij})$ is *increasing* over time. If the rate of increase in cumulative hazard is increasing, the hazard function must be increasing as well.

Similar strategies can be used with any cumulative hazard function. To illustrate, focus on Panel C, but this time, begin with cumulative hazard.

The diminishing slopes of the three line segments tangent to the curve indicate that the rate of increase is *decreasing* over time. If the rate of increase is decreasing, hazard must be decreasing as well, and this is precisely what we see in the upper panel, where $h(t_{ij})$ falls to a lower asymptote. $H(t_{ij})$ increases less rapidly because, in each successive instant, a *smaller* amount of risk is added to the prior cumulative level.

In Panel D, where hazard first rises and then falls, we infer the directional change through the *changing* rate of increase in cumulative hazard. When $h(t_{ij})$ rises slowly, the rate of increase in $H(t_{ij})$ is small; when $h(t_{ij})$ accelerates, the rate of increase in $H(t_{ij})$ is rapid; when $h(t_{ij})$ finally falls, the rate of increase in $H(t_{ij})$ finally diminishes. Whenever the rates of increase in a cumulative hazard function change in magnitude over time, we infer that the hazard function reaches either a peak (or a trough).

13.4.2 Estimating the Cumulative Hazard Function

There are two simple ways to estimate the cumulative hazard function: (1) the Nelson-Aalen method, which is based on Kaplan-Meier–type hazard estimates; and (2) the negative log survivor function method, which is based on Kaplan-Meier survivor function estimates. Although the latter method is more widely implemented in software, the former is more intuitive and for this reason, we begin there. Before doing so, however, notice that we do not work with either discrete-time or actuarial estimates for both are too coarse to yield adequate resolution.

The idea behind the Nelson-Aalen method is simple: if $H(t_{ij})$ cumulates together all the "hazard" that exists at all possible instants between t_0 and t_j, a reasonable estimate would total all the estimated hazard that exists between these points in time. This is where the Kaplan-Meier–type hazard estimates of section 13.3 come in. Each estimates the value of hazard at an "average instant" during its associated interval. To compute the *total* amount of hazard that exists during *all* instants in interval j, simply multiply the hazard estimate by its interval's width:

$$\text{Total hazard during interval } j = \hat{h}_{KM}(t_j)width_j.$$

The Nelson-Aalen estimator of cumulative hazard simply sums up these interval-specific estimates:

$$\hat{H}_{NA}(t_j) = \hat{h}_{KM}(t_1)width_1 + \hat{h}_{KM}(t_2)width_2 + \cdots \\ + \hat{h}_{KM}(t_j)width_j. \tag{13.7}$$

Equation 13.7 reinforces the graphical interpretation of cumulative hazard introduced in section 13.3.1. We estimate its value at time t_j by "summing up" the area under the associated hazard function from the beginning of time until the present.

The second method of estimation exploits a well-known mathematical relationship between the cumulative hazard and survivor functions. Because derivation of this relationship requires calculus, we simply assert the correspondence:

$$H(t_j) = -\ln S(t_{ij}). \tag{13.8}$$

Equation 13.8 tells us that the population cumulative hazard function is identical to the negative log of the population survivor function. This identity provides a simple alternative strategy for estimating the cumulative hazard function: apply equation 13.8 to the Kaplan-Meier estimate of the survivor function:

$$\hat{H}_{-LS}(t_j) = -\ln \hat{S}_{KM}(t_{ij}).$$

For obvious reasons, the results are labeled Kaplan-Meier estimates, *negative log survivor function* estimates, or simply the "negative log survivor function."

Nelson-Aalen and negative log survivor function estimates of the cumulative hazard function for the horn-honking data appear in figure 13.4. Because both estimate the same population function, it should come as no surprise that their values are similar. In general, the estimates will be most similar during early event times, when the risk set is large. As event occurrence decreases the risk set's size (to a small level if the original sample is small and hazard is high), the estimates will diverge. Because censoring takes a greater toll on the Nelson-Aalen estimates, the negative log survivor function estimates are more popular for descriptive analysis (although we use the Nelson-Aalen estimates in chapter 14 after fitting statistical models to continuous-time event occurrence data).

Regardless of estimation method, sample cumulative hazard functions suffer from a well-known visual problem apparent in figure 13.4: they tempt you into focusing on estimates in the upper right tail that are usually very unstable. Unless event occurrence is rare and many people remain at risk, later cumulative hazard estimates are based on small risk sets. You should therefore focus primarily on earlier sample estimates, de-emphasizing later values.

Jagged sample cumulative hazard functions are more difficult to interpret than were the smooth hypothetical functions in figure 13.3. To identify shifts in the underlying slope, look past the inevitable jigs and jags, and seek out the "big picture." As before, we suggest that you draw line segments at several different points in time to "guesstimate" the (potentially differing) rates of increase during selected periods of time.

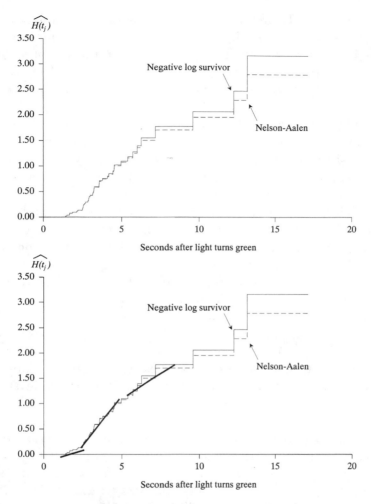

Figure 13.4. Sample cumulative hazard functions for the horn-honking data. The top panel presents both the negative log survivor function and Nelson-Aalen estimates. The bottom panel displays the estimated functions using faint lines and their approximate slopes using heavier line segments.

But because sample functions are not smooth, do not draw lines tangent at only two or three points in time. Instead, try to identify the "average" slope during several temporal periods.

We apply this approach in the bottom panel of figure 13.4, where we plot the two estimates of cumulative hazard using faint lines and we "guesstimate" three slopes using heavier lines. Comparing the varying slopes of these line segments, we find that the rate of increase in cumulative hazard is slowest between 1.5 and 2.5 seconds, fastest between 2.5

and 5.0 seconds, and somewhat slower again between 5.0 and 8.0 seconds. This suggests that hazard is initially low, increases until the fifth second, and then decreases again (a description consonant with that based upon the discrete-time and actuarial estimates examined earlier). We ignore the pronounced jumps after the eighth second as too few people remain at risk.

The cumulative hazard function provides a vital link between the hazard function—which we cannot estimate well in continuous time—and the survivor function, whose doubly bounded nature makes it insensitive to changes in hazard. We can clarify further the close nature of this link by invoking the simple mathematical principle that states that if two continuous functions are related by an identity, like $H(t_{ij})$ and $-\log_e S(t_{ij})$ in equation 13.8, the identity can be reexpressed using the *rates of change of the corresponding functions*. This tells us that the rate of change in $H(t_{ij})$ must be identical to the rate of change in $-\log_e S(t_{ij})$. But the rate of change in cumulative hazard—the object of study in figures 13.3 and 13.4—is simply the hazard function itself! We may therefore restate equation 13.8 as:

$$h(t_{ij}) = \text{rate of change in } \{-\ln S(t_{ij})\}. \qquad (13.9)$$

In other words, the hazard function assesses the rate of change *not* in the survivor function—as many people erroneously believe—but in the *negated logarithm* of the survivor function. This identity reinforces the utility of examining the negative log survivor function (in addition to the survivor function) for it is this transformation that is so closely related to hazard.

13.5 Kernel-Smoothed Estimates of the Hazard Function

A recurrent theme of this chapter is the difficulty of estimating well the continuous-time hazard function. Although the cumulative hazard function is informative, statisticians have devoted considerable energy to developing other methods for discerning the shape of the hazard function. In this section, we introduce one approach—known as *kernel smoothing*—which is becoming increasingly popular as it effectively converts any set of erratic point estimates into a smoother, well-behaved functional form (for a nontechnical description of kernel smoothing, see Fox, 1998).

The idea behind kernel smoothing is simple. At each of many distinct points in time, estimate a function's *average* value by aggregating together all the point estimates available within the focal time's temporal vicinity.

Conceptually, kernel-smoothed estimates are a type of moving average. They do not identify *precise* values of hazard at each point in time but rather *approximate* values based on the estimates nearby. Even though each smoothed value only approximates the underlying true value, a plot over time can help reveal the underlying function's shape.

Kernel smoothing requires a set of point estimates to smooth. For the hazard function, one way of obtaining these point estimates is by computing successive differences in the estimated cumulative hazard function from each observed event time until the next. Each difference acts as a pseudo-slope, a measure of the local rate of change in cumulative hazard during that period. Either Nelson-Aalen estimates or negative log survivor function estimates of cumulative hazard can be used; here, we use the latter.

Because each estimate of hazard is based on just two data points, its values are erratic. It is therefore unwise to let any individual value represent the sole estimate of hazard at that time. The logic of kernel smoothing is to *aggregate together* all the estimates that fall within a given temporal window on either side of a focal time and let the aggregated value estimate the *average* value of hazard in that temporal vicinity. To compute a smoothed value of hazard for the fifth second of the horn-honking data, for example, we aggregate all the point estimates of hazard near second 5. The definition of "near"—known as the *bandwidth*—is up to you. If you define "near" as ±1 second, you will aggregate together all point estimates between seconds 4 and 6. If you define "near" more expansively, as ±2 seconds, you will aggregate all point estimates between seconds 3 and 7.

Figure 13.5 presents kernel smoothed estimates of the hazard function for the horn-honking data obtained using bandwidths of 1, 2 and 3 seconds.[3] When examining kernel-smoothed estimates, be sure to explore a range of bandwidths, as we do here. Comparing results obtained using different bandwidths helps strike a balance between smoothness and precision. The use of Kernel smoothing is an art based on scientific principles. No bandwidth is "right," revealing the "true" shape of the underlying hazard function. Kernel-smoothed functions do not even estimate this population function, but rather its *average* value within the chosen bandwidth.

One hallmark of kernel smoothing is that as the bandwidth increases, the shape of the function becomes smoother. In figure 13.5, for example, the smoothed function obtained using a bandwidth of 3 (in the bottom panel) is almost linear, dropping steadily from its peak of .30 to a plateau of .10 around the ninth second. The smoothed function obtained using a bandwidth of 1 (in the top panel) is more erratic, displaying four local

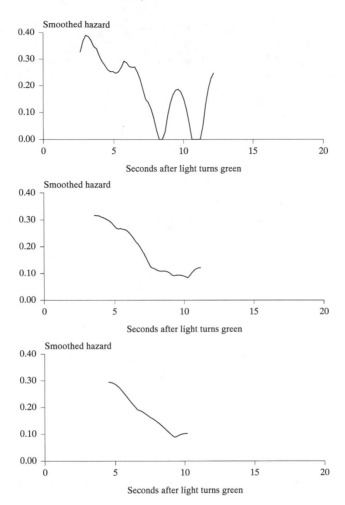

Figure 13.5. Kernel-smoothed hazard functions for the horn-honking data using three different bandwidths. Top panel: 1-second bandwidth. Middle panel: 2-second bandwidth. Bottom panel: 3-second bandwidth.

peaks (near seconds 3, 6, 9, and 12) and two periods of no measurable risk. The function in the middle, obtained using a bandwidth of 2, represents a compromise between these extremes. This display leads us to conclude that the hazard of a horn-honk drops steadily between seconds 3 and 7, after which time it levels off.

Widening the bandwidth to achieve a smoother appearance has two costs. First, each smoothed value does not estimate the population value of hazard at that point in time but rather the *average* of hazard in the temporal vicinity. Widening the bandwidth therefore weakens the link between each smoothed value and each specific point in time. To illustrate, consider what the smoothed value of hazard at second 6 represents. If we use a bandwidth of 1, it represents the "average" population value of hazard between seconds 5 and 7. But if we use a bandwidth of 2, it represents the "average" population value of hazard between seconds 4 and 8. Increasing the bandwidth causes the resultant value to estimate the "average" population value of hazard across a much wider span. This produces a smoother function, but blurs the meaning of its individual values. As the bandwidth widens, the discrepancy between the estimated and true values increase; in other words, widening the bandwidth increases the *bias*.

A second cost of widening the bandwidth is that it narrows the temporal region the smoothed function describes. In the upper panel, with a bandwidth of 1, the smoothed function runs from seconds 2 to 12, while in the bottom panel, with a bandwidth of 3, it runs from only seconds 4 to 10. Narrowing results from the fact that computation of the smoothed value requires equal amounts of time on both sides of the focal time.[4] So the *earliest* time that we can compute a smoothed value is one bandwidth *later* than the *minimum* observed event time, while the *latest* time that we can compute a smoothed value is one bandwidth *earlier* than the *maximum* observed event time. When the range of observed event times is limited, as it is here, the effect can be devastating. The temporal narrowing that results from using a wider bandwidth is also problematic in data sets in which it is important to ascertain values of hazard at early event times. Previous analyses of the horn-honking data, for example, have suggested that motorists give the blocked car a grace period, an initial moment in which to move. After this time, hazard skyrockets as motorists express their frustration. Yet we cannot discern these hypothesized phenomena in figure 13.5 for they occur too early. Lack of information about hazard at the extremes should never lead you to conclude that its value is 0 or that you can extrapolate values from the estimates you have. To discern the value of hazard at the beginning of time, you must simultaneously examine the sample cumulative hazard function as well, as we now discuss.

13.6   Developing an Intuition about Continuous-Time Survivor, Cumulative Hazard, and Kernel-Smoothed Hazard Functions

To describe the distribution of event times recorded in a continuous metric, we recommend that you simultaneously examine:

- *The Kaplan-Meier estimate of the survivor function.*
- *The negative log survivor function estimate, or Nelson-Aalen estimate, of the cumulative hazard function.*
- *The kernel-smoothed estimate of the hazard function.*

Although statistical packages can automatically produce the first two displays, you may need to obtain the kernel-smoothed estimates on your own. Although this obviously requires extra effort, we believe this effort is repaid by the additional insights that this display provides.

We now describe how to examine such displays, using data from four studies. Because no small set of examples can manifest all the possible patterns you might encounter in practice, our goal is to develop your intuition about how the functions behave and interrelate. This presentation is similar to that of section 10.3 (in discrete time). But here, we assume you are comfortable with survivor and hazard functions, allowing us to focus upon the new methods of this chapter.

Let us begin by describing briefly the four data sets summarized in figure 13.6. Panel A presents data from Cooney et al (1991), who tracked 89 recently treated alcoholics for two years to assess whether and, if so, when they first had a "heavy drinking" day. Panel B presents data from Zorn and Van Winkle (2000), who used archival data to assess the tenure (in years) of the 107 justices appointed to the U.S. Supreme Court between 1789 and 1980. Panel C presents data from Sorenson, Rutter, and Aneshensel (1991), who asked 2974 adults, ages 18 to 94, to recall whether and, if so, at what age (in years) they first experienced a depressive episode. Panel D presents data from Singer et al. (1998), who used administrative records to track, for up to 33 months, the length of employment (in days) of 2074 health care workers hired by community and migrant health centers. Kaplan-Meier estimates of the survivor function appear in the top row, negative log survivor function estimates of cumulative hazard appear in the middle row, and kernel-smoothed hazard functions appear in the bottom row. We do not present grouped estimates of the survivor or hazard functions for although they are simple to construct and easy to interpret, they are too coarse to provide the insights we seek.

Unlike discrete time, where we generally begin with the hazard function, in continuous time, we generally begin with the survivor function. We do so because it is the easiest to examine, in the sense that we: (1) can interpret its absolute level (which is difficult to do with cumulative hazard); and (2) have a value for all observed event times (which the smoothed hazard function cannot provide). We then use the smoothed

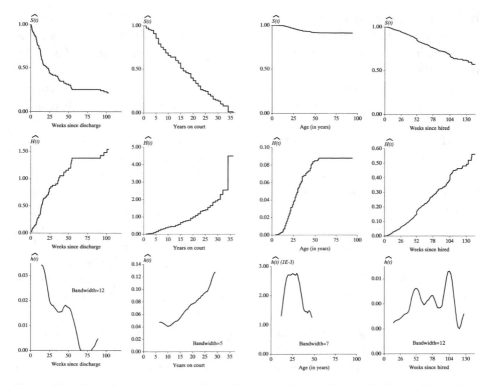

Figure 13.6. Sample survivor, cumulative hazard, and kernel-smoothed hazard functions for four different data sets. Panel A: time to first "heavy drinking day" among 89 recently treated alcoholics. Panel B: time to retirement or death among 107 U.S. Supreme Court Justices. Panel C: age at first depressive episode among 2974 adults. Panel D: employment duration among 2074 health care workers.

and cumulative hazard functions to identify periods of greater and lesser risk.

The four sample survivor functions in figure 13.6 display very disparate shapes. The first two—for the alcohol relapse study (Panel A) and the judicial career study (Panel B)—both drop quite low (albeit at very different paces). Because relapse is common among newly abstinent alcoholics, the estimated median lifetime is a mere 21.57 weeks (151 days) and the final estimated survival probability is only .2149 at 100.86 weeks (706 days). Event occurrence is also common in the judicial study because all justices must eventually retire or die. Its sample survivor function actually reaches 0 (something few do) and the estimated median lifetime is 16 years. Because event occurrence is less common in the other two studies, their sample survivor functions never even reach the halfway point (rendering median lifetime estimation impossible). For the depression onset data in Panel C, the lowest estimated survival probability is .9159 at 54 years, while for the health care employment data of Panel D, it is .5709 at 138.57 weeks (970 days).

These varying levels of survivorship have the expected relationship to hazard: a survivor function will be low only if the hazard is high during at least some of the time. But beyond this general statement, we cannot link the varying patterns of survivorship in figure 13.6 to specific *levels* of hazard because the studies clock time using different metrics. A link to hazard levels requires identical time metrics, which we have only for Panels A and D and Panels B and C, respectively. Comparing Panels B and C, where time is clocked in years, we see that hazard is generally higher for the judicial data (where the survivor function reaches 0) than for the depression data (where the survivor function never even reaches .90). Similarly, comparing Panels A and D, where time is clocked in weeks, we see that hazard is generally higher for the alcohol relapse data (where the survivor function nears 0) than for the health care employment data (where the survivor function never even reaches .50).

The need to attend to the metric for time is reinforced by the realization that we can only compare hazard rates across studies that use an identical time metric. Because a hazard rate assesses risk *per unit time*, a change of time metric produces a corresponding change of interpretation. If we clock time in days instead of weeks, the absolute levels of hazard in Panels A and D will drop because the risk of event occurrence *per day* (a smaller unit) must be lower. Similarly, if we clock time in months instead of weeks, the absolute levels of hazard will rise because the risk of event occurrence *per month* (a larger unit) must be higher.

This discussion highlights an important practical point: *the metric for analysis can differ from the metric of data collection*. Although both the alcohol relapse and health-care employment studies measured event occurrence in days, for example, we analyze their data using time clocked in weeks. This yields more meaningful statements because it is easier to understand a weekly hazard rate of relapse of .03 than a daily rate of $.03/7 = .0043$. Some might even argue that a more aggregated unit, like months, would be better still. If you convert finely measured event times into more aggregated units, be sure not to round the result. This discards potentially important information and increases the number of ties. (Lest you think that fractional values are difficult to handle, remember that the horn-honking study measured the timing of events to the nearest hundredth of a second.)

We now attempt to describe the shape of the underlying hazard functions. Because we have no direct estimates, we examine changes over time in both the absolute levels of smoothed hazard and the rate of increase in cumulative hazard. We typically begin with the smoothed hazard function because it directly depicts the changing level of risk. But because it cannot describe risk until one bandwidth later than the earliest observed event time, we also examine the changing slopes of the cumulative hazard function to gather information at these times. Unfortunately, the instability of the later values of cumulative hazard makes it difficult to learn much about risk at that end. For data sets with stable values—those with larger terminal risk sets such as the depression onset and health-care employment studies—we examine this end-stage behavior as well.

Begin with the alcohol relapse data shown in Panel A. The fairly regular decline in the smoothed hazard function suggests that the risk of relapse drops steadily after week 12. But what about the early weeks, just after release? To learn about this era, compare the cumulative hazard function's early and later rates of increase. Because the slope is shallower between weeks 0 and 12 than it is between weeks 12 and 18, we conclude that the risk of relapse is *lower* immediately after release than it is months later. This suggests that the former alcoholics go through a "safe" period—an initial phase when the risk of relapse is low. Safe periods are common in studies of relapse as newly committed individuals—be they former drinkers, smokers, or drug users—are able to abstain. As the novelty, support, and commitment diminish, the risk of relapse climbs, as in Panel A. This illustrates the importance of examining the cumulative hazard function. Had we examined only the smoothed hazard function, we would not have uncovered this phenomenon.

Next turn to the judicial data of Panel B, where the smoothed hazard function suggests that after 10 years of service, the risk of leaving office rises over time. Once again, this function cannot describe what happens during the early years, making the plateau between years 5 and 10 difficult to interpret. The cumulative hazard function, once again, provides the needed perspective. Comparing the rates of change in cumulative hazard before and after year 10, we see an approximately linear rise during the early years that is much less sharp than that of the later years. We conclude that hazard is low and relatively constant for approximately the first 10 years on the bench and then climbs steadily over time.

Panel C displays a common pattern for depression onset data. As in a similar study by Wheaton and colleagues (displayed in figure 12.5), risk peaks between ages 15 and 30. Before and after, hazard is low. Reinforcing this statement is the behavior of the cumulative hazard function. Its rates of change are fairly steady during the early and later phases of life, suggesting that the risk of onset is low during childhood and middle adulthood, peaking during adolescence and young adulthood.

Finally, examine the undulating smoothed hazard function in Panel D. This function differs dramatically from the others in that it peaks *several* times. If you examine when these peaks occur, you will notice they correspond to "anniversaries," approximately one year, 18 months, and two years after hire. A health-care worker is at much greater risk of leaving near these anniversaries. For some, the departure may result from the end of a contract; for others, it may reflect a "review process" initiated at these pre-specified times by either the worker or a supervisor. As the researchers could not distinguish voluntary and involuntary departures, nor did they have access to data about contractual obligations, we cannot know what causes these anniversary effects, yet their presence is of great interest. To describe the levels of risk immediately after hire, we turn to the cumulative hazard function. Its early shallow slope in comparison to its later steeper rise suggests that most employees give their job an initial chance.

We end with a word of caution about kernel smoothing. Although it can provide invaluable insight into the profile of risk over time, remember that its smoothed values do not actually estimate the population hazard function. The discrepancy between the smoothed hazard function and the unobservable full hazard function can be illustrated by examining the displays for the alcohol relapse data in Panel A. Notice that: (1) the cumulative hazard function plateaus around 52 weeks because no one relapsed for nearly one additional year; but (2) the smoothed hazard function decreases steadily during much of this time (between 52 and 68 weeks). Despite no events during the period, the smoothed hazard function drops because it uses values estimated at earlier and later times. The smoother provides insight into the profile of hazard over time, but its absolute values do not estimate the specific values.

14

Fitting Cox Regression Models

The future is something which everyone reaches at the rate
of 60 minutes an hour.

—C. S. Lewis

Having explored whether and, if so, when events occur in continuous time, as usual, we now examine whether variation in the risk of event occurrence varies systematically with predictors. In doing so, we focus exclusively on the most popular of possible methods: *Cox regression analysis* (often labeled the *proportional hazards model*).

Our goal in this chapter is to describe the conceptual underpinnings of the Cox regression model and to demonstrate how to fit it to data. Like the discrete-time hazard model, the Cox model expresses a transformation of hazard as a linear function of predictors, but the presence of a continuous-time metric demands that we change the way we posit the model, estimate parameters, interpret results, and communicate findings. We begin, in section 14.1, by developing the Cox model specification itself, demonstrating why it is a sensible representation. In section 14.2, we describe—conceptually and computationally—how the model is fit. In section 14.3, we examine the results of model fitting, showing how to interpret parameters, test hypotheses, evaluate goodness-of-fit, and summarize effects. We close, in section 14.4, by presenting strategies for displaying results graphically. Then, in chapter 15, we extend the basic model, relaxing its assumptions and demonstrating how to incorporate time-varying predictors into the specification.

14.1 Toward a Statistical Model for
Continuous-Time Hazard

We represent the population relationship between continuous-time hazard and predictors in much the same way as we represent the population relationship between discrete-time hazard and predictors. But because continuous-time hazard is a *rate*, not a probability, we treat its *logarithm*, not its logit, as the dependent variable. As you might expect from our earlier presentation, the new continuous-time model expresses log hazard as the sum of two components:

- A *baseline function*, the value of log hazard when the values of all predictors are 0.
- A *weighted linear combination of predictors*, whose parameters assess the shift in log hazard associated with unit differences in the corresponding predictor.

Owing to its origins in D. R. Cox's 1972 seminal paper, "Regression models and life tables," this representation is often known as the *Cox regression model*.

In this section, we introduce the Cox regression model and demonstrate why it offers a sensible strategy for analyzing continuous-time event occurrence data. Our presentation uses data collected by Henning and Frueh (1996), who tracked the criminal histories of 194 inmates released from a medium security prison to determine—to the nearest day— whether and, if so, when, the former inmates were re-arrested. During the period of data collection, which ranged from one day to three years, 106 former inmates (54.6%) were re-incarcerated. To develop the Cox regression model in a simple context, we first focus on the effect of a single dichotomous predictor, *PERSONAL*, which identifies the 61 former inmates (31.4%) with a history of person-related crimes (those who had more than one previous conviction for offenses such as simple assault, aggravated assault, or kidnapping). In later sections, we incorporate other predictors into the model, both categorical and continuous.

14.1.1 Plots of Within-Group Sample Functions

Plots of sample functions computed separately within groups distinguished by predictor values remain invaluable exploratory tools. Although it is easy to plot discrete-time and actuarial estimates, these displays are usually too coarse to be helpful. As explained in section 13.6, we therefore recommend that you plot: (1) Kaplan-Meier estimates of the survivor function; (2) negative log survivor estimates of the cumulative hazard function; and (3) kernel-smoothed estimates of the hazard function.

Figure 14.1 presents such plots for the recidivism data. Faint and bold lines represent groups of former inmates with and without a history of committing personal crime. Although event occurrence was recorded in

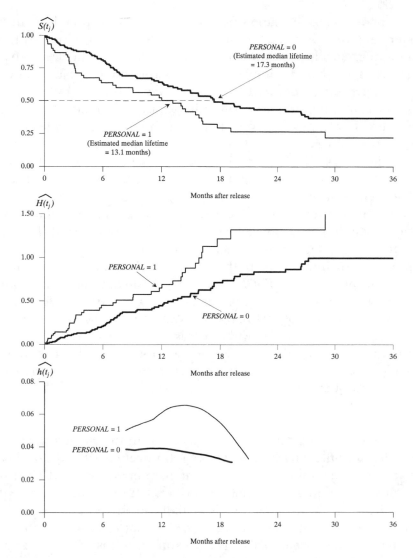

Figure 14.1. Estimated survivor functions, cumulative hazard functions, and kernel-smoothed hazard functions for the recidivism data, by *PERSONAL*, an indicator of whether the former inmate had committed more than one person-related offense prior to his focal incarceration.

days, we mark the time axis in months. To express the event times in days on a monthly scale, we divided each time by the length of an "average" month (365.25 ÷ 12), retaining fractional components. We believe that changing the time scale makes the metric more meaningful, but no temporal precision is lost because we have not rounded.

As in discrete time, we examine plots of within-group sample functions to learn about: (1) the shape of each group's hazard function; and (2) whether and how its level, or shape, differs across groups. Although kernel-smoothed hazard functions provide some insight, we focus primarily on survivor and cumulative hazard functions because they describe the distribution of event occurrence across the full range of event times. We can compute kernel-smoothed hazard estimates only during the middle of each group's event time distribution: one bandwidth later than the first observed event time and one bandwidth earlier than the last. In figure 14.1, for example, the first kernel-smoothed estimate appears at eight months, after 30 to 40% of the former inmates have already been re-arrested.

As recommended in section 13.6, we begin with the subsample survivor functions, which here drop well below .50 during the period of observation, indicating a high rate of recidivism in both groups. The growing separation of the functions with the passage of time highlights the importance of the predictor, *PERSONAL*. Estimated median lifetimes confirm the impressive magnitude of this effect. On average, an offender with a history of committing personal crime was arrested 13.1 months after release, whereas an offender without this history was arrested over four months later (17.3 months after release). Although recidivism is common in both groups, those with a history of person-related offenses are clearly at greater risk of re-arrest.

We next examine how the rate of increase in cumulative hazard changes over time to assess when the underlying hazard functions rise or fall. (When examining the middle range of event times, we also examine the kernel-smoothed hazard estimates.) Both cumulative hazard functions are approximately linear immediately after release, but each soon accelerates, beginning at somewhat different times. This suggests that each underlying hazard function is initially steady and then rises. For offenders with a history of committing personal crimes (the top function), acceleration occurs early, two to four months after release. For offenders without this history, acceleration occurs somewhat later, four to eight months after release. The subsequent deceleration of both cumulative hazard functions suggests that members of each group who survive these high-risk periods then face a lower risk. Although former inmates with a history of personal crime may face a second high risk period 12 to 15 months after release, so few people are still at risk that we do not emphasize this apparent acceleration.

Finally, we compare the *level* of cumulative hazard across the groups to evaluate just how different the underlying hazard functions are. Beyond the minor differences in the location of each group's initial acceleration, their *levels* of cumulative hazard differ markedly. The cumulative hazard function for those with a history of personal crime is consistently higher; the two functions never cross and the distance between them widens. A similar differential in risk is apparent in the kernel-smoothed hazard estimates, although we can see the discrepancy only in the middle range of event times. This suggests, once again, that the predictor *PERSONAL* is related to the risk of recidivism.

14.1.2   What Type of Statistical Model Do These Graphs Suggest?

It is difficult to move directly from the sub-sample graphs in figure 14.1 to an appropriate statistical model for hazard in the population because we even lack a full picture of hazard's values over time in the sample. We therefore take a different route, working first with the cumulative hazard functions in the middle panel. After developing a model that seems reasonable for cumulative hazard, in the next section we transform *the entire model* into an equivalent form expressed in terms of raw hazard. Although seemingly circuitous, we take this path because it reinforces the link between the sample functions we introduced for descriptive analyses and the statistical models we soon fit.

To specify an appropriate statistical model for cumulative hazard, we must first attend to its semi-bounded nature: bounded from below by 0, but without bound above. The transformations used in discrete time—logit and clog-log—are not appropriate here because they are undefined for quantities greater than 1. The logarithmic transformation, in contrast, lends itself nicely to the task for it is defined for any positive value. Taking logarithms of cumulative hazard yields a new unbounded function, the *log cumulative hazard function*, which can vary fully between $-\infty$ and $+\infty$. Because many software packages label the cumulative hazard function the "negative log survivor function," you often find its logarithmically transformed version labeled the "log negative log survivor function," the "log-log survivor function," or "log($-\log(S(t))$)."

The top panel of figure 14.2 displays log cumulative hazard functions for the recidivism data. Don't be alarmed by the negative numbers; log cumulative hazard is negative whenever cumulative hazard is less than 1. Comparing this graph and its raw version in figure 14.1 (middle panel) shows how taking logarithms expands the vertical distance between small values and compresses the vertical distance between large values. Because cumulative hazard is smallest immediately after release and grows over time, transformation further distinguishes initial values and renders later values more similar. This makes it easier to discern early between-group differences (which are more precise) and it diminishes differences in

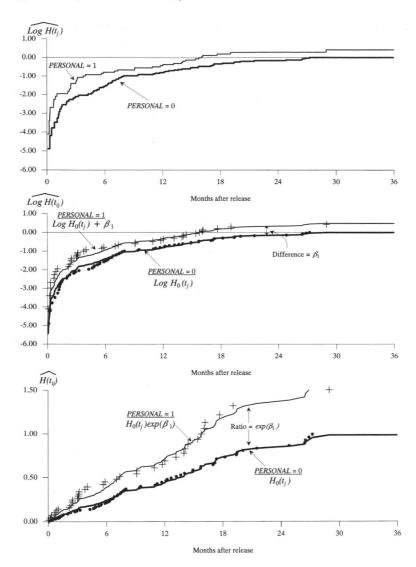

Figure 14.2. Developing the hypothesized Cox regression model. Top panel presents estimated log cumulative hazard functions for the predictor *PERSONAL* in the recidivism data. Middle panel presents estimated log cumulative hazard values as explicit points and hypothesized log cumulative hazard functions as smooth curves. Bottom panel is identical to middle panel except for a change of scale.

later values (which are generally more erratic). As an added benefit, transformation often stabilizes the distance between the functions over time, as it does here. So, not only does taking logarithms solve the boundedness problem, it has other salutary effects as well.

Now examine the top panel of figure 14.2 and ask: What kind of sta-

tistical model would provide a reasonable representation of the population relationship between log cumulative hazard and its predictors (here,
just *PERSONAL*)? As in discrete time, a dual partition makes sense, one
that expresses the entire log cumulative hazard function as the sum of a
baseline (log cumulative hazard) function and a weighted linear combination of predictors. And for the predictor *PERSONAL*, at least, the
relative distance between the sample log cumulative hazard functions is
relatively constant over time, suggesting the soundness of associating a
constant shift—say β_1—per unit difference in its values.

The real dilemma is the specification of the baseline. For this data set—
and most others—a straight line won't do. Although it is tempting to say
that we need "a curve," it is impossible to select, a priori, an appropriate
curve from the thousands of possible options. Different data sets will yield
log cumulative hazard functions with different shapes, and a curve suitable for one is unlikely to be suitable for another. In discrete time, we
resolved this dilemma by specifying a *completely general* baseline function,
using a series of dummy variables as predictors. In continuous time, a
completely general baseline also makes sense, but we cannot use a finite
set of dummy variables to specify its shape because we have an *infinite*
number of instants when the event could occur. Cox resolved this
dilemma by postulating the existence of a completely general baseline, a
function whose shape is entirely unconstrained. This baseline can take
on any shape necessary to describe the distribution of event occurrence
in the population adequately. Unlike the discrete-time hazard model,
however, we do not offer *any* specification for its shape, indicating only
that it has *some shape* and is a continuous function. Although this vague
characterization may seem problematic for parameter estimation, we will
soon demonstrate in section 14.2 that it is anything but.

Letting $\log H_0(t_j)$ represent the completely general baseline log cumulative hazard function, and using the same kind of logic that we employed
when setting up our earlier discrete-time hazard models, we can write a
Cox regression model for the recidivism data as:

$$\log H(t_{ij}) = \log H_0(t_j) + \beta_1 PERSONAL_i$$

where we add the subscript i to the variable *PERSONAL* to indicate that
the predictor takes on the appropriate value for each individual. We use
the subscript 0 to identify the baseline function because it represents the
value of the outcome (here log cumulative hazard) when all predictors
in the model (here, just *PERSONAL*) are set to 0.

One way of understanding what a Cox regression model stipulates
about the population relationship between event occurrence and predictors is to substitute in the predictor's two values:

when $PERSONAL = 0$: $\log H(t_{ij}) = \log H_0(t_j)$
when $PERSONAL = 1$: $\log H(t_{ij}) = \log H_0(t_j) + \beta_1$.

For offenders without a history of person-related crime, the model postulates that their log cumulative hazard function is represented by the completely general baseline, $\log H_0(t_j)$. For offenders with a history, the model postulates the existence of another log cumulative hazard function, $\log H_0(t_j) + \beta_1$. This second function is just a vertically shifted version of the first, where β_1, the parameter associated with the predictor, PERSONAL, measures the size of the vertical displacement.

Another way of understanding this model is to map its algebraic representation onto a graph of the sample log cumulative hazard functions. We take this approach in the middle panel of figure 14.2, where we use +'s and •'s to denote estimated subsample values. (Eliminating the line segments linking the estimates streamlines the display.) Each superimposed curve represents the hypothesized value of that group's population log cumulative hazard function. As hypothesized functions, we do not expect either to perfectly coincide with the sample values (although they will be similar if the model fits well). The lower curve is the hypothesized baseline, $\log H_0(t_j)$, the value of log cumulative hazard when PERSONAL $= 0$. The upper curve is identical to this function but shifted vertically by the constant amount, β_1.

The vertical distance between the functions assesses the magnitude of the predictor's effect. In postulating that this distance is always β_1, we stipulate that its effect is constant over time. The effect of the predictor does not depend on how long the offender has been out of prison. The model stipulates that its effect is identical for men who left prison yesterday and those who have been arrest-free for years.

Having written a Cox model in terms of log cumulative hazard, it is easy to re-express the model in terms of *raw* cumulative hazard. Taking the antilog of both sides, we have:

$$H(t_{ij}) = H_0(t_j)e^{(\beta_1 PERSONAL_i)}.$$

Notice what happens when we transform the model. The right side, which previously was a linear function, now includes both multiplication and exponentiation—it is nonlinear. To explore what this nonlinear representation means in practice, we apply the same two strategies used to explore the model in its original form: (1) substituting in different values of the predictor; and (2) superimposing the resultant hypothesized functions on a graph of sample estimates.

Substituting in the two values for the predictor PERSONAL we have:

$$\text{when } PERSONAL = 0: H(t_{ij}) = H_0(t_j)$$
$$\text{when } PERSONAL = 1: H(t_{ij}) = H_0(t_j)e^{(\beta_1)}.$$

For offenders with no history of personal crime, the hypothesized cumulative hazard function is still the completely general baseline, $H_0(t_j)$. But for offenders who *do* have a history of person-related crime, their hypothesized cumulative hazard function is $H_0(t_j)e^{(\beta_1)}$, which is *not* a linearly shifted version of the baseline. This second function is the *product* of the baseline and the antilog of the parameter estimate. This means that the two postulated cumulative hazard functions are *not* equidistant.

The varying gap between the two postulated cumulative hazard functions can be seen clearly in the bottom panel of figure 14.2. This graph is identical to that in the middle panel, except for a change of scale for the Y axis—from log cumulative hazard to raw cumulative hazard. Transformation converts the *constant* vertical distances between the functions in the middle graph into *increasing* vertical distances in lower graph. The cumulative hazard functions are anything but equidistant: in fact, the upper is a *magnification* of the lower. The magnitude of the vertical distance between the two, at any given moment, t_j, is:

$$H_0(t_j)e^{(\beta_1)} - H_0(t_j) = H_0(t_j)[e^{(\beta_1)} - 1].$$

This vertical distance depends not only on β_1 but also on the value of the baseline cumulative hazard function at that time, $H_0(t_j)$. In the scale of cumulative hazard, then, the shift associated with unit differences in a predictor is *relative* to the baseline hazard function at that point in time— it is *not* absolute.

Despite the varying vertical distance between the two postulated cumulative hazard functions, we still say that the effect of the predictor *PERSONAL* is constant over time. One reason for retaining this terminology is semantic: the effect of the predictor in the transformed model must be identical to the effect in the original model because the two models are the same. To move from one to the other, we have simply transformed both sides. The underlying model *has not changed*.

Beyond this semantic justification, it is important to understand how a predictor's effect can be constant over time if the vertical distance between hypothesized functions varies. To resolve this paradox, remember that there is more than one way for a predictor to have a time-constant effect. One way is for the *absolute* difference between two functions to be identical over time (as in the middle panel of figure 14.2). Another way is for the *ratio* of two functions—their *relative* distance—to be identical over time. Computing the ratio of the two postulated cumulative hazard functions for this model, we have:

$$\frac{H_0(t_j)e^{(\beta_1)}}{H_0(t_j)} = e^{(\beta_1)}.$$

The baseline cumulative hazard function, $H_0(t_j)$, cancels out of both numerator and denominator, leaving the antilog of the coefficient for *PERSONAL*, $e^{(\beta_1)}$. Because this ratio does not depend upon the baseline cumulative hazard function—or any other function of time, for that matter—it is appropriate to say that the effect of the predictor is constant over time.

In essence, then, the fundamental difference between the two representations of the Cox model surrounds the metric in which we interpret the time-constant effect: (1) $e^{(\beta_1)}$ measures the effect of the predictor on raw cumulative hazard; (2) β_1 measures the effect of the predictor on log cumulative hazard. In both cases, the effect of the predictor is assumed to be time-constant; all that changes is the reference function used to interpret the effect.

We close this section by extending these simple models to the general case of multiple predictors, both continuous and categorical. Letting X_1 through X_P represent P generic predictors, we may write the general Cox regression model as:

$$H(t_{ij}) = H_0(t_j)e^{[\beta_1 X_{1ij} + \beta_2 X_{2ij} + \cdots + \beta_P X_{Pij}]}. \tag{14.1}$$

where we add the subscript j to each substantive predictor so that they may, if necessary, take on time-varying values. Similarly, we may write the Cox model in terms of log cumulative hazard as:

$$\log H(t_{ij}) = \log H_0(t_j) + [\beta_1 X_{1ij} + \beta_2 X_{2ij} + \cdots + \beta_P X_{Pij}]. \tag{14.2}$$

When we invoke these more general representations in the context of other data sets, remember that all the properties just articulated for the recidivism data apply to these situations as well.

14.1.3 A Hazard Function Representation for the Cox Model

The cumulative hazard formulation of the Cox model is not the only representation possible. Mathematical identities allow us to express this model in terms of hazard directly. Because a detailed explanation of the re-expression requires the use of calculus, in this section we illustrate *graphically* what happens when the model is transformed.

The left side of figure 14.3 presents a pair of hypothesized population cumulative hazard functions displayed on a log scale (top row) and raw scale (bottom row). Unlike the hypothesized functions in the middle

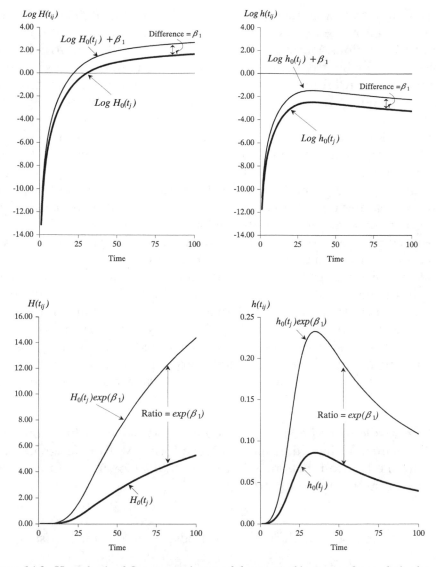

Figure 14.3. Hypothesized Cox regression model expressed in terms of cumulative hazard and raw hazard. Left panel is on a cumulative hazard scale (log and raw). Right panel is on a hazard scale (log and raw).

panel of figure 14.2, these are totally smooth. This regularity is no accident. As discussed in section 13.4.1, we can express a cumulative hazard function in terms of its corresponding hazard function only if both result from identical population assumptions about the distribution of event occurrence over time. These cumulative hazard functions are *idealized* representations of what these functions might look like if the distribution of event occurrence in a population conforms perfectly to a set of parametric assumptions. Working in this idealized arena allows us to construct a plot unattainable from real data: the hazard functions upon which the cumulative hazard functions are based. These are displayed on the right side of figure 14.3, using both a log scale (top row) and raw scale (bottom row).

In addition to reflecting specific assumptions about the distribution of event occurrence in the population, each graph on the left side of figure 14.3 was constructed so as to perfectly conform to the assumptions of the Cox model: the log cumulative hazard functions are separated by a constant *absolute* amount, β_1; the raw cumulative hazard functions are separated by a constant *relative* amount, $e^{(\beta_1)}$.

Now compare the vertical distances between the hypothesized hazard functions on the right side of figure 14.3 to the vertical distances between the cumulative hazard functions on the left side. When doing so, notice that although the scale for the raw functions in the bottom row differs (if it didn't, we could not provide adequate graphical resolution), the scale for the log functions is identical. Close comparison of the two sets of plots shows that: (1) both *log* functions are separated by a constant *absolute* amount, β_1; and, (2) both *raw* functions are separated by a constant *relative* amount, $e^{(\beta_1)}$. In other words, we find identical distances between each pair of functions whether they assess cumulative hazard or hazard!

These equivalencies suggest that the Cox model takes on an identical form, regardless of whether the outcome (and baseline) functions represent cumulative hazard or hazard. More formally, if we designate $h_0(t_j)$ the baseline hazard function, we may rewrite the Cox model in equation 14.1 by substituting $h(t_{ij})$ for $H(t_{ij})$:

$$h(t_{ij}) = h_0(t_j)e^{[\beta_1 X_{1ij} + \beta_2 X_{2ij} + \cdots + \beta_P X_{Pij}]}. \tag{14.3}$$

Taking the antilog of both sides, we can also write this model as:

$$\log h(t_{ij}) = \log h_0(t_j) + [\beta_1 X_{1ij} + \beta_2 X_{2ij} + \cdots + \beta_P X_{Pij}]. \tag{14.4}$$

Comparing equations 14.3 and 14.4 with equations 14.1 and 14.2 reveals that the two sets of models are identical, except for the substitution of hazard for cumulative hazard. This direct equivalence may not be intuitive, but it certainly is invaluable.

The representations in equations 14.3 and 14.4 allow us to articulate the assumptions of the Cox model using the metric of hazard (not cumulative hazard as we did before). For simplicity, let us consider a model with just one substantive predictor. In this context, the Cox model assumes:

1. *For each value of the predictor, there is a postulated log hazard function.* If the predictor is dichotomous, we postulate two log hazard functions. If the predictor is continuous, we postulate that there are as many log hazard functions as there are values of the predictor. (If the model includes two or more predictors, we postulate that there are as many log hazard functions as there are *combinations* of predictor levels.)

2. *Each of these log hazard functions has an identical shape,* although we don't place any constraints on the specification of that shape. The shape of each log hazard function is constrained to be the same across all predictor values. Within this constraint, the function can take on *any* form necessary to adequately describe the distribution of event occurrence in the population.

3. *The distance between each of these log hazard functions is identical at every possible instant.* Regardless of the common *shape* of the postulated log hazard functions, the difference in their *level* is constant over time. We do not allow the gap to be smaller during some periods of time and larger during others. This means that the effect of the predictor on log hazard is constant over time.

In chapter 15, we relax some of these assumptions, specifying more general models. For this chapter, however, we assume that these assumptions hold.

The hazard function formulations in equations 14.3 and 14.4 are the most popular ways of representing the Cox model. Their appeal undoubtedly stems from the fact that they are stated in the most intuitively appealing metric in which we can work: hazard. Yet we cannot overstate the importance of their equivalence with the cumulative hazard representation, which allows us to:

- *Directly interpret parameters from the Cox model in terms of a predictor's effect on hazard.* This allows us to make statements in a meaningful metric, even though we have no estimate of the hazard function itself.
- *Use sample cumulative hazard functions as our visual window on the effects of predictors.* When we examine the vertical distances between sample log cumulative hazard functions (as in the top panel of figure 14.2), we are assessing the very distances that we would observe were we able to plot the sample log hazard functions themselves (which we cannot).

This allows us to (1) move effortlessly between the cumulative hazard metric for exploratory work (the one that is more easily implemented) and the hazard metric for interpretation (the one that is more intuitively meaningful); and (2) circumvent our inability to estimate hazard well because our data analytic tools provide us with the very same graphical information that we would have, were we able to examine the underlying hazard functions directly.

We close this section by discussing why Cox's model is often labeled the *proportional hazards model.* As shown in the right side of figure 14.3, a constant vertical distance in the log hazard scale converts to a constant ratio (a constant multiple) in the hazard scale. (This scale conversion is identical to that discussed in section 14.1.2 in the context of cumulative hazard.) Whenever the ratio of two functions is constant over time, we say that the two functions are *proportional.* Since the Cox model assumes that the population hazard functions are proportional, the phrase "proportional hazards model" has become a common way of referencing this specification.

Despite its popularity, we do not routinely use the term "proportional hazards model" to refer to Cox's model. Although we currently invoke a proportional hazards assumption, we show (in chapter 15) that it is easy to relax this assumption and fit Cox models in which the hazard functions are *not* proportional. If we referred to the Cox model as the "proportional hazards model," we would soon find ourselves fitting "nonproportional proportional hazards models"! We therefore use the more neutral term "Cox regression model" throughout, even though we recognize, and remind you, that the term "proportional hazards model" is more widespread in use.

14.2 Fitting the Cox Regression Model to Data

Postulating a statistical model is easy; fitting it to data can be hard. Cox's landmark 1972 paper is remarkable because it offered both a compelling model and an ingenious new method of estimation. Originally labeled *conditional maximum likelihood* estimation, Cox (1975) expanded the method into a broader class known as *partial maximum likelihood* estimation. Although partial maximum likelihood estimation is not without limitations—most notably, it does not provide direct estimates of the baseline functions—its estimates share many of the asymptotic properties associated with other ML estimates: consistency, efficiency, and normality.

Below, we describe what happens when the partial maximum likelihood estimation method is used to fit a Cox regression model to data. In section 14.2.1, we provide an overview of the mathematics of the method—we expect that most readers will skip this subsection on first reading. In section 14.2.2, we identify three important practical consequences of using the method that all empirical researchers need to understand. Please be sure to read this latter subsection even if you skip over the mathematical details.

14.2.1 The Partial Likelihood Method of Estimation

Like full ML estimation, partial ML estimation uses a two-step process in which we (1) construct a partial likelihood function—an equation that expresses the probability of observing the sample data as a function of the unknown parameters; and (2) numerically examine the relative performance of alternative estimates of the unknown parameters until those values that maximize the partial likelihood are found. The fundamental difference between the full and partial ML methods is that we construct a full likelihood function by asking: What is the probability that individual i experiences his or her observed event time? Whereas, we construct a partial likelihood function by "conditioning" on the observed event times and asking: Given that *someone* experienced an event at time t_j, what is the probability that it was *individual i*?

This conditional argument has an important effect on the partial likelihood function: only those individuals who actually experience the target event contribute an explicit term. This means that the likelihood function will include only as many explicit terms as there are individuals with observed event times. This contrasts markedly with what happens under full ML, in which *every* person contributes an explicit term, regardless of whether his or her event time is observed or censored.

The first step in constructing the partial likelihood is to evaluate each person's contribution: the conditional probability that individual i experiences the event at time t_j, given that someone still at risk does. We compute this probability by dividing individual i's hazard at time t_j by the sum of all the contemporaneous hazards faced by everyone (including individual i) at risk at that time:

$$\text{Individual } i\text{'s contribution to the partial likelihood at } t_{ij}^* = \frac{h(t_{ij}^*)}{\sum\limits_{\text{risk set at } t_{ij}^*} h(t_{ij}^*)},$$

$$(14.5)$$

where t_{ij}^* identifies the time (t_j) when individual i experiences the target event. Someone who faces a high hazard when he or she experiences the event will contribute a large term; someone who faces a low hazard will contribute a small term. Notice that even though continuous-time hazard is not a probability, the ratio of an individual's hazard at a given time to the total amount of contemporaneous hazard *is*.

Because each individual's contribution to the partial likelihood is computed when he or she experiences the event, only individuals with non-censored event times have an explicit term. This is not to say that these people contribute only to this one term, or that censored individuals never contribute at all. Both contribute *indirectly* through their participation in the *denominator* of the explicit contributions for any noncensored individual whose event time is less than (or equal to) their observed (or censored) event time. The person with the 12th earliest observed event time has one *explicit contribution*, when he or she experiences the event, and 11 implicit contributions, in the partial likelihoods of each of the 11 individuals with earlier observed event times. The person with an event time censored between the 11th and 12th observed event times will similarly affect the contributions of the 11 people who previously experienced the event.

To construct the partial likelihood, we multiply together all the individual contributions:

$$\text{Partial likelihood} = \prod_{\substack{noncensored \\ individuals}} \frac{h\left(t_{ij}^*\right)}{\sum_{\substack{risk\ set\ at\ t_{ij}^*}} h\left(t_{ij}^*\right)}.$$

Like all likelihood functions, this one is the product of probabilities. It expresses the probability of observing the data actually observed (the event times) as a function of unknown population parameters. In its present form, the unknown parameters are the population values of hazard at each observed event time among the individuals who experienced the event and those still in the risk set.

Where do the unknown parameters of interest—the baseline hazard function and the β's—come in? Reflection on the model in equation 14.4 reveals their presence through the expression for the population values of hazard. To make their appearance explicit, let us substitute equation 14.4 into the partial likelihood to write:

$$\text{Partial likelihood} =$$

$$\prod_{\substack{noncensored \\ individuals}} \frac{h_0(t_j)e^{\left(\beta_1 X_{1ij} + \beta_2 X_{2ij} + \cdots + \beta_P X_{Pij}\right)}}{\sum_{\substack{risk\ set\ at\ t_{ij}^*}} h_0(t_j)e^{\left(\beta_1 X_{1ij} + \beta_2 X_{2ij} + \cdots + \beta_P X_{Pij}\right)}}. \tag{14.6}$$

Both the baseline hazard function $h_0(t_j)$ and the β's have appeared, demonstrating that the partial likelihood can be written solely using a combination of predictors and unknown parameters.

The elegance of Cox's approach becomes apparent when we notice that every term in equation 14.6 includes $h_0(t_j)$, the unknown baseline hazard. Dividing throughout by $h_0(t_j)$ we have:

$$\text{Partial likelihood} = \prod_{\substack{noncensored \\ individuals}} \frac{e^{(\beta_1 X_{1ij} + \beta_2 X_{2ij} + \cdots + \beta_P X_{Pij})}}{\sum_{risk\ set\ at\ t_{ij}^*} e^{(\beta_1 X_{1ij} + \beta_2 X_{2ij} + \cdots + \beta_P X_{Pij})}}. \tag{14.7}$$

All traces of the baseline hazard function have disappeared. Invoking the proportionality assumption allows us to use the partial likelihood method to fit the model without specifying *anything* about the shape of the baseline hazard. The fact that the baseline postulated is totally vague poses no problem because we do not even bother to estimate it. (And because we can extend this approach to *non*-proportional hazard models—using strategies discussed in chapter 15—we can actually use this method in an even wider array of situations.)

As with all ML methods, the next step is to simplify the estimation task by taking logarithms:

$$\text{Partial log likelihood} = \sum_{\substack{noncensored \\ individuals}} \left[(\beta_1 X_{1ij} + \cdots + \beta_P X_{Pij}) - \log \sum_{risk\ set\ at\ t_{ij}^*} (\beta_1 X_{1ij} + \cdots + \beta_P X_{Pij}) \right]. \tag{14.8}$$

Using iterative numerical methods, a computer can then obtain those estimated β's that maximize this function.

14.2.2 Three Important Practical Consequences of Using Cox's Method

Taken together, the Cox regression model and the method of partial likelihood constitute an elegant statistical feat and provide a powerful strategy for examining the effects of predictors on continuous time hazard. But Cox's approach has some hidden features that have practical implications for data analysis. Although knowledge of the mathematical underpinnings of the method helps identify the *source* of these implications, you need no such knowledge to understand their *consequences*. Below, we outline three important consequences of the Cox regression model that all empirical researchers should understand: (1) that the shape of the baseline hazard function is irrelevant; (2) that the precise event times are irrelevant, too (only their *order* counts); and (3) that ties can create analytic difficulties.

The Shape of the Baseline Hazard Function Is Irrelevant

When we postulated the Cox regression model in section 14.1, we discussed the concept of a completely general baseline hazard function–one that was as complex as was necessary for describing the distribution of risk in the population. At the time, it seemed that the use of a completely general baseline hazard function might pose problems for analysis, in that it is difficult to imagine how we could estimate such a function without exhausting most of the degrees of freedom available in any data set. In practice, however, when we fit a Cox model to data, we evaluate only the predictors' effects—we do not investigate the shape of the baseline hazard function at all because Cox's method of partial likelihood invokes a conditioning argument that eliminates it entirely from consideration. (For those who read the mathematical presentation in the previous section, notice that $h_0(t_j)$ does not appear in the partial log-likelihood function in equation 14.8.)

The intentional irrelevance of the baseline hazard function in a Cox regression analysis is the raison d'être of Cox's approach. One important implication is that we need not stipulate *anything* about the baseline hazard function when we fit a Cox model because we never even consider its level or its shape. We can fit a Cox regression model regardless of whether the baseline hazard function is high or low, flat or peaked, undulating or steady. We can fit a Cox model even if the baseline has 100 irregularly spaced peaks! No function is too simple or too complex. All we need assume is that the effect of predictors on the unknown baseline hazard function is identical on every observed occasion of measurement. If we are willing to invoke this proportionality assumption—a small price to pay, it seems, as it often holds—we can fit a Cox regression model by the method of partial likelihood and ignore the shape of the baseline hazard function with impunity.

This flexibility has made the Cox model the most popular method for analyzing continuous time-event occurrence data. Other methods—and there are many—invoke parametric assumptions about the shape of the baseline hazard function that may, or may not, be met in practice. When Cox proposed his new model and method in 1972, critics suggested that it might be inferior to parametric methods. This concern was largely laid to rest when Efron (1977) demonstrated that the method is nearly as efficient even if the underlying hazard function has a known identifiable shape, and is far more efficient if it does not. With this positive evaluation, empirical use of Cox regression analysis mushroomed. In fields such as medicine, it is the most popular statistical method in use today.

The irrelevance of the baseline hazard function in Cox regression analysis does have one unfortunate consequence: a fitted model does not provide predicted values of hazard. To offset this shortcoming, the best we can do is use nonparametric strategies to "recover" survivor and cumulative hazard functions (using techniques described in section 14.4). So, although Cox's method provides a strategy for evaluating the effects of predictors on the hazard function—effects that are the primary interest of most researchers—it does not provide a way of estimating the hazard function itself. Most empirical researchers find this problem a small price to pay.

The Precise Event Times Are Irrelevant: Only Their Rank Order Matters

Unlike most methods for the statistical modeling of continuous random variables, Cox's method does not use the precise values of the event times to compute parameter estimates, standard errors, or goodness-of-fit statistics. Instead, it uses the values of each person's *predictors* at each observed event time. The observed event times play a role only through their *relative ranking*. To estimate parameters, we compare a weighted linear combination of each person's predictor values to a similar weighted linear combination among everyone still in the risk set when that person experiences the target event. (Those who read the mathematical explanation in section 14.2.1 can confirm this observation by noticing that the partial log likelihood function in equation 14.8 does not include the observed event times.)

The consequences of this observation can be startling. The actual event times—the very data that you have taken great pains to gather in as precisely a fashion as possible—are irrelevant. You need only know the *rank order* of the event times: Who came first, who came second, and where the censored observations fall in between. You would obtain identical parameter estimates, standard errors, and goodness-of-fit statistics whether you analyzed time assessed in days, weeks, months, or years. You could even add, or subtract, a constant to each event time! You would even obtain identical results were you to *increase* some event times and *decrease* others, *as long as you do not alter their ranking*! To convince yourself of this unusual property, we suggest that you try these manipulations in an actual data set.

Because Cox's method depends only on the ranks of the observed event times, not their actual values, you might conclude that the model is *nonparametric*. This is not strictly true because the model does indeed make several parametric assumptions. It assumes, for example, that the effect of each predictor on hazard is identical over time. It also assumes that the shift in log hazard corresponding to unit differences in each predictor is identical across the full predictor range. Because the model invokes assumptions about the functional form that links predictors to (log) hazard, the *nonparametric* label is overstated. The more accurate term that statisticians prefer to use to describe the model is *semi-parametric*.

Ties Can Create Analytic Difficulties

The Cox regression model has one serious Achilles' heel—its sensitivity to *ties*. Even though the actual event times are unimportant to model fitting, their *ranked* values must be extremely precise lest imprecision lead to duplicates. In section 12.1, we showed that truly continuous data should contain no observed ties. But we also asserted that most data sets do contain ties, often because of rounding. In the recidivism data, for example, where event occurrence was recorded to the nearest day, one event time is shared by 28 people (everyone censored at three years), ten event times are shared by 2 individuals, and one event time is shared by 3 people. If time were measured using a truly continuous metric—an assumption inherent in the partial likelihood method—observed ties would not exist.

Because ties are inevitable in real data, statisticians have developed several modifications to Cox's basic approach. For instance, if a censored time happens to coincide with an observed time, we assume that the event *precedes* the censoring. (This is identical to assumptions made when constructing a Kaplan-Meier estimate of the survivor function.) This places the censored individual in the observed individual's risk set (making it possible to evaluate that person's contribution to the partial likelihood).

What if two or more individuals share an *observed* event time? In the recidivism data, two people were arrested on each of five different days: 9, 77, 178, 207, and 528 days after release. Although the precise time of arrest surely differed within pairs, only the *day* was recorded, thereby creating the ties.

Of the various available methods for dealing with ties, the most compelling approach is the *exact* method (Peto, 1972; Kalbfleisch & Prentice,

1980). Under this approach, we calculate each tied observation's contribution to the partial likelihood by evaluating its value under all possible underlying orderings that might exist. If two people are tied, this requires two additional calculations, but as the number of people tied at any specific event time increases, the number of calculations escalates exponentially, making this computationally intensive method impractical for everyday use.

Improvements in computation will eventually eliminate such pragmatic concerns, but until then, most statistical packages implement one of two approximations to the exact solution:

- *Breslow (1974)–Peto (1972) approximation.* Instead of considering *all* possible underlying orderings that might exist, assume that the observed ties occurred sequentially.
- *Efron (1977) approximation.* Consider all possible underlying orderings that might exist, but use a numeric *approximation* to simplify the computations.

There is also a third approach (due to Kalbfleisch & Prentice, 1973), but owing to its poor performance in comparison to these alternatives, we do not consider it here.

How should you handle the ties that appear in your data? When more powerful computers and sophisticated software eliminate the need for approximations, the exact approach will be the method of choice. Even today, if the number of ties is small—and, more important, if the number of cases tied at each time is smaller still—you should consider the exact method if it is available in your software package and you find it to be computationally feasible. Otherwise, simulation studies suggest that Efron's approximation is superior to Breslow's and Peto's (Hertz-Picciotto & Rockhill, 1997). For this reason, we use Efron's approximation for all the models fitted in this chapter and the next. In addition, remember that if your data contain many ties—if, for example, there are more *tied* than *unique* event times—none of these continuous time strategies works well. When this happens, we recommend—as does Cox himself—the discrete-time methods of chapters 11 and 12.

14.3 Interpreting the Results of Fitting the Cox Regression Model to Data

Most statistical packages include a routine for fitting the Cox regression model to continuous-time event history data. Although packages differ in many ways—their strategies for handling ties, the availability of residual diagnostics, the ease with which time-varying predictors can be included, and the ability to recover baseline functions—they are essentially identical when it comes to fitting the basic models with time-invariant predictors discussed in this chapter. The differences that do exist surround the choice—and form—of statistics displayed in the post-analysis output. Regardless of the particular package you choose, you should be able to obtain all summary statistics described in this section through simple numeric manipulation.

Table 14.1 presents the results of fitting four Cox regression models to the recidivism data. These analyses investigate the effects of *PERSONAL* and two other predictors: *PROPERTY*, a dummy variable indicating whether the former inmate was previously convicted for a property-related offense; and *AGE*, which assesses the impact of the former inmate's age (in years) at the time of release. For reasons described in section 14.3.4, we center age around its sample mean (30.7) and so predictor *AGE* represents the *difference* in age upon release between each individual and the average inmate under study. Models A, B, and C contain each predictor individually; Model D includes all three simultaneously. Below, we discuss strategies for: interpreting parameter estimates (section 14.3.1); evaluating model goodness-of-fit (section 14.3.2); testing statistical hypotheses (section 14.3.3); and using risk scores to communicate the effects of several predictors simultaneously (section 14.3.4).

14.3.1 Interpreting Parameter Estimates

When we fit a Cox regression model to data, we obtain parameter estimates that assess and summarize predictors' effects. We do not obtain parameter estimates that describe the baseline hazard function because, as discussed in section 14.2.2, this function is ignored entirely when the model is fitted by the partial likelihood method. Although we can ultimately recover an estimate of the baseline function using nonparametric methods described in section 14.4, most empirical researchers focus on the numeric summaries of predictors' effects, as we do here.

Corresponding to the two representations of the Cox model in equations 14.3 and 14.4, each parameter estimate can be interpreted in two interrelated ways:

- Each raw coefficient describes the effect of a one-unit difference in the associated predictor on log hazard.
- The antilog of each raw coefficient, $e^{(coefficient)}$, describes the effect of a one-unit difference in the associated predictor on *raw* hazard.

Table 14.1: Results of fitting four Cox regression models to the recidivism data ($n = 194$, $n\ events = 106$)

	Model A	Model B	Model C	Model D
Parameter Estimates and Asymptotic Standard Errors				
PERSONAL	0.4790*			0.5691**
	(0.2025)			(0.2052)
PROPERTY		1.1946***		0.9358**
		(0.3493)		(0.3509)
AGE			−0.0681***	−0.0667***
			(0.0156)	(0.0168)
Hazard Ratios and Their Asymptotic Standard Errors				
PERSONAL	1.6144***			1.7659***
	(0.3268)			(0.3624)
PROPERTY		3.3022***		2.5482**
		(1.1535)		(0.8941)
AGE			0.9342***	0.9355***
			(0.0146)	(0.0157)
Goodness-of-fit				
LL	−492.04	−486.60	−483.22	−475.22
−2LL	984.08	973.20	966.43	950.44
LR statistic	5.32	16.20	22.97	38.96
n parameters	1	1	1	3
p	0.0210	<0.0001	<0.0001	<0.0001
AIC	986.08	975.20	968.44	956.44
BIC	988.74	977.86	971.09	964.43
Likelihood-ratio Hypothesis Tests				
H_0: $\beta_{PERSONAL} = 0$	5.32* (1)			7.28(1)**
H_0: $\beta_{PROPERTY} = 0$		16.20(1)***		9.15(1)***
H_0: $\beta_{AGE} = 0$			22.97*** (1)	18.32(1)***
Wald Hypothesis Tests				
H_0: $\beta_{PERSONAL} = 0$	5.59* (1)			7.69(1)**
H_0: $\beta_{PROPERTY} = 0$		11.70(1)***		7.11(1)**
H_0: $\beta_{AGE} = 0$			19.00*** (1)	15.81(1)***

$\sim p < .10$; $*p < .05$; $**p < .01$; $***p < .001$.

Note: Efron method for ties.

Because the former approach is simpler, we begin with it. We then examine the latter approach, which allows us to express parameter estimates as *hazard ratios*.

Each raw parameter estimate assesses the estimated vertical separation—on a log hazard scale—associated with a one-unit difference in the associated predictor (controlling for all other predictors in the model). For a 0/1 dichotomy like *PERSONAL*, the coefficient of 0.4790 in Model A indicates that the log hazard function for individuals with a history of person-related offenses is 0.4790 units higher than that for individuals with no such history. For a continuous predictor like *AGE*, the coefficient of −0.0681 in Model C indicates that for every one-year increment in age at release, the estimated log hazard function is 0.0681 units lower.

Many researchers find it difficult to develop an intuition about such vertical separations in log hazard space because they lack good estimates of hazard itself, on either a raw or logarithmic scale. This is where the equivalence between the raw and cumulative hazard representations of the Cox model becomes useful. The vertical distances just ascribed to log hazard can be ascribed just as easily to log *cumulative* hazard. This means that we can relate these estimated vertical distances to the observed vertical distances on the sample log cumulative hazard function plots, as shown, for example, in the top panel of figure 14.2. The parameter estimate for *PERSONAL* in Model A indicates that we estimate the average vertical distance between the associated population log cumulative hazard functions to be 0.4790.

As in discrete-time hazard modeling, it is also common to interpret parameter estimates by taking their antilog. For the Cox model, antilogged coefficients—shown in the second panel of table 14.1—are known as *hazard ratios*—the ratio of hazard functions that correspond to unit differences in the value of the associated predictor. This kind of transformation is so helpful that most statistical packages output the hazard ratios alongside the Cox regression coefficients, and some even output hazard ratios *instead* of the coefficients. Automatic re-expression is helpful, but it means that you must read documentation and pay careful attention to output labeling, making sure that you know the metric in which results are displayed. If the raw coefficients have already been transformed into hazard ratios (erroneously labeled "risk ratios" by some statistical packages), you must again take their logarithm to obtain the graphical interpretation just discussed.

To understand why antilogged coefficients can be interpreted as hazard ratios, return to the Cox model in equation 14.3 and consider its representation for a single generic time-invariant predictor X:

$$h(t_{ij}) = h_0(t_j)e^{(\beta X_i)}.$$

Now consider what this model tells us about the population hazard functions for two individuals with different values of X. Let one person have a value of c, and the other have a value of $c + 1$. We use generic values of X, rather than specific values so that interpretation can apply equally, regardless of c, for both continuous and categorical predictors. Substituting these predictor values into the Cox model we have:

$$\text{when } X = c: \quad h(t_{ij}) = h_0(t_j)e^{(\beta c)}$$
$$\text{when } X = c+1: h(t_{ij}) = h_0(t_j)e^{(\beta(c+1))} = h_0(t_j)e^{(\beta c)}e^{(\beta)}.$$

If we take the ratio of these two hazard functions, both $h_0(t_j)$ and $e^{(\beta c)}$ cancel out, leaving the coefficient's antilog, $e^{(\beta)}$.

This algebraic exercise tells us that $e^{(\beta)}$ is the hypothesized constant hazard ratio for *any* one-unit difference in the predictor, X. Because this hazard ratio is solely a function of β—neither the predictor, nor time, appears—it measures the effect of the predictor regardless of: (1) c, the initial value of the predictor chosen as a point of comparison; and (2) the moment in time when the comparison is made.

Hazard ratios are easiest to interpret when they are very large or very small. The large hazard ratio of 3.30 for *PROPERTY* in Model B, for example, tells us that the estimated hazard of recidivism among offenders with a history of property offenses is three times that of those for whom *PROPERTY* = 0. If a hazard ratio is very small, much closer to 0 than to 1, we can adopt a similar interpretation by taking its reciprocal (that is, by computing 1/hazard ratio). This changes the reference group for the comparative statement, converting the very small hazard ratio into a very large one (as described in detail in sections 11.4.2 in the context of discrete time). If a hazard ratio is 0.25, for example, this tells us that the estimated hazard function for one group is four times (1/0.25) that for the other.

Hazard ratios can be linguistically awkward when they are near 1, as often happens for continuous predictors (because a one-unit difference in a predictor with a wide numeric range is likely to enjoy only a small effect). In Model C, for example, the hazard ratio for *AGE* (0.9342) indicates that with each extra year in age at release, the estimated hazard of recidivism is .93 times that for subjects one year younger. Another way to interpret this relative difference is to convert it into a statement about the *percentage* difference in hazard associated with a one-unit difference in the value of the predictor. We obtain this interpretation by subtracting the hazard ratio from 1 and multiplying by 100 (that is, by computing 100*(hazard ratio − 1)). Applying this strategy to the hazard ratio for *AGE* in Model C, we have 100(.9342 − 1) = −6.58%. This tells us that the hazard of recidivism is 6.6% lower for each additional year of age upon release.[1]

As this discussion suggests, we can interpret the effects of predictors in Cox regression models using strategies similar to those used in discrete time. Yet there is one fundamental way in which the Cox model differs.

Because we do not estimate the baseline hazard function, we can make only *comparative*, not *absolute*, statements about hazard. We can say that the hazard for one group is three times that of another, but we cannot say how high, or low, either function is. Even a large hazard ratio (like 3.30, for *PROPERTY*) could potentially be making statements about a small value of hazard because it only multiplies the risk of event occurrence from an infinitesimal level to a minuscule one.

In essence, then, Cox's regression model and the method of partial likelihood invoke a compromise: we trade our ability to estimate the actual values of the baseline hazard function for the ability to estimate the effects of predictors on the baseline hazard function. Some investigators equate this tradeoff with a decision to treat the baseline hazard function as a "nuisance" parameter—a feature that is present but discarded as nonessential. Knowing how informative knowledge of the baseline hazard function can be, it is difficult for us to accept the notion that it is simply a "nuisance." Yet we are willing to accept this tradeoff for it allows us to model the effects of predictors without invoking potentially inappropriate and constraining parametric assumptions. (In addition, using data-analytic strategies presented in section 14.4, we can recover post-model-fitting nonparametric estimates of the baseline hazard function, thereby ultimately providing a reassuring sense of the overall level of risk.)

14.3.2 Evaluating Goodness-of-Fit

Because the Cox regression model is fit using a maximum likelihood method, the log-likelihood statistic continues to provide the basis for assessing model fit. Even though the partial log likelihood in equation 13.8 is not a "true" likelihood, Cox (1975) demonstrated that it is reasonable to treat it as if it were. The sample log likelihood statistic (LL) is obtained by substituting the estimated parameters from the Cox model into the partial log-likelihood function and it can be interpreted in much the same way as the LL for the discrete time hazard model.[2]

The first row of the third panel of table 14.1 presents LL statistics for the four Cox regression models fit to the recidivism data. Notice that the LL statistics increase (become less negative) across the four models, suggesting that each model fits better than the previous one and that Model D, which includes all three predictors simultaneously, fits the best. In the next row of the table, we multiply the log-likelihood statistic by −2, but notice that we label the result −2LL, not deviance (as we have done before). We use this more modest label because "deviance" refers to a comparison with a "fully saturated" model that reproduces the sample data perfectly. When we use the method of partial likelihood, there is no fully saturated model that reproduces the sample data perfectly because the method of estimation does not use the actual event times. Since there exists no set of parameters that can reproduce the sample data perfectly, there is no deviance statistic for the Cox regression model.

Despite the lack of a deviance statistic, however, we can still use *differences* in the −2LL statistics to conduct likelihood ratio tests that compare alternative nested models as is our usual practice. (This makes sense because the common fully saturated model that would appear—were we able to compute deviance statistics—would simply cancel out, leaving the −2LL statistics.) As you would expect, we typically compare −2LL statistics for a given Cox model to either the *null model* with *no* substantive predictors (that is, one that contains just the baseline hazard function) or to simpler models that exclude one or more predictors whose contribution we would like to evaluate. The first tests the null hypothesis that *all* parameters in the current model are 0; the second tests the null hypothesis that all the *focal* parameters are 0. As usual, a difference in −2LL statistics between nested models has a χ^2 distribution with as many degrees of freedom as there are additional parameters. By judiciously fitting nested models, you can test the individual and joint contribution of any number (or type) of predictors.

In broad outline, the strategies for comparing −2LL statistics for the Cox regression model are identical to those for comparing deviance statistics for the discrete-time hazard model. Yet there is one important exception: the fitting of the null model. In discrete time, we can explicitly fit this model (as in Model A of table 11.3). In continuous time, we fit this model only implicitly because it does not include any explicit predictors or parameters. All statistical packages provide the necessary LL (or −2LL) statistic, either routinely as part of model fitting or as an option that the user may specify. (It is usually the initial value of the LL statistic (or −2LL statistic) before any iterations have occurred.) For the recidivism data, the LL statistic for the null model is −494.701, yielding a −2LL statistic of 989.402.

The difference between the −2LL statistic for the null model and that for each of the four models in table 14.1 is presented in the third row of the third panel. This difference is often labeled a *likelihood ratio* (LR) statistic because the difference between two −2LLs is identical numerically to −2 times the logarithm of the *ratio* of the two likelihoods. To determine whether a current model represents an improvement over the null, we compare its likelihood ratio statistic to a χ^2 distribution with degrees of freedom equal to the number of parameters in the model (because the null model, with no predictors, has 0 parameters).

The small *p*-values associated with each of these tests (shown in the next row of table 14.1) allow us to reject the individual null hypotheses that each parameter in Models A through C is separately 0 and the simultaneous null hypothesis that *all* three parameters in Model D are jointly 0.

Differences in −2LL statistics between nested models can also be used to test hypotheses about individual parameters (or subsets of parameters). As Models A through C include a single predictor, the only test possible is numerically identical to those just discussed. For Model D, though, we can compare its −2LL statistic to that of three nested models, each of which includes only two of the three predictors. By systematically *excluding* the focal predictor and comparing the −2LL statistic for that model and Model D, we obtain a statistical test on the effect of that predictor, controlling statistically for the presence of the others. Although we do not present the results of fitting these additional models, the fourth panel of table 14.1 presents the tests that follow from these comparisons, labeled *likelihood ratio tests*. Notice that each test also has a corresponding low *p*-value, indicating that the effect of each predictor is statistically significant, even on control for the other two.

If you wish to compare the goodness-of-fit of alternative models that are not nested, then AIC and BIC statistics continue to provide a useful guide. The last two rows of the third panel of table 14.1 present these statistics for each model. As in discrete-time survival analysis, we—and most software packages—use the number of observed events, (here, 106), to represent the sample size when computing BIC. As we have already seen, these information criteria suggest that each model in the table fits better than the last, and that Model D fits better still. Because we can use formal hypothesis tests for comparing these nested models, we do not belabor these comparisons here. When comparing *nonnested* models, however, AIC and BIC statistics can be very useful.

14.3.3 Drawing Inferences Using Asymptotic Standard Errors

We recommend testing statistical hypotheses about the effects of predictors by comparing −2LL statistics for nested models. On occasion, however, you may want to use asymptotic standard errors to conduct Wald hypothesis tests—based on the ratio of a parameter estimate to its asymptotic standard error—both because these tests are printed out routinely by computer packages and because the asymptotic standard errors upon which they are based can be used to construct asymptotic confidence intervals.

The Cox regression routines available in all statistical packages provide asymptotic standard errors for each of the parameters, Wald hypothesis tests (in the form of either χ^2 or z statistics), and associated p-values. In the top panel of table 14.1, we present asymptotic standard errors in parentheses; in the bottom panel, we present Wald tests computed as χ^2 statistics. (As discussed in section 11.7.1, z-statistics would be the square root of the accompanying χ^2 statistics.)

When sample sizes are large, the Wald and likelihood ratio tests will suggest similar conclusions (because the two are asymptotically equivalent). When sample sizes are modest—as they are here ($n = 194$)—they may not agree completely. Here, we find that while the results for *PERSONAL* are similar (5.59 vs. 5.32 in Model A; 7.69 vs. 7.28 in Model D), those of *PROPERTY* and *AGE* are not. The χ^2 statistics for the effect of *PROPERTY*, for example, are 11.70 versus 16.20 in Model B and 7.11 versus 9.15 in Model D, while those for the effect of *AGE* are 19.00 versus 22.97 in Model C and 15.81 versus 18.32 in Model D. Although these discrepancies do not lead to completely different conclusions, it is easy to imagine that discrepancies of even this magnitude could change conclusions somewhere. Because tests based on −2LL statistics are superior in small samples (when the tests are most likely to diverge), we recommend using likelihood ratio tests whenever possible.

Asymptotic standard errors serve an important role, however, because they provide a strategy for constructing asymptotic confidence intervals (ACIs) for parameter estimates. As you might expect, we construct a 95% ACI for the raw coefficients in the Cox regression model by computing:

$$\hat{\beta} \pm 1.96 ase\left(\hat{\beta}\right). \tag{14.9}$$

For example, the 95% ACI for $\beta_{PERSONAL}$ in Model A is $0.4790 \pm 1.96(0.2025) = (0.0821, 0.8759)$, while the 95% ACI for $\beta_{PROPERTY}$ in Model B is $1.1946 \pm 1.96(0.3493) = (0.5100, 1.8792)$. We construct asymptotic confidence intervals for hazard ratios by antilogging these ACI limits:

$$e^{\left(\hat{\beta} \pm 1.96 ase(\hat{\beta})\right)}. \tag{14.10}$$

For example, the 95% ACI for the hazard ratio for *PERSONAL* in Model A is $(e^{(0.0821)}, e^{(0.8759)}) = (1.09, 2.40)$, while the 95% for the hazard ratio for *PROPERTY* in Model B is $(e^{(0.5100)}, e^{(1.8792)}) = (1.66, 6.55)$.

How do you form an asymptotic confidence interval if a one-unit difference in a predictor is not substantively meaningful? All you need do is follow the two-step process outlined above, working with a c-unit increment in the predictor. To form a 95% ACI for a coefficient associated with a c-unit increment in a predictor, simply multiply both terms in equation 14.9 by c. For example, a 95% ACI for a ten-year increment in the predictor *AGE* in Model C is $10(-0.0667) \pm 1.96(10)(0.0168) = (-0.9963, -0.3377)$. You can then apply equation 14.10, antilogging these limits, to find the ACI for the hazard ratio, which here is $(0.37, 0.71)$.

Notice that the asymptotic confidence interval for a hazard ratio is not symmetric around its point estimate. Although initially unsettling, asymmetry has no effect on the interval's interpretation or utility. Notice, too, that we do not compute an ACI for a hazard ratio using its own asymptotic standard error (as presented, for example, in the second panel of table 14.1). Adding and subtracting 1.96 times the hazard ratio's asymptotic standard error from the hazard ratio *does not yield* a 95% asymptotic confidence interval. To construct the appropriate interval—which *must* be asymmetric—you must antilog the symmetric limits for the raw coefficient itself as shown in equation 14.10. This realization reinforces the need for care when retrieving coefficients from computer output. Be sure you know which coefficients and asymptotic standard errors your package presents—raw regression parameters or their associated hazard ratios. Incorrect interpretations will result from incorrect labels.

14.3.4 Summarizing Findings Using Risk Scores

Another strategy for communicating the results of fitting a Cox regression model is to use the concept of a *risk score*. Instead of summarizing the effect of a single predictor, risk scores summarize the effects of several predictors simultaneously. Risk scores are therefore especially attractive when you have fitted complex models with multiple predictors whose findings may not be easily explained.

To understand how to compute and interpret risk scores, examine the Cox model in equation 14.4 and consider how you might compare the risk of event occurrence for each individual in the sample to that of the "baseline individual"—the person who has value 0 for every predictor in the model. With time-invariant predictors, the hazard function for individual i is $h_0(t_j)e^{(\beta_1 X_{1i}+\beta_2 X_{2i}+\cdots+\beta_P X_{Pi})}$. The hazard function for the baseline individual is just $h_0(t_j)$, regardless of the number and type of predictors. Dividing the former function by the latter and substituting in parameter estimates for population values, the baseline hazard function cancels out, leaving:

$$\text{risk score}_i = e^{[\hat{\beta}_1 X_{1i}+\hat{\beta}_2 X_{2i}+\cdots+\hat{\beta}_P X_{Pi}]}. \tag{14.11}$$

Each individual's risk score compares the level of his or her fitted hazard function to that of the baseline hazard function.

Risk scores are measured in *relative*, not *absolute*, terms. The risk score for individual i does not assess that person's *absolute* level of hazard, but rather his or her *relative* level in comparison to that of the baseline hazard function. When contemplating the use of risk scores in an empirical analysis, it is therefore essential to think carefully about who the baseline hazard function represents. All Cox regression models possess a baseline hazard function, as a mathematical necessity, even if 0 is not a viable value for every predictor. Desire for a baseline hazard function that is substantively meaningful was the motivation for our decision to center predictor *AGE* by subtracting its sample mean. (Because 0 is a valid value for the other predictors, we felt that no further rescaling was necessary for them.) But, centering *AGE* ensured that the baseline hazard function for Model D was substantively plausible and conceptually appealing: it represents the hazard function for a former inmate with no history of personal or property offenses who was of "average" age upon release.

Because risk scores do not depend on an individual's event time, they can be calculated for every member of the sample, censored or not. Using Model D of table 14.1, table 14.2 presents predictor values and risk scores for eight former inmates in the recidivism study. Like hazard ratios, risk scores are positive. The higher the risk score, the higher the individual's predicted level of risk. Individuals who face no elevated risk will have risk scores of approximately 1 (inmates 22 and 8). Individuals who face *greater* comparative risk will have risk scores greater than 1 (inmates 187, 26 and 5). Individuals who face *lower* comparative risk will have risk scores less than 1 (inmates 130, 106 and 33).

We chose the eight individuals in table 14.2 to illustrate how risk scores of varying size can be interpreted to advantage. The risk score for the first entry in the table (inmate 22, who has a risk score of 0.98) indicates that his predicted hazard function is virtually identical to that of the baseline. This makes sense as this former inmate was of average age upon

Table 14.2: Risk scores estimated from Model D of table 14.1 for selected individuals in the recidivism study

ID	PERSONAL	PROPERTY	CENTERED AGE	Risk score	DAY	MONTHS	CENSOR
22	0	0	0.258	0.98	52	1.7084	1
8	1	1	22.451	1.01	19	0.6242	1
187	1	0	−7.200	2.86	1065	36.0000	1
26	0	1	−7.302	4.15	72	2.3655	0
5	1	1	−7.165	7.26	9	0.2957	0
130	0	1	22.391	0.57	486	15.9671	1
106	0	0	16.203	0.34	356	11.6961	0
33	1	0	27.061	0.29	85	2.7926	1

release with no history of either personal or property crime. But the second entry in the table, inmate 8, illustrates that this predictor profile is not the only way of attaining a risk score of 1. He also has a risk score of approximately 1 (1.01) despite his history of both personal and property crime. He attained this relatively low value because he was 22.45 years older than the average former inmate upon release (age 53). His age counterbalances his criminal history, placing him at identical risk as someone of average age with a less problematic record.

Each of the next three individuals have comparatively higher risk scores. Youth is one source of their increased risk (all were approximately 24 upon release, seven years younger than the average sample member). But their history of personal and property crime also plays a role. Inmate 187, who committed more than one person-related crime, is nearly three times more likely than a baseline individual to be rearrested. Despite this relatively high risk, he managed to remain arrest-free for three years. Inmates 26 and 5 were not so fortunate. The model predicts that each faces a greater risk of recidivism—inmate 5 was over seven times as likely as an individual at baseline to be rearrested—and both returned to jail within three months of release.

The remaining three individuals are at *lower* risk of recidivism in comparison to an individual at baseline. One reason for their lower risk is their "advanced" age upon release—all were in their late 40s or 50s. Once again, age mitigates criminal history, as the risk score for even inmate 130, who has a history of property crime, is nearly half that of the baseline. For two of these people—inmates 130 and 33—this lower predicted risk is born out by the fact that neither had been arrested by the end of data collection (they are both censored). But inmate 106, who had no history of personal or property crime, was rearrested just before his one-year anniversary (despite the model's prediction of very low risk).

Risk scores are invaluable for demonstrating that there is more than one way to attain a given level of risk. Each predictor's effect operates in the context of the others; some people who may appear to have a high risk because of the values of one or two predictors can attain lower levels of risk because of other factors. Other individuals who appear to have low levels of risk may attain higher levels because of other factors. Although each predictor has the controlled effect summarized by its single associated parameter, it is the *combination of predictor values* in association with the combination of parameter estimates that determines each person's overall level of risk. Few predictors have such large effects that they dominate all others. In essence, then, risk scores illustrate the inevitable push and pull of all countervailing effects.

Although risk scores are valuable summaries of analytic results, they are *not* immutable characteristics of individuals. Risk scores are a joint consequence of an individual's predictor values *and* a particular fitted model. If we fit another model to the recidivism data using just two of these three predictors, the risk scores would change. Risk scores are model dependent even though predictor values are not. We find it helpful to think of risk scores as predicted, or fitted, values based on a particular model and sample. But because the baseline hazard function is ignored in the calculation of risk scores, the predicted (or fitted) values are measured in relative, not absolute, terms.

If you want to use risk scores to communicate your findings, take the time to create a useful and substantively plausible baseline. The strategy we use here—centering continuous predictors on their sample mean and leaving dichotomous predictors alone—is just one approach. You can also center *all* predictors on their sample means. If we had adopted this strategy for Model D, for instance, the baseline hazard function would represent the risks for an "average" former inmate: someone released at an "average" age with an "average" history of personal crime and an "average" history of property crime. Although this "average' individual cannot exist, he can provide a valuable artifice for presentation. For further ideas, we suggest you read the discussion of centering in section 4.5.4. Our point is simple: by suitable construction of predictors, the interpretation of the baseline hazard function is entirely up to you, dictated by the kinds of descriptive statements you would like to make.

14.4 Nonparametric Strategies for Displaying the Results of Model Fitting

Throughout this chapter, we have stated repeatedly that Cox regression analysis provides no information about the shape of the baseline hazard function. Given that we *never* estimate the hazard function well in continuous time, this limitation seems unremarkable. But what about the *cumulative hazard function*? Can we discern its shape from a fitted model? If so, might we be able to recapture a fitted survivor function and perhaps even deduce the shape of the hazard function?

Soon after the publication of Cox's 1972 paper, statisticians began addressing these challenging questions. Reasoning that risk scores offer valuable information, Kalbfleisch and Prentice (1973, 1980) and Breslow (1974) developed *nonparametric* methods for *recovering* the baseline cumulative hazard function from a model fit using time-invariant predictors. These recovery methods do not produce *predicted* functions in the classical sense; in other words, they do not yield *model-based* estimates of the underlying population functions. Instead, they yield nonparametric estimates of these functions based on the *risk scores* of sample members.

Derivation of these algebraically complex and computationally intensive methods is beyond the scope of this book. Our goal is to demystify the process, explaining how you can recover baselines—even though we have ignored them entirely during the process of model fitting—and put the recovered values to good use. In section 14.4.1, we describe conceptually how the methods work and examine their values from Model D for the recidivism data. In section 14.4.2, we use these baselines for a second purpose: to obtain fitted functions at selected predictor values.

14.4.1 Recovered Baseline Functions

The recovery methods are creative extensions of the principles of Kaplan-Meier estimation. As described in section 13.3, we obtain Kaplan-Meier estimates by: (1) dividing continuous time into intervals; (2) computing the conditional probability of event occurrence in each interval; (3) multiplying the complements of these conditional probabilities together to estimate the survivor function; and (4) computing the negative log of the survivor function to estimate the cumulative hazard function. Central to this approach, then, are the conditional probabilities of event occurrence. If we can estimate them, we can compute the other summaries.

The key task for baseline recovery is estimation of these conditional probabilities for the model's baseline group—those individuals who take on the value 0 for every predictor in the model. You might think it logical to start the process by computing a risk score for each person under the assumption that he had the value 0 for every predictor in the model. But if we do this, all the risk scores would be 1 (since every parameter would be multiplied by 0, and taking the antilog of 0 yields 1). This strategy effectively eliminates information about predictors from the calculation, making it an unwise choice. Instead, we invoke the opposite logic: begin with each person's *observed* risk score because this compares his level of risk to that of the baseline individual. In essence, risk scores provide an alternative metric for assessing the size of the risk set, scaled in the metric of *risk scores*, not *people*. This allows us to estimate the conditional probability of event occurrence using the concept of the "amount of remaining risk," as we now explain.

The "amount of remaining risk" at the beginning of any interval is the sum of the risk scores among everyone still at risk. For Model D of the recidivism data, the amount of remaining risk at t_0 is 626.778—the sum of the observed risk scores among all 194 individuals. As people leave the risk set, either because of event occurrence or censoring, the amount of remaining risk diminishes. After the second interval, when the risk set drops to 193, the total amount of remaining risk drops to 621.742, because the one individual arrested (on day 2) has a risk score of 5.036. After the third interval, when the risk set drops to 192, the total amount of remaining risk drops to 616.612, because the one individual arrested (on day 4) has a risk score of 5.130. After the fourth interval, the total amount of remaining risk drops to 610.561. After the fifth interval, the amount of remaining risk drops more precipitously, to 597.869, because the *two* people arrested on day 9 have a combined risk of 12.692. After day 882 (the last observed event time), when the risk set falls to 31, the total level of remaining risk hits its floor, 90.702, the sum of the risk scores among the 31 individuals censored at the end of data collection. In any data set, the last observed event time is the last moment for which we can recover a baseline function.

Both recovery methods use the total amount of remaining risk to compute baseline conditional probabilities. The methods differ not in their underlying assumptions, but rather in the approximations invoked to increase computational feasibility. The simpler Breslow method substitutes the amount of remaining risk for the size of the risk set in the denominator of the standard formula for conditional probability estimation in equation 13.3. The more elaborate Kalbfleish-Prentice method (1973, 1980) goes further and alters the numerator to account for the amount of risk faced by those individuals who experience the target event in the interval. When there are ties, this computationally intensive method requires numeric iteration. Because the methods are asymptotically equivalent and virtually identical in practice, you are free to use either. For display purposes in what follows, we use the Kalbfleisch and Prentice method (sometimes labeled the product limit method in software packages).

The left panel of figure 14.4 presents baseline survivor, cumulative hazard, and kernel-smoothed hazard functions for Model D of table 14.1. To interpret these functions, you must first identify precisely who the baseline represents. Baselines are not necessarily "averages"; this interpretation accrues only if every predictor in a model—including dichotomies—is centered on its sample mean (which is not the case for Model D). Instead, an individual who takes on the value 0 for every predictor in Model D is someone of average age upon release from prison with *no* history of either personal or property crime. Because only *AGE* is centered, this baseline is someone at much lower risk (owing to lack of a history of personal or property crime). Reflecting this low risk, the baseline survivor function drops to only 0.6694 at 29 months. (Notice that

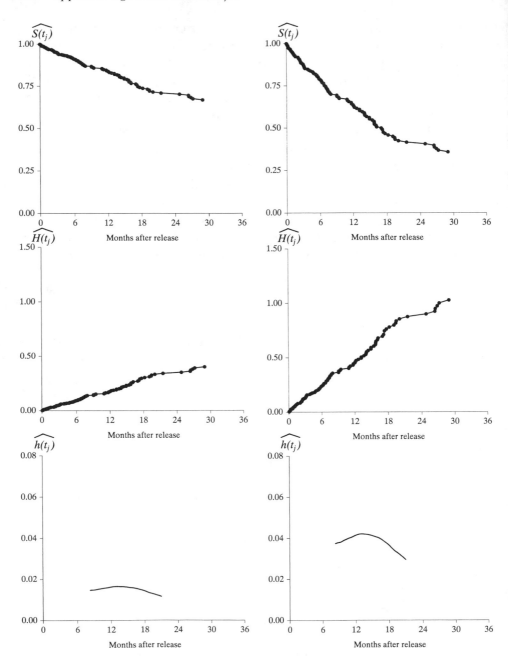

Figure 14.4. Recovered baseline survivor function, cumulative hazard function, and kernel-smoothed hazard function for Model D of table 14.1. The baseline in the left panel is a former inmate of average age upon release who had *no* history of either property or personal crime. The baseline in the right panel is a former inmate of average age upon release who had an "average" history of both property and personal crime.

the fitted baseline stops at this moment, which is the last observed event time.) In fact, we predict recidivism to be so rare for this group that we cannot even estimate a median lifetime. This stands in stark contrast to the sample functions in figure 14.1 according to values of the predictor, *PERSONAL*. Both groups in that display are at greater risk, even the group with no history of personal crime. If you think this is illogical—that the baseline should fall between the sample functions—remember that this baseline has no history of personal crime *and* no history of property crime. Such an individual is at much lower risk of recidivism than either group in figure 14.1.

If you want recovered baselines for an "average" individual, the easiest thing to do is to center all predictors—including dichotomies—before model fitting. Some software packages suggest that you routinely center all predictors because doing so improves numerical accuracy. Unlike the multilevel model for change, where centering affects parameter inter-pretation, centering has virtually no impact on the proportional hazards model. All statistics in table 14.1—parameter estimates, standard errors, and goodness-of-fit statistics—will be identical whether or not you center. (To confirm this result, try fitting identical models with raw and centered predictors.) What does change when you center is: (1) everyone's risk scores; and (2) the meaning of the baseline individual. The risk scores change not because the parameter estimates change (they remain the same) but rather because the *predictor values* change—they are now centered on the sample means. This changes the baselines, allowing recovered functions to represent model predictions for an "average" individual.

The right panel of figure 14.4 illustrates what happens if we use Model D to recover an "average" baseline for the recidivism data. We obtained these functions by fitting a model identical to Model D in table 14.1 with one important change: in addition to centering *AGE* at its sample mean, we also centered *PERSONAL* and *PROPERTY* on their sample means (0.3144 and 0.8144, respectively). This allows us to interpret the recov-ered baselines on the right side of figure 14.4 as predicted functions for an "average" former inmate *after adjusting statistically for the predictors AGE, PERSONAL and PROPERTY*. In contrast to the recovered baselines in the left panel, these fall near the middle of the sample functions in figure 14.1. We expect this behavior because these recovered baselines repre-sent the model's expectation for an average individual.

The ability to recover baseline functions increases the Cox model's interpretability. Without recovered functions, you may find yourself fitting model after model, getting further removed from your data. We suggest that you examine recovered baselines periodically to clarify the link between the functions you examine during exploratory analysis and the results of model fitting. This link can be further clarified if you use the recovered baselines to go one step further and compute predicted functions at interesting combinations of predictor values, as we now explain.

14.4.2 Predicted Survivor and Cumulative Hazard Functions

Having recovered the baseline survivor and cumulative hazard functions, it is simple to use these to display fitted models not just for "baseline" individuals but also for prototypical individuals who enjoy any combination of predictor values. To see how this works, imagine using equation 14.11 to compute a risk score that reflects predictor values for a prototypical individual. To recover a cumulative function for this prototype, simply substitute that risk score into equation 14.3 to find:

$$\hat{H}(t_{ij}) = \hat{H}_0(t_j) \text{risk score}_i. \tag{14.12}$$

The recovered fitted cumulative hazard function for each prototypical individual is just that risk score *multiple* of the baseline cumulative hazard function.

To recover a survivor function for a prototypical individual, we need an extra step that exploits the mathematical identity between cumulative hazard and the negative log survivor function in equation 13.8. Because of this identity, we can substitute the negative log survivor function into both the left and right hand sides of equation 14.12 to write:

$$-\log \hat{S}(t_{ij}) = -\log \hat{S}_0(t_j) \text{risk score}_i.$$

Canceling the two negative signs, we take the antilog of both sides to find:

$$\hat{S}(t_{ij}) = \hat{S}_0(t_j)^{\text{risk score}_i}. \tag{14.13}$$

The recovered survivor function for any combination of predictor values is just a risk score *power* of the baseline survivor function. You can then obtain the remaining functions from these values.

All statistical packages include some provision for computing fitted cumulative hazard and survivor functions corresponding to user-specified predictor values. If your package limits you to recovered baseline functions, you can calculate functions for prototypical individuals using equations 14.12 and 14.13. The major conceptual task in this process is the selection of the prototypical people (a topic discussed at length in section 4.5.3 in the context of the multilevel model for change). If you use the sample means for each predictor, you obtain the *adjusted* baseline

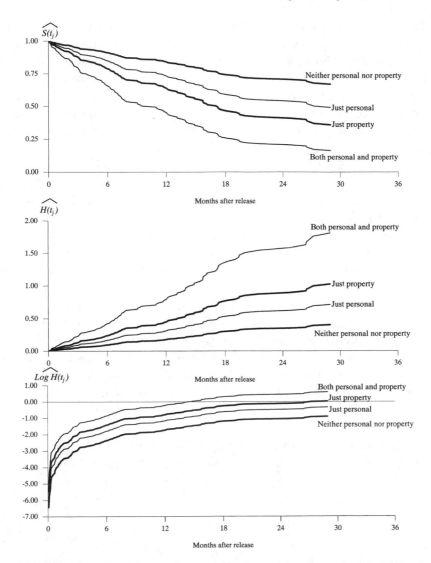

Figure 14.5. Recovered survivor and cumulative hazard functions, for Model D of table 14.1, obtained by setting *AGE* to its sample mean and varying the values of *PERSONAL* and *PROPERTY*.

functions presented in the right side of figure 14.4. More often, we select substantively interesting combinations of predictors—prototypical individuals whose behavior we would like to describe. We computed the functions in figure 14.5 by combining these strategies: *AGE* is set to its sample mean and we display results for all possible combinations of *PROPERTY* and *PERSONAL*.

Coupled with the numerical results in table 14.1, figure 14.5 provides compelling evidence about the large (and statistically significant) effects of a former inmate's criminal history on risk of recidivism. Controlling statistically for age at release, former inmates who committed no more than one personal crime and no property crime face the lowest risk of re-arrest. A history of either type of crime bodes poorly for remaining arrest-free, although the effect of property crime alone (hazard ratio of 2.55) is much larger than the effect of personal crime alone (hazard ratio of 1.77). Those with a history of *both* types of offenses are at especially great risk of recidivism. Controlling statistically for their age at release, their risk is 4.50 times ($e^{(0.5691+0.9358)} = 4.50$) the risk of someone with no history of either crime.

We close by drawing your attention to the bottom panel of figure 14.5, where we display the predicted cumulative hazard functions on a logarithmic scale. We display these functions instead of the kernel-smoothed hazard functions to highlight an important feature of the Cox model we want to re-emphasize: under the proportional hazards assumption, the predicted cumulative hazard functions will be equidistant when plotted on a logarithmic scale. Equidistance is an *assumption* of the model, not a statement about reality. In postulating the model, we assume that unit differences in the value of each predictor, controlling for all other predictors in the model, have the same effect at all points in time—here, regardless of whether the former inmate was just released from prison or whether he had been out for many months. Although this proportional hazards assumption is often tenable, you should always evaluate its appropriateness. We offer strategies for doing so in the next chapter.

15

Extending the Cox Regression Model

Time is nature's way of keeping everything from happening
at once.

—Woody Allen

Were the Cox regression model applicable only to time-invariant predictors with time-constant effects, it might not enjoy the popularity that it
does today. But just as we can extend a linear regression model in a variety
of ways, so, too, can we extend the Cox regression model. Some extensions—for example, the inclusion of nonlinear effects and statistical interactions—are identical to extensions we can make to any statistical model.
Rather than discuss familiar topics like these, we devote this chapter to a
set of novel extensions.

We begin, in section 15.1, by describing how to include time-varying
predictors in the Cox regression model. Although our approach resembles the one we used in discrete-time hazard modeling, the data demands
in continuous time complicate model fitting. We then introduce two
methods for relaxing the proportionality assumption. In section 15.2, we
present the stratified Cox regression model, which stipulates that while
the effects of each predictor are identical across strata, the baseline
hazard functions can *differ*. In section 15.3, we present an alternative strategy that closely mirrors the approach used in discrete time: the inclusion
of interactions with time as predictors in the model. In section 15.4, we
introduce a range of regression diagnostics useful for examining the
underlying assumptions of the Cox model. In section 15.5, we discuss
what you should do when modeling "competing risks"—multiple events
that compete to terminate an individual's lifetime. We close, in section
15.6, by describing what to do when you have not observed the beginning of time for everyone in your sample and there are so-called late
entrants to the risk set.

15.1 Time-Varying Predictors

In principle, you need no additional special strategies to include time-varying predictors in a Cox regression model. The representations in section 14.1 (see equations 14.1 through 14.4) allow for this possibility in that every predictor already includes a subscript j indexing time. When we fit the model to data in chapter 14, we restricted attention to time-invariant predictors so that we could better describe parameter interpretation in a simple context. Now, we reintroduce the possibility that some predictors may vary over time.

Using equation 14.3, for example we can specify a Cox regression model with one time-invariant predictor (X_1) and one time-varying predictor (X_{2j}) as:

$$h(t_{ij}) = h_0(t_j)e^{(\beta_1 X_{1i} + \beta_2 X_{2ij})}. \tag{15.1}$$

In writing this model, we postulate that individual i's hazard at time t_j is the product of the baseline hazard function at that time and individual i's contemporaneous true risk score [$e^{(\beta_1 X_{1i} + \beta_2 X_{2ij})}$]. Taking logarithms of both sides, we have:

$$\log h(t_{ij}) = \log h(t_j) + \beta_1 X_{1i} + \beta_2 X_{2ij}. \tag{15.2}$$

This representation demonstrates the indigenous similarity between a Cox model that contains time-varying predictors and a discrete-time hazard model with time-varying predictors (as shown, for example, in equation 12.5).

Given how straightforward this appears, you might question the need for this section. We have already discussed interpretive difficulties that time-varying predictors raise, specifically the problems of reciprocal causation and rate- and state-dependence (see section 5.3.4 and section 12.3.3). We have also discussed how to fit the model, as if every predictor in section 14.2.1 did in fact include a subscript j. In equation 14.7, for example, individual i's contribution to the partial likelihood is the ratio of his or her true risk score *at his or her moment of event occurrence* [$e^{(\beta_1 X_{1i} + \cdots + \beta_P X_{Pi})}$] to the sum of the contemporaneous true risk scores among everyone still at risk.

Yet time-varying predictors do pose a problem: they present an enormous—sometimes insurmountable—data requirement. For each time-varying predictor, you must know *its value, for everyone still at risk, at every moment when someone experiences the target event*. If your data set includes 100 unique event times, you need to know the time-varying predictor's values—for everyone still at risk—at each of those 100 moments. If your data set includes 1000 unique event times, you need to know its values at each of those *1000* moments. Although it may seem novel, we have seen this data requirement before. In discrete time, however, it was unremarkable because: (1) the number of unique event times was small; and (2) event occurrence *and* predictors are usually assessed using an identical temporal schedule.

The practical consequences of this data requirement can be formidable. Even if resources allow you to assess event times precisely, it can be prohibitively expensive—if not logistically impossible—to measure the value of all time-varying predictors at every observed event time. Researchers typically gather predictor data on a subset of occasions that may—or more often, may not—coincide with event occurrence. (Few of us are so prescient that we can tie data collection to unknown event times!) Even noncensored individuals rarely provide the data you desire most: their predictor values at their moment of transition. And if data collection is retrospective, information about temporal variation in time-varying predictors may be unobtainable. Perhaps the only relief in this bleak situation is that you need not know the predictor values at every moment when an event *could* occur but only at those moments when events *do* occur.

Examining the effects of time-varying predictors, therefore, requires advance planning and subsequent creativity. Research design is crucial in this regard, but once the data are in, your task is analytic. The complexity of the task depends on the information needed to construct a full predictor history. Below, we discuss what happens under an increasingly complex set of circumstances. In section 15.1.1, we discuss the handling of nonreversible dichotomies, variables whose values indicate a unidirectional transition (e.g., from never married to married). In section 15.1.2, we discuss potentially reversible dichotomies (e.g., from employed to unemployed and back) and continuous predictors that can change at will. Under these circumstances, you can rarely generate temporal histories that are as fine as you might like, which often leads to imputation, the topic of section 15.1.3.

15.1.1 Nonreversible Dichotomies

Time-varying predictors that describe an individual's nonreversible status—for example, high school student versus graduate—are the simplest to deal with because you can construct someone's entire predictor history using just one piece of information: his or her date of "transition." Before this date, the individual occupies state A; after, the individual occupies state B. Not everyone must experience the transition, but no one can "go backwards" and revert again to state A.

We illustrate strategies for examining the effects of time-varying predictors that are nonreversible dichotomies using data from Burton et al. (1996), who studied the precursors of cocaine initiation among a random sample of Americans interviewed twice, 11 years apart (in 1974 and 1985). Among the 1658 White men studied, 382 (23.0%) reported using cocaine for the first time between ages 17 and 41. Each also provided a detailed drug-use history—with ages of initiation—which allowed the researchers to determine that while at risk of initiating cocaine use: (1) 61.4% previously used marijuana; (2) 9.5% previously sold marijuana; (3) 32.6% previously used one or more other drugs (including amphetamines, psychedelics, depressants, opiates, or heroin); and (4) 17.0% previously used two or more other drugs. These data allowed us to construct six drug-use history variables:

- Two time-invariant predictors—*EARLYMJ* and *EARLYOD*—that indicate whether the respondent initiated marijuana use (7.2% of the sample) or other drug use (3.7%) so early that he could be classified as a previous user at t_0 (age 17).
- Four time-varying predictors—$USEDMJ_j$, $SOLDMJ_j$, $USEDOD_j$, and $MOREOD_j$—which identify, at each age t_j, whether the respondent had previously used marijuana, sold marijuana, used another drug, or used two or more other drugs.

To account for the well-known societal changes that occurred during the era under study (1961 to 1985), we also include a time-invariant predictor *BIRTHYR* as a control variable.

Each drug-use predictor assesses an individual's *prior* status. We use this approach because *contemporaneous* drug-use status—perhaps the more "obvious" variable—is an *internal* time-varying predictor. As discussed in section 12.3.3, internal time-varying predictors raise interpretive dilemmas because of the possibility of state dependence (if someone's decision to use another drug depends upon his cocaine-use status) and rate dependence (if someone's decision to use another drug depends upon his hazard of cocaine use). Both concerns are valid. Cocaine users may be more likely to sell marijuana (to make money to support their habit) and men at greater risk of using cocaine may be more likely to use other drugs as well. Were we to find an association between contemporaneous use of other drugs and the hazard of cocaine initiation, it would be difficult to argue the other drug led to cocaine use. A critic could reason-ᵇly argue that the causal arrow was reversed (or, at best, its direction

'ᵈilemma, we, like the original researchers, *lag* the values
ᵈictors so that they describe an individual's status

in the immediately prior year. We use a one-year lag because this is the shortest possible increment when time is assessed in whole years. Computationally, individual i's value of each time-varying predictor at age t_j reflects his drug use status at age t_{j-1}. For a man who started using marijuana at age 21, for example, $USEDMJ_j$ takes on the value 0 at all times up to *and including* age 21, switching to 1 at age 22 and beyond. If a man started selling marijuana at age 25, $SOLDMJ_j$ takes on the value 0 at all times up to and including age 25, switching to 1 at age 26 and beyond. Lagging makes it difficult to argue that an individual's initiation of cocaine use in one year influenced his decision to use marijuana (or another drug) in a previous year.[1]

All Cox regression routines available in standard statistical packages have facilities for fitting a model with time-varying predictors. The routines differ markedly, however, in how they are implemented. Broadly speaking, there are two distinct approaches:

- The *single record method*, in which each person has one line of data and you write computer code to indicate which variables contain the appropriate values of the time-varying predictor at each point in time; and,
- The *counting process method*, in which you construct a person-period data set with the time periods chosen so that all time-varying predictors are constant within each record. Unlike other person-period data sets, the discretization need not be constant across cases. Each individual's discretization is determined by his or her pattern of temporal variation in the time-varying predictors.

Some statistical packages allow you to use either approach; others offer just one. For time-varying predictors that change only once, the single record method is easier. For time-varying predictors that change periodically, the counting process method may be easier. Once you become comfortable with either method, the distinction is relatively unimportant because the analytic results are identical. As a practical matter, then, the choice is yours (constrained, in part, by your statistical package).

Table 15.1 presents the results of fitting four Cox regression models to the cocaine initiation data. Let us begin with Model A, which includes only the three time-invariant predictors. The three large statistically significant effects indicate that the risk of cocaine initiation is higher among men who: (1) were born later; (2) used marijuana at an early age; and (3) used other drugs at an early age. Holding constant the effects of early drug use, we estimate that the hazard of cocaine initiation is $100(e^{(0.1551)} - 1) = 16.8\%$ higher for each year later the respondent was

Table 15.1: Including the effects of time-varying predictors in a Cox regression model

	Model A	Model B	Model C	Model D
Parameter Estimates, Asymptotic Standard Errors, and Deviance-Based Hypothesis Tests				
BIRTHYR	0.1551***	0.1074***	0.0849***	0.0835***
	(0.0199)	(0.0215)	(0.0218)	(0.0226)
Marijuana use				
EARLYMJ	1.2171***			0.0753
	(0.1640)			(0.1709)
USEDMJ$_j$		2.5518***	2.4592***	2.4525***
		(0.2810)	(0.2836)	(0.2843)
SOLDMJ$_j$			0.6899***	0.6789***
			(0.1226)	(0.1250)
Other drug use				
EARLYOD	0.7912***			−0.0803
	(0.1962)			(0.2033)
USEDOD$_j$		1.8539***	1.2511***	1.2543***
		(0.1292)	(0.1566)	(0.1572)
MOREOD$_j$			0.7604***	0.7638***
			(0.1307)	(0.1322)
Goodness-of-fit				
−2LL	5277.228	4669.096	4580.537	4580.311
AIC	5283.228	4675.096	4590.537	4594.311
Δ −2LL	247.830***	55.962***	88.559***	0.226 (ns)
(df)	(3)	(3)	(2)	(2)
Comparison	Null	Null	Model B	Model C

$\sim p < .10$; $* p < .05$; $** p < .01$; $*** p < .001$.

These are the results of fitting four Cox models to the cocaine initiation data ($n = 1658$, n *events* = 382). Each model includes the respondent's birth year as a predictor as well as a combination of time-varying predictors assessing his drug use history. Model A includes time-invariant predictors assessing drug use prior to age 17. Model B includes time-varying predictors assessing whether the respondent used marijuana or other drugs, while Model C includes additional time-varying predictors assessing whether the respondent *sold* marijuana or other drugs. Model D simultaneously assesses the effects of all drug use predictors.

Note: Efron method for ties. −2LL for null model is 5525.059.

born. Holding constant this cohort effect, we estimate that the hazard of cocaine use among men who were early marijuana users is $e^{(1.2171)} = 3.38$ times the hazard for men who were not users, and among men who were early users of other drugs is $e^{(0.7912)} = 2.21$ times the hazard for men who were not.

In Model B, we replace the two time-invariant measures of drug use with their time-varying cousins USEDMJ$_j$ and USEDOD$_j$. Once again, the risk of cocaine initiation is higher among men who: (1) were born later; (2) previously used marijuana; and (3) previously used other drugs. But in contrast to Model A, which assesses the effect of *early* drug use, Model B assesses the effect of *previous* drug use regardless of age of initiation. These redefined predictors yield a model with larger parameter estimates and a smaller AIC. (We do not compare −2LL statistics because A is not nested within B.) Holding constant the cohort effect, we now estimate that the hazard of cocaine initiation is $e^{(2.5518)} = 12.83$ times the hazard for men who had not, among men who previously used marijuana and among men who previously used other drugs the hazard of cocaine use is $e^{(1.8539)} = 6.38$ times the hazard for men who had not. Few empirical researchers find hazard ratios this large, emphasizing not only the statistical significance of these findings but their practical significance as well.

Do time-varying predictors always yield larger effects and better fitting models? Although this is often true—indeed, this hope motivates most researchers who investigate the effects of time-varying predictors—superior performance is not guaranteed, but depends on the substantive mechanisms underlying the life histories. If *early* experiences are more important precursors of event occurrence, time-invariant predictors will be superior. One appealing feature of this data set is that we can examine this question empirically because the values of the time-varying drug use predictors at the beginning of time are identical to their time-invariant values (i.e., at time t_0, *EARLYMJ* = *USEDMJ$_j$* and *EARLYOD* = *USEDOD$_j$*). As the parameter estimates for Model B are over twice as large as those in Model A, the time-varying predictors appear superior. To address this question fully, however, we should include both sets of predictors in a single model. We will take this step in one moment, after adding the two further drug use predictors to the model.

In Model C, we add *SOLDMJ$_j$* and *MOREOD$_j$* to Model B. Each variable is interesting in its own right, but combined with the two other drug use predictors they are even more informative. This is because each new predictor can take on the value 1 only if its corresponding initial use predictor is also 1. Each pair of predictors—(*USEDMJ$_j$*, *SOLDMJ$_j$*) and (*USEDOD$_j$*, *MOREOD$_j$*)—therefore represents an ordinal variable that contrasts no drug use with two escalating levels. The large and statistically significant effects of these new predictors coupled with the persistent effects of *USEDMJ$_j$* and *USEDOD$_j$* suggest that escalating use is indeed associated with an increased risk of cocaine initiation. Controlling for birth cohort and the previous use of marijuana and one other drug, the hazard of cocaine initiation among men who also sold marijuana or who used a second other drug is twice ($e^{(0.6899)} = 1.99$; $e^{(0.7604)} = 2.14$) the hazard for men who had not. And lest you think that none of these effects is really "that large," antilog the sum of the four parameter estimates in Model C (i.e., compute $e^{(2.4592+0.6899+1.2511+0.7604)} = e^{(5.1606)} = 174.27$) to see that the estimated hazard of cocaine initiation among men who previously used and sold marijuana and who also used two or more other drugs is *174 times* the hazard for men who had not. (This is likely one of the largest hazard ratios you will ever encounter!)

With Model D, we ask: Do the time-varying predictors capture everything important about an individual's drug history or is early drug use of consequence in its own right? Comparison with Model C shows that neither *EARLYMJ* nor *EARLYOD* is statistically significant once all four time-varying predictors are included. Both parameter estimates are indistinguishable from 0 and the difference in −2LL statistics is trivial. For these data at least, we are drawn to a model with the time-varying predictors.

Let us now examine the effects of *BIRTHYR* as the time-varying drug-use predictors are added. Between Model A and C, its estimate halves— from 0.1551 to 0.0849—a precipitous drop especially when considered in light of its uncontrolled estimate (0.2026, not shown). This is but one illustration of a phenomenon often observed when adding time-varying predictors to statistical models: the magnitude of the parameter estimates for the time-invariant predictors diminish. *Substitution effects* like these arise when the values of the time-varying predictors themselves change in response to the time-invariant predictors. In this data set, men in later cohorts are not only more likely to try cocaine (as seen by the persistent effect of *BIRTHYR*), they are also more likely to try the "predictor" drugs—marijuana, opiates, and the like. When we add the time-varying drug-use predictors to the model, the effect of *BIRTHYR* diminishes because some of the effect previously attributed to cohort is now attributed to drug history. Although substitution effects can appear in any model with correlated predictors, a special feature of longitudinal analysis is that the values of the time-varying predictors themselves can *change* in response to the other predictors.

Substitution effects can even occur in settings less prone to inferential problems, like randomized trials. Consider an experiment evaluating the risk of stroke among individuals randomized to one of two drugs, each designed to lower blood pressure. If the experimental drug is more effective than the standard at lowering the risk of stroke, an uncontrolled comparison would yield a statistically significant treatment effect. But if we include a time-varying predictor assessing blood pressure, the treatment effect might disappear. Why? If all of the treatment effect operates through its regulation of blood pressure, there may be no additional variation in stroke risk to explain once blood pressure is taken into account. If so, the blood pressure effect *masks* (or *mediates*) the treatment effect.

It is not that the treatment has no effect—it does—it is that the blood pressure effect substitutes for the treatment effect.

We conclude this subsection by highlighting a purposeful omission: a complete lack of graphical displays. For example, we did not examine sample survivor and cumulative hazard functions by levels of the predictors. Although we could—and did—examine these functions for the full sample, subgroup plots are virtually impossible to construct for time-varying predictors. The best we can usually do is examine estimates for subgroups who have (1) constant values; or (2) specific patterns of temporal variation. The first approach is unsatisfying unless a constant pattern is common in the sample and the second approach is unsatisfying because it does not reflect the statistical model that was fit (which links event occurrence to contemporaneous predictor values). Notice, too, that we have we not displayed fitted survivor or cumulative hazard functions. The difficulty here is that when a model includes time-varying predictors, derivation of fitted values requires numerical integration routines not widely available in most statistical computer packages. The result is that when including time-varying predictors, we must generally content ourselves with numeric summaries.

15.1.2 Complex Patterns of Temporal Variation

Many time-varying predictors display more complex patterns of temporal variation than the nonreversible dichotomies discussed above. When studying readmission patterns among adolescents released from inpatient psychiatric facilities, for example, Foster (1999) examined the effectiveness of four forms of after-care: outpatient therapy, case management, placement in an intermediate care facility, and placement in a residential treatment center. Among the 204 adolescents, 59 (28.9%) were readmitted within the year; this demanded that Foster collect data on after-care services on each of the 45 unique days of readmission. Using administrative records, he compiled *day-by-day* treatment histories for each teen, which supported the construction of four time-varying predictors. Foster was then able to document the ineffectiveness of after-care in reducing the risk of readmission, after controlling for an adolescent's family background and mental health history.

Few researchers are able to gather data on time-varying predictors at such a fine level of temporal precision. Predictors are commonly assessed using a schedule coarser than that used to assess event occurrence. Event histories might be recorded in days, but the values of time-varying predictors are assessed once a week. Events might be recorded weekly, but time-varying predictors are assessed once a month. When the data collection for time-varying predictors is not temporally aligned with event times, you cannot analyze their effects directly because you do not know their values at every moment that an event occurs.

There are three ways of dealing with this type of information shortfall:

- *Round the event times so that they reflect the periodicity of the time-varying predictors.* In their study of U.S. Supreme Court justices' careers, Zorn and VanWinkle (2000) assessed event occurrence (retirement or death) in years, even though they could have done so in weeks or months. Doing so enabled them to use time-varying predictors summarizing the justices' annual productivity level and the match between the justice's political party and the president's.
- *Impute predictor values for the intermediate event times.* In their four-year study of employment duration, Dickter, Roznowski, and Harrison (1996) had weekly data on job turnover, but annual assessments of regional unemployment. Their analyses used the most recent value of each time-varying predictor, carrying forward its value to each of the next 52 weeks.
- *Use only time-invariant predictors constructed from baseline data.* In their two-year study of relapse to alcohol use, Cooney and colleagues (1991) measured event occurrence to the nearest day but sociopathy levels at three points in time—0, 6, and 18 months after release. But with only two follow-up assessments, they ultimately analyzed only the effects of the predictor's *initial* values rather than extrapolating the two time-varying values over long periods of time.

None of these strategies is uniformly preferable. None has any formal justification. To make an informed choice, consider your analytic purposes and the biases that each strategy may introduce. We find it helpful to ask four related questions:

- *How precise are the event times anyway?* If they are based on retrospective recall gathered on the same temporal schedule as the time-varying predictors, they may not be as precise as you think.
- *Would rounding increase the number of ties?* If most event times remain unique, rounding sacrifices little because Cox regression depends only on the rank order of the event times.
- *Is imputation feasible and reasonable?* You can treat this dilemma as a missing-data problem (Little & Rubin, 1987). Is there a defensible imputation strategy?

- *What, precisely, are you trying to accomplish?* Which strategy would be
  more convincing to your audience? Analyses conducted using
  time-invariant baseline data impervious to issues of rate- and state-
  dependence may be more credible than analyses conducted using
  imputed time-varying values or rounded event times.

If you decide to use time-invariant versions of time-varying predictors, no
additional steps are necessary: you simply fit a regular Cox regression
model. If you decide to discretize event times into a very small number
of periods, you can fit a discrete-time hazard model. And if discretization
is not that coarse, but it does allow you to construct a complete predictor history for each individual, you can apply the methods just described
in section 15.1.1. But if you decide to try imputation, your analyses will
be more complex because then you must select an imputation strategy.
It is this topic that we discuss next.

15.1.3 Imputation Strategies for Time-Varying Predictors

We discuss imputation strategies for time-varying predictors using data
from Hall et al. (1990), who followed 104 newly abstinent cocaine users
for up to 12 weeks to determine whether and, if so, when the former
addicts began using cocaine again. Everyone completed an intake interview during their last week of treatment and follow-up interviews every
week thereafter so that the researchers could gather data on a wide range
of predictors that might be associated with relapse. We focus here on two
of them: (1) *NEEDLE*, a time-invariant predictor that identifies the 35
individuals (33.7%) who used cocaine intravenously; and (2) $MOOD_j$, a
time-varying predictor assessing the respondent's positive mood score
using a standardized questionnaire. A total of 62 participants (59.4%)
began using cocaine within 12 weeks after release, some as soon as the
next day, others after weeks of abstinence.[2]

The presence of 38 unique event times requires that we generate
predictor histories that provide near daily information. But follow-up
interviews were conducted only once a week. Before imputing the required predictor values, we must craft a suitable strategy. Of the dozens
of approaches possible, the three below—cited with brief justifications—
are the most popular:

- *Use the most recent assessment.* Carry forward each *MOOD* assessment
  until the next one is available because this corresponds most
  closely to the approach that someone tracking the respondent's
  behavior would use.

- *Compute a moving average based on the most recent and several past assessments.* If the time-specific *MOOD* scores are just selected realizations of an underlying continuous process, the last value may not be the most meaningful. A more stable estimate of the underlying value may be obtained by averaging the most recent value with several previous ones.
- *Interpolate between adjacent assessments.* Imputed values depend on both past and future *MOOD* scores. Interpolating between adjacent assessments will provide more accurate values than those based solely on measures gathered in the past.

These are but three alternatives; we could create many more. However, we suggest that you resist the temptation to design sophisticated imputation algorithms, because: (1) all are ad hoc; (2) difficulties can arise when imputing values at the beginning of time, as there are rarely adequate prior data; (3) difficulties can arise when imputing values at the end of time, as event occurrence often terminates data collection; (4) difficulties can arise when some assessments are missing, creating a need for further imputation; and (5) the need to lag can complicate implementation. The more "creative" your imputation strategy, the easier it is for a critic to argue that your results are an artifact of your approach!

The challenge, then, is to come up with a feasible and credible strategy. In many ways, carrying forward past values nicely achieves this balance. Not only is it intuitively appealing—we update the predictor only when we get new data—it protects against state- and rate-dependence as well. This approach is especially attractive for categorical time-varying predictors because all other strategies yield (impossible) fractional values. On the down side, the resultant predictor histories behave erratically, changing in staccato bursts. Moving averages can soften this pattern, dampening the amplitude of the shifts. This retains the advantages of the last value approach and ameliorates some of the disadvantages. But if you find interpolation tempting, we caution that its use is controversial. Allison (1995) strongly questions whether it is *ever* appropriate to use later data to impute earlier values. Altman and de Stavola (1994) argue that interpolation can be appropriate if your goal is to describe the event history process, not to identify factors that could be identified *a priori* as potentially causal.

In practice, you may want to explore several options, compare results, and note how the findings change. We illustrate this strategy in table 15.2, which presents three Cox regression models fitted to the cocaine relapse data. Each includes the time-invariant predictor *NEEDLE* and a different variable representing the predictor for mood. Model A uses *BASEMOOD*,

Table 15.2: Comparing alternative imputation strategies for time-varying predictors

	Model A	Model B	Model C
Parameter Estimates, Asymptotic Standard Errors, and Deviance-Based Hypothesis Tests			
NEEDLE	1.0207**	1.0796***	1.1208***
	(0.3141)	(0.3157)	(0.3170)
Positive mood score			
BASEMOOD	−0.0037		
	(0.0147)		
WEEKMOOD$_j$		−0.0349*	
		(0.0139)	
DAYMOOD$_j$			−0.0544***
			(0.0149)
Goodness-of-fit			
−2LL	515.680	509.576	502.664
Δ −2LL from model that includes only NEEDLE	.065 (ns)	6.169*	13.081***

$\sim p < .10$; $* p < .05$; $** p < .01$; $*** p < .001$.

These are the results of fitting three Cox regression models to the cocaine use data ($n = 104$, n events $= 62$). Each model includes the respondent's mode of administration (NEEDLE) as well as a predictor describing his or her positive mood score. Model A includes BASEMOOD, the value at baseline; Model B includes WEEKMOOD, the value in the immediately prior week; Model C includes DAYMOOD, a linearly interpolated estimate of his or her mood on the immediately prior day.

Note: −2LL for null model is 528.186; −2LL for model with only predictor NEEDLE is 515.745. Efron method for ties.

a time-invariant predictor assessing the respondent's mood score just before release from treatment. Model B uses WEEKMOOD$_j$, a time-varying predictor assessing the respondent's mood score in the immediately prior week. This predictor lags the weekly mood data by one week—associating, for example, the baseline mood data with all days in the first week and the first week's mood data with all days in the second week. Model C uses DAYMOOD$_j$, a time-varying predictor assessing a respondent's mood score on the immediately prior day. We constructed DAYMOOD$_j$ by linearly interpolating between adjacent weekly values to yield daily values, and then assigning to each given day the mood value we imputed for the immediate prior day.

In comparing these models, first notice that the findings for NEEDLE remain virtually unchanged throughout. If our goal was to describe the effects of this variable, our conclusions would be unaffected by our selected version of the mood predictor, suggesting that the time-invariant BASEMOOD, which happens to provide the most conservative estimate, may be best. But if research interest focuses on the relationship between mood and relapse, this option is unsatisfying because *BASEMOOD* is not statistically significant. The inability to detect an effect tells us nothing about the effect of mood were its values allowed to vary. The specification of the mood predictor in Model B, using the previous week's data, *is* associated with risk of relapse. And the effect of mood is even greater when we use time-varying values imputed to the previous *day* (Model C).

Which predictor specification would we recommend? Although there is no single answer, the strong finding for $DAYMOOD_j$ argues persuasively for interpolation. But it raises issues of state- and rate-dependence because its values in any particular week depend upon future values (actually contemporaneous values, given that $DAYMOOD_j$ is lagged by one week). A critic would argue that the larger impact of this predictor may be due to temporal circularity. $WEEKMOOD_j$ is less susceptible to this problem, suggesting that its more conservative estimate may represent a good compromise. As this discussion makes clear, however, time-varying predictors can increase the predictive power of a model but often at a cost. Some difficulties can be resolved through design, but others will remain under even watchful vigilance. The ability to update predictor values is an attractive option, but data are not enough. Identification of links between data and outcomes demands careful research design, variable construction, and model parameterization. For further discussion of the use of time-varying predictors in Cox regression, we recommend that you consult the helpful papers by Altman and de Stavola (1994), Aydemir, Aydemir, and Dirschedl (1999), and Fisher and Lin (1999).

15.2 Nonproportional Hazards Models via Stratification

A Cox regression model invokes a *proportionality assumption*, that the hazard function for each individual in the population is a constant multiple of a common baseline function. Although this assumption often holds, you may encounter data sets in which it does not. If exploratory analyses, theory, or regression diagnostics (discussed in section 15.4) suggest that subgroups of individuals have *different* baseline hazard functions, you have two options: (1) fit a stratified model, which posits explicitly the existence of the multiple baseline hazard functions; or (2) fit a model that includes an interaction with time as a predictor, to represent the time-varying effect. Below, we describe former approach; in section 15.3, we describe the latter.

We illustrate the rationale underlying the stratified model solution using a simple hypothetical example. Imagine a randomized experiment that compares the effectiveness of two programs designed to reduce

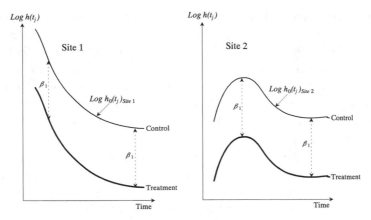

Figure 15.1. When is a stratified Cox regression model appropriate? Hypothesized (log) hazard functions, within two sites, by treatment status.

recidivism among inmates released from minimum security prison: (1) a standard program, comprising the current vocational, educational, and social services; and (2) an innovative program, comprising enhanced services. To increase generalizability and sample size, you conduct the study at two prisons. Although implementation is identical across sites, the inmates come from different geographic areas and they return to different areas upon release. The underlying pattern of recidivism may therefore differ across prisons.

Figure 15.1 presents (admittedly idealized) population (log) hazard functions for the treatment and control groups at each site. We have drawn these functions so that they illustrate the two statistical properties essential for proper use of a stratified model. Using the generic term "strata" to refer to sites, we see that:

- *Each stratum has its own baseline (log) hazard function.* At site 1, hazard is initially high and then drops steadily toward an asymptote; at site 2, hazard is initially low, rises to a peak, and then drops back toward a different asymptote.
- *The predictor's effect is identical across strata.* Although the baseline hazard functions differ, treatment lowers the log hazard of recidivism at each site by a constant amount, β_1, *at all points in time.*

Because the baseline (log) hazard functions differ across sites, it is inappropriate to fit a Cox regression model to the data pooled across sites. Doing so would knowingly violate the proportionality assumption. Including a dummy variable for site does not resolve the violation because, as shown in figure 15.1, the baseline (log) hazard functions differ across strata. Men from site 1—regardless of group—are at greatest risk of recidivism soon after release. Many men from site 2 forestall this initial risk, but soon, they, too, succumb. Including a statistical interaction between site and treatment as a predictor will not resolve the violation either because these predictors do not interact. It is not the effect of treatment that differs across sites (it is constant), but rather it is the baseline hazard functions themselves.

The stratified Cox regression model is designed for this type of situation. Instead of positing the existence of a single population baseline hazard function, it posits the existence of *multiple* baseline hazard functions, one per stratum. Using the subscript s to denote stratum, we can extend equation 14.4, for example, by writing:

$$\log h(t_{ij}) = \log h_{0s}(t_j) + \beta_1 X_{1ij} + \beta_2 X_{2ij} + \cdots + \beta_k X_{kij}. \qquad (15.3)$$

Adding subscript s to the baseline log hazard function allows each stratum to possess its own baseline hazard function. Omitting subscript s in the remainder of the model constrains the effect of each predictor to be identical across strata.

Most survival analysis routines can fit stratified Cox regression models. Computationally, the algorithm is simple. Instead of computing individual i's contribution to the partial likelihood by taking the ratio of his or her risk score at the moment of event occurrence to the sum of the contemporaneous risk scores among *everyone* still at risk (as in equation 14.5), revise the denominator so that it sums contemporaneous risk scores among those people still at risk in individual i's stratum. To compute the total partial likelihood across the full sample, multiply together these s stratum-specific partial likelihoods. This two-step process ensures that we compare individual i only to members of his or her own stratum, but that we constrain the parameters representing the effects of predictors—the β's—to be identical across strata.

We illustrate the use of a stratified Cox regression model by returning to the cocaine initiation study introduced in section 15.1. There, we saw that birth year and previous drug use were statistically significant predictors of the risk of cocaine initiation. Suspecting that the shape of the hazard function might differ by the respondent's residence, we examined sample (log) cumulative hazard functions for two groups of men: those who lived in urban and suburban areas ($n = 1316$) and those who lived in rural areas ($n = 342$). We display these sample functions in figure 15.2, where we use smooth curves, not step functions, to join the point estimates. We find that it is often easier to identify potential differences in shape if we use smooth curves instead of step functions in exploratory

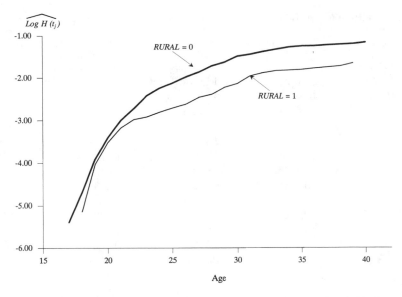

Figure 15.2. Sample log cumulative hazard functions for the cocaine initiation data by geographic region.

graphs like these. Here we find that the log cumulative hazard functions are nearly coincident when the men are young (between ages 17 and 22) and diverge thereafter. The varying vertical separation of the log cumulative hazard functions suggests a violation of the proportionality assumption, and the inappropriateness of the basic Cox model.

Table 15.3 presents the results of fitting four models to these data, each using the five predictors highlighted at the end of section 15.1.1: $BIRTHYR$, $USEDMJ_j$, $SOLDMJ_j$, $USEDOD_j$, $MOREOD_j$. The first column presents the fitted unstratified model, which is identical to Model C in table 15.1. Although figure 15.2 suggests the unsuitability of this model, we present it here for comparison. The second column documents what happens when we fit a model stratified by $RURAL$ to allow for the existence of two (unknown) baseline hazard functions. Comparing parameter estimates and standard errors for these two models, we see that stratification has little effect in this case. Both the unstratified and stratified models lead to identical conclusions, making the stratified model an attractive option that addresses the potential proportionality violation.

Of course, the stratified Cox model also has assumptions. Most notably, the stratified model is appropriate only if the predictors' effects are identical across strata—that is, if stratifier and predictors do not interact. For this data set, this implies that the effects of birth cohort and drug use

Table 15.3: Fitting a stratified Cox regression model

	Unstratified (Model C)	Stratified by RURAL	NON-RURAL ($n = 1316$)	RURAL ($n = 342$)
Parameter Estimates, Asymptotic Standard Errors, and Deviance-Based Hypothesis Tests				
BIRTHYR	0.0849***	0.0854***	0.0813***	0.1098~
	(0.0218)	(0.0219)	(0.0236)	(0.0584)
Marijuana use				
USEDMJ$_j$	2.4592***	2.4579***	2.4370***	2.5180***
	(0.2836)	(0.2837)	(0.3155)	(0.6488)
SOLDMJ$_j$	0.6899***	0.6847***	0.7151***	0.4542
	(0.1226)	(0.1228)	(0.1313)	(0.3530)
Other drug use				
USEDOD$_j$	1.2511***	1.2529***	1.2727***	1.1456**
	(0.1566)	(0.1567)	(0.1716)	(0.3843)
MOREOD$_j$	0.7604***	0.7468***	0.6925***	1.1050**
	(0.1307)	(0.1323)	(0.1410)	(0.3523)
Goodness-of-fit				
−2LL	4580.537	4271.899	3809.358	460.957

$\sim p < .10$; $* p < .05$; $** p < .01$; $*** p < .001$.

These are the results of fitting four Cox regression models to the cocaine initiation data ($n = 1658$, $n\ events = 382$). Each model includes the five predictors in Model C of table 15.1. The unstratified model reproduces that model for comparison purposes. The second column presents the results of fitting the identical model *stratified* by *RURAL*. The remaining columns present the results of fitting the identical model *separately* for the two levels of *RURAL*.

Note: All models fit using Efron's method for handling ties.

must be identical for men from rural areas and men from elsewhere. One way of evaluating this assumption is to examine the results of fitting *separate* Cox models to the data in each stratum. By comparing the results within each stratum to each other, and to those of the stratified model, we can evaluate the tenability of this noninteraction assumption.

The final two columns of table 15.3 present separate within-strata results for the cocaine initiation data. The effects of *BIRTHYR*, *USEDMJ$_j$* and *USEDOD$_j$* are similar across groups, but the effects of the secondary drug variables—*SOLDMJ$_j$* and *MOREOD$_j$*—are not. Moreover, the *direction* of the difference between estimates is inconsistent: the effect of *SOLDMJ$_j$* is larger in the nonrural areas (indeed, it is not statistically significant in the rural areas), whereas the effect of *MOREOD$_j$* is larger in the rural areas (although it *is* statistically significant in the nonrural areas). This suggests that the sale of marijuana may be a precursor to cocaine initiation only in urban and suburban areas and that the magnitude of the risk associated with additional use of other drugs may be larger in rural areas. These differentials suggest the possibility of statistical interactions between *RURAL* and *SOLDMJ$_j$* and between *RURAL* and *MOREOD$_j$*, each of which would represent a violation of the noninteraction assumption required for the stratified model.

How can you determine whether the observed variation in parameter estimates across strata is great enough to confirm a violation of the stratification assumption? Visual examination is insufficient because the standard errors for parameter estimates differ across strata. And within-strata estimates are typically less precise than full sample estimates because they are computed on smaller samples. Fortunately, it is easy to conduct a formal test of the null hypothesis that within-strata parameters are identical across strata, and identical to their counterparts in the stratified model itself.[3] You simply compare the $-2LL$ statistic for the stratified model to the sum of the $-2LL$ statistics for separate models fitted within each stratum. In our example, we compute the relevant test statistic by subtracting the sum ($3809.358 + 460.957$) from 4271.899, which yields 1.584. If the model includes p predictors and s strata, under the null hypothesis that the set of parameters is identical across the s strata this test statistic is distributed as χ^2 on $(s - 1)p$ degrees of freedom (here, $(2 - 1)5 = 5$ *d.f.*). Because the resulting test statistic is so small here, we conclude that variation in the set of parameter values across strata is inconsequential and hence that the stratified model is suitable. (Had we rejected the null hypothesis, we would have concluded that the stratified model was unsuitable, and we would have presented Cox regression models fitted within each stratum separately.)

Having conducted a formal test comparing the fits of the stratified model and the sum of the separate within-stratum models, you might wonder whether there is a similar test of the need for stratification that could be conducted by comparing the fits of the unstratified and stratified models. Unfortunately, such a test is not appropriate. The $-2LL$ statistic for the stratified model will always be smaller than that of the unstratified model simply because we postulate the existence of multiple baselines. Because the multiple baselines drop out of the stratified partial likelihoods, we do not estimate their values, and hence we cannot formally evaluate whether the resultant improvement in fit is statistically significant.

How, then, should you determine whether stratification is necessary? The paramount criterion is the tenability of the proportionality assumption. If it is violated, you must address it, either by implementing the stratified model (or by including an interaction with time as a predictor in the model, as we will soon discuss). If it is not clear whether the assumption is violated, Kalbfleisch and Prentice (1980) suggest that you have little to lose by choosing the stratified model because this choice generally has little effect on the efficiency of estimation. Choosing the stratified model has computational advantages as well, in that: (1) it decreases the number of observed ties (as tied individuals are inevitably spread across strata); and, (2) it decreases computational time (as each individual is compared only to members of the same stratum). In large data sets with many ties, these advantages can be considerable.

But the stratified model solution has a potentially profound cost: because you do not model the effect of the stratifier, you cannot describe its effect explicitly in your findings! The stratifier is a "hidden" variable, incorporated into the model but with its effect not explicitly modeled. Proponents of stratification argue that the inability to comment on the effect of the stratifier is no loss. They reason that if the shape of the baseline hazard functions differs across strata, it is inappropriate to make global statements about the stratifier's effect anyway because it varies over time. Others argue, and we share their view, that the inability to describe the stratifier's effect is a high price to pay to resolve a model violation that can be addressed through other mechanisms (as we show in section 15.3). And because choosing a stratified model requires grouping, you must weigh its benefits against the potential information loss that can result from categorization. From this perspective, then, choosing a stratified model is useful primarily when: (1) the stratifier is a "nuisance" variable, of no analytic interest; or (2) the baseline hazard functions in the different strata are so distinct that you cannot model their differences explicitly and easily. As full discussion of this issue involves contrasting the stratified model solution with an alternative approach—the inclusion of interactions with time as predictors in the model—it is to this topic that we now turn.

15.3 Nonproportional Hazards Models via Interactions with Time

Many researchers treat violations of the proportionality assumption as an analytic nuisance. The argument for the stratified model solution introduced in section 15.2 reflects this position in that it offers a strategy for addressing the violation that precludes saying anything about the predictor's effect. In discrete-time hazard modeling, we argued that violations of the proportionality assumption are often substantively interesting. To support this view, in section 12.5, we reframed a concern about a model violation into a positive statement that the predictor's effect *varying over time*. In this section, we extend these principles by postulating, fitting, and interpreting Cox regression models in which the effects of predictors are permitted to vary over time.

15.3.1 Alternative Representations of Interactions with Time

To include an interaction with time as a predictor in a Cox regression model, you must first create a time-varying variable, or set of time-varying variables, that represent the interaction. As we show below, the representation of the interaction can be continuous (if you hypothesize that the effect of the predictor varies smoothly with time) or categorical (if you hypothesize that the effect of the predictor differs in a piecewise fashion with time).

Using equation 14.4 as a starting point, we can specify a Cox regression model in which the effect of a single time-invariant predictor X varies linearly over time as:

$$\log h(t_{ij}) = \log h_0(t_j) + \beta_1 X_i + \beta_2 X_i (TIME_{ij} - c). \tag{15.4}$$

In this model, β_1 represents the vertical displacement in log hazard associated with a one-unit difference in X at time c and β_2 indicates how much this vertical displacement is increased (if β_2 is positive) or decreased (if β_2 is negative) with each one-unit increase in time. Subtracting the constant c from the value of $TIME$ facilitates interpretation. Common choices for c are: (1) the "beginning of time," so that $e^{(\beta_1)}$ is the hazard ratio when the clock "starts;" and (2) the estimated median lifetime, so that $e^{(\beta_1)}$ is the hazard ratio at the "average" event time. Another popular option is to associate β_2 with the interaction of X and the logarithm of $TIME$:

$$\log h(t_{ij}) = \log h_0(t_j) + \beta_1 X_i + \beta_2 X_i \log(TIME_{ij} - c).$$

However, in making this choice, you must be careful to select a value for the constant c that prevents the occurrence of undesirable infinities (recall that the log of zero is negative infinity).

To allow a predictor's effect to vary piecewise across epochs, you need a different model. The basic idea in this approach is to divide continuous time into k contiguous epochs, each represented by one of k time-indicators, $D_1 - D_k$. If time is measured in days, the epochs might be weeks; if time is measured in months, the epochs might be years. All epochs need not be the same length—it is more important to equalize the number of events per epoch—but taken together they must cover all observed event times. We can then posit a model such as:

$$\log h(t_{ij}) = \log h_0(t_j) + \beta_1 X_i D_{1ij} + \beta_2 X_i D_{2ij} + \cdots + \beta_k X_i D_{kij}. \qquad (15.5)$$

In this model, β_1 is the difference in log hazard associated with a one-unit difference in X during the first epoch, β_2 is the difference in log hazard associated with a one-unit difference in X during the second epoch, and so on. Although a piecewise specification like this typically uses more parameters than one that is smooth, its virtue is that it places no constraints on the functional form of the time-varying effect.

Notice the resemblance between these models and their discrete-time equivalents in equations 12.12 and 12.10. The primary difference is that here we use: (1) a *log*, not a *logit*, transformation of hazard; and (2) a smooth, not discrete, baseline. Those portions of each model that specify the predictor's time-varying effect are identical across metrics. In essence, then, we specify time-varying effects in Cox models using strategies similar to those used in discrete-time.

15.3.2 Fitting "Interactions with Time" Models to Continuous-Time Data

To fit an interaction with time model to continuous-time data, you must first construct the time-varying variable(s) that represent the interaction. Most data sets already include everything you need: the value of the predictor (which is fixed if it is time-invariant, or which you already had to cope with if it is time-varying) and *TIME* (which is known at every observed event time). Statistical packages that use the single record method for handling time-varying predictors are the most flexible in that they can accommodate any type of time-varying effect. All you need do is write computer code that creates the interaction variables, be they smooth, piecewise, or a combination thereof.

Statistical packages that exclusively use the counting process method are less flexible in that you must construct the person-period data set so that it appropriately includes the new variables that represent the interaction with time. Because all time-varying predictors—including these new interaction terms—must be constant within a record, the data set's periods must reflect the type of time-varying effect you would like to explore. This complicates data set construction in that individual *i*'s discretization of time depends not only upon the observed pattern of temporal variation for any time-varying predictors (as discussed in section 15.1) but also on your chosen specification for the time-varying effects. Researchers working with such programs may therefore find themselves drawn to piecewise specifications (like equation 15.5) so that their data set's periods can correspond to epochs. If you would like to allow the effect of a predictor to vary smoothly over time (as in equation 15.4), you will need to construct a data set that includes many separate periods. At the extreme, to represent a perfectly smooth interaction with time, you need a person-period record for each unique event time!

We demonstrate strategies for exploring interactions between predictors and time using data from Foster (2000). As part of a larger evaluation of the effects of an innovative system of delivering mental health services to children and adolescents with emotional and behavioral problems, Foster studied whether and, if so, when 174 individuals admitted to a psychiatric hospital were discharged. One half ($n = 88$) had traditional coverage ($TREAT = 0$), which required authorization for long-term inpatient care, provided outpatient care on a co-payment basis, and did not reimburse services provided in intermediate settings. The other half ($n = 86$) had an innovative plan ($TREAT = 1$), which offered coordinated mental health services at no cost, regardless of setting (inpatient, outpatient, or in between). Foster's primary research question was whether provision of comprehensive services, regardless of setting, reduced the length of inpatient stays.

To address this question, Foster initially fit a Cox regression model that included the single dichotomy $TREAT$. As shown in Model A of table 15.4, this variable has no main effect. Suspecting that its effect might vary over time, Foster investigated whether the treatment effect depended on how long the child had been hospitalized. The sample log cumulative hazard functions in the top panel of figure 15.3 confirm this suspicion. During the first few weeks after admission, children covered by the innovative plan are much more likely to be discharged. Over time, this differential diminishes, so that by day 32, the functions cross, indicating that an equal fraction of each group remains. Eventually, the risk of discharge is somewhat higher among those in the traditional plan (most likely because long-term hospitalization for this group must be authorized by a utilization coordinator).

The graph in the top panel of figure 15.3 depicts a classic "interaction with time." It is not that the treatment has no effect; it is that its effect varies over time. Fitting a Cox regression model that constrains its effect to be constant is both inappropriate (because the proportionality assumption is violated) and unsuccessful (because *no* effect is found). Stratification will not help because the violation concerns $TREAT$, the predictor of greatest interest. This leads us to specify Cox regression models in which the effect of $TREAT$ is permitted to vary over time.

We begin graphically, by plotting—in the bottom panel of figure 15.3— the *difference in sample log cumulative hazard functions* for the two groups. We constructed this graph by subtracting, at each point in time, the

Table 15.4: Fitting a nonproportional hazards Cox regression model

	Model A	Model B	Model C	Model D
Parameter Estimates, Asymptotic Standard Errors and Deviance-Based Hypothesis Tests				
TREAT	0.1457	0.7061*		2.5335***
	(0.1542)	(0.2924)		(0.7603)
TREAT × (TIME-1)		−0.0208*		
		(0.0092)		
TREAT1			1.5711*	
			(0.6406)	
TREAT2			0.5678	
			(0.4929)	
TREAT3			0.8497	
			(0.3621)	
TREAT4			−0.3499	
			(0.3641)	
TREAT5			−0.7660~	
			(0.4161)	
TREAT6+			−0.0993	
			(0.3111)	
TREAT × L2(TIME)				−0.5301**
				(0.1619)
Goodness-of-fit				
−2LL	1436.628	1431.374	1417.730	1423.062
n parameters	1	2	6	2
AIC	1438.628	1435.374	1429.730	1427.062

~$p < .10$; *$p < .05$; **$p < .01$; ***$p < .001$.

These are the results of fitting four Cox regression models to the length of hospital stay data ($n = 174$, n events = 172). Model A includes the main effect of treatment (TREAT). Model B allows the effect of treatment to vary linearly over time. Model C allows the effect of treatment to differ week by week. Model D allows the effect of treatment to vary linearly with the logarithm (to base 2) of time.

Note: −2LL for null model is 1437.520. Efron's method for handling ties.

sample log cumulative hazard function for the comparison and treated groups. We recommend that you construct such graphs when evaluating the proportionality assumption for time-invariant predictors. Although easier to construct if the predictors are categorical, you can construct a similar graph for continuous predictors via judicious categorization. (Another useful strategy is a plot of Schoenfeld residuals, as described in section 15.4.3.)

To understand why this graph is so useful, remember what each parameter in the Cox model represents: the difference in log (cumulative) hazard corresponding to unit differences in the value of the predictor.

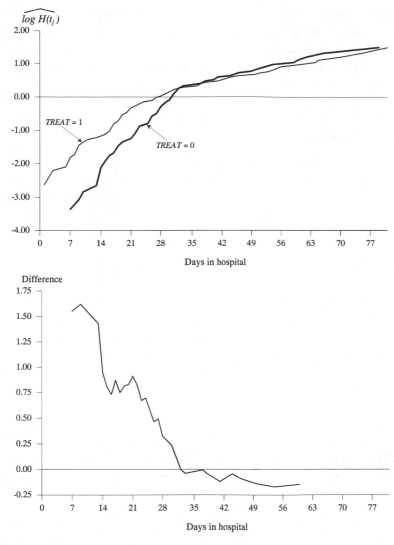

Figure 15.3. Sample log cumulative hazard functions and the difference in log cumulative hazard functions for the inpatient length of stay data, by treatment group.

Under the proportionality assumption, this difference should be identical at every point in time. If the model is appropriately specified, this diagnostic graph will be flat. Although it may "wiggle" owing to sampling variation, we should not find any consistent trend over time. If the graph is not flat, the proportionality assumption is violated. The shape of this function will reflect the predictor's time-varying effect. Ascertaining which algebraic function best describes its shape usually suggests a parameterization for the interaction with time. A steady equal-increment decline suggests an interaction with linear time. A stepwise shift in broad epochs suggests a piecewise specification. Here, the steep decline that we observe during the first few weeks, dropping to a lower asymptote, suggests a logarithmic specification.

Rather than move immediately to the logarithmic specification, let us proceed incrementally by examining three possible models, each with a different specification for the interaction with time:

- *Model B postulates that the effect of* TREAT *varies linearly over time.* Although we suspect that this model will not be ideal, many people find it easiest to routinely explore this specification first. To facilitate interpretation, we subtract 1 from *TIME*, so that the main effect of *TREAT* represents the treatment effect on the first day of hospitalization.

- *Model C postulates that the effect of* TREAT *differs piecewise across periods.* We defined six weekly time periods. The first represents days 1–7, the intermediate periods are 8–14, 15–21, 22–28, and 29–35 days; the last represents day 36 on. The decision to collapse together the data after day 35 was based on the graphs in figure 15.3, the declining number of events during the later days, and an examination of the parameter estimates from a finer classification.

- *Model D postulates that the effect of* TREAT *varies linearly with the* logarithm *of time.* To facilitate interpretation, we take logs to the base 2, so that the interaction term represents the change in the treatment effect as length of stay doubles. We do not center time on the first day of hospitalization, in order to avoid an infinite value for the logarithm of time on this day.

Although we could easily theorize the existence of other parameterizations, for these data we restrict ourselves to these three for they span a suitable range of alternatives.

Let us begin with the linear specification in Model B. We begin here not because this model is best but rather because we often use this specification as an initial screen for interactions with time. Here, the significant difference in fit from Model A ($\chi^2 = 5.254$, 1 *d.f.*, $p < .05$) demonstrates that, in this data set at least, we have uncovered cause for concern. As popular and easy as this screen is, however, we caution that it can miss many violations of the proportionality assumption. (We discuss still other ways of identifying these violations in section 15.4.3.)

The piecewise specification in Model C provides a better fit. Comparison with the main effects model (A) reveals a statistically significant difference in fit between models (χ^2 = 18.898, 5 *d.f.*, *p* < .01) and although we cannot directly compare its −2LL statistic to that for Model B, comparison of AIC statistics between models B and C suggests the superiority of the latter model. Examining parameter estimates, we can quantify how the effect of *TREAT* diminishes over time. In weeks 1, 2, and 3, the estimated hazard ratios are $e^{(1.5711)}$ = 4.81, $e^{(0.5678)}$ = 1.76, and $e^{(0.8497)}$ = 2.34. By week 6 and beyond, the treatment effect has all but disappeared ($e^{-(0.0993)}$ = 0.91).

This brings us to Model D, a parsimonious representation that fits better than the main effects model (χ^2 = 13.56, 1 *d.f.*, *p* < .001) and has the lowest AIC statistic of all the models displayed. Equally important, it allows us to go beyond the general statement that the effect of *TREAT* "varies over time" to a characterization of how it varies.[4] Because the log of 1 is 0, antilogging the coefficient for *TREAT* in Model D— computing $e^{(2.5335)}$ = 12.60—yields the estimated hazard ratio on day 1. By taking logarithms to the base 2, the parameter for log(*TIME*) indicates the drop in log hazard associated with a doubling of length of stay as it goes from 1 to 2, 2 to 4, 4 to 8, and so on. As length of stay doubles, the estimated log hazard for *TREAT* declines by 0.5301. By substitution, you can estimate a hazard ratio for any given day. For day 8, we have $e^{(2.5335+\log_2(8)\times(-.5301))}$ = 2.56. For day 32, $e^{(2.5335+\log_2(32)\times(-.5301))}$ = 0.89. Logarithmic specification of time in interactions with time are especially useful when the observed event times vary widely (because the transformation draws in the upper tail of the event time distribution).

Does the parameterization in Model D *fully* characterize the interaction with time? In discrete time, we could answer this question by comparing its fit with that of a fully interactive model (in which the predictor's effect was allowed to differ in each discrete time period). In continuous time, no such fully interactive model exists. All we can fit are smooth representations, which are parsimonious but constrained, and epoch-to-epoch representations, which are unconstrained but not parsimonious. There is no model in which the effect of a predictor can be permitted to differ at every observed event time. To select among models, then, we must rely on a combination of common sense and model checking. For this data set, we have a strong preference for Model D.

We now return to the question first raised at the end of section 15.2: given two distinct strategies for handling violations of the proportionality assumption—choosing a stratified model and fitting an interaction with time model—which is preferable? Although there are no definitive rules, we offer the following principles to guide you:

- *If the predictor that violates the proportionality assumption is not of explicit research interest,* and it is either discrete or can be categorized with little loss of information, adopting a stratified Cox regression model is a simple low-cost strategy.
- *If the predictor that violates the proportionality assumption is of explicit research interest,* you should adopt the "interaction with time" strategy because fitting a stratified model will prevent you from learning precisely what it is you want to know.

Note, too, that most computer programs can stratify the Cox model only by time-invariant predictors. If the "offending" predictor is *continuous, time-varying,* or both, you may have little choice.

We close this section by returning to a paradox in nomenclature mentioned briefly in section 14.1. Many people refer to the Cox regression model as a *proportional hazards* model. But as we have just seen, the Cox model need not invoke a proportional hazards assumption. Indeed, we could label the Cox regression models in this section as "nonproportional proportional hazards models." Before expressing alarm at this terminology, remember that linear regression models can include nonlinear effects. So although the *proportional hazards* label is useful, it should not be interpreted literally or narrowly. The Cox regression model is equally well suited to data sets in which the proportionality assumption does not hold.

15.4 Regression Diagnostics

Practicing data analysts know the value of residuals and influence statistics. Diagnostics akin to these are especially useful in Cox regression analysis. We make this assertion not because the model is excessively sensitive to violations but because the strategies available for exploratory analysis are so coarse and limited. When fitting a multilevel model for change, you can examine empirical growth trajectories to detect individuals with unusual temporal patterns of change (outliers) or with extreme predictor values (high leverage cases), or just to investigate interesting transformations of the data. In survival analysis, you do not plot data points, but rather *summary* statistics—the sample survivor and cumulative hazard functions—computed at some group level. To estimate these, you collapse your data twice: once across individuals, which precludes any insight into the behavior of specific cases; and a second time across predictor values, which limits your insight into event occurrence at specific levels of the predictor.[5] When fitting Cox models, the associated regression diagnostics allow you to learn things about your data that are unobservable by other means.

Yet the Cox regression model poses a thorny statistical problem: How should diagnostics be defined? In linear regression analysis, for example, individual i's residual is $y_i - \hat{y}_i$, the difference between the observed and predicted outcome values. To extend this same idea to the Cox model, we need to: (1) choose a quantity to focus upon (e.g., the event time, the cumulative hazard function, or something else?); and (2) develop a strategy for dealing evenhandedly with censoring. These dilemmas have led statisticians to create many different diagnostics, each with a specific purpose in mind. In the sections below, we focus on four diagnostics that we find particularly useful: martingale residuals (section 15.4.1), deviance residuals (section 15.4.2), Schoenfeld residuals (section 15.4.3), and score residuals (section 15.4.4). We conclude, in section 15.4.5, with an overview of how to use these diagnostics efficiently.

15.4.1 Martingale Residuals

Martingale residuals are helpful in Cox regression analysis for selecting a particular functional form to be used for a continuous predictor. Their name derives from extensions of the Cox model rooted in martingale theory, an advanced branch of probability theory. Developed for situations in which individuals can experience a target event more than once, martingale theory allows you to model the hazard function for repeatable events (Anderson, Borgan, Gill, & Keiding, 1993; Fleming & Harrington, 1991). Martingale residuals assess the quality of a model's predictions by comparing, for each individual, the number of events *actually* experienced to the number of events *predicted* to happen:

$$\hat{M}_i = \text{observed}(n \text{ events})_i - \text{expected}(n \text{ events})_i. \qquad (15.6)$$

This definition yields a residual measured in a familiar metric—the discrepancy between observed and expected values—expressed in terms of the *number of events*. Like residuals from linear regression analysis, martingale residuals sum to 0 across individuals and in large samples, are approximately uncorrelated.

In a Cox regression model—in which each person can experience either only one, or no, event—martingale residuals assess the relative magnitude of an individual's *event time* in comparison to what we would predict on the basis of the fitted model. To understand how we can make this statement, let us examine equation 15.6 and identify the conditions that produce positive and negative values for the residual. For $\hat{M}_i$ to be positive, individual i must have experienced "more events" than expected on the basis of his or her predictor values and the estimated parameters. This means that he or she must have experienced the event, and that it must have happened *before* it was expected—it must have occurred "too soon." For $\hat{M}_i$ to be negative, individual i must have experienced "fewer events" than expected. This can only happen if he or she: (1) experienced *no* event (i.e., the event time was censored); or (2) experienced the event, but *later* than expected. Either way, the event occurred "too late." As a result, positive residuals indicate that the event time is *earlier than expected*—that the model "overpredicts." Negative residuals indicate that the event time is *later than expected*—that the model "underpredicts."

There are two ways of using martingale residuals to discern the appropriate functional form for a continuous predictor.

- *Plot martingale residuals versus an omitted predictor.* You construct these displays *before* including the predictor of interest in the model. The fitted model for which you estimate the residuals can: (a) *already include other predictors*, so that the martingale residuals then suggest a functional form for the new predictor after statistically controlling for the existing predictors; or (b) *include no predictors*, so that the residuals suggest a functional form in the absence of statistical controls.
- *Plot martingale residuals versus an included predictor.* You construct these displays *after* including the predictor of interest in the model, to determine whether the chosen functional form is appropriate. If no pattern emerges in the plot, you can conclude that the chosen form is fine; if a pattern emerges, the form chosen is probably inappropriate.

As martingale residual plots are more difficult to interpret than regular scatterplots, we suggest that you superimpose a smoothed nonparametric summary of the residual/predictor relationship in the plot, such as a lowess smooth or a cubic spline.

We illustrate these strategies in figure 15.4, which plots martingale residuals from two fitted Cox models for the recidivism data of chapter 14. Recall that the data set included three predictors: *PROPERTY* and *PERSONAL*, describing the former inmate's criminal history, and *AGE*, assessing the difference in years between this individual's age at release and the "average" former inmate in the sample. In the top panel, we plot martingale residuals from a null model versus *AGE*. In the bottom panel, we plot the martingale residuals for a model that includes *AGE*, *PERSONAL*, and *PROPERTY* versus *AGE* (Model D of table 14.1). In both panels, the • symbol identifies individuals with observed event times; the

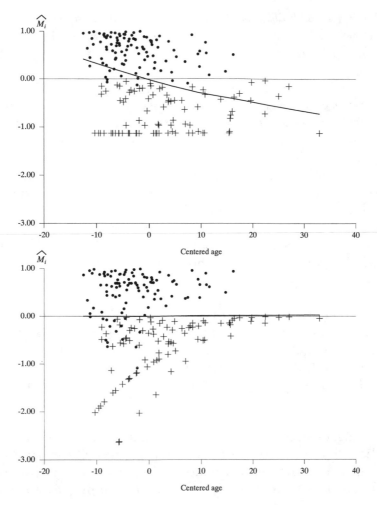

Figure 15.4. Martingale residuals for the recidivism data. The top panel displays residuals from a null model (with no predictors) vs. the continuous predictor *AGE*. The bottom panel displays residuals from a model with three predictors (Model D of table 14.1) vs. the included predictor *AGE*. A • identifies individuals with observed event times; a + identifies individuals with censored event times. The trend lines are lowess smooths.

+ symbol identifies individuals with censored event times. The superimposed trend lines are lowess smooths, computed across all observations in the sample.

These displays exhibit some patterns common to all martingale residual plots and other patterns specific to this particular data set. We begin with three general observations:

- *Martingale residuals have a maximum of 1 and are skewed toward negative values.* This is a consequence of their definition in equation 15.6, which subtracts from either 0 or 1, a quantity that must be positive.
- *Individuals with observed event times have larger martingale residuals, on average, than those with censored event times.* This, too, is a consequence of the definition in equation 15.6, and makes it difficult to use martingale residuals to identify poorly predicted cases.
- *If many people are censored at a single late event time, their martingale residuals will display predetermined patterns.* For a null model, they will concentrate at a single negative value (here, a value of −1.1). For a model that contains predictors, the martingale residuals will form bands that reflect their particular combination of predictor values (as shown in the bottom panel of figure 15.4.

Taken together, these properties highlight the need for care when interpreting martingale residual plots, because some patterns (like the line of +'s in the top panel or the swath of +'s in the bottom panel) are just a consequence of the residual's definition.

How, then, should you interpret these plots? We recommend that you concentrate on the trend revealed by a superimposed smooth. The approximately linear negative trend with *AGE* in the top panel, for example, suggests the appropriateness of adding linear *AGE* as a predictor in the Cox regression model. A jagged or a nonlinear pattern would suggest an alternative representation. The virtually nonexistent trend with *AGE* in the bottom panel confirms the appropriateness of this decision.

Martingale residuals are a powerful data analytic tool for detecting candidate predictors and selecting an appropriate functional form when examining continuous predictors for potential inclusion in a Cox model. Unlike model-based strategies (described in section 12.4.2 in discrete time), you need not categorize a predictor nor compare "shot-in-the-dark" representations. Because they provide such direct information, it is good data analytic practice to compute martingale residuals for a null model and add their values to your basic data set. This allows you to plot their values versus each potential predictor to help discern an appropriate functional form for that predictor. If the relationship is approximately linear, you can feel comfortable analyzing the effect of the predictor in its raw state. If the relationship you detect is non-linear, you can quickly transform the predictor as appropriate.[6]

15.4.2 Deviance Residuals

Deviance residuals are helpful for identifying individuals whose outcome is poorly predicted. Although you might hope that we could use martingale residuals for this purpose—eliminating the need for another diagnostic—their inherent skewness makes them ill-suited to the task. Without documenting their complex algebraic computation, which all the major statistical packages implement, suffice it to say that deviance residuals are a transformation of martingale residuals designed to symmetrize their distribution around 0. Under light censoring (say, less than 25%), their distribution will be approximately normal. Under heavier censoring (more than approximately 40%), their distribution will still be approximately symmetric, but the normal approximation will be less accurate because there will be a larger proportion of small values (Therneau & Grambsch, 2000).

You interpret deviance residuals in ways similar to the martingale residuals upon which they are based. A positive residual suggests the event occurred earlier than expected; a negative residual suggests the event occurred later than expected (or not at all). Large deviance residuals of *either sign* identify potential outliers. You can identify individuals whose outcomes are predicted poorly using one of two strategies:

- *Examine the distribution of the deviance residuals.* If censoring is light and the model appropriate, approximately 5% will be greater than ±2. An excessive percentage of large values suggests a problem, and individuals with very large values (say, >±2.5 or 3.0) warrant scrutiny.
- *Plot the deviance residuals* versus *risk scores.* A risk score summarizes the information contained across all predictors in a model for each individual. If the model is appropriate, the deviance residuals should be unrelated to the risk scores (indicating that the model predicts equally well at all levels of risk). Concentrations of deviance residuals at particular levels of risk suggest a poor fit at these levels.

Because the second plot is useful only if risk scores are continuous, you should construct this display only for models that include either: (1) one or more continuous predictors; or (2) at least several categorical predictors. For other models, simply examine the residuals' distribution.

We illustrate these strategies in figure 15.5, which presents deviance residuals from a Cox regression model containing all three predictors for the recidivism data. In the top panel, we present a univariate

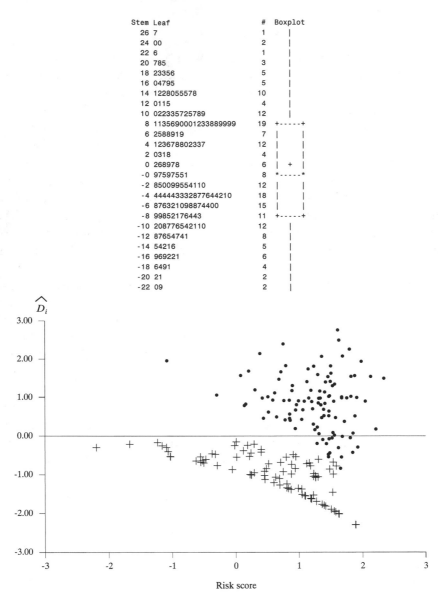

Figure 15.5. Deviance residuals for the recidivism data. The top panel displays their univariate distribution; the bottom panel plots them vs. risk scores from a model with three predictors (Model D of table 14.1). A • identifies individuals with observed event times; a + identifies individuals with censored event times.

(stem-and-leaf) display of the residuals; in the bottom panel, we plot the deviance residuals $\hat{D}_i$ versus the risk scores (once again using symbol • for cases with observed and + for cases with censored event times). To facilitate inspection, the risk score plot includes a horizontal reference line at 0.

In a moderately censored data set like this—in which 88 of the 194 cases (45.4%) remained arrest-free—we do not find the approximate normality that the deviance transformation is designed to achieve. Instead, the stem-and-leaf display contains larger concentrations of deviance residuals near +1 and −0.5. But when examining the distribution of the deviance residuals, we are not trying to determine whether its shape is approximately normal. The purpose of the plots is to identify "extreme" values. So in a stem-and-leaf display, we focus on ascertaining how many individuals have deviance residuals greater than ±2. As only 11 former inmates meet this criterion (5.6% of the full sample), leading us to conclude that the fraction of poorly predicted cases is within expected limits.

Plots of deviance residuals versus risk scores are tricky to interpret because our eye is inevitably drawn to an irrelevant predetermined pattern: the concentration of individuals with observed event times (the •'s) in the upper half and the fan spread of individuals with censored event times (the +'s) in the lower half. These patterns result directly from the residual's definition. Of greater interest is whether the poorly predicted cases are concentrated at particular levels of risk. In figure 15.5, we see that some of the larger values of $\hat{D}_i$ appear at the highest levels of risk (risk scores around 2.0). This does not suggest that the model predicts poorly for everyone at high-risk, for the plot includes many high-risk individuals whose outcomes are precisely predicted (with points on, or near, the line of 0). What it does suggest is that most of the poorly predicted cases for this model are former inmates who face a high risk of recidivism.

What should you do if your data set includes individuals with extreme deviance residuals? As when identifying potential outliers after fitting any statistical model, the first thing you should look for are: (1) transcription errors; and (2) omitted predictors that might help explain the poor predictions. You can also set aside the cases and refit the model. If the results remain unchanged, retaining the extreme observations causes little harm. If the results do differ substantially, further analyses are needed to ensure that the small number of highly deviant cases does not unduly influence the findings. In no case would we set the problematic cases aside without considerable investigation, as the resulting improvement in fit is unlikely to be justifiable.

15.4.3 Schoenfeld Residuals

Schoenfeld residuals (also known as *partial* residuals) are useful for identifying violations of the proportionality assumption. Unlike martingale and deviance residuals, which compare the observed and expected number of events, Schoenfeld residuals compare observed and expected *predictor values*. If a Cox model includes more than one predictor, there is a set of Schoenfeld residuals for each.

To concretize this definition, let us first explain what we mean by the "expected value" of a predictor. In general, the term "expected value" refers not just to predicted values but also to "average" values—here, the "average" value of the predictor at individual i's event time. To compute a Schoenfeld residual for a predictor, X, we compare individual i's value of X to the "average" value of X among everyone still in the risk set when i experiences the event. Although we could also compute this average as a simple sample mean, Schoenfeld (1982) showed that it is more useful to *weight* each person's value of X by his or her risk score. For each predictor X, then, we define the Schoenfeld residual for individual i as:

$$\hat{S}_i(X) = x_i - expected_i(X), \qquad (15.7)$$

where the second term on the right side of the equation refers to the expected value of X among everyone in the risk set when individual i experiences the target event. Because Schoenfeld residuals compare predictor values at observed event times, they are defined only for individuals who actually experience the event. Censored individuals contribute to the computations when they fall in an observed person's risk set, but they do not have Schoenfeld residuals of their own.

The simplicity of the definition in equation 15.7 makes it easy to interpret Schoenfeld residuals. Positive values identify individuals whose predictor values are greater than those of their peers at risk at their event time; negative values identify individuals whose predictor values are smaller than those of their peers at their event time. Schoenfeld residuals, therefore, assess the relative magnitude of an individual's predictor value in comparison to what we expect given his or her event time. For samples that are sufficiently large, Schoenfeld residuals for each predictor will sum to 0 across individuals and will be approximately uncorrelated.

The most important property of Schoenfeld residuals is their behavior under the proportionality assumption. If the assumption holds, the Schoenfeld residuals will be unrelated to time. This means that for each predictor in a model, a plot of Schoenfeld residuals versus observed event times should reveal no consistent trend. If the Schoenfeld residuals tend to increase (or decrease), this suggests that the effect of the predictor is not constant over time—as the proportionality assumption suggests—but rather that it increases (or decreases) over time.

To discern whether Schoenfeld residuals are related to time, you can:

- *Plot the Schoenfeld residuals versus time and superimpose a smooth summary.* If the proportionality assumption holds, no nonhorizontal trend should appear. Any discernible trend suggests a model violation (Schoenfeld, 1982).
- *Compute the simple correlation between Schoenfeld residuals and time.* If the correlation is zero and not statistically significant, then the validity of the proportionality assumption is confirmed. A statistically significant non-zero correlation suggests the opposite (Harrell, 2001).

Because the Cox model is *semi-parametric*, based on not the actual event times but their *ranking*, we usually implement these approaches using the *rank* of the event times (computed across the full sample, including observed and censored cases). If either analysis suggests a violation of the proportionality assumption, you can follow up by testing explicitly for interactions with time.

We illustrate these strategies in figure 15.6, which presents Schoenfeld residuals from Model D for the recidivism data versus ranked event time. The top and middle panels display residuals for the two dichotomous predictors, *PERSONAL* and *PROPERTY*; the bottom panel displays residuals for the continuous predictor, *AGE*. As Schoenfeld residuals are defined only for individuals with observed event times, we need no plotting symbols to distinguish observations. To facilitate pattern detection, each plot displays a lowess smooth and the sample bivariate correlation of the Schoenfeld residuals with ranked event times.

Schoenfeld residuals for dichotomous predictors will always appear in two bands, one for each predictor value. When interpreting these displays, we therefore rely heavily on the summary information in the lowess smooth and the correlation. For *PERSONAL*, the potential importance of the slight downward trend is mitigated by the small statistically nonsignificant correlation coefficient ($r = -0.09$). The plot for *PROPERTY* is difficult to interpret because only 9 of the 106 former inmates with observed event times had no property crime history. This proportion is so small that the lowess smooth is flat. This suggests no consistent trend over time, a conclusion supported by the small statistically nonsignificant correlation ($r = -0.07$). For both dichotomous predictors, then, we have no evidence of a proportionality violation.

Plots of Schoenfeld residuals for continuous predictors are easier to interpret. In the bottom panel for *AGE*, we find a distinct upward trend,

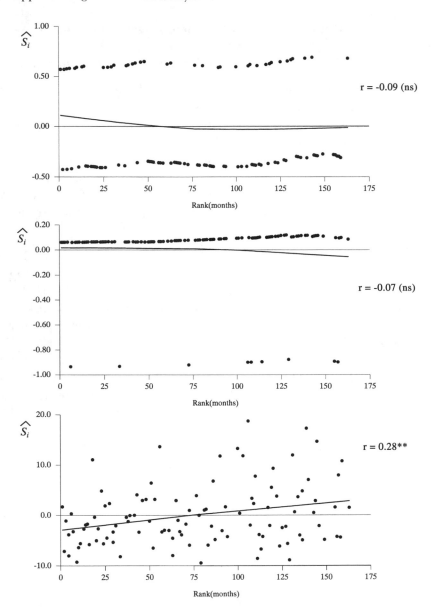

Figure 15.6. Schoenfeld residuals for the recidivism data vs. ranked event times. The top panel is for *PERSONAL*; the middle panel is for *PROPERTY*; the bottom panel is for *AGE*. The trend lines are lowess smooths.

suggesting that the proportionality assumption may not be met for this predictor. The statistically significant non-zero correlation (r = .28, p = .0039) associated with this plot reinforces this suspicion. If you find a similar pattern in any Schoenfeld residual plot, you should fit a Cox regression model that incorporates an interaction between the predictor and time, using the trend in the residual plot to inform the interaction's parameterization. Here, the steady rise with ranked time suggests a smooth specification. Had a step-function emerged, an epoch-by-epoch representation would have been explored. When we include an interaction between linear time and *AGE* in the Cox model, we find that the interaction *is* indeed non-zero and statistically significant (χ^2 = 7.52, 1 *d.f.*, p < .01). This tells us that the proportionality assumption does not hold for *AGE* (after controlling for *PERSONAL* and *PROPERTY*). The effect of inmate age upon recidivism—which we have described previously as large and negative—diminishes over time.

Schoenfeld residuals are a powerful tool for examining the tenability of the proportionality assumption, especially for continuous predictors. For dichotomous predictors, you can often spot a model violation by examining plots of differences in log cumulative hazard functions by levels of the predictor. For continuous predictors, construction of these plots requires categorization. Schoenfeld residuals are invaluable in this latter case because you need not categorize nor specify a parametric form for the interaction up-front. Another advantage of Schoenfeld residuals is that they provide a graphical vehicle for exploring proportionality when there are other predictors in the model.

15.4.4 Score Residuals

Score residuals are helpful for identifying individuals who exert a great influence on model fit as a consequence of their combined predictor values. Although you might hope that we could use the Schoenfeld residuals for this purpose—eliminating the need for yet another diagnostic— the fact that Schoenfeld residuals are defined only for individuals with observed event times limits their utility for this purpose. So, too, although you might wish to assess influence by comparing parameter estimates computed with, and without, each observation, the iterative nature of the partial likelihood method of estimation makes this computationally intensive approach less attractive. Although some statistical packages include an option for computing the changes in parameter estimates that result from the systematic deletion of each individual case, they more commonly include an approximation to this quantity known as the *score residual.*

Like Schoenfeld residuals, there is one set of score residuals for each predictor in a model. Without delving into the details, suffice it to say that for each predictor, for individual i, the score residual is a function of the approximate change in the parameter estimate that would result if individual i were deleted from the sample. A positive score residual indicates that the parameter estimate would increase; a negative score residual indicates that the parameter estimate would decrease. But when interpreting their values, be sure not to take their magnitude too literally. Score residual values do not *actually* assess the changes in parameter estimates that would result from the deletion of the case.

Because score residuals are a type of *case deletion diagnostic*, we examine their behavior using methods that we use for all such diagnostics (Cook & Weisberg, 1982). Of the many possible alternatives, we recommend two simple approaches:

- *Examine the distribution of the score residuals.* Excessively large values identify individuals who exert a great influence on the fit.
- *Plot the score residuals* versus *ranked event times.* This plot should help discern whether high-influence cases are concentrated at specific event times.

When examining these displays, remember that the primary purpose is to identify extreme values so that you may determine whether model fit is unduly affected by high influence cases.

In figure 15.7, we use the second of these strategies to present plots of score residuals for the three-predictor Cox regression model fitted to the recidivism data (once again using a • for observed event times and a + for censored event times). To facilitate visual examination, the plots include a reference line drawn horizontally at 0. Examination of the three plots shows that each includes a small number of individuals with large score residuals. In the bottom of the top panel, we find eight individuals with large score residuals for *PERSONAL*. In the lower right corner of the middle panel, we find several individuals with large score residuals for *PROPERTY*. In the upper portion of the bottom panel, we find a handful of individuals with large score residuals for *AGE*. All these cases warrant closer examination.

15.4.5 Using Regression Diagnostics as an Integral Part of Data Analysis

No Cox regression model-building task is complete without a thorough examination of regression diagnostics. To encourage their use and interpretation, table 15.5 presents a brief overview of each. We offer this rubric in the hope that the potentially dizzying array of diagnostics does not

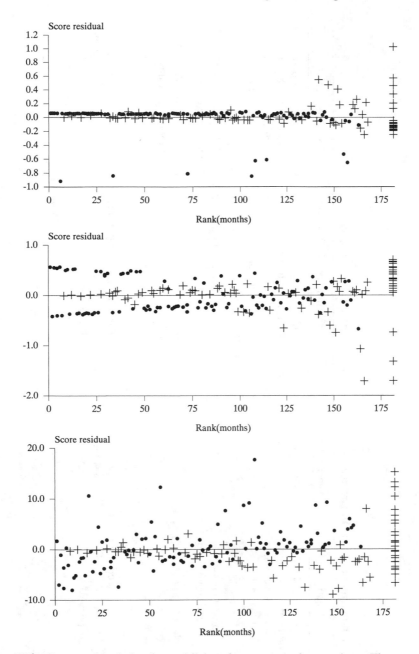

Figure 15.7. Score residuals for the recidivism data vs. ranked event times. The top panel display is for *PERSONAL*; the middle panel is for *PROPERTY*; the bottom panel is for *AGE*. A • identifies individuals with observed event times; a + identifies individuals with censored event times.

Table 15.5: Regression diagnostics useful in Cox regression analysis—definition, interpretation, and uses of residuals

	Martingale residual	Deviance residual	Schoenfeld residual	Score residual
Definition	Difference between the observed and expected "number of events" for individual i	Martingale residual "standardized" to be an approximately standard Normal deviate	Difference between the value of X for individual i and the weighted average of X among everyone still at risk at i's event time	Difference in parameter estimate for X were individual i removed from the sample
Best at detecting	The functional form for a continuous predictor	Cases not predicted well by the model	Violations of the proportionality assumption	Influential and high leverage cases
Number of diagnostics per individual	1	1	One per predictor in model, but only for individuals with observed event times	One per predictor in model, regardless of whether individual i is censored
Interpretation of a large positive value	i experienced the event "too soon" (earlier than expected)	i experienced the event "too soon" (earlier than expected)	i has a relatively high value of X (a higher predictor value than his event time suggests)	Parameter estimate would increase without i

Interpretation of a large negative value	i experienced the event "too late" (later than expected)	i experienced the event "too late" (later than expected)	i has a relatively low value of X (a lower predictor value than his event time suggests)	Parameter estimate would decrease without i
Two most important things to do with the diagnostics (and what to look for)	Plot vs. *omitted* X to identify appropriate functional form. An added "smooth" is especially helpful	Examine distribution. Check cases with extremely large values	Plot vs. ranked event time and add a "smooth" summary. No pattern should emerge	Examine distribution. Large values identify potentially influential cases
	Plot vs. *included* X. If functional form is correct, no discernible pattern should emerge	Plot vs. risk score. Should predict equally well at all risk levels	Compute simple correlations with ranked event time. The correlations should be statistically non-significant	Plot vs. ranked event times. Influential observations should not be concentrated at particular event times
Useful for categorical predictors?	Not especially	Can be, if there are other predictors in the model	Yes, but plots are somewhat more difficult to interpret	Yes

deter you from using them. We have found that once you become comfortable with these strategies, you can quickly learn how to exploit the insights they offer and improve the quality of your analyses.

For each potential predictor, we recommend that you initially explore its functional form and its behavior under the proportionality assumption. For continuous predictors, you can examine: (1) martingale residuals from a null model; and (2) Schoenfeld residuals from a model that includes the predictor in the chosen functional form. For categorical predictors, the issue of functional form is moot, but the proportionality assumption should be examined both using Schoenfeld residuals and the graphical and model-based approaches of section 15.3. Deviance and score residuals may be helpful at this stage to identify potential outliers and high-leverage cases, but given limited time, we usually emphasize the other diagnostics.

After identifying a small number of potential "final" models, we recommend that you examine the behavior of all four types of residuals. If your exploratory analyses have been thorough (and if statistical control does not substantially alter your conclusions), you can likely devote these latter analyses to confirming that the model fit behaves as expected and that no individual cases (or set of cases) unduly affects your findings. The hope is that if your initial and periodic residual screenings were successful, you will not encounter any rude surprises at the end.

15.5 Competing Risks

In a standard survival analysis, all sample members begin in one state (employed, virgin, well) and we investigate whether and, if so, when a single target event occurs (unemployment, sexual intercourse, depression). In a *competing-risks* survival analysis, all sample members also begin in one state but we study whether and, if so, when one of *several* events occurs. Students may leave school either by graduating or by dropping out. Employees can leave an employer by quitting, by being laid off, or by being fired. Cardiac patients can die of a heart attack, of noncardiac diseases, or in an accident. Because only one of these events can occur first, statisticians say that the multiple events *compete* to end an individual's lifetime.

In this section, we demonstrate how to describe (section 15.5.1) and model (section 15.5.2) competing-risks data. Although our presentation is framed in terms of continuous-time event occurrence, these strategies can be extended into discrete time by conducting similar analyses using discrete-time methods. To illustrate the general approach, we analyze the

Table 15.6: Structure of a competing-risks survival analysis data set: career histories for three of the 109 justices who served on the U.S. Supreme Court between 1789 and 1999

Justice	YEARIN	YEARLEFT	HOW LEFT	TENURE	DEAD	RETIRE	LEAVE
John Jay	1789	1795	retired	6	0	1	1
John Marshall	1801	1835	died	34	1	0	1
Clarence Thomas	1991	—	—	8	0	0	0

length of time on the U.S. Supreme Court among the 109 justices appointed between 1789 and 1994. By 1999, when data collection ended, 100 had left the bench; 9 were still in office. When we first introduced these data in section 13.6, we did not discriminate between the two ways a justice's career could end: death ($n = 47$) or retirement ($n = 53$). Now, we distinguish sharply between these competing events and examine the effects of two predictors on risk: *YEAR*, the year the justice took office and *AGE*, the age of the justice at that time.

15.5.1 Describing Competing Risks Data

The major difference between a competing-risks data set and a standard survival analysis data set is that it includes additional variables identifying the particular competing event that terminated each individual's lifetime. Table 15.6 illustrates the structure for three justices in the Supreme Court data set: (a) John Jay, who retired after 6 years; (b) John Marshall, who died after 34 years; and (c) Clarence Thomas, who was still in office when data collection ended in 1999, 8 years after appointment. The event time, *TENURE*, and the two event-type indicators, *DEAD* and *RETIRE*, contain the necessary data. For comparison, the data set also includes a global event indicator, *LEAVE*, which collapses the occurrence of both competing events together.

Even though interest lies in the competing risks, we usually begin by examining the distribution of event occurrence regardless of type. Here, we analyze variables *TENURE* and *LEAVE*, noting that 0, not 1, identifies censored cases. Panel B of figure 13.6 presents the sample functions, so we do not reproduce them here. As discussed in section 13.6, the risk of leaving office is low during the first 10 years and then rises steadily over time. The estimated median career duration is 16 years.

We now consider the reality that death and retirement are hardly equivalent ways of ending a career. Not only might the distributions of risk

differ, so, too, might the predictors. We begin to investigate these possibilities by examining the *event-specific hazard rate*. Like the overall hazard
rate defined in equation 13.2, the event-specific hazard rate assesses the
instantaneous risk of experiencing that event, given that an individual
is at risk of doing so. In a competing-risks context, however, sample
members can leave the risk set not just by experiencing the target event
or by censoring but also by experiencing a competing event. Justices who
die in office are no longer at risk of retiring. Justices who retire are no
longer at risk of dying in office. The event-specific risk set at each
observed event time includes people who are not yet censored and
people who have not yet experienced any competing event. Conceptually, then, the occurrence of a competing event acts like a form of censoring—it removes an individual from the risk set for all other events.

This logic yields a simple strategy for estimating the event-specific
hazard rate: for each competing event, analyze the event times as you
would in a standard survival analysis, but redefine the censoring indicator so that it identifies both:

- *The fully censored*—everyone who did not experience *any* event
  during data collection.
- *The event censored*—those who experienced a competing event
  before this target event.

This strategy allows us to use a single event time variable (here, *TENURE*)
to analyze all competing risks. The only thing that changes across the
multiple analyses is the censoring indicator. For the event of retirement,
we censor 56 justices—the 9 still in office and the 47 who died. For the
event of death, we censor 62 justices—the 9 still in office and the 53 who
retired. Notice that our data set already includes the necessary censoring
indicators in the guise of the event indicators, *DEATH* and *RETIRE*.
Notice our reverse coding for these variables: 0, not 1, identifies the censored cases (as it did for *LEAVE*).

Use of multiple censoring indicators makes it easy to estimate sample
survivor, cumulative hazard, and kernel-smoothed hazard functions
for each event. Figure 15.8 presents the results for the Supreme Court
data, where bold lines represent the event of death and faint lines
represent the event of retirement. All major statistical packages can
produce plots like these, although you must merge together output files
containing event-specific estimates to obtain the overlaid displays shown
here.

Let us begin with the sample cumulative hazard and kernel-smoothed
hazard functions, for they are easiest to interpret. During the first five
years on the bench, both cumulative hazard functions are low and flat,

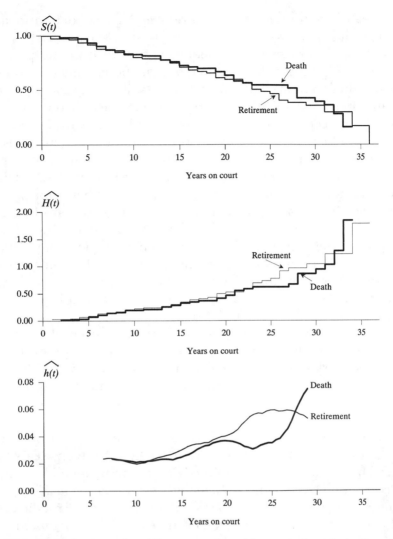

Figure 15.8. Preliminary analyses for competing risks survival analysis. For the U.S. Supreme Court data, sample survivor functions, cumulative hazard functions, and kernel-smoothed hazard functions for the two competing events: death and retirement.

indicating the rarity of either death or retirement. During the next ten years, both hazard functions rise steadily, although neither smoothed value ever exceeds .04. Between years 15 and 25, the two profiles of risk diverge: the hazard of retirement increases while the hazard of death decreases (or remains steady). After 25 years, only 24 justices remain, rendering later sample estimates for both events imprecise. It appears, however, that the risk of retirement abates while the risk of death rises, making death, not retirement, the more likely cause of departure among longer term justices.

To understand why survivor functions are difficult to interpret in the context of competing risks, remember that these functions assess the probability that an individual will "survive"—will *not* experience an event. When multiple events compete, the nonoccurrence of one event does not imply nonoccurrence of *all* others—it only refers to the nonoccurrence of this *specific* event. Someone who does not experience this event may well have experienced another, so the event-specific survivor function does not assess the overall probability of survival. Instead, it describes the probability that people survive given that they have not previously experienced this, *or any other*, competing event. This convoluted interpretation has led Pepe and Mori (1993) to suggest that event-specific survivor functions never be estimated when multiple events compete. Certainly, we cannot use them to estimate a median lifetime, and hence the importance of the global survivor function estimated without regard to event type. If you decide to examine these event-specific functions, be sure interpret them with a grain of salt.

By now, you may be wondering whether it is even "legitimate" to analyze competing-risks data by redefining the censoring indicators. Are the redefined risk sets representative of everyone who would have been at risk had a competing event not occurred? In practice, we can never know what would have happened to sample members had a competing event not occurred. Would Thurgood Marshall, who died two years after retirement, had lived for decades had he remained on the Court? Would Benjamin Cardozo, who died in office, have lived for many years had he retired? These questions speak directly to the fundamental assumption of the competing risks method: *that the occurrence of one event tells us nothing about the risk of occurrence of all others.* Contingent on all predictors in a model (which we will soon add), the occurrence of each competing event must be *noninformative* for all others.

We first encountered the concept of noninformativeness in section 9.3.2, in which we introduced the concept of censoring. A noninformative censoring mechanism operates independent of event occurrence and the risk of event occurrence. When censoring is due to design, the assumption seems reasonable. When censoring is due to a competing event, the assumption needs critical evaluation. Because there is no formal test, you must evaluate its tenability in context. Ask yourself what would have happened to those people who experienced the competing event had they not done so. Would their subsequent risk of all other events be identical, on average, to their peers who did not experience the competing event? If so, the competing events are noninformative.

To illustrate how you might evaluate this assumption, consider the event of accidental death when studying human lifetimes. For postsurgical survival, accidental deaths are likely noninformative for disease-related deaths because we have no reason to believe: (1) that people at greater risk of postsurgical complications are more (or less) likely to die accidentally; or (2) that people who die from accidental deaths would follow a different hazard function for disease-related deaths had they lived. But the same argument does not apply to a study of psychiatric patients. Because some types of accidental death (e.g., a drowning, an overdose) may be suicides, we could not blindly accept the noninformativeness assumption in this context.

All is not lost if competing events appear informative. Recall that the noninformative assumption must hold *conditional* on predictors. We can often identify predictors that allow us to treat the competing risks "as if" they were noninformative. Although we rarely have the predictors we might like, proxies may offer decent alternatives. When analyzing deaths due to heart disease and lung cancer, for example, we would include predictors assessing exercise, diet, and smoking history. We invoke a similar argument for the Supreme Court data. Although death is likely noninformative for retirement (as it is usually beyond an individual's control), retirement may not be noninformative for death. This would happen if some justices choose to retire because of poor health (because of their increased risk of death). Lacking health data, we use two proxies—age and year of nomination. Although imperfect, it seems reasonable to expect that younger justices, and more recent justices, will enjoy better health. If true, the retirement decision of a justice of a given age and era will operate independently of his or her risk of death.

The importance of the noninformativeness assumption cannot be overstated. When met, it is what allows us to analyze competing-risks data using standard survival analysis methods. It also has a practical consequence that allows you to target resources. If the competing events are noninformative, you need not analyze every competing event. It is perfectly legitimate to study whatever subset of events you find interesting. When studying student careers, for example, you might focus on graduation; when studying employee careers, you might focus on quitting. As long as the nonmodeled competing events are noninformative for the modeled competing events, each analysis will be valid.

We conclude this section by noting that the competing risks method is not appropriate when studying all types of multiple events. The method is appropriate only if everyone is simultaneously at risk of every possible outcome. The event process should not involve a sequential mechanism in which people are first at risk of an initial event, whose occurrence sets in motion the selection among a set of alternatives. For example, consider a study of whether and, if so, when newly licensed drivers purchase their first motor vehicle. The beginning of time is the date of licensure; the event occurs when the vehicle is bought. Would it be reasonable to frame this process as a competing-risks problem, where sedan, sport utility vehicle, minivan, and truck represent "competing" alternatives? Although these vehicles compete for a driver's dollars, the event-occurrence process is probably not one of competing risks. Instead, it is likely sequential: an individual decides to purchase a vehicle and then selects a particular type. Researchers modeling sequential processes should refer to Hachen (1988).

15.5.2 Statistical Models for Competing Risks Data

Armed with an event-time variable and multiple censoring indicators, model fitting is easy. All you do is fit the same model to the same data set several times, once for each censoring indicator. Under the assumption of noninformativeness, the likelihood functions for each event are separable, so you can estimate parameters using separate, but parallel, analyses. Although you may be tempted to include a different set of predictors in the multiple models, we caution against this approach. Use of identical predictors increases the tenability of the noninformativeness assumption and facilitates comparison of estimates, as we describe below.

Table 15.7 presents results for the Supreme Court data. In addition to the two event-specific fitted models, the table also includes a third

Table 15.7: Conducting a competing-risks survival analysis

	Model A Death only	Model B Retirement only	Model C Retirement or death
Parameter Estimates, Asymptotic Standard Errors, and Deviance-Based Hypothesis Tests			
AGE AT	0.0671**	0.1061***	0.0861***
NOMINATION	(0.0239)	(0.0258)	(0.0177)
YEAR	−0.0116***	0.0007	−0.0052**
NOMINATED	(0.0029)	(0.0026)	(0.0019)
Goodness-of-fit			
−2LL	331.447	373.497	711.765

$\sim p < .10$; $* p < .05$; $** p < .01$; $*** p < .001$.
These are the results of fitting three Cox regression models to the Supreme Court Justice data ($n = 109$, n retirements = 53, n deaths = 47). Model A is for the event of death only; Model B is for the event of retirement only; Model C is for the combined events of retirement or death.

Note: Efron's method for handling ties.

"global" model that does not distinguish between event type. Each model includes the same two predictors, *AGE* and *YEAR*, which together allow us to address research questions about secular changes in the risk of termination over time and the role that age at nomination plays in this process.

Let us first examine the event-specific models, comparing the absolute magnitude of their individual coefficients and their statistical significance as well. For *AGE*, the coefficient in the death model is 0.06711, while it is much higher (.10611) in the retirement model. Expressed as hazard ratios, we find that if we compare two justices whose ages differ by one year, the older justice faces a risk of death that is 6.9% higher and a risk of retirement that is 11.1% higher. If these hazard ratios appear small, remember that they represent the predicted *annual* elevation in risk associated with one year of age. Following the approach outlined in section 14.3.1, we can get a better sense of their magnitude by computing hazard ratios for justices separated by a greater age distance, say ten years. Computing $100(\text{hazard ratio}^{10}-1)$, we find that, in any given year, the justice who is ten years older is nearly twice as likely to die (a 95.6% increase in the risk of death) and three times as likely to retire (a 190% increase in the risk of retirement). The effect of age is not only statistically significant, it is practically significant as well. Older justices are at much greater risk of dying and they face an even greater risk of retiring.

Have the risks of dying or retiring changed over time? The data set includes appointees whose tenures cover a span of over two hundred years. Examining the model predicting death, we see clear evidence of a declining risk. Taking the antilog of the coefficient for *YEAR*, the hazard ratio is .9885. Because secular changes have a long gestational period, it is easier to interpret the magnitude of this effect using a larger temporal gap, of say 10, 50, or even 100 years. Comparing two justices of the same age a century apart, we find that the hazard of death for the recent appointee is 68.7% lower than that of his predecessor 100 years ago.

Having fit separate, but parallel, models, we now raise the obvious question: Are the effects of the same predictors different across the competing events—that is, are the coefficients for each predictor identical across the set of event-types (here, just two)? We can address this question by testing either:

- The *compound* null hypothesis that all coefficients associated with each predictor are identical across event-types
- The *targeted* null hypothesis that one specific set of coefficients is identical across event-types

Let us discuss each type of test in turn.

We begin by testing the compound null hypothesis that every coefficient in one event-specific model is identical to its counterpart(s) in all other event-specific model(s). Because the likelihood functions for the event-specific models are separable, we can test this null hypothesis by comparing the sum of the goodness-of-fit statistics for the separate models to that of the global model that does not distinguish between event types. Under the null hypothesis that the set of coefficients is identical across event types, this difference in −2LL statistics is distributed as χ^2 on $p(k-1)$ degrees of freedom, where p is the number of predictors and k is the number of competing event-types. For the models in table 15.7 we compute:

Value of the −2LL statistic for the global model: = 711.765
Sum of −2LL statistics for the event-specific models = 704.944
    (331.447 + 373.497)
Difference in −2LL statistics =   6.821

which exceeds the 0.05 critical value of 5.99 for a χ^2 distribution on $2(2-1) = 2$ degrees of freedom. We reject the null hypothesis that each predictor has identical coefficients across the events of death and retirement. Death and retirement are not equivalent ways of leaving office.

Having established that some parameters differ across models, we next test the set of targeted null hypotheses that assess whether the population parameter for each predictor is identical across event types. We can compute a separate test statistic for every predictor in the set of competing-risks models. In the general case of k event types, we must test this null hypothesis using a generalized Wald statistic (Long, 1997). In the common situation of only two event types (as here), the Wald test statistic simplifies to:

$$\chi_1^2 = \frac{\left[\hat{\beta}_1 - \hat{\beta}_2\right]^2}{\left[se(\hat{\beta}_1)\right]^2 + \left[se(\hat{\beta}_2)\right]^2}.$$

Conducting this test for AGE, we obtain an observed test statistic of 1.23 and we fail to reject the null hypothesis that the coefficient on AGE in the model for death is identical to the coefficient on AGE in the model for retirement. Conducting this test for $YEAR$, however, we obtain an observed test statistic of 9.97, which exceeds the 0.01 critical value of 6.63 for a χ^2 distribution on one degree of freedom. We conclude that the effect of $YEAR$ differs across the two events. Combined with the non-significance of the coefficient for $YEAR$ in the retirement model, we conclude that justices of any era were equally likely to step down, but that the risk of dying in office has declined.

As this discussion demonstrates, it is simple to implement a competing-risks survival analysis. Once the appropriate censoring indicators have been constructed, all you need to do is fit an identical Cox model several times. Yet competing risks analyses are not without problem. Although implementation is easy, establishing the validity of the noninformativeness assumption is hard. There is no test for evaluating its tenability, and sample data cannot be used to martial an argument. It is therefore impossible to know for certain that the assumption has been met. Seeking shelter from criticism, some researchers choose *never* to distinguish between competing events, analyzing only the global event indicator (like *LEAVE*). But, this may mask important differences across event types, making it an unattractive option. We therefore end this discussion on a pragmatic note. If you think you may be studying a set of competing risks, be sure to act as your own devil's advocate before data collection, or at least before data analysis. If you can anticipate potential criticisms, you may be able to collect data on a set of predictors whose inclusion might facilitate a conditional noninformativeness argument.

15.6 Late Entry into the Risk Set

A fundamental tenet of research design, articulated in chapter 9, is that you should try to track each person's event history from a single common starting point, designated t_0. If your definition of t_0 provides immediate access to participants, tracking is easy. You might follow hospitalized patients from their day of admission or newborn babies from their day of birth. Although t_0 need not be 0—in the cocaine initiation study of section 15.1.1, for example, it was age 17—it should represent the same "point in time" for each study participant. This ensures that everyone remains the same "age" throughout, so event times assess event occurrence in a meaningful and comparable metric.

In some studies, you may find yourself unable to track everyone from a single common starting point. This is a familiar predicament when studying *stock samples*—age-heterogenous groups of people who *already* occupy the initial state when data collection begins (Lancaster, 1990). Although this design may seem unusual, it is more common than you might think. Suppose you conducted a ten-year prospective study of depression onset among a group of individuals who had never been depressed. If the participants vary in initial age, the ten-year period of data collection does not cover the same ten-year period of risk. The data for people who are initially 30 describe what happens between ages 30 to 39 while the data for people who are initially 40 describe what happens between ages 40 to 49. Even though each person is tracked during the same chronological period (the same ten years), their "time at risk," which should be clocked in the metric of age, differs. To model depression onset appropriately, you need a second random variable, *ENTRY*, which assesses the time—here, the participant's age—when he or she is first observed as being at risk of event occurrence. Statisticians refer to this random variable as an individual's *late*, or *delayed, entry time.*

In this section, we describe how to fit Cox regression models to data when some individuals have late entry times. In section 15.6.1, we present the basic strategy; in section 15.6.2, we show how this strategy can be profitably applied to research contexts in which the late entrants may be less apparent. In doing so, we demonstrate how the possibility of late entry is related to another analytic issue that arises when fitting Cox models: the choice of an appropriate metric for time.

15.6.1 Fitting a Cox Model with Late Entrants

We illustrate how to fit Cox regression models to data sets with late entrants by examining how Singer et al. (1998) tracked the careers of physicians working in community and migrant health centers (CMHCs) to fulfill medical school loan obligations. Although the researchers originally wanted to track a random sample of physicians hired since the loan forgiveness program began, poor national record-keeping systems rendered this impossible. Even within individual centers, unreliable record-keeping systems made it impossible to reconstruct the careers of physicians who left long ago retrospectively. Reasoning that center directors could reliably provide the names and dates of hire (and departure) if they were asked to focus on physicians working during a recent brief measurement window, the researchers ended up using a stock sample design.

The top panel of figure 15.9 illustrates how their sampling plan worked. Designating the period, January 1, 1990–September 30, 1992, as the measurement window, the researchers asked each director to list every physician who worked during that period. Center records were then used to identify the dates of hire for these 812 physicians (and the dates of departure for the 396 who left before the measurement window closed). In the figure, a diamond indicates the date of departure and an arrowhead indicates the last known date of employment (the last day of data collection) for physicians who were still employed. A solid line identifies that portion of the career that occurred during the measurement window; a dashed line identifies that portion that occurred earlier. Physicians hired the day the window opened (January 1, 1990)

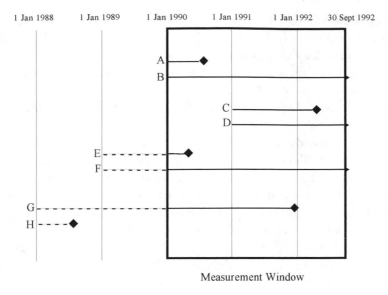

Figure 15.9. Understanding the consequences of late entry into the risk set. Hypothetical career histories for 8 individuals in the physician career study.

were tracked until they left the center (A) or were censored (B). Physicians hired *after* the window opened were also tracked until they left the center (C) or were censored (D). Physicians working during the measurement window who were hired *before* it opened were tracked from their date of hire until they left (E and G) or were censored (F). Colleagues who were hired *and left* before the window opened were excluded from the sample (H).

To analyze these data, we must compute entry and event times by repositioning each physician's trajectory on a new time axis, one that begins at t_0, the day of hire. The bottom panel of figure 15.9 illustrates this process where the resulting event times, $EVENT_i$, are displayed above each event or censoring symbol. The panel also presents sample values for the random variable, $ENTRY_i$, which identifies the day—expressed in the metric of "days since hire"—when the physician first entered the measurement window (the late, or delayed, entry time). The data set collects this information together:

PHYSICIAN	ENTRY	EVENT	CENSOR
A	0	180	0
B	0	1000	1
C	0	455	0
D	0	635	1
E	366	455	0
F	366	1370	1
G	731	1430	0

For physicians A, B, C, and D, who were hired on or after January 1, 1990, *ENTRY* is 0: they entered the measurement window the day they were hired. For everyone else, *ENTRY* identifies the point they were at in their career when the measurement window opened. This corresponds to the day when their trajectory switched from a dashed to a solid line. For physicians E and F, who were hired on January 1, 1989, *ENTRY* is 366: they entered the window on the first day of their second year. For physician G, who was hired on January 1, 1988, *ENTRY* is 731: she entered the window on the first day of her third year. Notice that we display a trajectory for physician H only for illustration; no corresponding record appears in the data set. Because this physician was hired, and left, before data collection began, we do not know either his event times or entry times (nor those of his colleagues who had also already left).

Even though each individual now has an event time and a censoring indicator we still cannot use a standard approach to fitting a Cox model. It is not that a statistical package will not compute parameter estimates, standard errors, and goodness-of-fit statistics for some specified model—unfortunately, it will—it is that the results will be incorrect! To understand the problem, examine the bottom panel of figure 15.9 and ask: Who should be included in the risk set for each observed event time? To concretize the question, focus on day 180, the first event time shown. Physicians A, B, C, and D are clearly "at risk" because we have monitored their careers from t_0 to this day. But are physicians E, F, and G also "at risk?" Even though they survived *past* this day, they are not members of the day 180 risk set because: (1) we did not observe their careers on this day; and (2) we lack comparative data on their peers hired at the same time (like H), who left before data collection began. Physicians E, F, and G are not representative of all physicians hired on January 1, 1988 (or 1989), who were still working on day 180. They are the "longer lived" physicians—the ones who persisted long enough to be present when data collection began. Because physician H and his unobserved colleagues cannot be included in the day 180 risk set—as we have no data for them—we should not include E, F, or G, either.

One way of addressing this concern is to limit analyses to the 290 physicians who began their careers during the measurement window. Although not invalid, this approach is unattractive for at least three reasons: (1) it focuses exclusively on early event times—here, the first 2.75 years of a physician's career—which may not be those of greatest risk or substantive interest; (2) it increases the proportion of censored cases—here, to 80.3%—because individuals hired later have had less time to leave; and (3) it sets aside valuable data—here, for the 522 physicians already employed on 1 January 1990. Taken together, these factors decrease statistical power. In this data set, power is so reduced that we cannot find a single statistically significant predictor of event occurrence (although we will soon show that there are some).

How, then, can we simultaneously analyze the data for all sampled individuals? If we are willing to make a strong, but often reasonable, assumption—that the entry and event times are conditionally independent—we can use the following common sense approach:

- Restrict the risk set at each observed event time to those individuals actually observed at that moment.
- Compute individual i's contribution to the likelihood at that moment by comparing his or her true risk score $[e^{(\beta_1 X_{1ij} + \beta_2 X_{2ij} + \cdots + \beta_P X_{Pij})}]$ to the sum of the contemporaneous true risk scores among this revised risk set.

From a computational perspective, this means that at time t_j, individual i is a member of the risk set if, and only if, he or she satisfies the inequality:

$$ENTRY_i \leq t_j \leq EVENT_i$$

In other words, to be in the risk set at time t_j, individual i must have entered the measurement window, but he or she must not have already experienced the target event.

Any Cox regression routine that can handle time-varying predictors can handle late entrants. Just as the packages differ markedly in the way time-varying predictors are specified, so, too, do they differ markedly in how the delayed entrants must be identified. Easiest to work with are those statistical packages that use the counting process notation. (Notice that this differs from our preference in section 15.1 when specifying time-varying predictors.) All you need do is ensure that each record for each person includes an entry time, exit time, and censoring indicator (as shown above). If you identify these variables appropriately, the package will adjust the risk set accordingly. You need not create additional data records, as when modeling time-varying predictors, because under the counting process system, the computer automatically screens each data record using the inequality above.

If the only way of including time-varying predictors is through computer code, you can still include late entrants in your model, but you must do so by "manually" manipulating the risk set at each observed event time. Although this may seem onerous, it is easy to do using a simple "trick": using the delayed-entry variable *ENTRY*, create a time varying predictor that is *missing* during any earlier event time. Whenever the predictor's value is missing, the computer will remove individual i from the risk set; whenever the predictor's value is not missing, individual i will appear. If your statistical model already includes a time-varying predictor, you can implement this strategy for that predictor. If your model is limited to time-invariant predictors, you must create a "phantom" time-varying predictor that sets the missing values.

Model A in table 15.8 presents results for the physician career data using the late entrants approach. The model includes three predictors: *PARTTIME*, a dummy variable identifying the 64 physicians who did not work full time; *AGE*, a continuous variable indicating the physician's age at hire; and *AGE* by *YEAR*, which represents the two-way interaction between *AGE* and time. (We include this latter term because preliminary analyses showed that the effect of *AGE* diminished over time. To facilitate its interpretation, we rescaled the time metric from days into years by dividing each event time and entry time by 365.25. This allows the

Table 15.8: Accounting for late entry into the risk set

	Model A (Appropriate late entrants model)	Model B (The "pure" sample— those hired during the measurement window)	Model C (Incorrect model— ignoring the presence of late entrants)
Parameter Estimates, Asymptotic Standard Errors, and Deviance-Based Hypothesis Tests			
PARTTIME	−1.3473***	−0.1705	−1.3394***
	(0.2764)	(0.5208)	(0.2765)
AGE	0.0478*	0.0398	0.0977***
	(0.0242)	(0.0455)	(0.0192)
AGE × YEAR	−0.0287**	−0.0039	−0.0370***
	(0.0091)	(0.0446)	(0.0078)
Goodness-of-fit			
−2LL	4214.515	558.391	4482.988
−2LL for null model	4264.335	560.760	4539.431
Difference in −2LL	49.820	2.369	56.443
n parameters	3	3	3

$\sim p < .10$; $* p < .05$; $** p < .01$; $*** p < .001$.
These are the results of fitting three Cox regression models to the physician career data ($n = 812$, n events = 396). Model A appropriately includes the late entrants. Model B is fit to only those physicians hired during the measurement window ($n = 290$, n events = 57). Model C is an *incorrect model*, shown for pedagogic purposes only—it ignores the presence of late entrants.

Note: Efron's method for handling ties.

interaction term to represent the decrement in the effect of *AGE* with each passing year, not day). We find that the estimated hazard of leaving a CMHC among physicians who work part time is one fourth of that for other physicians ($e^{-(1.3473)} = 0.26$). Immediately after hire, the older the physician, the greater the hazard of departure. For each extra year, the initial hazard is $100(e^{(0.0478)} - 1) = 4.9\%$ higher. By year 2, however, the age effect has disappeared (the estimated hazard ratio at this time is $e^{(0.0478) - 0.0287(2)} = 0.99$). And by year 4, the effect of *AGE* has reversed itself: older physicians are *less* likely to leave.

Table 15.8 also presents the results of fitting two other models to these data: (1) B, which focuses on the 290 people hired during the data collection window; and, (2) C, which analyzes the full data set without identifying the 522 physicians who entered the risk set late. We present these results not because we recommend either approach but to demonstrate how misleading they can be. As suggested earlier, restricting attention to physicians hired during the measurement window diminishes power so completely that we fail to find any statistically significant effects. We are clearly better served by including the 522 physicians hired before the measurement window opened. Yet, as we do in Model A, this is true only if we include them appropriately. If we do not adjust for late entry, the effect of *AGE* erroneously appears to be twice as large! Were we to have fit this model, we would have erroneously concluded that for each extra year of age, the initial hazard of leaving is $100(e^{(0.0977)} - 1) = 10.3\%$ higher. When some sample members enter the risk set late, you must appropriately account for their behavior lest you reach incorrect inclusions.

15.6.2 Using Late Entrants to Introduce Alternative Metrics for Clocking Time

The ability to include data from late entrants has important practical implications for analysis. Not only can you use this approach when you know, by design, that your sample includes individuals whose time at risk varies, you can use them when you decide, after data collection has ended, that the original metric for time may not be the best metric for analysis. As we show below, late entry methods can be used even when you *thought* you tracked everyone from a single common starting point, but you later decide that the t_0 you once thought "obvious" may not be the best.

To understand how you might find yourself in this situation, imagine designing a new study of physician careers, one that tracks every doctor from his or her first day of work. Most researchers would prefer this design to the stock sample approach of the previous section because it seemingly eliminates all late entrants. But late entrants disappear only if the best metric for time is the number of "days since hire." What if it was more appropriate to measure time using the metric of physician *age*. Perhaps a person's age, not his or her tenure on the job, is the more important predictor of the stay-or-leave decision. In this case, every physician in this new study is now a late entrant, even though each was tracked from hire.

Once you recognize that event occurrence can be clocked using more than one time metric, you may find alternative time metrics wherever you look. Should the efficacy of kidney transplantation be evaluated by clocking survival from an individual's date of eligibility for transplant or from his or her date of surgery? Should employee retirement decisions be modeled as a function of date of hire (as we did in section 15.1 for the Supreme Court justices) or is age more appropriate? Should the clock for evaluating a teen pregnancy prevention program begin at the girl's date of program entry, her date of birth (which leads to her age), or her date of menarche (when she is first at risk of getting pregnant)?

The existence of alternative metrics for time would be of little consequence if a switch had little analytic impact. But your conclusions can differ markedly depending on the metric for time. We demonstrate this impact using data from Ha, Kimpo, and Sackett (1997), who studied whether and, if so, when monkeys attain the classic Piagetian stage known as object recognition. According to Piaget, an infant's relationship with his or her external world progresses in an orderly fashion over time, beginning with object recognition (the infant is aware that visible objects exist) through object permanence (the infant knows that an object exists even if it is completely out of sight). As part of a larger study, Ha and colleagues assessed how long it took 123 monkeys to successfully demonstrate object recognition by completing what is known as a *plain reach* task. (In this task, the infant monkey must reach out and pick up a familiar object; here, a small rubber toy, rubbed with grape juice, that he or she had previously sucked). At the beginning of the study, the monkeys ranged in age from 11 to 38 days. After 37 days, all but one monkey had attained the milestone. Some did so in just 1 day; others took as long as 33 days.

Table 15.9 presents the results of fitting three Cox regression models to these data. Each includes the same three predictors: *FEMALE*; *BIRTH WEIGHT*, the decile equivalent of the monkey's birth weight in comparison to colony-wide sex-specific standards; and *INITIAL AGE*, the monkey's age at initial testing. The models differ, however, in the metric used to assess event occurrence:

- Model A uses session number, where time is measured as the number of days between initial testing and successful task completion (or censoring).
- Model B uses *AGE* in days, where time is measured as the monkey's age (in days) upon successful task completion (or censoring).
- Model C uses *AGE* in days, but adjusts for late entry, so that each monkey is included in the risk set for a given age *only if* he or she was at risk of event occurrence at that age. The late entry time is coded as the monkey's age on the initial day of testing (*INITIAL AGE*).

Before delving into specific findings, notice that the effects of *FEMALE* and *BIRTH WEIGHT* are virtually identical across models. Regardless of measurement metric, male and female monkeys are equally likely to master the plain reach task, and bigger monkeys are more likely than smaller monkeys to reach criterion earlier.

What is most fascinating about these three models is the way in which the effect of *INITIAL AGE* varies, from significantly positive in Model A, to

Table 15.9: Empirically comparing alternative metrics for clocking *TIME* in Cox regression analysis

	Model A (Session number)	Model B (*AGE* in days, no late entry)	Model C (*AGE* in days, with late entry)
Parameter Estimates, Asymptotic Standard Errors, and Deviance-Based Hypothesis Tests			
FEMALE	0.3720	0.2908	0.3118
	(0.1902)	(0.1888)	(0.1879)
BIRTH WEIGHT	0.1163***	0.1021***	0.1025***
	(0.0280)	(0.0277)	(0.0277)
INITIAL AGE	0.1064**	−0.0939**	−0.0189
	(0.0339)	(0.0301)	(0.0349)
Goodness-of-fit			
−2LL	919.898	915.073	896.822
−2LL for null model	944.449	944.449	914.654
difference in −2LL	24.551***	29.376***	17.8327***
n parameters	3	3	3

~$p < .10$; * $p < .05$; ** $p < .01$; *** $p < .001$.
These are results of fitting three Cox models to the Piagetian task data ($n = 123$, n events = 122). Model A clocks time using session number. Model B clocks time using age, but does not account for late entry into the risk set. Model C clocks time using age, but does appropriately account for late entry.

Note: Efron's method for handling ties.

significantly negative in Model B, to nonsignificant in Model C. To understand why this happens, and what this says about the effects of age, we must carefully determine what the parameters in each model represent.

Let us begin with Model A, which is based on the time metric that most researchers would initially adopt: number of testing sessions. This metric is appealing because monkeys are eligible to complete the plain reach task only after testing begins. The positive coefficient for *INITIAL AGE* tells us that monkeys who are initially older reach criterion earlier than those who are initially younger, a commonsense finding that conforms to Piagetian theory. The intuitive appeal of this model increases when we realize that we would obtain the identical parameter estimate if we substituted the time-varying predictor *AGE AT TESTING* for the time-invariant predictor *INITIAL AGE*. As a result, we can interpret the effect of *INITIAL AGE* in Model A by saying that older monkeys reach criterion earlier than younger monkeys.

But if age is such an important factor, might it be as appropriate—or more appropriate—to model task attainment as a function of the monkey's age, not the number of testing sessions? After all, Piaget argued that task attainment is inherently developmental: infants reach the stage of object recognition (and ultimately object permanence) as they mature. Training may have an effect—you can teach a child (or monkey) particular skills—but Piaget believed that regardless of training, all infants would progress through these stages in an orderly fashion. This suggests that age, not session number, might be a better metric for time.

We initially explore this possibility in Model B, which uses the monkey's age when he or she reached criterion (or censoring) as the outcome. Although the effects of *FEMALE* and *BIRTH WEIGHT* are virtually identical, the sign on the coefficient for *INITIAL AGE* is reversed! This model suggests that older monkeys have a *lower* hazard of reaching criterion. Before you rationalize this counterintuitive finding, remember that this analysis—and hence this finding—has a fundamental flaw: it does not correct for late entry into the risk set. We cannot change the metric for time (from session number to age) without simultaneously changing the way in which we specify event occurrence. We obtain a negative coefficient here because the older monkeys have less opportunity to reach criterion early (as they are not even tested during the earlier portions of their lives). The negative coefficient is an artifact; it does not present unbiased information about the relationship between age and task attainment.

Let us therefore move to Model C, in which we record event occurrence in terms of age, but we adjust for late entry into the risk set. Although the effect of *BIRTH WEIGHT* continues to be statistically significant, the effect of *INITIAL AGE* disappears. Before you conclude that age and task mastery are unrelated, remember what happens when we adopt this new metric for time. The effect of the monkey's age is now absorbed entirely into the baseline hazard function. Absorption prevents age from exerting any additional effect on the risk of task attainment; whatever effect it has appears in the baseline hazard function. By adopting this metric, we trade the ability to comment on age trends for the certainty that we have controlled fully for all of its possible effects. This ensures that Model C will provide the cleanest estimates of all other predictors' effects.

Which of the three models is best? Although Model B is flawed, both A and C are "correct." The difference lies in the research questions they address. If you want to focus on the developmental aspects of task mastery, Model C may be preferable because each monkey's task attainment is compared to peers *of an identical age*. If you want to focus on the training aspects of task mastery, Model A may be preferable because each monkey's task attainment is compared to peers with an *identical training level* (measured in days). And although you might be tempted to choose a model based on the associated −2LL statistics, you cannot do so because their null models differ, making numeric comparison inappropriate.

Although both models are valid, we have three reasons for preferring Model C (or rather a revised version of Model C without the two non-significant predictors, *FEMALE* and *INITIAL AGE*). The first reason is substantive, not statistical. Piagetian theory is inherently developmental; if infants master tasks at particular ages, it simply makes sense to use age as the metric for measuring task attainment. The second argument builds upon the *BIRTH WEIGHT* effect. Ha and colleagues suggest that *BIRTH WEIGHT* assesses the monkey's maturity at birth; it is a proxy for *gestational* age. This, too, suggests that age, not session number, is the key issue. (It also suggests that we might want to track time using a third metric, time from conception, if we had the data.) The third argument focuses on what it means to control for *INITIAL AGE* in Model A. If we did not include this predictor, older monkeys would have an "unfair advantage." The hope is that this predictor places all monkeys on a level playing field, allowing the model to evenhandedly assess other predictors' effects. But will the effect of age be entirely removed just by including its main effect? Further analyses, not shown, reveal that *INITIAL AGE* in Model A interacts with time. Its effect is largest at initial testing and disappears entirely by day 16. Although we could include this interaction as well, Model C achieves the same goal of controlling statistically for age, and it does it with certainty and far less effort. In Model C, *all* effects of age are eliminated, providing what we believe is the clearest vehicle for evaluating other predictors' effects. Model A is perfectly valid (as long as the assumptions are met), but Model C will always do a better job of controlling for the effects of age. This makes Model C preferable if the goal is to remove the effects of age and evaluate the effects of other predictors.

Not all studies lend themselves to the consideration of alternative time metrics. Researchers wishing to explore this possibility must be prescient enough to collect the data that will permit them to explore the other metrics. In the primate study, for example, hindsight suggests that assessing gestational age would have been prudent, allowing us to explore that third metric for time. For now, we simply remind you that the most "obvious" metric for time may not be the best. When possible, and more important, when theory suggests its appropriateness, you may want to explore alternative metrics.

Notes

CHAPTER 1

1. To be completely accurate, this restrictive statement must be qualified. Under certain conditions, growth modeling can be used to analyze change in noncontinuous outcomes, such as counts (e.g., number of days of school absence per month), and dichotomies (e.g., whether a crime has been committed in each of several weeks after a prisoner's release from prison).

CHAPTER 2

1. A method-of-moments estimate of the sampling variance of the OLS slope for individual i is obtained by replacing terms in the numerator and denominator of equation 2.1 by corresponding sample statistics.

CHAPTER 3

1. If the window of observation were wider or there were more waves of data, we might hypothesize a more complex trajectory. In the larger data set from which this sample was drawn, which included up to 12 waves of data per child, Burchinal and colleagues specified individual change as a cubic function of age.

2. With time-structured data, the subscript i on the temporal predictor (AGE) is redundant because we need not distinguish measurement occasions for different individuals. When this happens, you can eliminate this subscript. We preserve it here to emphasize the model's generality.

3. The situation is more complex than we admit here. When longitudinal data are balanced and time structured, with no missing data, and the same predictors used in each part of the level-2 submodel, statistical tests conducted via *restricted* ML (see section 4.3.2) are exact, not approximate, even in small samples (Raudenbush, 2002, private communication).

4. The estimates presented here are *full* maximum likelihood estimates. In section 4.3, we distinguish between *full* and *restricted* ML methods.

5. Under the conditions listed in note 3, all tests are *exact* and the label "*t*-statistic" is appropriate (Raudenbush, 2002, private communication).

CHAPTER 4

1. GLS minimizes a multivariate weighted sum of squares (Raudenbush, 2002, private communication).

2. Actually, ML and GLS may not be identical, even under normality, depending on how the weighted sum of squares that is minimized in GLS is constructed. If the weights are based on ML estimates of the residual variances and covariances, the GLS estimator will produce ML estimates (Raudenbush, 2002, private communication).

3. A further advantage of grand mean centering is that it allows the data analyst to evaluate the increased precision in the level-1 individual growth model that accrues from adding level-2 predictors by comparing the changing magnitude of the associated standard errors. In this example, notice that while the level-2 intercepts remain identical from Model B to G, their associated standard errors decline (from 0.105 to 0.080 for initial status and from 0.062 to 0.061 for rates of change). This demonstrates that the inclusion of *COA* and *PEER* in the multilevel model improves the precision with which we estimate the average individual growth trajectory.

4. S. W. Raudenbush (2002, private communication) states that methodologists disagree about using deviance statistics to test null hypotheses that constrain parameters to values at the boundary of the parameter space (i.e., testing that a variance component is zero). Then, a difference of deviances does not have the χ^2 distribution we think it has and the test is conservative, but can still be used (Pinheiro & Bates, 2000). Other authors disagree (Verbeke & Molenbergs, 2000).

CHAPTER 5

1. A fundamental advantage of the multilevel model for change is that it can be used under all three missingness assumptions. This stands in contrast to other longitudinal methods (including the generalized estimating equations (GEE) approach of Diggle, Liang, and Zeger (1994), which requires the MCAR assumption).

2. The terms defined, ancillary, and internal were first used by Kalbfleisch and Prentice (1980). The fourth category—contextual—builds upon ideas in Blossfeld and Rohwer (1995) and Lancaster (1990).

CHAPTER 6

1. As shown in section 6.1.2, this perspective has important implications when postulating level-2 submodels. Because these level-1 discontinuities are fundamental features of our model, we will allow these parameters to vary randomly at level-2. This stands in stark contrast to our recommendations about generic time-varying predictors offered in chapter 5.

2. Differentiating with respect to time, the true slope of the quadratic change trajectory is $\pi_{1i} + 2\pi_{2i}TIME_{ij}$, which then has value π_{1i} when $TIME_{ij} = 0$.

3. Actually, we oversimplify here. The number of waves of data needed also depends on whether you fix any of the individual growth parameters at

level-2. If you fix some of them, you may require fewer waves of data, depending on the temporal design of the data-collection (Rindskopf, 2002, private communication).

4. Strictly speaking, this argument about average curves requires the data to be balanced and time structured.

5. The slope of the hyperbolic trajectory is $(1/\pi_{1i}t^2)$, and so rate of change in the outcome is proportional to $(1/\pi_{1i})$, as implied in the text, but also depends on the value of time, as expected for a curved trajectory.

6. Specifically, $\pi_0 = e^{(-c\alpha)}$ and $\pi_1 = k\alpha$, where c is the constant of integration.

CHAPTER 7

1. Specifically, the identical blocks in the error covariance matrix in equation 7.8 require completely balanced data.

CHAPTER 8

1. In practical analysis, we would need to define a measurement metric for each of the underlying constructs perhaps by fixing one loading per construct in the Λ_x matrix to a constant value, such as 1 (see Bollen, 1989).

2. Again, in practical analysis, the metrics of the underlying endogenous constructs would need to be determined, perhaps by fixing selected loadings, one per construct, to a constant value, such as 1 (Bollen, 1989).

3. Taking expectations in equation 8.14 and substituting for μ_{η_1} from the first expression into the second, the population means of the endogenous constructs are $\mu_{\eta_1} = \alpha_1 + \gamma_{11}\mu_{\xi_1} + \gamma_{12}\mu_{\xi_2}$ and $\mu_{\eta_2} = \alpha_2 + \beta_{21}(\alpha_1 + \gamma_{11}\mu_{\xi_1} + \gamma_{12}\mu_{\xi_2})$.

4. If β_{12} and β_{21} were both present in **B**, then a hypothesis of "reciprocal causation" (in which each endogenous construct was simultaneously predicting the other) would be entertained.

5. Notice that this is a different research question than would be addressed by incorporating a time-varying variable as an extra predictor in the Level-1 submodel, as described in section 5.3 in the regular multilevel model for change.

6. With completely unstructured data, the number of required subgroups would become inhospitable and the sample size within subgroup too small to estimate the sample covariance matrices required for multigroup analysis.

CHAPTER 10

1. Strictly speaking, hazard is the conditional probability of event occurrence *per unit of time*. Because attention here is restricted to the case of discrete-time, where the unit of time is an "interval" assumed to be of length 1, we omit the temporal qualifier. When we move to continuous time, in chapters 13 through 15, we invoke a temporal qualifier (as well as an altered definition of hazard).

2. Owing to its genesis in modeling human lifetimes, the hazard function is also known as the *conditional death density function* and its realizations at any given time are known as the *force of mortality* (Gross & Clark, 1975). A comparison of these terms suggests that while the phrase "hazard function" has a

negative connotation, the valence of this word is far milder than the valence of the competing alternatives!

CHAPTER 11

1. In section 12.2, we introduce a third transformation, the complementary log-log transformation, which can also be used to specify a discrete-time hazard model.

2. Some readers may be more familiar with a specification that includes a stand-along intercept and excludes one of the time indicators. This alternative specification, although identical in fit to the specification we present, precludes the simple interpretation of the α's (as in equation 11.6). Our decision to specify the model without a single stand-alone intercept has implications for model fitting and estimation. We return to this topic in chapter 12.

CHAPTER 12

1. You might wonder whether all the comparisons in table 12.2 are really nested. Certainly, the constant model is nested within *all* other models as it includes only a single intercept, and all others include at least *that* intercept (for the polynomial specifications), or multiple intercepts, one per time period (for the general specification). So, too, each less complex polynomial is nested within all subsequent polynomials because the former can be obtained by setting higher order parameters in the latter to 0. But is each model also nested within the general specification? Although perhaps not obvious, each is indeed nested. Remember from section 11.6 that one model is nested within another if we can obtain the former by placing *constraints* on parameters in the latter. The constraints need not be setting parameters to 0. Because *TIME* takes on only *J* distinct values, each polynomial can be obtained by setting the multiple intercepts in the general specification to particular values (or multiples).

2. Be sure to take care when interpreting these observed proportions. Unlike the proportions examined in chapters 9 and 10, which could be interpreted as sample values of hazard, these proportions do not admit of such interpretations. This is because the denominators for consecutive proportions do not differ only because of event occurrence and censoring (as they do when computing a sample hazard function over time). Each denominator can include a totally distinct group of individuals, and the same person can appear in the denominator for both groups (albeit in different time periods).

3. Notice, too, that we do not suggest that the effect of the time-varying predictor is to *change* the risk of event occurrence as the predictor itself changes. All this parameter does is contrast individuals with different values of the predictor in each time period. Within each given time period, a time-varying predictor is *time-invariant*. As a result, we do not relate *changes* in time-varying predictors to *changes* in hazard; instead, we relate differences in the values of time-varying predictors to differences in the values of hazard.

4. We note that although the sib-size is time-varying, the data do not include information on the timing of each sibling's birth. In the analyses presented here, we therefore treat this predictor as time-invariant.

5. This correspondence can be confirmed by comparing equation 12.14 with the log-likelihood in equation 11.13. Notice that the term underneath the square

root in equation 12.14 is identical (except for the factor of −2) to the elements of the log-likelihood, with fitted hazard probabilities substituted for true hazard probabilities.

CHAPTER 13

1. The steady rise in conditional probabilities does not indicate that the underlying hazard function is rising (as some people erroneously conclude). Rather, it tells us that we lack sufficient data—for the infinite number of *unobserved* times between every pair of *observed* events—for the estimates to describe the underlying hazard function's shape.

2. In discrete time, we can define $H(t_{ij})$ by summing the j individual values of $h(t_{ij})$. In continuous time, summation no longer works because there are an infinite number of specific hazard values between t_0 and t_j. For this reason, we adopt the "cumulation" phrase included in equation 13.6. Readers familiar with calculus will note that we may write this definition as: $H(t_{ij}) = \int h(t_{ij}) \, dt$.

3. Only one statistical package (Egret) currently includes built-in routines for kernel-smoothing hazard functions. To facilitate their computation, we have written a SAS macro (based upon Allison, 1995) and STATA ado file (based upon McKnight, 1995) that computes and plots kernel smoothed estimates. Readers interested in learning more about kernel smoothing hazard functions should consult the seminal paper by Ramlau-Hansen (1983) or the excellent discussion in Klein and Moeschberger (1997).

4. Asymmetric kernel smoothers can address this problem. See Klein and Moeschberger (1997) for details.

CHAPTER 14

1. A similar strategy can be used for other increments. To estimate the percentage difference in hazard associated with a c-unit difference in a predictor, compute $100(\text{hazard ratio}^c - 1)$. A 10-year increase in the age of release, for example, is associated with a change of $100(.9342^{10} - 1) = -49.4\%$—a halving of—the hazard of recidivism.

2. If there are no ties, we calculate LL using equation 14.8. If there are ties, we calculate LL using a revised partial log-likelihood function that reflects the method used to handle the ties. The log-likelihood statistics in table 14.1 were computed using Efron's method for handling ties; if a different method is used, the obtained log-likelihood statistics will differ.

CHAPTER 15

1. What would have happened if we did not lag the time-varying predictors? In most data sets, as here, we would find even larger effects. Lagging protects against drawing unwarranted inferences that result from simultaneity. Of the 1018 men who used marijuana while at risk of initiating cocaine use, 367 (36.1%) eventually used cocaine; of the 640 who did not previously use marijuana while at risk of initiating cocaine use, only 15 (2.3%) used cocaine. The lagged estimate of marijuana's effect is conservative (albeit large) because all but 2 of these 15 actually *did* use marijuana, but they first did so in the same year as their cocaine initiation. By lagging the predictors, we do not "credit" these

individuals with marijuana use, rendering our estimates of the effect of marijuana more conservative.

2. In the authors' analyses, relapse was measured in weeks, which enabled them to construct a complete predictor history for each individual. To use these same data to illustrate imputation strategies, we jittered (Cleveland, 1994) the event times, converting the weekly relapse information into *days*. To jitter, we simply add some random "error" to a variable's values, effectively converting the event times from discrete to continuous. The traditional use of jittering is in the construction of graphic displays. We do not recommend this approach in routine analytic use. We have taken this liberty here simply so that we could use these data to illustrate the use of imputation techniques.

3. This test has another application: It can serve as a routine screen for uncovering interactions between predictors. The idea is simple. To test whether a set of predictors interacts with a focal predictor, make the focal predictor a stratifier and test whether the set of within-strata estimates are identical across strata. This approach should not replace the targeted approach of including relevant cross-products (as explained in section 12.4.1 in the context of discrete time) but it does provide a simple strategy that is easy to implement.

4. Had we not been drawn toward the logarithmic specification for the interaction with time from the graph in figure 15.3, we could have used the parameter estimates from the piecewise specification itself to point toward a similar conclusion. In Model C, we see an immediate drop in parameter estimates followed by a leveling off. This pattern also suggests a logarithmic specification that allows hazard ratios to diminish rapidly during the first few weeks after admission, but constrains the decreases to stabilize over time.

5. A number of case-specific graphical exploratory methods have been proposed (Goldman, 1992; Gentleman & Crowley, 1991; Schmueli & Cohen, 1999), but few have yet to "catch on" and none are currently implemented in today's statistical computer packages.

6. This idea can be extended to the detection of statistical interactions. If you suspect that the effect of one predictor may vary by levels of another, divide your sample by levels of one predictor and create separate martingale residual plots (vs. the other predictor) for each subgroup. If the smoothers in each display differ in form, direction or level, you know that an interaction is possible.

References

Aalen, O. O. (1988). Heterogeneity in survival analysis. *Statistics in Medicine, 7,* 1121–1137.

Abedi, J., & Benkin, E. (1987). The effect of students' academic, financial and demographic variables on time to the doctorate. *Research in Higher Education, 27,* 3–14.

Achenbach, T. M. (1991). Manual for the Child Behavior Checklist 4–18 and 1991 Profile. Burlington, VT: University of Vermont Press.

Akaike, H. (1973). Information theory as an extension of the maximum likelihood principle. In B. N. Petrov & F. Csaki (Eds.), *Second international Symposium on information theory* (pp. 267–228). Akademiai Kiado, Budapest, Hungary.

Allison, P. D. (1982). Discrete-time methods for the analysis of event histories. In S. Leinhardt (Ed.), *Sociological methodology* (pp. 61–98). San Francisco: Jossey-Bass.

Allison, P. D. (1984). *Event history analysis: Regression for longitudinal data* (Sage University paper series on quantitative applications in the social sciences, Number 07-046). Beverly Hills, CA: Sage.

Allison, P. D. (1995). *Survival analysis using the SAS System: A practical guide.* Cary, NC: SAS Institute.

Altman, D. G., & de Stavola, B. L. (1994). Practical problems in fitting a proportional hazards model to data with updated measurements of the covariates. *Statistics in Medicine, 13,* 301–341.

Anderson, P. K., Borgan, O., Gill, R. D., & Keiding, N. (1993). *Statistical models based on counting processes.* New York: Springer.

Arbuckle, J. L. (1995). Amos for Windows. Analysis of moment structures (Version 3.5) [Computer software]. Chicago: SmallWaters. Available at http://www. smallwaters. com

Aydemir, U., Aydemir, S., & Dirschedl, P. (1999). Analysis of time dependent covariates in failure time data. *Statistics in Medicine, 18,* 2123–2134.

Barnes, G. M., Farrell, M. P., & Banerjee, S. (1994). Family influences on alcohol abuse and other problem behaviors among black and white adolescents in a general population sample. *Journal of Research on Adolescence, 4,* 183–201.

Bayley, N. (1935). The development of motor abilities during the first three years. *Monographs of the Society for Research in Child Development*, 1.

Beck, N. (1999). Modelling space and time: The event history approach. In E. Scarbrough & E. Tanenbaum (Eds.), *Research strategies in social science*. New York: Oxford University Press.

Beck, N., Katz, J. N., & Tucker, R. (1998). Taking time seriously: Time-series-cross-section analysis with a binary dependent variable. *American Journal of Political Science, 42*, 1260–1288.

Bentler, P. M. (1995). *EQS: Structural equations program manual.* Encino, CA: Multivariate Software, Inc.

Bereiter, C. (1963). Some persisting dilemmas in the measurement of change. In C. W. Harris (Ed.), *Problems in the measurement of change* (pp. 3–20). Madison, WI: University of Wisconsin Press.

Berk, R. A., & Sherman, L. W. (1988). Police responses to family violence incidents: An analysis of an experimental design with incomplete randomization. *Journal of the American Statistical Association, 83*, 70–76.

Blossfeld, H.-P., & Rohwer, G. (1995). *Techniques of event history modeling.* Mahwah, NJ: Lawrence Erlbaum.

Bolger, N., Downey, G., Walker, E., & Steininger, P. (1989). The onset of suicide ideation in childhood and adolescence. *Journal of Youth and Adolescence, 18*, 175–189.

Bollen, K. (1989). *Structural equations with latent variables.* New York: Wiley.

Box, G. E. P., & Cox, D. R. (1964). An analysis of transformations. *Journal of the Royal Statistical Society, Series B, 26*, 211–252.

Breslow, N. E. (1974). Covariance analysis of censored survival data. *Biometrics, 30*, 88–99.

Brown, J. V., Bakeman, R., Coles, C. D., Sexson, W. R., & Demi, A. S. (1998). Maternal drug use during pregnancy: Are preterms and fullterms affected differently? *Developmental Psychology, 34*, 540–554.

Brown, M. (1996). Refining the risk concept: Decision context as a factor mediating the relation between risk and program effectiveness. *Crime and Delinquency, 42*, 435–455.

Browne, W. J., & Draper, D. (2000). Implementation and performance issues in the Bayesian and likelihood fitting of multilevel models. *Computational Statistics, 15*, 391–420.

Bryk, A. S., & Raudenbush, S. W. (1987). Application of hierarchical linear models to assessing change. *Psychological Bulletin, 101*, 147–158.

Burchinal, M. R., Campbell, F. A., Bryant, D. M., Wasik, B. H., & Ramey, C. T. (1997). Early intervention and mediating processes in cognitive performance of children of low income African American families. *Child Development, 68*, 935–954.

Burchinal, M. R., Roberts, J. E., Riggins, R., Zeisel, S. A., Neebe, E., & Bryant, D. (2000). Relating quality of center child care to early cognitive and language development longitudinally. *Child Development, 71*, 339–357.

Burton, R. P. D., Johnson, R. J., Ritter, C., & Clayton. R. R. (1996). The effects of role socialization on the initiation of cocaine use: An event history analysis from adolescence into middle adulthood. *Journal of Health and Social Behavior, 37*, 75–90.

Campbell, R. T., Mutran, E., & Parker, R. N. (1987). Longitudinal design and longitudinal analysis: A comparison of three approaches. *Research on Aging, 8,* 480–504.

Capaldi, D. M., Crosby, L., & Stoolmiller, M. (1996). Predicting the timing of first sexual intercourse for at-risk adolescent males. *Child Development, 67,* 344–359.

Cherlin, A. J., Chase-Lansdale, P. L., & McRae, C. (1998). Effects of divorce and mental health through the life course. *American Sociological Review, 63,* 239–249.

Cleveland, W. S. (1994). *The elements of graphing data.* Summitt, NJ: Hobart Press.

Coie, J. D., Terry, R., Lenox, K. F., Lochman, J. E., & Hyman, C. (1995). Peer rejection and aggression as predictors of stable risk across adolescence. *Development and Psychopathology, 7,* 697–713.

Collett, D. (1991). *Modeling binary data.* London: Chapman and Hall.

Collett, D. (1994). *Modeling survival data in medical research.* London: Chapman and Hall.

Conger, R. D., Ge, X., Elder, G. H., Lorenz, F. O., & Simons, R. L. (1994). Economic stress, coercive family process, and developmental problems of adolescents. *Child Development, 65,* 541–561.

Cook, R. D., & Weisberg, S. (1982). *Residuals and influence in regression.* London: Chapman and Hall.

Cooney, N. L., Kadden, R. M., Litt, M. D., & Getter, H. (1991). Matching alcoholics to coping skills or interactional therapies: Two-year follow-up results. *Journal of Consulting and Clinical Psychology, 59,* 598–601.

Cox, D. R. (1972). Regression models and life tables. *Journal of the Royal Statistical Society, 34,* 187–202.

Cox, D. R. (1975). Partial likelihood. *Biometrika, 62,* 269–276.

Cox, D. R., & Oakes, D. (1984). *Analysis of survival data.* London: Chapman and Hall.

Cronbach, L. J., & Furby, L. (1970). How we should measure "change"—or should we? *Psychological Bulletin, 74,* 68–80.

Curran, P. J., Stice, E., & Chassin, L. (1997). The relation between adolescent and peer alcohol use: A longitudinal random coefficients model. *Journal of Consulting and Clinical Psychology, 65,* 130–140.

Dempster, A. P., Laird, N. M., & Rubin, D. B. (1977). Maximum likelihood from incomplete data via the EM algorithm. *Journal of the Royal Statistical Society, Series B, 44,* 1–38.

Derogatis, L. R. (1994). SCL-90-R: Administration, scoring and procedural manual. Minneapolis: National Computer Systems, Inc.

Dickter, D. N., Roznowski, M., & Harrison, D. A. (1996). Temporal tempering: An event history analysis of the process of voluntary turnover. *Journal of Applied Psychology, 81,* 705–716.

Diekmann, A., Jungbauer-Gans, M., Krassnig, H., & Lorenz, S. (1996). Social status and aggression: A field study analyzed by survival analysis. *Journal of Social Psychology, 136,* 761–768.

Diekmann, A., & Mitter, P. (1983). The sickle-hypothesis: A time dependent poisson model with applications to deviant behavior. *Journal of Mathematical Sociology, 9,* 85–101.

Diggle, P. J., Liang, K-Y., & Zeger, S. L. (1994). *Analysis of longitudinal data.* New York: Oxford University Press.

Draper, D. (1995). Assessment and propagation of model uncertainty (with discussion). *Journal of the Royal Statistical Society, Series B, 57*, 45–97.

Efron, B. (1977). The efficiency of Cox's likelihood function for censored data. *Journal of the American Statistical Association, 72*, 557–565.

Efron, B. (1988). Logistic regression, survival analysis, and the Kaplan-Meier curve. *Journal of the American Statistical Association, 83*, 414–425.

Espy, K. A., Francis, D. J., & Riese, M. L. (2000). Prenatal cocaine exposure and prematurity: Developmental growth. *Developmental and Behavioral Pediatrics, 21*, 264–272.

Estes, W. K. (1950). Toward a statistical theory of learning. *Psychological Review, 57*, 94–107.

Estes, W. K. (1956). The problem of inference from curves based on grouped data. *Psychological Bulletin, 53*, 134–140.

Estes, W. K., &. Burke, C. J. (1955). Applications of a statistical model to simple discrimination learning in human subjects. *Journal of Experimental Psychology, 50*, 81–88.

Fahrmeir, L., & Wagenpfeil, S. (1996). Smoothing hazard functions and time-varying effects in discrete duration and competing risks models. *Journal of the American Statistical Association, 91*, 1584–1594.

Felmlee, D., & Eder, D. (1984). Contextual effects in the classroom: The impact of ability groups on student attention. *Sociology of Education, 56*, 77–87.

Fergusson, D. M., Horwood, L. J., & Shannon, F. T. (1984). A proportional hazards model of family breakdown. *Journal of Marriage and the Family, 46*, 539–549.

Fichman, M. (1989). Attendance makes the heart grow fonder: A hazard rate approach to modeling attendance. *Journal of Applied Psychology, 74*, 325–335.

Fisher, L. D., & Lin, D. Y. (1999). Time dependent covariates in the Cox proportional hazards regression model. *Annual Review of Public Health, 20*, 145–157.

Fleming, T. R., & Harrington, D. P. (1991). *Counting processes and survival analysis.* New York: Wiley.

Flinn, C. J., & Heckman, J. J. (1982). New methods for analyzing individual event histories. In S. Leinhardt (Ed.), *Sociological methodology* (pp. 99–140). San Francisco: Jossey-Bass.

Follmann, D. A., & Goldberg, M. S. (1988). Distinguishing heterogeneity from decreasing hazard rates. *Technometrics, 30*, 389–396.

Foster, E. M. (1999). Service use under the continuum of care: Do followup services forestall hospital readmission? *Health Services Research, 34*, 715–736.

Foster, E. M. (2000). Does the continuum of care reduce inpatient length of stay? *Evaluation and Program Planning, 23*, 53–65.

Fox, J. (1998). *Applied regression analysis, linear models, and related methods.* Thousand Oaks, CA: Sage.

Francis, D. J., Shaywitz, S. E., Stuebing, K. K., Shaywitz, B. A., & Fletcher, J. M. (1996). Developmental lag versus deficit models of reading disability: A longitudinal, individual growth curves analysis. *Journal of Educational Psychology, 88*, 3–17.

Frank, A. R., & Keith, T. Z. (1984). Academic abilities of persons entering and remaining in special education. *Exceptional Children, 51*, 76–77.

Fryer, G. E., & Miyoshi, T. J. (1994). A survival analysis of the revictimization of children: The case of Colorado. *Child Abuse and Neglect, 18,* 1063–1071.

Furby, L., Weinrott, M. R., & Blackshaw, L. (1989). Sex offender recidivism: A review. *Psychological Bulletin, 105,* 3–30.

Gail, M. H., Wieand, S., & Piantadosi, S. (1984). Biased estimates of treatment effect in randomized experiments with nonlinear regression and omitted covariates. *Biometrika, 71,* 431–444.

Gamse, B. C., & Conger, D. (1997). An evaluation of the Spencer post-doctoral dissertation program. Cambridge, MA: Abt Associates.

Gardner, W., & Griffin, W. A. (1988). Methods for the analysis of parallel streams of continuously recorded social behaviors. *Psychological Bulletin, 106,* 497–502.

Gehan, E. A. (1969). Estimating survival functions for the life table. *Journal of Chronic Diseases, 21,* 629–644.

Gelman, A., & Rubin, D. B. (1995). Avoiding model selection in Bayesian social research. *Sociological Methodology, 25,* 165–195.

Gentleman, R., & Crowley, J. (1991). Graphical methods for censored data. *Journal of the American Statistical Association, 86,* 678–683.

Gilks, W., Richardson, S., & Spiegelhalter, D. (1996). *Markov chain Monte Carlo in practice.* London: Chapman and Hall.

Ginexi, E. M., Howe, G. W., & Caplan, R. D. (2000). Depression and control beliefs in relation to reemployment: What are the directions of effect? *Journal of Occupational Health Psychology, 5,* 323–336.

Goldman, A. I. (1992). EVENTCHARTS: Visualizing survival and other timed-events data. *The American Statistician, 46,* 13–18.

Goldstein, H. (1995). *Multilevel statistical models,* 2nd ed. New York: Halstead Press.

Goldstein, H. (1998). MLwiN, available from http://multilevel.ioe.ac.uk

Goldstein, H., Healy, M. J. R., & Rasbash, J. (1994). Multilevel time series models with applications to repeated measures data. *Statistics in Medicine, 13,* 1643–1655.

Graham, S. E. (1997). *The exodus from mathematics: When and why?* Unpublished doctoral dissertation. Harvard University, Graduate School of Education.

Greenhouse, J. B., Stangl, D., & Bromberg, J. (1989). An introduction to survival analysis methods for analysis of clinical trial data. *Journal of Consulting and Clinical Psychology, 57,* 536–544.

Greenland S., & Finkle, W. D. (1995). A critical look at methods for handling missing covariates in epidemiologic regression analyses. *American Journal of Epidemiology, 142,* 1255–1264.

Greenwood, M. (1926). The natural duration of cancer. *Reports on Public Health and Medical Subjects. 33,* 1–26. London: Her Majesty's Stationery Office.

Grice, G. R. (1942). An experimental study of the gradient of reinforcement in maze learning. *Journal of Experimental Psychology, 30,* 475–489.

Gross, A. J., & Clark, V. A. (1975). *Survival distributions: Reliability applications in the biomedical sciences.* New York: Wiley.

Guire, K. E., & Kowalski, C. (1979). Mathematical description and representation of developmental change functions on the intra- and inter-individual levels. In J. Nesselroade & P. Baltes (Ed.), *Longitudinal research in the study of behavior and development.* New York: Academic Press.

Gulliksen, H. (1934). A rational equation of the learning curve based on Thorndike's law of effect. *Journal of General Psychology, 11,* 395–433.

Gulliksen, H. (1953). A generalization of Thurstone's learning function. *Psychometrika, 18,* 297–307.

Ha, J. C., Kimpo, C. L., & Sackett, G. P. (1997). Multiple-spell, discrete-time survival analysis of developmental data: Object concept in pigtailed macaques. *Developmental Psychology, 33,* 1054–1059.

Hachen, D. S., Jr. (1988). The competing risks model: A method for analyzing processes with multiple types of events. *Sociological Methods and Research, 17,* 21–54.

Hall, S. M., Havassy, B. E., & Wasserman, D. A. (1990). Commitment to abstinence and acute stress in relapse to alcohol, opiates, and nicotine. *Journal of Consulting and Clinical Psychology, 58,* 175–181.

Harrell, F. E. (2001). *Regression modeling strategies with applications to linear models, survival analysis and logistic regression.* New York: Springer.

Harris, E. K., & Albert, A. (1991). *Survivorship analysis for clinical studies.* New York: Marcel Dekker.

Harris, V., & Koepsell, T. D. (1996). Criminal recidivism in mentally ill offenders: A pilot study. *Bulletin of the American Academy of Psychiatry and Law, 24,* 177–186.

Harrison, D. A., Virick, M., & William, S. (1996). Working without a net: Time, performance, and turnover under maximally contingent rewards. *Journal of Applied Psychology, 81,* 331–345.

Harville, D. A. (1974). Bayesian inference for variance components using only error contrasts. *Biometrika, 61,* 383–385.

Hasin, D. S., Tsai, W.-Y., Endicott, J., Mueller, T. I., Coryell, W., & Keller, M. (1996). The effect of major depression on alcoholism: Five year course. *The American Journal on Addictions, 5,* 144–155.

Hauck, W. W., & Donner, A. (1977). Wald's test as applied to hypotheses in logit analysis. *Journal of the American Statistical Association. 72,* 851–853.

Heckman, J. J., & Singer, B. (Eds.). (1984). *Longitudinal analysis of labor market data.* New York: Cambridge University Press.

Hedeker, D. R., & Gibbons, R. D. (1996). MIXREG: A computer program for mixed-effects regression analysis with autocorrelated errors. *Computer Methods and Programs in Biomedicine, 49,* 229–252, available from http://www.uic.edu/~hedeker

Hedeker D. R., & Gibbons R. D. (1997). Application of random-effects pattern-mixture models for missing data in longitudinal studies. *Psychological Methods, 2,* 64–78.

Hedeker D. R., Gibbons R. D., & Flay B. R. (1994). Random regression models for clustered data: With an example from smoking prevention research. *Journal of Clinical and Consulting Psychology, 62,* 757–765.

Henning, K. R., & Frueh, B. C. (1996). Cognitive-behavioral treatment of incarcerated offenders: An evaluation of the Vermont Department of Corrections' cognitive self-change program. *Criminal Justice and Behavior, 23,* 523–541.

Hertz-Picciotto, I., & Rockhill, B. (1997). Validity and efficiency of approximation methods for tied survival times in Cox regression. *Biometrics, 53,* 1151–1156.

Hicklin, W. J. (1976). A model for mastery learning based on dynamic equilibrium theory. *Journal of Mathematical Psychology, 13,* 79–88.

Hodges, K., McKnew, D., Cytryn, L., Stern, L., & Klien, J. (1982). The child assessment schedule (CAS) diagnostic interview: A report on reliability and validity. *Journal of the American Academy of Child Psychiatry, 21,* 468–473.

Hofmann, D. A., & Gavin, M. B. (1998). Centering decisions in hierarchical linear models: Theoretical and methodological implications for organizational science. *Journal of Management, 24,* 623–641.

Hosmer, D. W., Jr., & Lemeshow, S. (1999). *Applied survival analysis: Regression modeling of time to event data.* New York: Wiley.

Hosmer, D. W., Jr., & Lemeshow, S. (2000). *Applied logistic regression* (2nd ed.). New York: Wiley.

Hsieh, F. Y. (1995). A cautionary note on the analysis of extreme data with Cox regression. *The American Statistician, 49,* 226–228.

Hu, X. J., & Lawless, J. F. (1996). Estimation from truncated lifetime data with supplementary information on covariates and censoring times. *Biometrika, 83,* 747–762.

Hull, C. (1939). Simple trial and error learning: An empirical investigation. *Journal of Comparative Psychology, 27,* 233–258.

Hull, C. (1943). *Principles of behavior.* New York: Appleton-Century-Crofts.

Hull, C. (1952). *A behavior system.* New Haven, CT: Yale University Press.

Hurlburt, M. S., Wood, P. A., & Hough, R. L. (1996). Providing independent housing for the homeless mentally ill: A novel approach to evaluating long-term longitudinal housing patterns. *Journal of Community Psychology, 24,* 291–310.

Huttenlocher, J., Haight, W., Bryk, A., Seltzer, M., & Lyons, T. (1991). Early vocabulary growth: relation to language input and gender. *Developmental Psychology, 27,* 236–248.

Joreskog, K. G., & Sorbom, D. (1996). *LISREL 8: User's reference guide.* Chicago: Scientific Software International.

Kalbfleisch, J. D., & Prentice, R. L. (1973). Marginal likelihoods based on Cox's regression and life model. *Biometrika, 60,* 267–279.

Kalbfleisch, J. D., & Prentice, R. L. (1980). *The statistical analysis of failure time data.* New York: Wiley.

Kaplan, E. L., & Meier, P. (1958). Nonparametric estimation from incomplete observations. *Journal of the American Statistical Association, 53,* 457–481.

Keats, J. A. (1983). Ability measures and theories of cognitive development. In H. Wainer and S. Messick (Eds.), *Principles of modern psychological measurement: A festschrift for Frederick M. Lord* (pp. 81–101). Hillsdale, NJ: Lawrence Erlbaum.

Keifer, N. M. (1988). Economic duration data and hazard functions. *Journal of Economic Literature, 26,* 646–679.

Keiley, M. K., & Martin, N. C. (2002). *Child abuse, neglect, and juvenile delinquency: How "new" statistical approaches can inform our understanding of "old" questions— a reanalysis of Widon, 1989.* Manuscript submitted for publication.

Keiley, M. K., Bates, J. E., Dodge, K. A., & Pettit, G. S. (2000). A cross-domain growth analysis: Externalizing and internalizing behavior during 8 years of childhood. *Journal of Abnormal Child Psychology, 28,* 161–179.

Killen, J. D., Robinson, T. N., Haydel, K. F., et al. (1997). Prospective study of risk factors for the initiation of cigarette smoking. *Journal of Consulting and Clinical Psychology, 65* (6): 1011–1016.

Klein, J. P., & Moeschberger, M. L. (1997). *Survival analysis: Techniques for censored and truncated data.* New York: Springer.

Kleinbaum, D. G., Kupper, L. L., & Morgenstern, H. (1982). *Epidemiological research: Principles and quantitative methods.* Belmont, CA: Lifetime Learning Publications.

Kreft, I. G. G., & de Leeuw, J. (1998). *Introducing multilevel modeling.* Thousand Oaks, CA: Sage.

Kreft, I. G. G., & de Leeuw, J., & Aiken, L. S. (1995). The effect of different forms of centering in hierarchical linear models. *Multivariate Behavioral Research, 30,* 1–21.

Kreft, I. G. G., de Leeuw, J., & Kim, K. S. (1990). *Comparing four different statistical packages for hierarchical linear regression: GENMOD, HLM, ML2, and VARCL* (Technical report 311). Los Angeles: Center for the Study of Evaluation, University of California at Los Angeles.

Lahey, B. B., McBurnett, K., Loeber, R., & Hart, E. L. (1995). Psychobiology of conduct disorder. In G. P. Sholevar (Ed.), *Conduct disorders in children and adolescents: Assessments and Interventions* (pp. 27–44). Washington, DC: American Psychiatric Press.

Laird, N. M. (1988). Missing data in longitudinal studies. *Statistics in Medicine, 7,* 305–315.

Laird, N., & Olivier, D. (1981). Covariance analysis of censored survival data using log-linear analysis techniques. *Journal of the American Statistical Association, 76,* 231–240.

Lancaster, T. (1990). *The econometric analysis of transition data.* New York: Cambridge University Press.

Lavori, P. W., Dawson, R., Mueller, T. I., Warshaw, M., Swartz, A., & Leon, A. (1996). Analysis of the course of psychopathology: Transitions among states of health and illness. *International Journal of Methods in Psychiatric Research, 6,* 321–334.

Lawless, J. F. (1982). *Statistical models and methods for lifetime data.* New York: Wiley.

Lee, E. T. (1992). *Statistical methods for survival data analysis* (2nd ed.). New York: Wiley.

Lewis, D. (1960). *Quantitative methods in psychology.* New York: McGraw-Hill.

Lilienfeld, A. M., & Stolley, P. D. (1994). *Foundations of epidemiology* (3rd ed.). New York: Oxford University Press.

Linn, R. L., & Slinde, J. A. (1977). The determination of the significance of change between pre- and posttesting periods. *Review of Educational Research, 47,* 121–150.

Little, R. J. A. (1995). Modeling the dropout mechanism in repeated-measures studies. *Journal of the American Statistical Association, 90,* 1112–1121.

Little, R. J. A., & Rubin, D. (1987). *Statistical analysis with missing data.* New York: Wiley.

Little, R. J. A., & Yau, L. (1998). Statistical techniques for analyzing data from prevention trials: Treatment of no-shows using Rubin's causal model. *Psychological Methods, 3,* 147–159.

Long, J. S. (1997). *Regression models for categorical and limited dependent variables.* Beverly Hills: Sage.

Longford, N. T. (1990). VARCL. Software for variance component analysis of data with nested random effects (maximum likelihood). Princeton, NJ: Educational Testing Service.

Longford, N. T. (1993). *Random coefficient models.* New York: Oxford University Press.

Longford, N. T. (1999). Standard errors in multilevel analysis. *Multilevel Modeling Newsletter, 11*(1), 10–13.

Lord, F. M. (1963). Elementary models for measuring change. In C. W. Harris (Ed.), *Problems in the measurement of change* (pp. 21–39). Madison, WI: University of Wisconsin Press.

MaCurdy, T., Mroz, T., & Gritz, R. M. (1998). An evaluation of the National Longitudinal Survey of Youth. *Journal of Human Resources, 33,* 345–436.

Mare, R. D. (1994). Discrete-time bivariate hazards with unobserved heterogeneity: A partially observed contingency table approach. *Sociological Methodology, 24,* 341–383.

Mason, W. M., Anderson, A. F., & Hayat, N. (1988). *Manual for GENMOD.* Ann Arbor: Population Studies Center, University of Michigan.

Masse, L. C., & Tremblay, R. E. (1997). Behavior of boys in kindergarten and the onset of substance use during adolescence. *Archives of General Psychiatry, 54,* 52–68.

McArdle, J. J. (1986a). Dynamic but structural equation modeling of repeated measures data. In J. R. Nesselrode & R. B. Cattell (Eds.), *Handbook of multivariate experimental psychology* (Vol. 2, pp. 561–614). New York: Plenum.

McArdle, J. J. (1986b). Latent variable growth within behavior genetic models. *Behavior Genetics, 16,* 163–200.

McArdle, J. J. (1989). A structural modeling experiment with multiple growth functions. In P. Ackerman, R. Kanfer, & R. Cudek (Eds.), *Learning and individual differences: Abilities, motivation, and methodology* (pp. 71–117). Hillsdale, NJ: Lawrence Erlbaum.

McArdle, J. J. (1991). Structural models of developmental theory in psychology. *Annals of Theoretical Psychology, 7,* 139–159.

McArdle, J. J., Anderson, E., & Aber, M. (1987). Covergence hypotheses modeled and tested with linear structural equations. *Proceedings of the 1987 public health conference on records and statistics* (pp. 351–357). Hyattsville, MD: National Center for Health Statistics.

McArdle, J. J., & Epstein, D. (1987). Latent growth curves within developmental structural equation models. *Child Development, 58,* 110–133.

McArdle, J. J., Hamagami, F., Elias, M. F., & Robbins, M. A. (1991). Structural modeling of mixed longitudinal and cross-sectional data, *Experimental Aging Research, 17,* 29–52.

McCullagh, P., & Nelder, J. A. (1989). *Generalized linear models* (2nd ed). London: Chapman and Hall.

Mead, R., & Pike, D. J. (1975). A review of response surface methodology from a biometric point of view. *Biometrics, 31,* 803–851.

Meredith, W., & Tisak, J. (1984). *"Tuckerizing" curves.* Paper presented at the annual meeting of the Psychometric Society, Santa Barbara, CA.

Meredith, W., & Tisak, J. (1990). Latent curve analysis. *Psychometrika, 55,* 107–122.

Miller, R. G. (1981). *Survival analysis.* New York: Wiley.

Miller, R. (1986). *Beyond ANOVA, Basics of applied statistics.* New York: Wiley.

Mojtabai, R., Nicholson, R. A., & Neesmith, D. H. (1997). Factors affecting relapse in patients discharged from a public hospital: Results from survival analysis. *Psychiatric Quarterly, 68,* 117–129.

Mosteller, F., & Tukey, J. W. (1977). *Data analysis and regression: A second course in statistics.* Reading, MA: Addison-Wesley.

Murnane, R. J., Boudett, K. P., & Willett, J. B. (1999). Do male dropouts benefit from obtaining a GED, postsecondary education, and training? *Evaluation Review, 23,* 475–502.

Murnane, R. J., Singer, J. D., & Willett, J. B. (1989). The influences of salaries and opportunity costs on teachers' career choices: Evidence from North Carolina. *Harvard Educational Review, 59,* 325–346.

Muthen, B. O. (1989). Latent variable modelling in heterogeneous populations. *Psychometrika, 54,* 557–585.

Muthen, B. O. (1991). Analysis of longitudinal data using latent variable models with varying parameters. In L. M. Collins & J. L. Horn (Eds.), *Best methods for the analysis of change: Recent advances, unanswered questions, future directions* (pp. 1–17). Washington, DC: American Psychological Assocation.

Muthen, B. O. (1992). *Latent variable modeling of growth with missing data and multilevel data.* Paper presented at the Seventh International Conference on Multivariate Analysis, Barcelona, Spain.

Muthen, B. O. (2001). MPLUS. Available at http://www.statmodel.com

Muthén, B. O., & Satorra, A. (1989). Multilevel aspects of varying parameters in structural models. In D. Bock (Ed.), *Multilevel analysis of educational data* (pp. 87–89). San Diego, CA: Academic Press.

Myers, K., McCauley, E., Calderon, R., & Treder, R. (1991). The 3-year longitudinal course of suicidality and predictive factors for subsequent suicidality in youths with major depressive disorder. *Journal of the American Academy of Child and Adolescent Psychiatry, 30,* 804–810.

Patterson H. D., & Thompson R. (1971). Recovery of inter-block information when block sizes are unequal. *Biometrika, 58,* 545–554.

Pepe, M., & Mori, M. (1993). Kaplan-Meier, marginal or conditional probability curves in summarizing competing risks failure time data. *Statistics in Medicine, 12,* 737–751.

Peto, R. (1972). Discussion of Professor Cox's paper. *Journal of the Royal Statistical Society, Series B, 34,* 205–207.

Peto, R., Pike, M. C., Armitage, P., Breslow, N. E., Cox, D. R., Howard, S. V., Mantel, N., McPherson, K., Peto, J., & Smith, P. G. (1976). Design and analysis of randomized clinical trials requiring prolonged observation of each patient: Introduction and design. *British Journal of Cancer, 34,* 585–612.

Pinheiro, J. C., & Bates, D. M. (1995). Approximations to the log-likelihood function in nonlinear mixed-effects models. *Journal of Computational and Graphical Statistics, 4,* 12–35.

Pinheiro, J. C., & Bates, D. M. (2000). *Mixed-effects models in S and S-PLUS.* New York: Springer-Verlag.

Pinheiro, J. C., & Bates, D. M. (2001). SPLUS' NLME, available at http://cm.belllabs.com/cm/ms/departments/sia/project/nlme/

Powers, D. A., & Xie, Y. (1999). *Statistical methods for categorical data analysis.* San Diego: Academic Press.

Prentice, R. L., & Gloeckler, L. A. (1978). Regression analysis of grouped survival data with application to breast cancer data. *Biometrics, 34,* 57–67.

Radloff, L. S. (1977). The CES-D scale: A self report major depressive disorder scale for research in the general population. *Applied Psychological Measurement, 1,* 385–401.

Raftery, A. E. (1995). Bayesian model selection in social research. *Sociological Methodology, 25,* 111–163.

Raftery, A. E., Lewis, S. M., Aghajanian, A., & Kahn, M. J. (1996). Event history modeling of World Fertility Survey data. *Mathematical Population Studies, 6,* 129–153.

Ramlau-Hansen, H. (1983). Smoothing counting process intensities by means of kernel functions. *Annals of Statistics, 11,* 453–466.

Rasbash, J., & Woodhouse, G. (1995). *MLn Command Reference. Multilevel Models Project.* London: Institute of Education.

Raudenbush, S. W., & Bryk, A. S. (2002). *Hierarchical linear models: Applications and data analysis methods* (2nd ed.). Thousand Oaks, CA: Sage.

Raudenbush, S. W., Bryk, A. S., Cheong, Y., & Congdon, R. (2001). HLM. Available from http://www.ssicentral.com

Raudenbush, S. W., Bryk., A. S., & Congdon, R. T. (1988). *HLM: Hierarchical linear modeling.* Chicago: Scientific Software International, Inc.

Raudenbush, S. W., & Chan, W. S. (1992). Growth curve analysis in accelerated longitudinal designs. *Journal of Research in Crime and Delinquency, 29,* 387–411.

Rayman, P., & Brett, B. (1995). Women science majors: What makes a difference in persistence to graduation? *Journal of Higher Education, 66,* 388–414.

Robertson, T. B. (1908). Sur la dynamique du systeme nerveux central. *Archives of International Physiology, 6,* 388–454.

Robertson, T. B. (1909). A biochemical conception of memory and sensation. *The Monist, 19,* 367–386.

Roderick, M. (1994). Grade retention and school dropout: Investigating the association. *American Educational Research Journal, 31,* 729–759.

Rogosa, D. R., Brandt, D., & Zimowski, M. (1982). A growth curve approach to the measurement of change. *Psychological Bulletin, 90,* 726–748.

Rogosa, D. R., & Willett, J. B. (1985). Understanding correlates of change by modeling individual differences in growth. *Psychometrika, 50,* 203–228.

Rothenberg, T. J. (1984). Hypothesis testing in linear models when the error covariance matrix is nonscalar. *Econometrica, 52,* 827–842.

Sargeant J. K., Bruce, M. L., Florio L. P., & Weissman, M. (1990). Factors associated with 1-year outcome of major depression in the community. *Archives of General Psychiatry, 47,* 519–526.

SAS Institute. (2001). *Statistical analysis system.* Available at http://www.sas.com

Schafer, J. (1997). *Analysis of incomplete multivariate data.* New York: Chapman and Hall.

Scheike, T. H., & Jensen, T. K. (1997). A discrete survival model with random effects: An application to time to pregnancy. *Biometrics, 53,* 318–329.

Schmueli, G., & Cohen, A. (1999). Analysis and display of hierarchical lifetime data. *The American Statistician, 53,* 140–146.

Schoenfeld, D. (1982). Partial residuals for the proportional hazards regression model. *Biometrika, 69,* 239–241.

Schukarew, A. (1907). Uber die energetischen grundlagen des Gesetzes von Weber-Sechner und der Dynamik des Gedachtnisses. *Annalen des Naturphilosophie, 6,* 139–149.

Schwarz, G. (1978). Estimating the dimensions of a model. *Annals of Statistics, 6,* 461–464.

Shanahan, M. J., Elder, G. H., Jr., Burchinal, M., & Conger, R. D. (1996). Adolescent paid labor and relationships with parents: Early work-family linkages. *Child Development, 67,* 2183–2200.

Siegfried, J. J., & Stock, W. A. (2001). So you want to earn a Ph.D. in economics?: How long do you think it will take? *Journal of Human Resources, 34,* 364–378.

Singer, J. D. (1993). Are special educators' careers special? *Exceptional Children, 59,* 262–279.

Singer, J. D. (1998). Using SAS PROC MIXED to fit multilevel models, hierarchical models, and individual growth models. *Journal of Educational and Behavioral Statistics, 25,* 323–355.

Singer, J. D. (2001). Fitting individual growth models using SAS PROC MIXED. In D. S. Moskowitz & S. L. Hershberger (Eds.), *Modeling intraindividual variability with repeated measures data: Method and applications.* Englewood Clifs, NJ: Erlbaum.

Singer, J. D., Davidson, S., Graham, S., & Davidson, H. S. (1998). Physician retention in community and migrant health centers: Who stays and for how long? *Medical Care, 38,* 1198–1213.

Singer, J. D., Fuller, B., Keiley, M. K., & Wolf, A. (1998). Early child care selection: Variation by geographic location, maternal characteristics, and family structure. *Developmental Psychology, 34,* 1129–1144.

Singer, J. D., & Willett, J. B. (1991). Modeling the days of our lives: Using survival analysis when designing and analyzing longitudinal studies of duration and the timing of events. *Psychological Bulletin, 110,* 268–290.

Singer, J. D., & Willett, J. B. (1993). It's about time: Using discrete-time survival analysis to study duration and the timing of events. *Journal of Educational Statistics, 18,* 155–195.

Singer, J. D., & Willett, J. B. (2001, April). *Improving the quality of longitudinal research.* Paper presented at the Annual Meeting of the American Educational Research Association, Seattle, Washington.

Snijders, T. A. B., & Bosker, R. J. (1994). Modeled variance in two-level models. *Sociological Methods and Research, 22,* 342–363.

Snijders, T. A. B., & Bosker, R. J. (1999). *Multilevel analysis: An introduction to basic and advanced multilevel modeling.* London: Sage.

Sorenson, S. B., Rutter, C. M., & Aneshensel, C. S. (1991). Depression in the community: An investigation into age of onset. *Journal of Consulting and Clinical Psychology, 59,* 541–546.

South, S. J. (1995). Do you need to shop around: Age at marriage, spousal alternatives, and marital dissolution. *Journal of Family Issues, 16,* 432–449.

South, S. J. (2001). Time-dependent effects of wives' employment on marital dissolution. *American Sociological Review, 66,* 226–245.

Stage, S. A. (2001). Program evaluation using hierarchical linear modeling with curriculum-based measurement reading probes. *School Psychology Quarterly, 16,* 91–112.

Stata Corporation. (2001). STATA. Available at http://www.stata.com

Strober, M., Freeman, R., Bower, S., & Rigali, J. (1996). Binge eating in anorexia nervosa predicts later onset of substance use disorder: A ten-year prospective, longitudinal follow-up of 95 adolescents. *Journal of Youth and Adolescence, 25,* 519–532.

Sueyoshi, G. T. (1995). A class of binary response models for grouped duration data. *Journal of Applied Econometrics, 10,* 411–431.

Svartzberg, M., Seltzer, M. H., Stiles, T. C., & Khoo, E. (1995). Symptom improvement and its temporal course in short-term dynamic psychotherapy—A growth curve analysis. *Journal of Nervous and Mental Disorders, 183,* 242–248.

Tanner, J. M. (1964). The human growth curve. In G. A. Harrison, J. S. Winer, J. M. Tanner, & N. A. Barnicot (Eds.), *Human biology* (pp. 299–320). New York: Oxford University Press.

Therneau, T. M., & Grambsch, P. M. (2000). *Modeling survival data: Extending the Cox model.* New York: Springer.

Thurstone, L. L. (1917). The learning curve equation. *Psychological Bulletin, 14,* 64–65.

Thurstone, L. L. (1930). The learning function. *Journal of General Psychology, 3,* 469–493.

Tisak, J., & Meredith, W. (1990). Descriptive and associative developmental models. In A. Von Eye (Ed.), *Statistical methods in longitudinal research* (Vol. 2, pp. 387–406). New York: Academic Press.

Tivnan, T. (1980). *Improvements in performance on cognitive tasks: The acquisition of new skills by elementary school children.* Unpublished doctoral dissertation. Harvard University, Graduate School of Education.

Tomarken, A. J., Shelton, R. C., Elkins, L., & Anderson, T. (1997). *Sleep deprivation and anti-depressant medication: Unique effects on positive and negative affect.* Poster session presented at the 9th annual meeting of the American Psychological Society, Washington, DC.

Tuma, N. B., & Hannan, M. T. (1984). *Social dynamics: Models and methods.* New York: Academic Press.

Turnbull, B. W. (1974). Non-parametric estimation of a survivorship function with doubly censored data. *Journal of the American Statistical Association, 69,* 169–173.

Turnbull, B. W. (1976). The empirical distribution function with arbitrarily grouped, censored and truncated data. *Journal of the Royal Statistical Society, 38,* 290–295.

Van Leeuwen, D. M. (1997). A note on the covariance structure in a linear model. *The American Statistician, 51,* 140–144.

Vaupel, J. W., Manton, K. G., & Stallard, E. (1979). The impact of heterogeneity in individual frailty on the dynamics of mortality. *Demography, 16,* 439–454.

Vaupel, J. W., & Yashin, A. I. (1985). Heterogeneity's ruses: Some surprising effects of selection on population dynamics. *The American Statistician, 39,* 176–185.

Verbeke, G., & Molenberghs, G. (2000). *Linear mixed models for longitudinal data.* New York: Springer-Verlag.

Wheaton, B., Roszell, P., & Hall, K. (1997). The impact of twenty childhood and adult traumatic stressors on the risk of psychiatric disorder. In I. H. Gotlib &

B. Wheaton (Eds.), *Stress and adversity over the life course: Trajectories and turning points* (pp. 50–72). New York: Cambridge University Press.

Willett, J. B. (1988). Questions and answers in the measurement of change. In E. Rothkopf (Ed.), *Review of research in education (1988–89)* (pp. 345–422). Washington, DC: American Educational Research Association.

Willett, J. B. (1989). Some results on reliability for the longitudinal measurement of change: Implications for the design of studies of individual growth. *Educational and Psychological Measurement, 49,* 587–602.

Willett, J. B., & Sayer, A. G. (1994). Using covariance structure analysis to detect correlates and predictors of change. *Psychological Bulletin, 116,* 363–381.

Willett, J. B., & Singer, J. D. (1991). From whether to when: New methods for studying student dropout and teacher attrition. *Review of Educational Research, 61,* 407–450.

Willett, J. B., & Singer, J. D. (1993). Investigating onset, cessation, relapse, and recovery: Why you should, and how you can, use discrete-time survival analysis to examine event occurrence. *Journal of Consulting and Clinical Psychology, 61,* 952–965.

Wolfinger, R. D. (1993). Covariance structure selection in general mixed models. *Communications in Statistics-Simulations, 22,* 1079–1106.

Wolfinger, R. D. (1996). Heterogeneous variance-covariance structures for repeated measures. *Journal of Agricultural, Biological, and Environmental Statistics, 1,* 205–230.

Wu, Z. (1995). Premarital cohabitation and postmarital cohabiting union formation. *Journal of Family Issues, 16,* 212–232.

Wu, L. L., & Martinson, B. C. (1993). Family structure and the risk of a premarital birth. *American Sociological Review, 58,* 210–232.

Xie, Y. (1994). The log-multiplicative models for discrete-time, discrete-covariate event history data. *Sociological Methodology, 24,* 301–340.

Xue, X., & Brookmeyer, R. (1997). Regression analysis of discrete time survival data under heterogeneity. *Statistics in Medicine, 16,* 1983–1993.

Yamaguchi, K. (1991). *Event history analysis.* Newbury Park, CA: Sage.

Yamaguchi, K., & Kandel, D. B. (1985). Dynamic relationships between premarital cohabitation and illicit drug use: An event history analysis of role selection and role socialization. *American Sociological Review, 50,* 530–546.

Yamaguchi, K., & Kandel, D. B. (1987). Drug use and other determinants of premarital pregnancy and its outcome: A dynamic analysis of competing life events. *Journal of Marriage and the Family, 49,* 257–270.

Young, M. A., Meaden, P. M., Fogg, L. F., Cherin, E. A., & Eastman, C. I. (1997). Which environmental variables are related to the onset of seasonal affective disorder? *Journal of Abnormal Psychology, 106,* 554–562.

Zorn, C. J., & van Winkle, S. R. (2000). A competing risks model of Supreme Court vacancies, 1780–1992. *Political Behavior, 22,* 145–166.

Index